Reviews in Fluorescence 2007

Editor

Chris D. Geddes
University of Maryland, Baltimore, MD, USA

For further volumes:
http://www.springer.com/series/6946

Chris D. Geddes
Editor

Reviews in Fluorescence 2007

Springer

Chris D. Geddes
Director of the Institute of Fluorescence
University of Maryland Biotechnology Institute
The Columbus Center
Suites 3017–21
701 East Pratt Street
Baltimore, MD 21202
USA

Joseph R. Lakowicz
Center for Fluorescence Spectroscopy
725 West Lombard Street
Balitmore, MD, 21201
USA

ISSN 1573-8086
ISBN 978-0-387-88721-0 e-ISBN 978-0-387-88722-7
DOI 10.1007/978-0-387-88722-7
Springer Dordrecht Heidelberg London New York

Library of Congress Control Number: Applied For

Printed on acid-free paper

Springer is part of Springer Science+Business Media (www.springer.com)

Preface

This is the fourth volume in the *Reviews in Fluorescence* series. To date, three volumes have been both published and well received by the scientific community. Several book reviews in the last few years have also favorably remarked on the series.

In this fourth volume we continue the tradition of publishing leading edge and timely articles from authors around the world. We thank the authors for their timely and exciting contributions. We hope you find this volume as useful as past volumes, which promises to be just as diverse with regard to fluorescence-based content.

Finally, in closing, I would like to thank to Aaron Johnson, formerly at Springer, for helping me to publish this book serial over the last four volumes. Thanks also go to Michael Weston at Springer for help in publishing this current volume.

Baltimore, Maryland — Chris D. Geddes

Contents

Contributors

Freek Ariese Analytical Chemistry and Applied Spectroscopy, Laser Centre, Vrije Universiteit Amsterdam, The Netherlands

Arjen N. Bader Analytical Chemistry and Applied Spectroscopy, Laser Centre, Vrije Universiteit Amsterdam, The Netherlands

Luis A. Bagatolli Membrane Biophysics and Biophotonics group/MEMPHYS – Center for Biomembrane Physics, Department of Biochemistry and Molecular Biology, University of Southern Denmark, Campusvej 55 DK-5230, Odense M, Denmark

Kankan Bhattacharyya Department of Physical Chemistry, Indian Association for the Cultivation of Science, Jadavpur, Kolkata-700 032, India

Nigel J.F. Blamey Nanoscale Biophotonics Laboratory, School of Chemistry, National University of Ireland-Galway, Galway, Ireland

Denis Boudreau Centre d'optique, photonique et laser (COPL) and Département de chimie, Université Laval, Québec, QC, Canada G1K 7P4

Harry G. Brittain Center for Pharmaceutical Physics, 10 Charles Road, Milford NJ 08848, USA

Ekaterina A. Bykova Department of Physiology and Membrane Biology, University of California, One Shields Avenue, Davis, California 95616, USA

Patrik R. Callis Department of Chemistry and Biochemistry, Montana State University, Bozeman MT 59717-3400, USA

Abhijit Chakrabarti Biophysics Division and Strcutural Genomics Section, Saha Institute of Nuclear Physics, 1/AF Bidhannagar, Kolkata 700064, India

Paulo J. G. Coutinho Centro de Física, Universidade do Minho, Campus de Gualtar, 4710-057 Braga, Portugal

Joost S. de Klerk Analytical Chemistry and Applied Spectroscopy, Laser Centre, Vrije Universiteit Amsterdam, The Netherlands

Kim Doré Centre de Recherche Université Laval Robert-Giffard (CRULRG), Unité de neurobiologie cellulaire, 2601 Ch. Canardière, Québec, QC, Canada G1J2G3

Christoph J. Fahrni School of Chemistry and Biochemistry and Petit Institute for Bioengineering and Bioscience, Georgia Institute of Technology, 901 Atlantic Drive, Atlanta, Georgia 30332, USA

John S. Fossey Department of Chemistry, University of Bath, Bath, BA2 7AY, UK

Xiaohu Gao Department of Bioengineering, University of Washington at Seattle, WA 98195, USA

Cees Gooijer Analytical Chemistry and Applied Spectroscopy, Laser Centre, Vrije Universiteit Amsterdam, The Netherlands

Martin Hof J. Heyrovský Institute of Physical Chemistry of the ASCR, v. v. i., Dolejškova 3, 182 23 Prague 8, Czech Republic

Katrin Hoffmann Working Group Optical Spectroscopy, Department I, Federal Institute for Materials Research and Testing (BAM), Richard-Willstaetter-Str. 11, D-12489 Berlin, Germany

Tony D. James Department of Chemistry, University of Bath, Bath, BA2 7AY UK

Lennart B.-Å. Johansson Departments of Chemistry, Biophysical Chemistry, Umeå University, S-901 87 Umeå, Sweden

Piotr Jurkiewicz J. Heyrovský Institute of Physical Chemistry of the ASCR, v. v. i., Dolejškova 3, 182 23 Prague 8, Czech Republic

Mario Leclerc Centre de recherche sur la science et l'ingénierie des matériaux (CERSIM), and Département de chimie, Université Laval, Québec, QC, Canada G1K 7P4

Therese Mikaelsson Departments of Chemistry, Biophysical Chemistry, Umeå University, S-901 87 Umeå, Sweden

Sudip Kumar Mondal Department of Physical Chemistry, Indian Association for the Cultivation of Science, Jadavpur, Kolkata-700 032, India

Agnieszka Olżyńska J. Heyrovský Institute of Physical Chemistry of the ASCR, v. v. i., Dolejškova 3, 182 23 Prague 8, Czech Republic

Dietmar Pfeifer Working Group Optical Spectroscopy, Department I, Federal Institute for Materials Research and Testing (BAM), Richard-Willstaetter-Str. 11, D-12489 Berlin, Germany

Ute Resch-Genger Working Group Spectroscopy, BAM Department I, Federal Institute for Materials Research and Testing (BAM), Richard-Willstaetter-Str. 11, D-12489 Berlin, Germany

Alan G. Ryder Nanoscale Biophotonics Laboratory, School of Chemistry, National University of Ireland-Galway, Galway, Ireland

Radek Šachl Departments of Chemistry, Biophysical Chemistry, Umeå University, S-901 87 Umeå, Sweden

Kalyanasis Sahu Department of Physical Chemistry, Indian Association for the Cultivation of Science, Jadavpur, Kolkata-700 032, India

Mark P. Sena Department of Bioengineering, University of Washington at Seattle, WA 98195, USA

Jian Yang Department of Chemical Engineering, South China University of Technology, Guangzhou 510640, China

Jie Zheng Department of Physiology and Membrane Biology, University of California, One Shields Avenue, Davis, California 95616, USA

Simple Calibration and Validation Standards for Fluorometry

Ute Resch-Genger, Katrin Hoffmann, and Dietmar Pfeifer

Abstract Physical and chemical standards for fluorometry are classified, and general and type-specific requirements on suitable standards are derived. To improve the comparability and reliability of fluorescence data, simple calibration and validation standards are presented that enable an instrument characterization under routinely used measurement conditions.

Keywords Fluorescence · Standard · Spectral correction · Emission · Glass · Quantum yield · Fluorescence intensity · Quality assurance

Introduction

Within the last decades, luminescence techniques developed into analytical tools widely used in material and life sciences [1–5]. The ever-growing interest in these methods, that yield analyte-specific quantities such as emission and excitation spectra, fluorescence quantum yields, fluorescence lifetimes, and emission anisotropies, is due to their comparable ease of use, non-invasive character, potential for combining spectrally, temporally, and spatially resolved measurements, and suitability for multiplexing and remote sensing. Moreover, fluorescence detection techniques can provide a sensitivity down to the single molecule level. General drawbacks of all fluorescence-based methods are, however, intensity-, wavelength-, polarization-, and time-dependent, instrument-specific contributions to otherwise dye- or analyte-specific signals, general difficulties in accurately measuring absolute fluorescence intensities, and the dependence of the spectroscopic properties of most chromophores on their microenvironment [6–15]. This limits the comparability of luminescence data across instruments and, for the same instrument, over time, especially if these instrument-specific effects are not properly removed. Simultaneously, these dependences render quantitation from measurements of relative

U. Resch-Genger (✉)
Working Group Fluorescence Spectroscopy, BAM Department I, Federal Institute for Materials Research and Testing, Richard-Willstaetter-Str. 11, D-12489 Berlin, Germany
e-mail: ute.resch@bam.de

C.D. Geddes (ed.), *Reviews in Fluorescence 2007*, Reviews in Fluorescence 2007,
DOI 10.1007/978-0-387-88722-7_1,

fluorescence intensities difficult. Critical with respect to the globalization-induced trends of harmonization of measurements, traceability, and accreditation [16, 17] is the comparatively small number of reliable standards and guidelines for instrument characterization and instrument performance validation (IPV). With the exception of colorimetry or surface fluorescence [18, 19], and in part flow cytometry [20], the standardization of fluorescence measurements is still in its infancy. This hampers many applications of fluorescence techniques at reach in strongly regulated areas such as medical diagnostics and drug discovery.

The often demanded improvement of quality assurance in fluorometry is directly linked to the availability of suitable, evaluated, and preferably certified instrument calibration and validation standards for all the instrument quantities and parameters that can affect the analyte-specific spectral position, spectral shape, and intensity of measured fluorescence signals. This includes, e.g., the (relative) spectral responsivity of the emission channel of a fluorescence instrument, the (relative) spectral irradiance at sample position (equaling the (relative) spectral radiance of the excitation channel reaching the sample), the wavelength accuracy of the instrument's excitation and/or emission channel, and the instrument's spectral resolution. In addition, evaluated and internationally accepted recommendations and guidelines for instrument qualification are needed which consider the current state-of-the-art of fluorescence techniques. However, despite the ever-increasing use of fluorescence measurements, only in 2004, a consensus document was published by the Clinical and Laboratory Standards Institute (CLSI) in the area of strongly regulated laboratory medicine [20]. This document that has regulatory implications focuses on quantitative fluorescence calibration and specifically addresses analysis of cells and microspheres by flow cytometry. For other fluorescence techniques, there currently exist only comparably few guidelines and recommendations for the characterization of the respective instrumentation and the performance and evaluation of fluorescence measurements [21–26]. These documents were developed mostly in the 1980s, e.g., by ASTM International, the UK-based Ultraviolet Spectrometry Group, and the International Union of Pure and Applied Chemistry (IUPAC). This includes for instance methods for the determination of the wavelength accuracy and spectral resolution of fluorescence instruments and their linear and dynamic range as well as the limit of detection for particular analytes introduced by ASTM [21, 22, 23] and methods for the evaluation of fluorescence lifetime measurements from IUPAC [27]. Also, only comparably few recommendations on emission and excitation standards as well as fluorescence quantum yield and lifetime standards are available [2, 4, 7, 10, 24, 25, 28]. The quality and reliability of these documents is, however, often limited. Frequently, these recommendations are based on non- or only in part evaluated literature data and only recently, requirements on and quality criteria for fluorescence standards have been derived [9]. In addition, many standards are not fully characterized with respect to all the parameters that can affect their calibration-relevant properties and their application. This can result in considerable calibration and measurement uncertainties.

Aiming at an improved quality assurance for fluorescence techniques, in this chapter, a large variety of chemical and physical transfer standards commonly used in fluorometry is introduced and classified according to their scope. General and application- or type-specific quality criteria for and requirements on instrument calibration and validation standards are defined, and simple strategies and methods are presented for instrument characterization and performance validation under application-relevant measurement conditions. In addition, approaches toward the adaptation of fluorescence standards to different fluorescence techniques are illustrated. Special emphasis is dedicated to steady state fluorometry and newly developed, simple chromophore-based calibration and validation tools.

Overview of Fluorescence Standards

Instrument-Specific Quantities Affecting Fluorescence Signals

Each luminescence technique yields signals $I_m(\lambda_{ex},\lambda_{em})$ that are determined by both instrument- and analyte-specific quantities, see eq. (1) [29]. λ_{ex} and λ_{em} represent the excitation and the emission wavelength, respectively:

$$I_m(\lambda_{ex}, \lambda_{em}) = f(\lambda_{ex}) \times F_\lambda(\lambda_{ex}, \lambda_{em}) \times E_\lambda(\lambda_{ex}) \times s(\lambda_{em}) \times G \qquad (1)$$

Instrument-specific quantities include the spectral irradiance E_λ reaching the sample and the spectral responsivity s [30] of the emission or detection channel [9, 10, 24, 31]. The wavelength dependence of E_λ, i.e., $E_\lambda(\lambda_{ex})$, is controlled by the spectral radiance of the excitation light source and the transmittance of optical components such as lenses, mirrors, filters, monochromator gratings, beam splitters, and polarizers in the excitation channel. The wavelength dependence of s, i.e., $s(\lambda_{em})$, is determined by the transmittance of the optical components in the emission channel and by the spectral responsivity of the detector. In addition to their wavelength dependences illustrated in Fig. 1, E_λ and s are polarization-dependent and, due to aging of optical and optoelectronical instrument components, they are also time-dependent. The geometry factor G in eq. (1), which accounts for the ratio of the solid angle of fluorescence emission, the solid angle of detection, and the size of the illuminated volume, depends on both the instrument and the sample and needs to be determined only for measurements of absolute fluorescence intensities [6, 32].

Analyte-specific quantities that determine measured fluorescence signals $I_m(\lambda_{ex},\lambda_{em})$ from the sample side are the analyte's absorption factor (λ_{ex}) α at the excitation wavelength and its spectral fluorescence yield $F_\lambda(\lambda_{ex},\lambda_{em})$ [1–5, 9, 10, 24, 34]. Both α and F_λ are typically sensitive to dye microenvironment. α is nonlinearly linked to absorbance and thus to the concentration by the Beer–Lambert law. $F_\lambda(\lambda_{ex},\lambda_{em})$ reveals the spectral shape of the fluorescence spectrum of the analyte. For a single nonaggregated chromophore in a homogeneous matrix displaying simple photophysics [35], F_λ does not depend on excitation wavelength. F_λ is linked to the integral quantum yield of photoluminescence Φ_l that represents the number of emitted photons per number of absorbed photons [36, 37]. Φ_l is one of the most important parameters in luminescence analysis. It characterizes a radiative transition

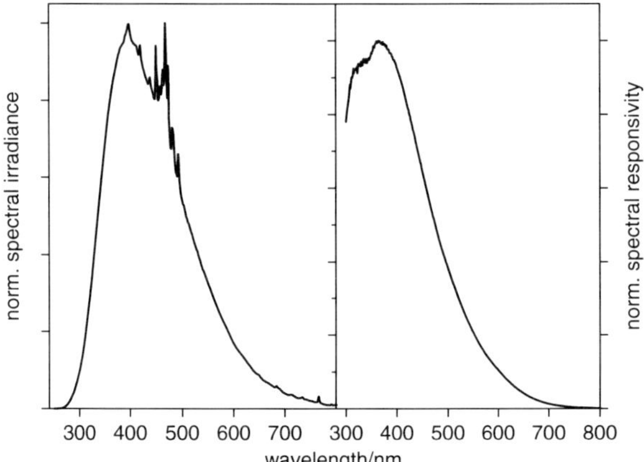

Fig. 1 Wavelength dependences of the spectral irradiance reaching the sample E_λ (*left*) and the spectral responsivity s of the emission channel (*right*) of common fluorometers equipped with a xenon excitation light source and a photomultiplier tube (PMT). $E_\lambda(\lambda_{ex})$ was determined with a calibrated detector at sample position and $s(\lambda_{em})$ with a calibrated light source and a calibrated white standard at sample position [33]

in combination with the luminescence lifetime and the luminescence spectrum and determines the sensitivity for the detection of a certain analyte from the sample side together with the analyte's molar absorption coefficient at the excitation wavelength.

The influence of the instrument-specific wavelength dependences of E_λ and s depicted in Fig. 1 on measured fluorescence data is exemplarily illustrated in Fig. 2

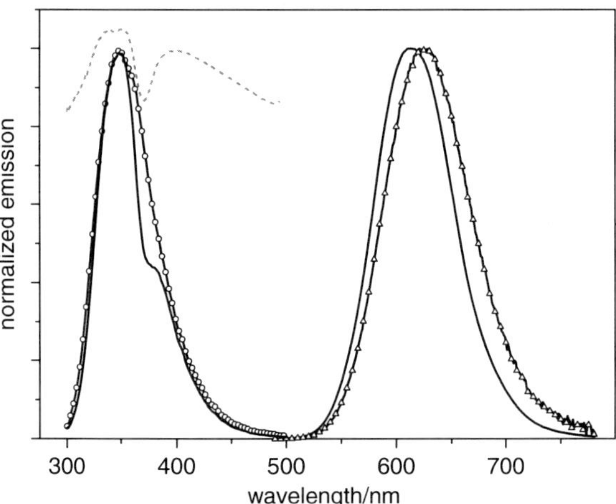

Fig. 2 Effect of $s(\lambda_{em})$ of a typical fluorometer on the measured emission spectra of tryptophan (*circles*) and an red emitting fluorophore dye Y (*triangles*): uncorrected emission spectra (*solid line*) vs. corrected emission spectra (*symbols*). The indentation at ca. 370 nm in the uncorrected spectrum of tryptophan results from diffraction effects (*Wood anomalies*) of the instrument's monochromator gratings. As this indentation naturally appears also in the wavelength dependence of the instrument's spectral responsivity (*dashed line*), it can be corrected for, see indentation-free corrected spectrum

for the spectral responsivity and for the emission spectra of two chromophores, tryptophan and dye Y, emitting in the UV and the long wavelength region. Tryptophan is a frequently measured, naturally occurring fluorescent biological molecule, and dye Y is a representative example of a bioanalytically relevant red-emitting fluorescent probe. Such red-emitting probes and labels are of increasing importance in fluorescence analysis. The obvious deviations between the measured instrument-specific or so-called uncorrected fluorescence emission spectra and the emission spectra (spectrally) corrected for the wavelength (and polarization) dependences of s underline the need for spectral correction as a prerequisite for fluorescence data that are comparable across instruments and, for the same instrument, over time [33, 38–40]. Similar instrument-specific distortions result for excitation spectra, here related to the wavelength (and polarization) dependence of E_λ.

Types of Fluorescence Standards

The comparability of fluorescence data between instruments and for a single instrument, over time, and the minimization of fluorescence measurement-inherent sources of error require a reliable instrument characterization and a regular performance validation following evaluated and overall accepted procedures with suitable fluorescence standards. Depending on the desired application, such standards can be of physical or chemical nature. Classical examples for physical transfer standards are calibrated lamps or detectors. Chemical transfer standards are liquid or solid chromophore-based reference materials.

Fluorescence standards can be divided into three general types depending on their scope and application: (i) instrument calibration standards, (ii) instrument validation standards, and (iii) application-specific fluorescence standards. *Instrument calibration standards* are standards for the determination and correction of instrument bias. The scope of these standards, which can be of either physical or chemical nature, is to rule out instrumentation as a major source of variability and to yield instrument-independent comparable fluorescence data. Typical instrument calibration standards are spectral fluorescence standards for the determination of the wavelength accuracy of wavelength-selecting optical components or the spectral characteristics of fluorescence instruments [9].

Instrument validation standards represent tools for the periodic check of instrument performance [9]. Such standards can be of either physical or chemical type. Instrument validation standards can be identical with instrument calibration standards or can, e.g., belong to the class of fluorescence intensity standards depending on the instrument parameter(s) to be checked. Typical examples of instrument validation standards are day-to-day intensity standards, which test the instrument's day-to-day performance and long-term stability and may even enable correction for these variabilities.

Application-specific fluorescence standards include the class of fluorescence intensity standards as well as fluorescence lifetime and fluorescence polarization standards [2, 41]. Fluorescence lifetime and fluorescence polarization standards are beyond the scope of this article. Application-specific fluorescence standards are based on routinely measured or closely related [42] chromophores and consider the fluorescence properties of typically measured samples. The scope of these reference materials is, e.g., the quantitation from measured fluorescence intensities [43, 44], i.e., to relate chemical concentration to instrument response or the validation of methods involving fluorescent samples. The class of fluorescence intensity standards also includes fluorescence quantum yield standards [9] that can be used to link fluorescence intensity to an instrument-independent, comparable, relative scale.

General Requirements on Fluorescence Standards

Every fit-for-purpose qualification of a fluorescence instrument with minimum uncertainty has to be performed under commonly used measurement conditions, i.e., application-relevant measurement conditions [4, 9]. Solely, this guarantees its reliability and suitability for the consideration of instrument-specific effects and instrument drift. Exceptions are here, e.g., the validation of the wavelength accuracy of fluorescence instruments, which is typically performed at maximum spectral resolution, and measurements aiming at the comparability of fluorescence signals between different instruments. In the latter case, instrument settings are to be chosen that can be employed for a broad variety of different instruments. Accordingly, suitable fluorescence standards must be not only well characterized with respect to their calibration-relevant properties, but measurable with routinely used instrument settings such as spectral bandpass or spectral resolution, detector voltage, polarizer settings as well as measurement geometry. For measurements with pulsed light sources, delay, gate, and (integration or scanning) time also need to be considered. The latter parameters can influence the emission spectra of materials that contain a mixture of chromophores displaying long and different luminescence lifetimes [9]. In addition, knowledge of the linear range of the instrument's detection system is mandatory, since nonlinearities of the detection system can result in distorted signals and can thus lead to considerable measurement uncertainties [33]. Fluorescence instruments equipped with a reference channel, which accounts for fluctuations of the excitation light intensity (flux), need to be operated within their linear range.

The stringent requirement of use under application-relevant measurement conditions is best met with chemical transfer standards due to their chromophore nature that guarantees emission characteristics comparable to those of typically measured fluorescent samples [33]. The fulfillment of this criterion can be critical for physical transfer standards, especially, e.g., for source-based standards, the spectral radiances or emission intensities of which exceed those of common fluorophores by several orders of magnitude [32].

International Equivalence of Measurements

The use of similar instrument settings for instrument characterization and the actual fluorescence measurements is also the prerequisite for *traceable fluorescence measurements*. Traceability is a metrological requirement that is closely related to the globalization-induced general need for the harmonization and comparability of measurements. It is documented in ISO/IEC 17025 relevant, e.g., for laboratory accreditation [45]. ISO/IEC 17025 demands an instrument qualification to be traceable to the relevant SI units, if possible, or to other accepted primary standards via an unbroken chain of comparisons with given uncertainties [45, 46].

Traceability in the case of fluorometry implies at least linking of fluorescence spectra or (integral) fluorescence intensities (fluxes) to radiometric quantities, i.e., to the *spectral radiance scale* and/or the *spectral responsivity scale* with the aid of accordingly characterized physical or chemical transfer standards [31, 32, 33, 47]. In the case of physical transfer standards such as lamps or detectors, this denotes calibration against the source- and detector-based primary standards via working standards and with a quoted uncertainty. These primary standards, which are realized by National Metrology Institutes (NMIs) from different countries, are a high-temperature black body radiator in the vis/NIR spectral region, and an electron storage ring in the UV spectral region in the case of the spectral radiance, or a cryogenic radiometer for the spectral responsivity. The calibration of physical transfer standards is typically performed by NMIs or accordingly accredited companies equipped with NMI-calibrated working standards. In the case of chemical transfer standards, the calibration-relevant fluorometric properties, such as the corrected fluorescence spectra, need to be measured with a reference fluorometer traceably characterized with calibrated physical transfer standards with a quoted uncertainty. At present, this is only fulfilled by fluorescence standards certified and released by the National Institute of Standards and Technology (NIST), USA and the Federal Institute for Materials Research and Testing (BAM), Germany.

To guarantee the international equivalence of the source- and detector-based primary standards used for the dissemination of these scales between the NMIs of different countries, in particular as signatories of the CIPM (Comité International des Poids et Mesures) Mutual Recognition Arrangement (MRA) of 1999, regular comparisons of measurement capabilities are performed [48–50]. Due to the importance of comparable fluorescence data, the comparability and equivalence of the fluorescence measurements of NIST and BAM and other NMIs active in the area of high-precision spectrofluorometry, i.e., the National Research Council (NRC), Canada and the Physikalisch Technische Bundesanstalt (PTB), Germany, have been recently evaluated by a Round Robin test.

Relative vs. Absolute Instrument Characterization

Depending on their calibration, physical transfer standards, such as lamps and detectors, can enable an absolute instrument characterization and thus the determination of the absolute values of $E_\lambda(\lambda_{ex})$ and $s(\lambda_{em})$ in eq. (1). Chromophore-based reference materials typically allow only a relative characterization of the instrument's spectral characteristics such as the determination of the relative spectral shape of the spectral responsivity and the spectral irradiance reaching the sample. This is sufficient for the comparable measurement of fluorescence spectra and for traceable fluorescence measurements. However, the comparability of fluorescence intensities over time and between instruments is more challenging. For the broad users of fluorescence techniques, comparable fluorescence intensities can be best achieved with the use of fluorescence intensity standards providing (relative) reference systems [8, 9] in addition to a relative characterization of the instrument's spectral characteristics. When an absolute scale for fluorescence intensities is needed, this can be realized with a suitable fluorescence instrument in conjunction with a very sophisticated absolute instrument characterization that includes the very sophisticated determination of G in eq. (1) [7, 32, 47], see also "Measurement of Absolute Fluorescence".

General Requirements on the Characterization of Standards

The quality of a standard is determined by its properties and the wealth of information provided with the standard. As the calibration-relevant radiometric or fluorometric properties of a standard are affected by the instrument settings and measurement conditions chosen for their determination, see "General Requirements on Fluorescence Standards", generally, not only the standard's calibration-relevant properties and its scope and limitations for use as well as the recommended recalibration intervals or the shelf life (stability) must be reported, but also all the parameters that can influence them and their uncertainty [9,51]. Solely this enables the evaluation of the standard's suitability. Additional application- or type-specific requirements are discussed in the forthcoming sections.

These general quality criteria are typically fulfilled for physical standards, such as lamps and detectors, but scarcely for chromophore-based standards. Similar to the commonly measured fluorophores, the absorption and emission spectra, molar absorption coefficients, and fluorescence quantum yields of chemical transfer standards can also depend on microenvironment. This includes solvation, solvent polarity, and, in aqueous media, pH value and ionic strength as well as viscosity and temperature. Moreover, the fluorescence quantum yields and lifetimes of these standards can be sensitive to the presence of fluorescence quenchers such as oxygen [52]. Accordingly, for well-characterized chromophore-based fluorescence standards—in addition to the report of the instrument settings and measurement geometry employed—not only the chromophore purity [53], chromophore concentration, and solvent or matrix (type and purity) should be provided, but also the information on, e.g., concentration-dependent absorption and emission properties, an excitation wavelength-dependent spectral shape of the emission spectrum, and excitation wavelength-dependent fluorescence quantum yields [54–56]. Additionally, solid materials need to be tested for the homogeneity of the fluorophore content [57]. Here, the suitability can also depend on the spatial resolution of the respective fluorescence technique. Furthermore, the standard's fluorescence anisotropy (r) should be given as the size of this quantity of the standard's emission that determines whether the standard can be used without polarizers. As a rule of thumb, nearly isotropic emitters with r ca. ≤ 0.05 render polarizers dispensable, whereas for partly or strongly anisotropic materials, such as organic fluorophores embedded into a solid matrix [33], the use without polarizers results in an enhanced calibration uncertainty, the size of which increases with increasing anisotropy of the standard's emission.

For commercial chromophore-based standards, in addition the criteria for the production of reference materials as stated in ISO Guide 34 [58] and ISO Guide 35 [59] should be fulfilled. NMIs certify their reference materials according to the requirements imposed by the respective ISO guides and typically provide a wealth of information on their certified standards and reference materials in their certificates and certification reports [60–62] including uncertainty budgets. Also, these fluorescence standards are traceable. For chromophore-based fluorescence standards from non-NMI sources, if not stated, the production must not necessarily follow the respective ISO guides. Also, helpful information on parameters affecting calibration-relevant properties are often missing, at least partly, and uncertainties are generally not provided. To the best of our knowledge, at present, reference materials from non-NMI sources are not traceable.

Spectral Fluorescence Standards

Spectral fluorescence standards are devices or reference materials, the wavelength dependences of the spectral radianceor spectral responsivity (physical transfer standards) or the instrument-independent corrected fluorescence spectra (chemical transfer standards) of which are known and preferentially certified with a quoted uncertainty. The standard's scope determines the requirements of the spectral shape and structure of the spectra.

Standards for the Validation of the Wavelength Accuracy

Wavelength standards are physical and chemical standards suitable for the validation of the wavelength accuracy of the wavelength-selecting optical components incorporated into fluorescence instruments. In addition to the general requirements of fluorescence standards derived in "General Requirements on Fluorescence Standards" and "General Requirements on the Characterization of Standards", such wavelength standards must emit a multitude of very narrow emission bands in the UV/vis/NIR spectral region at known spectral positions with a given uncertainty [63]. The spectral resolution of the instrument to be characterized determines the acceptable width of the spectral lines of the wavelength standard. The validation of the wavelength accuracy is the first step of the qualification of most fluorescence instruments [21]. Also, the instrument's spectral resolution can be obtained with the aid of wavelength standards.

For the calibration of the wavelength scale of high-precision spectrofluorometers, where typically an accuracy of about 20 cm^{-1} ($\pm$ 0.5 nm at 500 nm) is desired, the most common and best choice are atomic discharge lamps that display extremely narrow emission lines, see Fig. 3 (top). To cover the UV/vis/NIR spectral region, such lamps often contain mixtures of gases such as mercury, argon, and neon [64–69]. As the spectral position of these emission bands is affected by gas pressure, this parameter should be reported by the standard's manufacturer and supplier. Since atomic discharge lamps exhibit typically a very large spectral radiance (emission intensity) as compared to fluorescent samples, the use of an attenuator such as a white standard or diffuse scatterer is often mandatory to avoid detector saturation.

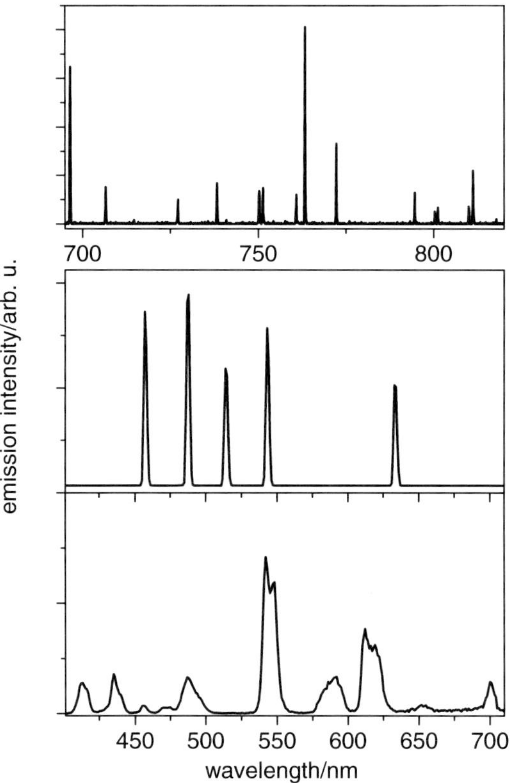

Fig. 3 Methods and standards for the determination and control of the wavelength accuracy of different fluorescence instruments. *Top*: emission spectrum of an atomic discharge lamp containing a mixture of mercury and argon for the validation of the wavelength scale of high-precision spectrofluorometers. *Middle*: emission lines of different lasers, here a multiline Ar^+-laser and a HeNe-laser (543 nm and 633 nm, respectively), normally used for checking the wavelength accuracy of confocal spectral imaging systems in conjunction with a mirror. *Bottom*: emission spectrum of a fluorescent glass doped with a multitude of rare earth (RE) metal ions for the validation of the wavelength accuracy of fluorescence measuring systems with low spectral resolution. Excitation was at 365 nm

For instruments with a lower spectral resolution, which are equipped with different lasers as excitation sources such as confocal spectral imaging systems, the emission lines of the lasers can be used for the control of the wavelength accuracy of the emission channel in conjunction with, e.g., a mirror, see middle panel of Fig. 3. Confocal spectral imaging systems, such as confocal laser scanning microscopes (CLSM), are typically operated with a fixed spectral bandpass of at the minimum 2 or 2.5 nm or more common, 5 or 11 nm, with the applicable spectral bandpass depending on the instrument's design.

An easy-to-use chemical alternative for the determination of the wavelength accuracy are chromophore-based wavelength standards, particularly for fluorescence instruments with a reduced spectral resolution such as confocal spectral imaging systems or certain microplate readers [9, 63] where the high accuracy provided by atomic discharge lamps is not necessary. Examples of chemical wavelength standards include, e.g., $Y_{3-x}Dy_xAl_5O_{12}$, a dysprosium-activated yttrium garnet [70, 71], adaptation of a reflectance standard such as SRM 2036 [72] to fluorescence measurements, and glass-based materials currently developed by NIST and BAM [4]. In addition, although questionable, the potential of luminescent nanocrystals or the so-called quantum dots has been tested [73]. Different manufacturers of steady state fluorometers also recommend the determination of the wavelength accuracy via scanning of the emission lines of the instrument's xenon excitation source or the transmission minima in the spectrum of a solution of holmium perchlorate. The former procedure is restricted to the spectral region of ca. 450–550 nm where the xenon spectrum displays narrow lines, see also Fig. 1. The latter method is hampered by the necessary use of a specific calibration accessory and measurement geometry.

The corrected emission spectrum of a very promising candidate wavelength standard currently tested by BAM [9, 63] is shown in the bottom of Fig. 3. This material, a glass doped with a mixture of rare earth (RE) ions, will be eventually provided in a variety of formats for different fluorescence techniques such as, e.g., cuvette-shaped for spectrofluorometry and slide-shaped for fluorescence microscopy with corrected emission spectra certified for different application-relevant excitation wavelengths [74], temperatures, and different spectral bandpasses/spectral resolutions of the BAM reference fluorometer. Since the luminescence intensity of such a material can be controlled by dopant concentration, not only the shape and size of its radiating volume, but also its spectral radiance or fluorescence intensity can be adapted to that of commonly measured samples.

Standards for the Determination of the Relative Spectral Responsivity

The wavelength dependence of the spectral responsivity of the emission channel of a fluorescence instrument, see $s(\lambda_{em})$ in eq. (1), can be obtained with physical or chemical spectral radiance transfer standards. Suitable devices or materials must emit a known broad and unstructured spectrum within the application-relevant wavelength range, typically the UV/vis/NIR spectral region [6, 7, 10, 12, 24, 75]. Fulfillment of this requirement is mandatory to minimize the dependence of the shape of the standard's spectrum on instrument resolution/spectral bandpass. Knowledge of $s(\lambda_{em})$ is the prerequisite for the correction of emission spectra for instrument-specific effects and the determination of fluorescence quantum yields.

Physical spectral radiance transfer standards are tungsten ribbon lamps or integrating sphere-type radiators [10, 24]. Advantageous of these devices are their very broad unstructured emission spectra that cover the UV/vis/NIR spectral region [31]. General drawbacks are a tedious alignment, regular and expensive recalibrations [76], restrictions on measurement geometry, and a considerable size that can hamper their application for compact fluorescence instruments. Moreover, their spectral radiances exceed those of typical fluorescent samples by at least four (tungsten ribbon lamp) to two (integrating sphere radiator) orders of magnitude [32, 33]. Accordingly, operation of these devices under application-relevant measurement conditions within the linear range of the instrument's detection system requires sophisticated attenuation procedures that do not introduce additional spectral effects increasing the calibration and, subsequently, the measurement uncertainty [33]. Also, the previously calibrated emission channel of a fluorescence instrument, see "Standards for the Determination of the Relative Spectral Irradiance", principally represents an equivalent of a calibrated lamp [33]. Generally, the use of physical source-based transfer standards requires a solid knowledge of optics and is only recommended for NMIs and expert laboratories.

Chemical equivalents of physical spectral radiance transfer standards are commonly referred to as emission standards. Only with these simple and easy-to-use calibration tools, the broad community of fluorescence users can be reached. The close match of the spectral radiance and the size and shape of the radiating volume of both standard(s) and samples enables a straightforward determination of the instrument's relative spectral responsivity under application-relevant conditions [33]. Type-specific requirements on emission standards include, e.g., minimum overlap between absorption and emission for a moderate influence of dye concentration and measurement geometry on spectral shape, moderate to high fluorescence quantum yields to enhance the signal-to-noise ratio and to reduce the influence of stray light, solvent emission, and fluorescent impurities on the spectral shape of the standard's fluorescence spectrum, sufficient thermal and photochemical stability under application-relevant conditions, and a small fluorescence anisotropy (r), e.g., $r \leq 0.05$ within the analytically relevant room temperature region. The latter circumvents additional material-related polarization effects and reduces uncertainties for use without polarizers [33, 77, 78], see also "General Requirements on the Characterization of Standards".

A number of emission standards have been recommended in the literature [9, 10, 24, 25, 79–85] and several are available from commercial (non-NMI) sources [86, 87]. However, none of these materials is supplied with traceable and certified, corrected fluorescence data and wavelength-dependent uncertainties, and most of the general and type-specific quality criteria discussed in "General Requirements on Fluorescence Standards" and "General Requirements on the Characterization of

Standards" are not fulfilled [9, 33]. This renders the reliability of the supplied spectra questionable. The only emission standards that provide the necessary reliability and traceability are Standard Reference Material (SRM) 936 *Quinine Sulfate Dihydrate* from NIST [60, 61] and BAM-F001 to BAM-F005 (equaling the organic dyes A–E in earlier publications) recently certified by BAM [62], see Fig. 4, all used as solutions. Most likely, this will also be true for the glass-type emission standards SRMs 2940 and 2941 soon to be released by NIST [88, 89]. The slightly structured emission spectrum of SRM 2941, however, may be problematic as this can introduce a considerable dependence on spectral bandpass and spectral resolution.

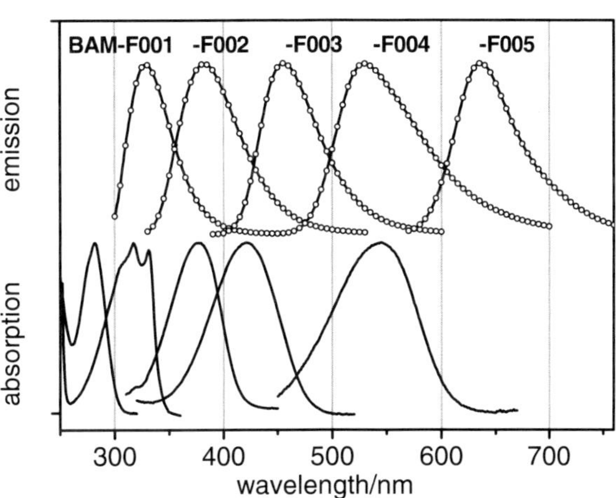

Fig. 4 Absorption (*bottom*) and corrected emission (*top*) spectra of the BAM Calibration Kit *Spectral Fluorescence Standards* BAM-F001 to BAM-F005 in ethanol. Excitation of each dye was at its longest wavelength absorption maximum. BAM-F001 to BAM-F005 equal the organic dyes A–E in earlier publications

A drawback of otherwise straightforward emission standards is the, compared to a tungsten ribbon lamp or an integrating sphere-type radiator, small spectral region covered by the emission spectrum of a typical organic or inorganic chromophore. For instance, quinine sulfate dihydrate that reveals a very broad emission spectrum for a fluorophore covers only the spectral region of 395–565 nm [90,91]. Accordingly, a dye-based emission calibration within the UV/vis/NIR region can only be realized with a set of emission standards. However, although such a chromophore-based strategy has been first applied by Lippert [82], fit-for-purpose combinations of emission standards are still very rare. For such sets, not only each component must be suitable, but the spectra of the set dyes must cross at points of sufficient fluorescence intensity, e.g., at least at 20% of the relative maximum fluorescence intensity for a reasonable calibration uncertainty [33, 83]. To the best of our knowledge, at present, the only set of emission standards that meets all the stringent requirements on and quality criteria for emission standards is the Calibration Kit *Spectral Fluorescence Standards* BAM-F001 to BAM-F005 [92–95] covering the spectral range from 300 to 770 nm. The absorption and emission spectra of the Kit components are revealed in Fig. 4.

The working principle of BAM-F001 to BAM-F005 is illustrated in Fig. 5. For each spectral standard F00*x* (*x* equaling 1–5), the quotients $Q^{F00x}(\lambda_{em})$ of

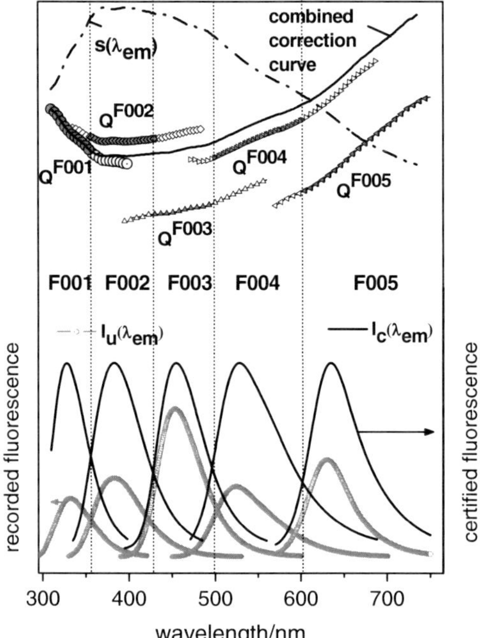

Fig. 5 Determination of $s(\lambda_{em})$ of a typical fluorescence instrument with the Calibration Kit *Spectral Fluorescence Standards* BAM-F001 to BAM-F005. *Bottom*: normalized certified corrected emission spectra $I_c(\lambda_{em})$ of BAM-F001 to BAM-F005 (*solid black lines*) and corresponding spectrally uncorrected (*background-corrected*) emission spectra $I_u(\lambda_{em}$; *grey symbols*) measured with the instrument to be calibrated. *Middle*: quotients $Q^{F00x}(\lambda_{em}) = I_c(\lambda_{em})/I_u(\lambda_{em})$ obtained for each dye F00*x* with closed *symbols* representing quotients being considered for the calculation of the combined correction *curve* and *open symbols* equaling neglected quotients, respectively. *Top*: weighted combination of $Q^{F00x}(\lambda_{em})$ to the combined correction curve equaling $1/s(\lambda_{em})$ (*solid line*) and its reciprocal $s(\lambda_{em})$ (*dash-dotted line*). Multiplication of measured spectra with the correction curve yields corrected spectra

the certified normalized corrected emission spectra $I_c(\lambda_{em})$ (Fig. 5, bottom, solid lines) and the spectrally uncorrected (background-corrected, [96]) emission spectra $I_u(\lambda_{em})$ (Fig. 5, bottom, grey symbols; measured with the instrument to be calibrated) are calculated. These quotients $Q^{F00x}(\lambda_{em})$, shown in the middle part of Fig. 5, represent the individual inverse spectral responsivities $1/s(\lambda_{em})$ within the emission regions of each dye F00*x*. The global inverse relative spectral responsivity $1/s(\lambda_{em})$ of the instrument to be calibrated (Fig. 5, top) follows from the statistically weighted combination of $Q^{F00x}(\lambda_{em})$. $1/s(\lambda_{em})$ equals the combined emission correction curve. A critical step of the determination of the global emission correction curve is the reliable and comparable linkage of $Q^{F00x}(\lambda_{em})$. To achieve this for the broad majority of users, a software that performs this linking procedure is mandatory. Accordingly, we developed the software *LINKCORR* for data evaluation to minimize standard- and calibration-related uncertainties in conjunction with the BAM reference materials.

As the corrected emission spectra of the Kit components were determined with a traceably (to the spectral radiance scale) calibrated reference fluorometer, the use of these calibration tools provides the basis for traceable emission measurements [97, 98]. The corrected emission spectra of BAM-F001 to BAM-F005 and the determination of a dye-based emission correction with the BAM Calibration Kit and the BAM software *LINKCORR* were both validated in an interlaboratory comparison of the NMIs NIST, NRC, PTB, and BAM. In addition, the performance of these reference materials and the data evaluation procedure was tested by selected laboratories from academia and industry comparing the uncorrected spectra and Kit-based corrected spectra of three test dyes. Results from these comparisons yield an excellent agreement of Kit dye-based corrected emission spectra with relative deviations of at maximum 5% for signals $\geq$ 10% of the maximum intensity. This provides a hint for the achievable comparability of emission spectra upon a widespread use of this new calibration tool.

Standards for the Determination of the Relative Spectral Irradiance

The wavelength dependence of the spectral irradiance reaching the sample can be determined with physical detector-based transfer standards and chemical transfer standards. Type-specific requirements on such standards are, e.g., either a known spectral responsivity or a known corrected fluorescence excitation spectrum. Knowledge of $E_\lambda(\lambda_{ex})$, see eq. (1), is the prerequisite for the correction of excitation spectra for instrument-specific effects and the comparison of (integral) emission intensities measured at different excitation wavelengths.

The most common method for the determination of the relative spectral shape of E_λ is the use of a calibrated spectral responsivity transfer standard such as a calibrated detector, typically a silicon photodiode (simple or integrating sphere-type trap detector [99]). This is also pursued by us. Alternatively, a pyroelectric detector [100] can be used, that is, however, much less sensitive. Also, the previously characterized emission channel can be applied as equivalent of a calibrated detector in a synchronous scan of the excitation and emission channel with a white standard at sample position [33].

Chemical alternatives to these physical transfer standards represent excitation standards with known corrected excitation spectra [10, 24, 25, 33], quantum counters [36, 80, 101, 102], and actinometers [103, 104]. Also, the absorption and excitation spectrum of a chromophore can simply be compared [33]. Advisable is neither the use of a quantum counter that is prone to polarization and geometry effects, nor the application of an actinometer that relies on the wavelength-independent quantum yield of a photochemical reaction, yielding a measurable and well-characterized product. Also, the comparison of the absorption and excitation spectrum of a chromophore can lead to a comparatively high calibration uncertainty [105], if the dye photophysics is not very well known and, e.g., the fluorescence quantum yield of the dye depends on excitation wavelength for excitation at two different electronic transitions.

The tempting use of excitation standards that must similarly fulfill the type-specific requirements detailed for emission standards in "Standards for the Determination of the Relative Spectral Irradiance" is hampered by the lack of certified excitation standards and generally the limited reliability of literature data on excitation standards [10, 24, 33]. In addition, for excitation standards, the use of dilute dye solutions is mandatory as the proportionality of fluorescence intensity to the absorption factor results in a concentration dependence of the spectral shape of excitation spectra and introduces a dependence on measurement geometry [33]. For instance, for a 0°/90° measurement geometry and a 1-cm cell, the chromophore absorbance should commonly not exceed 0.05. This imposes stringent requirements on dye concentration, optical pathlength in the cuvette, and measurement geometry. For dyes that display two optical transitions with considerable differences in oscillator strength, this implies the measurement of excitation spectra at two different concentrations. Accordingly, the coverage of a broad spectral region with excitation standards requires the combination of several dyes to a set (different dyes or reduced number of dyes and different concentrations). The first set of excitation standards that was only recently presented by us [33] is shown in Fig. 6. Principally, instrument characterization with such excitation standards can be performed similarly as described for BAM-F001 to BAM-F005 in "Standards for the Determination of the Relative Spectral Responsivity".

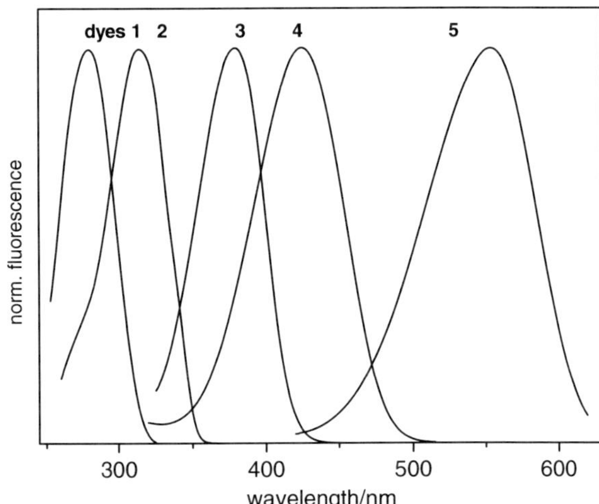

Fig. 6 Corrected excitation spectra of a set of liquid excitation standards developed by BAM

Fluorescence Intensity Standards

The quantitation of fluorophores from measurements of fluorescence intensities is hampered by two facts: the very challenging determination of absolute fluorescence intensities and the sensitivity of the spectroscopic properties of most chromophores to their microenvironment. In addition to the consideration of the instrument's spectral characteristics, the determination of absolute fluorescence intensities requires the knowledge of both the absolute fraction of incident photons per time in the sample, that are absorbed by the analyte, and the collection efficiency of the instrument which together control the fraction of the fluorescence photons detected. For the majority of instruments and users of fluorescence techniques, such a characterization that is detailed in "Measurement of Absolute Fluorescence" is not practicable. The tedious measurement of absolute fluorescence intensities can be elegantly circumvented by the determination of the relative spectral shape of s and/or E_λ with spectral fluorescence standards as described in "Spectral Fluorescence Standards" in conjunction with the straightforward use of fluorescence intensity standards. Such fluorescence intensity standards link the measured fluorescence intensity of a sample to that of a standard, thereby defining a relative intensity scale or reference system. Based on such a scale, chemical concentration can be linked to instrument response, (relative) fluorescence quantum yields can be determined, and measurements of fluorescence intensities can be made comparable between instruments and laboratories as well as over time. This provides also a tool for the validation of the instrument performance as described in "Instrument Validation Standards" .

Standards to Relate Chemical Concentration to Instrument Response

Standards that relate chemical concentration to instrument response compare the fluorescence intensity of a sample to that of a standard of known fluorophore content under identical measurement conditions to quantify the concentration or number of fluorophores in the sample. Though often overlooked, this is a common approach pursued in many analytical techniques. This type of intensity standard typically relies on the same fluorophore(s) as to be quantified. Problematic can be here the inherent sensitivity of spectroscopic properties such as molar absorption coefficient and fluorescence quantum yield to dye microenvironment. Accordingly, for quantitation with minimum uncertainty, the chromophore has to be specified, and the standard has to be in the same microenvironment as the analyte. Only this guarantees identical absorption spectra, fluorescence spectra, molar absorption coefficients, and especially fluorescence quantum yields. This is, e.g., fulfilled for many applications of separation techniques with fluorescence detection such as high-performance liquid chromatography (HPLC) where the fluorescence intensities from free, i.e., unbound fluorophores in solutions of identical or at least very similar chemical composition are compared [106, 107]. To eliminate errors in quantitation, the dye purity in the reference material also needs to be reported [53, 108]. An example represents here SRM 1932 *Fluorescein Solution*, a solution of high and well-characterized purity of the most widely used fluorescent label, fluorescein, dissolved in a borate buffer at pH 9.5, certified for concentration by NIST in 2004 [109].

For the majority of fluorescence techniques, the criterion of matching microenvironments between sample and standard can only be fulfilled to a certain degree.

This is particularly true for the comparison of free and immobilized fluorophores, i.e., dyes attached to beads, particles, or macro- and biomolecules, e.g., polymers, proteins, antibodies, and DNA where considerable changes of at least the fluorescence quantum yield are to be expected. Accordingly, different concepts for fluorescence intensity standards have been developed that all aim at the consideration and minimization of the effect of dye microenvironment on fluorophore quantitation. For flow cytometry, which has become an increasingly popular technique in clinical diagnostics, the concept of molecules of equivalent soluble fluorophore (MESF) has been developed [20, 110–112] as an exemplary scheme for the reliable quantitation from measured integral fluorescence intensities in complex microenvironments. The calibration of the fluorescence intensity scale or fluorescence channel number of flow cytometers that can be only carried out with labeled cells or particles and not with simple fluorophores is performed with sets of microbeads with different amounts of surface-bound fluorophores with assigned MESF units. These units express the fluorescence intensity of these calibration beads in terms of the number of the corresponding free fluorophores in the same matrix under similar measurement conditions. In a second step, e.g., suspensions of cells labeled with the same fluorophore are measured.

The MESF concept represents a straightforward (relative) intensity scale comparable across instruments, laboratories, and over time, but does not enable to derive the absolute number of fluorophores in the sample. This concept requires well-characterized bead-type MESF standards for common fluorescent labels. MESF standards of different quality are available from various commercial (non-NMI) sources. At present, the only certified concentration standard for the assignment of MESF values is SRM 1932 *Fluorescein Solution* [44, 109]. As the MESF concept relies on matching fluorescence spectra of the standard and the sample, standard-related contributions to the overall measurement uncertainty are directly linked to the degree of spectra matching of the standard and the sample [9, 20]. Accordingly, the availability of corrected emission spectra of representative standards used in flow cytometry and typical samples in application-relevant microenvironments could be helpful to estimate the size of these contributions to the overall uncertainty of quantitation in flow cytometry. In addition, the MESF concept needs to be extended from fluorescein to other more advanced fluorophores and to the simultaneous measurement of multiple dyes.

Fluorescence Quantum Yield Standards

Fluorescence quantum yield standards are used as a reference for the determination of the fluorescence quantum yield of an analyte. The fluorescence quantum yield is an instrument-independent property inherent to a chromophore that characterizes a radiative transition in combination with the luminescence lifetime, the emission anisotropy and the luminescence spectrum, see "Instrument-Specific Quantities Affecting Fluorescence Signals". Contrary to fluorescence intensity standards that relate chemical concentration to instrument response, fluorescence quantum yield standards are typically not based on the same fluorophore as the sample and do not require matching of spectra. However, they rely on spectrally corrected emission spectra of both the standard and the sample. In addition to the general quality criteria derived in "Overview of Fluorescence Standards", the most stringent requirements on fluorescence quantum yield standards are reliable fluorescence quantum yields with quoted uncertainties. Furthermore, standard and sample should absorb and emit within comparable spectral regions. Currently, the main problem related to fluorescence quantum yield standards is the very limited reliability of literature data with deviations of up to 50% being not uncommon, even for well-established chromophores such as coumarins. Accordingly, there is a considerable need for well-characterized fluorescence quantum yield standards for the UV/vis/NIR spectral region with high - and low -fluorescence quantum yields. This takes into account that not only the absorption and emission spectra of standard and sample should lay within a similar spectral region, but also the size of their quantum yields should be preferably close to avoid problems related to nonlinearities of the detection system or dilution errors.

Measurement of Absolute Fluorescence

One of the future trends in fluorometry is the enhanced interest in measuring absolute fluorescence intensities, e.g., for the characterization of the luminescence quantum yields of LED and OLED materials [113–117] and NIR chromophores [56]. The latter are increasingly applied in the life sciences. Moreover, only absolute fluorescence quantum yields can provide the necessary basis for the evaluation of literature data on quantum yield standards [25, 56, 118, 119] with the prerequisites of careful and reported method evaluation and supply of uncertainties.

The measurement of absolute fluorescence requires the additional determination of the geometry factor G in eq. (1) [32]. G depends strongly on the solid angle for the detection of fluorescence and the numerical apertures for excitation as well as on the type of sample (e.g., translucent or nontranslucent with emission from the surface) and the anisotropy of its emission [6, 47, 120], i.e., the angular distribution of its luminescence [77]. Also, for instance, the wavelength dependence of the refractive index of the solvent or matrix and of the container walls play a role. Accordingly, all these properties have to be known. Absolute fluorescence intensities of dilute solutions and cuvette-shaped transparent solid materials, such as glasses, can be measured either with an accordingly designed and characterized spectrofluorometer [47], or with an integrating-sphere setup [121]. In the case of solid anisotropic and scattering or strongly absorbing materials, typically an integrating-sphere setup is used [115, 116, 117, 122].

To improve the reliability of fluorescence quantum yield standards and generally of measurements of fluorescence quantum yields, the performance of interlaboratory comparisons on absolute quantum yields of application-relevant chromophores should be encouraged. The ultimate goal here should be to establish a set of fluorescence quantum yield standards for the UV/vis/NIR spectral region and to identify suitable methods for the measurement of the photoluminescence quantum yield of different types of samples such as nearly isotropic, anisotropic, scattering, and/or strongly absorbing materials.

Instrument Validation Standards

A stringent prerequisite for quality assurance in fluorometry is the regular validation of instrument performance at fixed (comparable) application-relevant measurement conditions with physical and/or chemical instrument validation standards. The aim here is to detect changes of the instrument's optical and optoelectronical components on time and to account for these changes, especially in the case of fluorescence intensity measurements. Only this ensures the necessary measurement accuracy and the ability to carry out meaningful comparisons of data acquired from multiple instruments as well as from the same instrument over time. Especially for clinical trials, not only instrument calibration, but also instrument performance validation is mandatory [11, 20]. Despite its obvious importance, however, to the best of our knowledge, there exist no overall accepted procedures for the broad majority of fluorescence techniques at present.

Instrument validation standards can either be identical with instrument calibration standards or represent standards exclusively employed for periodic controls such as day-to-day intensity standards. Such instrument validation standards do not necessarily need to closely match commonly measured samples, but should be measurable with typical instrument settings to guarantee the reliability of the instrument performance under routinely used measurement conditions. In addition to application-specific quality criteria discussed in the previous sections, the most stringent requirement on instrument validation standards is either a sufficient, well-characterized stability under applicable conditions, or, for single-use standards, an excellent reproducibility, preferably in combination with an assigned uncertainty. For chemical standards, stability and reproducibility are both closely linked to dye purity, and in the case of solutions, also to the purity of the solvent.

The choice of suitable instrument validation standards depends on the instrument parameters to be checked and thus, to a certain extent, on the respective fluorescence technique. Also, the desired accuracy of the IPV and the ease-of-use of a standard need to be taken into account. Generally, the assignment of changes in instrument performance to certain instrument parts requires tools for the independent measurement of $s(\lambda_{em})$ and $E_{\lambda}(\lambda_{ex})$. This enables at least a clear distinction between drifts arising from changes of the excitation and the emission channel. For frequent use to capture any drift in instrument performance under application-relevant conditions, simple standards are recommended to reduce costs and measurement time.

For steady state fluorometry that is the focus of this chapter and for the huge majority of fluorescence techniques, the regular check of the wavelength accuracy of the excitation and/or the emission channel and the wavelength dependence of the (relative) spectral responsivity and spectral sensitivity of the emission channel are strongly recommended. In many cases, the control of the (relative) spectral irradiance of the excitation channel at sample position is not that critical. Often, only emission spectra or integral fluorescence intensities at a fixed excitation wavelength are measured and short-term fluctuations of the spectral radiance of the excitation channel are typically considered for each fluorescence measurement via a reference detector. For the reliable correction of excitation spectra and the comparison of fluorescence intensities measured at different excitation wavelengths, however, the regular determination of $E_{\lambda}(\lambda_{ex})$ is mandatory.

The validation of the wavelength accuracy is typically performed with the same wavelength standards as used for instrument calibration. Due to its ease of use, the BAM wavelength standard shown in Fig. 3 (bottom) has most likely considerable potential here. The long-term stability of the spectral characteristics of the emission channel can be, e.g., determined by regular measurement and comparison of the (uncorrected) emission spectra of certified or in-house emission standards. BAM-F001 to BAM-F005 are suitable for this application. The long-term stability of $E_{\lambda}(\lambda_{ex})$ is best controlled with a detector standard. The comparability of measured fluorescence intensities over time can be achieved with the aid of fluorescence intensity standards such as quantum yield standards as detailed in "Fluorescence Intensity Standards" or with day-to-day intensity standards.

Day-to-Day Intensity Standards

Most users of fluorescence techniques are not particularly interested in the measurement of absolute fluorescence intensities, but in the determination of the long-term stability of their fluorescence instrument and aging-induced changes in its spectral sensitivity as prerequisites for the comparability of fluorescence intensities over time. Of additional interest is the comparability of fluorescence intensities across instruments. Thus, standards for day-to-day and instrument-to-instrument intensity are highly requested to establish a comparable fit-for-purpose intensity scale for both spectrally resolved and integral fluorescence measurements. The significance of such standards is related not only to the ever-increasing use of fluorescence measurements and according need for the qualification of fluorescence instruments, but also to the application of portable fluorometers, microplate, and microarray readers, as well as fluorescence microscopes in drug discovery and clinical diagnostics.

The major requirements on day-to-day and instrument-to-instrument intensity standards have been derived in the previous sections. Additional prerequisites are known corrected spectra, if, e.g., their intensities are to be compared with those of other fluorophores or between instruments with different spectral bandpasses. At present, the most popular day-to-day and instrument-to-instrument intensity standard of excellent stability is water in combination with the so-called Raman test incorporated into the software of many spectrofluorometers [9, 123, 124]. This simple test that measures the intensity of the Raman band from nonfluorescent water (typically provided in a sealed cell to prevent the uptake of impurities) at an excitation wavelength of 350 nm and a detection wavelength of 397 nm responds to both changes in the instrument's excitation and emission channel and is very convenient for daily use. Unfortunately, this method is limited to the UV region due to the λ^{-4}-dependence of the intensity of scattered light that results in already very weak signals in the vis region. Principally, solutions of fluorophores offer a greater flexibility with respect to the spectral region of excitation and emission and advantageous is also the close spectral match to typical fluorescent labels, but their often limited stability renders the regular displacement of these dye solutions necessary. However, many users of fluorescence techniques favor solid materials with an excellent long-term stability (desired shelf lives of at least 2 years) over liquid standards to be regularly replaced. As prerequisites for ease-of-use, such solid materials should be transparent and measurable without the use of polarizers. This is best met by inorganic fluorophores in a glass or polymer matrix [9, 63, 70, 125], since organic fluorophores embedded into polymers typically display an anisotropic emission that can make the use of polarizers mandatory. Also the photostability of solid materials doped with organic chromophores is reduced compared to that of, e.g., glasses containing inorganic chromophores such as RE ions or transition metal ions. Some day-to-day intensity standards are already commercially available, but their corrected fluorescence spectra and other relevant information such as data on their photostability are typically not supplied [9, 71, 125].

The significance of day-to-day and instrument-to-instrument intensity standards encouraged NIST and BAM to develop such materials [4]. Attractive candidates currently tested by both institutes are inorganic glasses doped with inorganic fluorophores displaying either broad emission spectra, such as SRMs 2940 and 2941 [88, 89], or narrow line-shaped spectra covering the UV/vis/NIR spectral region [9, 63]. The emission spectrum of a BAM candidate reference material is shown in Fig. 7. This versatile calibration tool not only represents a wavelength standard and a fluorescence intensity standard for a very broad spectral region, but also communicates changes in $s(\lambda_{em})$ via changes in the intensity pattern of its uncorrected emission spectrum. This is exemplarily illustrated in Fig. 7 by a comparison of the uncorrected spectrum (left panel) and the spectrally corrected spectrum (right panel) of this material.

These metal ion-doped glasses are very robust and do not photodegrade when excited by a fluorometer's conventional light source, even over long periods of time. Since many metal ion-based dopants have comparably long emission lifetimes, on the order of milliseconds in many cases [126], when pulsed light sources are used, the emission properties of these dopants can be affected by measurement parameters such as pulse duration, delay, and gate [9]. Accordingly, reference materials containing such dopants can be used as day-to-day intensity standards only in conjunction with a constant set of these measurement parameters. The suitability of these materials as instrument-to-instrument intensity standards is directly linked to the applicability of identical measurement conditions for the fluorescence instruments to be compared. For instruments equipped with continuous (non-pulsed) excitation

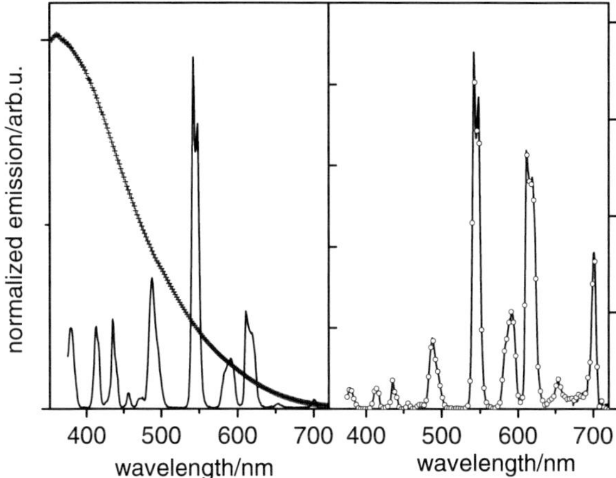

Fig. 7 Effect of $s(\lambda_{em})$ (*left panel, symbols*) on the measured emission spectrum of a fluorescent glass doped with RE ions: uncorrected emission spectrum (*left panel, solid line*) vs. corrected emission spectrum (*right panel*). Excitation was at 365 nm. This versatile BAM candidate reference material represents a potential wavelength standard and an instrument validation standard, i.e., a day-to-day intensity standard and a standard for the check of $s(\lambda_{em})$

sources, this is typically not critical. However, care has to be taken for instruments with pulsed excitation sources.

Adaptation to Different Fluorescence Techniques

Equation (1) in "Overview of Fluorescence Standards" is valid for each method measuring photoluminescence. Accordingly, it is very tempting to transfer and adapt evaluated and established procedures for instrument characterization and IPV including suitable standards to many other fluorescence techniques. Eventually, this may even enable a comparison of fluorescence data between different fluorescence techniques. Such an adaptation can be only successful if method-inherent requirements on standards are properly considered as well as method-specific limitations. This includes for instance adaptation of measurement geometry, sample/standard format, excitation wavelength(s), and (photochemical and thermal) stability [9]. Also, the luminescence lifetime of a potential standard needs to be carefully examined as this parameter controls the standard's suitability for techniques that use pulsed excitation light sources as derived in "General Requirements on Fluorescence Standards" and "Day-to-Day Intensity Standards" or that employ short measurement or integration times (pixel times) such as fluorescence microscopy.

A representative example for a method-adaptable standard is highlighted in Fig. 8. This figure displays the wavelength dependences of the (normalized) spectral

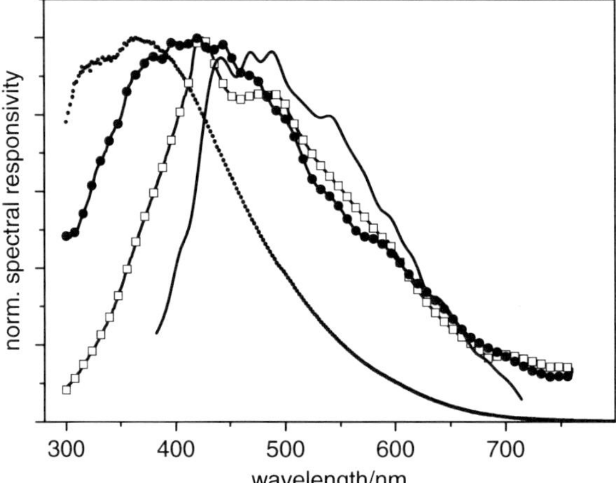

Fig. 8 Wavelength dependences of the normalized spectral responsivity of different fluorescence measuring systems determined with the Calibration Kit *Spectral Fluorescence Standards* BAM-F001 to BAM-F005: three spectrofluorometers from different instrument manufacturers (*dotted line, symbols*) and one confocal spectral imaging system (CLSM; *solid line*). In the latter case, BAM-F003 to BAM-F005 were used in conjunction with a microchannel device

responsivity of different fluorescence measuring systems such as spectrofluorometers and a confocal spectral imaging system determined with the spectral fluorescence standards BAM-F001 to BAM-F005. For the calibration of the microscope's emission channel, solutions of F003, F004, and F005 were filled into a microchannel device [127]. BAM-F001 to BAM-F005 offer a unique flexibility with respect to measurement geometry, format, and type of fluorescence instrument to be calibrated due to their use as dye solutions, the minimum spectral overlap between absorption and emission, their broad absorption spectra, the minimum dependence of the emission spectra on excitation wavelength, and their excellent photostability. Meanwhile, these reference materials have been successfully employed, e.g., for the characterization of the spectral characteristics of the emission channel of different confocal spectral imaging systems, monochromator-based microplate readers, and a colorimeter in front face, i.e., 45°/0° measurement geometry [9, 33, 127].

Similarly, the potential of the BAM glass standards is currently being evaluated for different fluorescence techniques ranging from conventional fluorometry over microfluorometry to fluorescence microscopy and Raman spectroscopy [9, 63, 128, 129] using different formats such as cuvette- or slide-shaped. Recently performed studies with these glass reference materials and different CLSM revealed not only an excellent chromophore homogeneity suitable for fluorescence microscopy, but also an encouraging long-term photostability under UV and vis laser excitation/illumination.

Conclusion and Outlook

To improve the quality assurance in fluorometry and the comparability of fluorescence data between instruments and laboratories, reliable and fit-for-purpose fluorescence standards with certified, calibration-relevant properties are mandatory, along with evaluated and internationally accepted guidelines for instrument calibration and performance validation. This encourages and requires a close interaction and collaboration between NMIs, instrument manufacturers, and regulatory agencies. In addition, it underlines the importance of the determination of uncertainties for representative fluorescence measurements for different fluorescence techniques. One approach here could be the performance of interlaboratory comparisons between National Metrology Institutes, expert laboratories, and routine users. The ultimate goal here should be to establish the achievable repeatability and accuracy of analyte determinations and assay results that can be eventually considered in recommendations and guidelines. Additionally, workshops and special training courses on drawbacks and sources of uncertainties of fluorescence techniques and instrument characterization—e.g., jointly organized by NMIs—may help to broaden the understanding for the need of an improved quality assurance in fluorometry within the fluorescence community.

As a contribution to these very challenging goals, we derived quality criteria for the different classes of fluorescence standards in need to help users of fluorescence techniques with the choice of suitable fluorescence standards. To reach the broad community of fluorescence users, we started to develop simple and robust liquid and solid chromophore-based tools for the characterization and performance validation of fluorescence instruments. A first set of emission standards for the spectral region of 300–770 nm was certified and released by BAM in March 2006. As a prerequisite for their straightforward use and the reliability of the instrument characterization performed with these transfer standards, these chromophore-based instrument calibration and validation standards have been designed for application under typically employed measurement conditions thereby avoiding, e.g., tedious and erroneous attenuation procedures. To provide the necessary basis for the traceability to radiometric units, the calibration-relevant fluorometric properties of these fit-for-purpose transfer standards were determined with a properly characterized reference spectrofluorometer with a known measurement uncertainty thereby eventually linking fluorescence measurements to the spectral radiance scale with the aid of these standards.

Acknowledgment Financial support from the Federal Ministry of Economics and Technology (BMWi) and the German Ministry of Education and Research (BMBF) is gratefully acknowledged. We express our gratitude to Professor D. M. Jameson for critically reading the manuscript and to Ms. C. Heinrich for technical assistance with the manuscript.

References

1. Lakowicz, J. R. (1999) *Principles of Fluorescence Spectroscopy*, 2nd edn. Kluwer Academic/Plenum Press, New York.

2. Lakowicz, J. R. (ed) (1992–2004) *Topics in Fluorescence Spectroscopy Series*, Vol. 1–8. Plenum, New York. (1992–2004); Valeur, B. (ed) (2002) *Molecular Fluorescence: Principles and Application*, Wiley-VCH, Weinheim; Schulman, S. G. (ed) (1985–1993) *Molecular Luminescence Spectroscopy, Parts 1–3*. Wiley, New York.
3. Wolfbeis, O. S. (ed) (2001–2004) *Springer Series on Fluorescence, Methods and Applications*, Vol. 1–3. Springer, Berlin.
4. Resch-Genger, U. (ed) (2008) *Standardization and Quality Assurance in Fluorescence Measurements I: Techniques*, vol. 5, Springer-Verlag, Berlin-Heidelberg; Resch-Genger, U. (ed) (2008) *Standardization and Quality Assurance in Fluorescence Measurements II: Bioanalytical and Biomedical Applications*, vol. 6, Springer-Verlag, Berlin-Heidelberg.
5. Mason, W. T. (1999) *Fluorescent and Luminescent Probes for Biological Activity*, 2nd edn. Academic Press, San Diego.
6. Mielenz, K. D. (1982) *Optical Radiation Measurements*. Vol. 3, *Measurement of Photoluminescence*. Academic Press, New York.
7. Burgess, C. and Jones, D. G. (1995) *Spectrophotometry, Luminescence and Colour.* Elsevier B. V., Amsterdam.
8. Gaigalas, A. K., Li, L., Henderson, O., Vogt, R., Barr, J., Marti, G., Weaver, J. and Schwartz, A. (2001) The development of fluorescence intensity standards. J Res Natl Inst Stand Technol 106, 381–389.
9. Resch-Genger, U., Hoffmann, K., Nietfeld, W., Engel, A., Ebert, B., Macdonald, R., Neukammer, J., Pfeifer, D. and Hoffmann, A. (2005) How to improve quality assurance in fluorometry: Fluorescence-inherent sources of error and suited fluorescence standards. J Fluoresc 15, 337–362.
10. Parker, C. A. (1968) *Photoluminescence of Solutions*, Elsevier, Amsterdam.
11. Marin, N. M., MacKinnon, N., MacAulay, C., Chang, S. K., Atkinson, E. N., Cox, D., Serachitopol, D., Pikkula, B., Follen, M. and Richards-Kortum, R. (2006) Calibration standards for multicenter clinical trials of fluorescence spectroscopy for in vivo diagnosis. J Biomed Opt 11, 014010-1-014010-14.
12. Jameson, D.M., Croney, J.C. and Moens, P.D.J. (2003) Fluorescence: Basic concepts, practical aspects, and some anecdotes. Methods Enzymol 360, 1–43.
13. Nickel, B. (1996) Pioneers in photochemistry: From the Perrin diagram to the Jablonski diagram. EPA Newsl 58, 9–27.
14. Nickel, B. (1997) Pioneers in photophysics: From the Perrin diagram to the Jablonski diagram. Part 2. EPA Newsl 61, 27–60.
15. Nickel, B. (1998) Pioneers in photophysics: From Widemann's discovery to the Jablonski diagram. EPA Newsl 64, 19–72.
16. Saunders, G. and Parkes, H. (1999) *Analytical Molecular Biology: Quality and Validation*. RSC, Cambridge.
17. ISO/IEC 17025, International Organization for Standardization (2005) *General Requirements for the Competence of Testing and Calibration Laboratories*, 2nd edn., Geneva.
18. CIE Publication 15.2 (1986) (Commission internationale de l'éclairage) Colorimetry, 2nd edn.
19. Rich, D. C. and Martin, D. (1999) Improved model for improving the inter-instrument agreement of spectrocolorimeters. Anal Chim Acta 380, 263–276.
20. Clinical and Laboratory Standards Institute (CLSI) (2004) Approved Guideline for Fluorescence Calibration and Quantitative Measurements of Fluorescence Intensity, USA.
21. ASTM E 388-04 (2004, original version 1972) Spectral bandwidth and wavelength accuracy of fluorescence spectrometers. In: Annual Book of ASTM Standards, vol. 03.06.
22. ASTM E 578-01 (2001, original version 1983) Linearity of fluorescence measuring system. In: Annual Book of ASTM Standards, vol. 03.06.
23. ASTM E 579-04 (2004, original version 1984) Limit of detection of fluorescence of quinine sulfate. In: Annual Book of ASTM Standards, vol. 03.06.
24. Miller, J. N. (ed) (1981) *Techniques in Visible and Ultraviolet Spectrometry*, Vol. 2, *Standards in Fluorescence Spectrometry*. Chapman and Hall, New York.

25. Eaton, D. F. (1988) Reference compounds for fluorescent measurements. Pure Appl Chem 60, 1107–1114.
26. Chapman, J. H., Förster, T., Kortüm, G., Parker, C. A., Lippert, E., Melhuish, W. H. and Nebbia, G. (1963) Proposal for the standardization of methods of reporting fluorescence emission spectra. J Am Chem Soc 17, 171.
27. Eaton, D. F. (1990) Recommended methods for the fluorescence decay analysis. Pure Appl Chem 62, 1631–1648.
28. Velapoldi, R. A. and Epstein, M. S. (1989) Luminescence standards for macro- and microspectro-fluorometry. In: M. C. Goldberg (ed.) *Luminescence Applications in Biological, Chemical, Environmental and Hydrological Sciences.* ACS Symposium Series 383, American Chemical Society, Washington, DC, pp. 97–126.
29. Equation (1) assumes very dilute solutions, negligible inner filter effects, and validity of the Beer–Lambert law as is typically fulfilled for the majority of fluorescence measurements.
30. Due to radiometric conventions, s implies s_{λ}. The subscript λ denotes per nanometer or spectral.
31. Hollandt, J., Taubert, D. R., Seidel, J., Resch-Genger, U., Gugg-Helminger, A., Pfeifer, D. and Monte, C. (2005) Traceability in fluorometry: Part I, Physical standards. J Fluoresc 15, 301–314.
32. Monte, C., Resch-Genger, U., Pfeifer, D., Taubert, R. D. and Hollandt, J. (2006) Linking fluorescence measurement to radiometric units. Metrologia 43, S89–S93.
33. Resch-Genger, U., Pfeifer, D., Monte, C., Pilz, W., Hoffmann, A., Spieles, M., Rurack, K., Taubert, D. R., Schönenberger and B., Nording, P. (2005) Traceability in fluorometry: Part II. Spectral fluorescence standards. J Fluoresc 15, 315–336.
34. Nighswander-Rempel, S. P. (2006) Quantum yield calculation for strongly absorbing chromophores. J Fluoresc 16, 483–485.
35. This is valid only for a chromophore with simple photophysics and, e.g., the absence of dye-dye interactions and exciplex formation.
36. Verhoeven, J. W. (1996) Glossary of terms used in photochemistry. Pure Appl Chem 68, 2223–2286.
37. Melhuish, W. H. (1984) Nomenclature, symbols, units and their usage in spectrochemical analysis. VI: Molecular luminescence spectroscopy. Pure Appl Chem 56, 231–245.
38. Spectrally corrected fluorescence spectra are occasionally also referred to as technical spectra, see, e.g., [37]. However, terminology is often not very consistent here.
39. Spectral correction typically does not include correction for uncertainties introduced by the sample itself such as pre- and post- or so-called inner filter effects, quenching by oxygen, and refraction at the sample boundaries (refractive index of the matrix). Such effects need to be considered by the proper choice of measurement conditions and additional corrections.
40. Mielenz, K. D. (1978) Refraction correction for fluorescence spectra of aqueous solutions. Appl Opt 17, 2875–2876.
41. Thompson, R. B., Gryczynski, I. and Malicka, J. (2002) Fluorescence polarization standards for high-throughput screening and imaging. Biotechniques 32(1), 34–41.
42. "Closely related" refers to the spectral behavior, i.e., matching or similar absorption and emission spectra.
43. Schwartz, A., Gaigalas, A. K., Wang, L., Marti, G. E., Vogt, R. F. and Fernandez-Repollet, E. (2004) Formalization of the MESF unit of fluorescence intensity. Cytometry 57B(1), 1–6.
44. RM 8640, National Institute of Standards and Technology (NIST).
45. ISO (2005) ISO/IEC 17025, 2nd edn. International Organization for Standardization, Geneva.
46. The calculation of uncertainties is described in the *ISO Guide to the Expression of Uncertainty in Measurement* (GUM).
47. Monte, C., Pilz, W. and Resch-Genger, U. (2005) Linking fluorescence to the scale of spectral sensitivity – The BAM reference fluorometer. Proc SPIE 5880, 588019-1-588019-6.

48. A National Metrological Institute (NMI) is an institute designated by national decision to develop and maintain national measurement standards for one or more quantities.
49. In October 1999, the CIPM (CIPM (Comité Internationale des Poids et Mesures) Mutual Recognition Agreement (MRA) for national standards and calibration and measurement certificates issued by National Metrological Institutes was signed. By the end of 2003, NMIs of 44 Signatory States of the Metre Convention, 2 international organizations and 13 associates of CGPM had signed CIPM MRA.
50. Howarth, P. and Redgrave, F. (2003) *Metrology in Short*. 2nd edn. MKom Aps., Denmark.
51. International Organization for Standardization (ISO) (1995) *Guide to the Expression of Uncertainty in Measurement*. ISBN 92-67-10188-9, 1st edn., Geneva.
52. Melhuish, W. H. (1965) Quantum efficiencies of fluorescence of organic substances: Effect of solvent and concentration of the fluorescent solute. J Phys Chem 65, 229–235.
53. DeRose, P. C. and Kramer, G. W. (2005) Bias in the absorption coefficient determination of a fluorescent dye, standard reference material 1932 fluorescein solution. J Lumin 113, 314–319.
54. Chauvin, A.-S., Gumy, F., Imbert, D. and Bünzli, J.-C. (2004) Europium and terbium *tris* (dipicolinates) as secondary standards for quantum yield determination. Spectrosc Lett 37, 517–532.
55. Values of the fluorescence anisotropy should be provided as this quantity determines the need for the use of polarizers and the size of polarization-related contributions to the overall uncertainty.
56. Benson, R. C. and Kues, H. A. (1977) Absorption and fluorescence properties of cyanine dyes. J Chem Eng Data 22, 379–383.
57. The homogeneity of the dopant also needs to be considered as part of the photochemical stability studies to determine if local photodegradation effects are significant.
58. International Organization for Standardization (ISO) (2000, corrigendum 2003) *ISO: General Requirements for the Competence of Reference Material Producers*, 2nd edn., Geneva.
59. International Organization for Standardization (ISO) (2006) Reference Materials—General and Statistical Principles for Certification, Geneva.
60. National Bureau of Standards (NBS) (1979) Certificate of analysis. Standard Reference Material 936, quinine sulfate dihydrate. National Bureau of Standards, Gaithersberg.
61. Velapoldi, R. A. and Mielenz, K. D. (1980) A fluorescence standard reference material: Quinine sulfate dihydrate. NBS Spec Publ 260-64, PB 80132046, Springfield, VA.
62. Federal Institute for Materials Research and Testing (BAM) (2006) Certificates of analysis, Certified Reference Material BAM-F001, BAM-F002, BAM-F003, BAM-F004, and BAM-F001. Spectral fluorescence standard for the determination of the relative spectral responsivity of fluorescence instruments within its emission range. Certification of emission spectra in 1 nm-intervals.
63. Hoffmann, K., Monte, C., Pfeifer, D. and Resch-Genger, U. (2005) Standards in fluorescence spectroscopy: Simple tool for the characterization of fluorescence instruments. G.I.T. Lab J 6, 29–31.
64. Salit, C. J. M. L. et al. (1996) Wavelengths of spectral lines in mercury pencil lamps. Appl Opt 35, 74.
65. http://www.physics.nist.gov/PhysRefData/Handbook/index.html.
66. Harrison, G. R. (1982) *MIT Wavelength Tables*, Vol. 2, *Wavelengths by Element*. MIT Press, Cambridge, MA.
67. Zaidel, A. N., Prokofev, V. K., Raiskii, S. M., Slavnyi, V. A. and Shreider, E. Y. (1970) *Tables of Spectral Lines*. Plenum Press, New York.
68. Calibration light source CAL-2000, MIKROPACK GmbH (http://www.mikropack.de) or Ocean Optics Inc (http://www.oceanoptics.com).
69. Certain commercial equipments, instruments, or materials are identified in this chapter to foster understanding. Such identification does not imply recommendation or endorsement by the Federal Institute for Materials Research and Testing (BAM), nor does it imply that the materials or equipment identified are necessarily the best available for the purpose.

70. Lifshitz, I. T. and Meilman, M. L. (1989) Standard sample for calibrating wavelength scales of spectral fluorometers. Sov J Opt Technol 55, 487–492.
71. Photon Technology International Inc. (DYAG) FA-2036.
72. National Institute of Standards and Technology (NIST) (2003) Certificate of Analysis, Standard Reference Material 2036, Near-infrared wavelength/wavenumber reflection standard.
73. Knight, A., Gaunt, J., Davidson, T., Chechik, V. and Windsor, S. (2004) Evaluation of the suitability of quantum dots as fluorescence standards. NPL report DQL-AS 007.
74. Mixtures of different fluorophores such as rare earth ions always display excitation wavelength-dependent emission spectra.
75. Bartholomeusz, D. and Andrade, J. D. (2002) Photodetector calibration method for reporting bioluminescence measurements in standardized units. In: Bioluminescence & Chemiluminescence: Progress & Current Applications, Proc. Symposium on Bioluminescence and Chemiluminescenc, World Scientific Publishing Co. Pte. Ltd., Singapore, pp. 189–192.
76. Eppeldauer, G. (1998) Spectral response based calibration method of tristimulus colorimeters. J Res Natl Inst Stand Technol 103, 615–619.
77. Mielenz, K. D., Cehelnik, E. D. and McKenzie R. L. (1976) Elimination of polarization bias in fluorescence intensity measurements. J Phys Chem 64, 370–374.
78. Azumi, T. and McLynn, S. P. (1962) Polarization of the luminescence of phenanthrene. J Chem Phys 37, 2413–2420.
79. Velapoldi, R. A. and Tonnesen, H. H. (2004) Corrected emission spectra and quantum yields for a series of fluorescent compounds in the visible spectral region. J Fluoresc 14, 465–472.
80. Hofstraat, J. W. and Latuhihin, M. J. (1994) Correction of fluorescence-spectra. Appl Spectr 48, 436–447.
81. Kortüm, G. and Finckh, B. (1941–1944) Eine photographische Methode zur Aufnahme quantitativer vergleichbarer Fluoreszenzspektren. Spectrochim Acta 2, 137–143.
82. Lippert, E., Nägele, W., Seibold-Blankenstein, I., Staiger, U. and Voss, W. (1959) Messung von Fluoreszenzspektren mit Hilfe von Spektralphotometern und Vergleichsstandards., Z Analyt Chem 170, 1–18.
83. Gardecki, J. A. and Maroncelli, M. (1998) Set of secondary emission standards for calibration of the spectral responsivity in emission spectroscopy. Appl Spectr 52, 1179–1189.
84. Thompson, A. and Eckerle, K. L. (1989) Standards for corrected fluorescence spectra. Proc SPIE-Int Soc Opt Eng 1054, 20–25.
85. National Institute of Standards and Technology (NIST) (1989) Certificate of Analysis, Standard Reference Material 1931, fluorescence emission standards for the visible region. This set of four solid spectral fluorescence standards in a cuvette format that is no longer available was restricted in measurement geometry and certified using polarizers.
86. See for Instance Invitrogen or Former Molecular Probes, Starna GmbH, Matech Precision Dynamics Coorp., Labsphere Inc., and LambdaChem GmbH.
87. Certain commercial equipment, instruments, or materials are identified in this chapter to foster understanding. Such identification does not imply recommendation or endorsement by the National Institute of Standards and Technology, nor does it imply that the materials or equipment identified are necessarily the best available for the purpose.
88. National Institute of Standards and Technology (NIST) (2006) Certificate of Analysis, Standard Reference Material 2940, Relative intensity correction standard for fluorescence spectroscopy: Orange emission. Certification of emission spectra in 1 nm-intervals. (http://ts.nist.gov/ts/htdocs/230/232/232.htm).
89. National Institute of Standards and Technology (NIST) (2006) Certificate of Analysis, Standard Reference Material 2941, Relative intensity correction standard for fluorescence spectroscopy: Green emission. Certification of emission spectra in 1 nm-intervals. (http://ts.nist.gov/ts/htdocs/230/232/232.htm).

90. For instance, the emission spectrum of quinine sulfate dihydrate in 0.105 M perchloric acid that is very broad for an organic dye can be used for spectral correction only from ca. 395 to 565 nm where the emission intensity is at least 10% of the intensity at the emission maximum.
91. Typically, for the determination of the spectral responsivity of fluorescence instruments with a reasonable uncertainty, only fluorescence intensities $\geq$ 10% of the intensity at the emission maximum are used. Otherwise the signal-to-noise ratio can be poor.
92. Pfeifer, D., Hoffmann, K., Hoffmann, A., Monte, C. and Resch-Genger, U. (2006) The Calibration Kit Spectral Fluorescence Standards: A simple tool for the standardization of the spectral characteristics of fluorescence instruments. J Fluoresc 16, 581–587.
93. Federal Institute for Materials Research and Testing (BAM) (2006) Certificate of Analysis, Certified reference materials BAM-F001 - BAM-F005, Calibration Kit Spectral Fluorescence Standards for the determination of the relative spectral responsivity of fluorescence instruments. Certification according to ISO guides 34 and 35 in 1 nm-steps for three different spectral bandpasses of the BAM fluorometer.
94. BAM-F001, BAM-F002, BAM-F003, BAM-F004, and BAM-F005 ready-made from Sigma-Aldrich GmbH (former Fluka GmbH) are available from BAM or from Sigma-Aldrich. The corresponding product numbers from Sigma-Aldrich are 97003-1KT-F for the Calibration Kit and 72594, 23923, 96158, 74245, and 94053 for BAM-F001, BAM-F002, BAM-F003, BAM-F004, and BAM-F005, respectively.
95. Resch-Genger, U. and Pfeifer, D. (2006) Certification report, Calibration Kit Spectral Fluorescence Standards BAM-F001 – BAM-F005, BAM.
96. Removal of background signals, such as scattering and fluorescence from the solvent and dark counts at the detector from measured fluorescence spectra by subtraction of a background spectrum, $I_b(\lambda_{ex},\lambda_{em})$ that was recorded under identical measurement conditions for a blank solvent sample $I_m(\lambda_{ex},\lambda_{em})$ yields spectrally uncorrected spectra. $I_u(\lambda_{ex},\lambda_{em})$: $I_u(\lambda_{ex},\lambda_{em}) = I_m(\lambda_{ex},\lambda_{em}) - I_b(\lambda_{ex},\lambda_{em})$.
97. Ejder, E. (1969) Methods of representing emission, excitation, and photoconductivity spectra. J Opt Soc A 59,223–224.
98. For the determination of fluorescence quantum yields, the energy of the emitted photons needs to be taken into account as the fluorescence quantum yield is the ratio of the number of emitted to absorbed photons and not a ratio of radiated fluxes or powers. Accordingly, corrected emission spectra referenced to the spectral radiance have to be multiplied by the wavelength prior to integration on a wavelength scale to consider the photonic nature of the emitted light.
99. Fox, N. P. (1991) Trap detectors and their properties. Metrologia 28, 197–201.
100. A pyroelectric detector measures the energy of absorbed photon with a wavelength independent responsivity (grey detector), but with a drastically reduced sensitivity and accuracy compared to, for instance, a Si photodiode.
101. Melhuish, W. H. (1975) Modified technique for determining the wavelength-sensitivity curve of a spectrofluorometer. Appl Opt 14, 26–27.
102. Hart, S. J. and Jones, P.J. (2001) Fiber-optic quantum counter for incident excitation correction in fluorescence measurements. Appl Spectrosc 55, 1717–1724.
103. Demas, J. N. and Crosby, G. A. (1971) The measurement of photoluminescence quantum yields. A review. J Phys Chem 75, 991–1024.
104. Mielenz, K. D., Velapoldi, R. A. and Mavrodineanu (1977) *Standardization in Spectrophometry and Luminescence Measurements*. NBS Special Publication 466, Gaithersfield, MD.
105. Prerequisites for this method are a pure compound and an excitation wavelength-independent emission spectrum and quantum yield, and thus a straightforward excited-state photochemistry can yield a comparatively high calibration uncertainty.
106. Wise, S. A., Sander, L. C. and May, W. E. (1993) Determination of polycyclic aromatic hydrocarbons by liquid chromatography. J Chromatogr 642, 329–349.
107. National Institute of Standards and Technology (NIST). SRM 1647b.

108. Duewer, D. L., Parris, R. M., White, V. E., May, W. E. and Elbaum, H. (2004) NIST Spec Pub 1012. An approach to the metrologically sound traceable assessment of the chemical purity of organic reference materials. U.S. Government Printing Office, Washington, DC.
109. National Institute of Standards and Technology (NIST) (2004) Certificate of Analysis, Standard Reference Material 1932, Fluorescein solution. (http://ts.nist.gov/ts/htdocs/230/232/232.htm)
110. Schwartz, A., Gaigalas, A. K., Wang, L., Marti, G. E., Vogt, R. F. and Fernandez-Repollet, E. (2004) Formalization of the MESF unit of fluorescence intensity. Cytometry 57B, 1–6.
111. National Institute of Standards and Technology (NIST) (2004) Report of Investigation, Reference Material 8640, Microspheres with Immobilized Fluorescein Isothiocyanate.
112. Gaigalas, A. K., Wang, L., Schwartz, A., Marti, G. E., Vogt, R. F. (2005) Quantitating fluorescence intensity from fluorophore: assignment of MESF values. J Res Natl Inst Stand Technol 110, 101–114.
113. LED: light emitting diode; OLED: organic light emitting diode.
114. Greenham, N. C., Samuel, I. D. W., Hayes, G. R., Phillips, R. T., Kessener, Y. A. R. R., Moratti S. C., Holmes, A. B. and Friend, R. H. (1995) Measurement of absolute photoluminescence quantum efficiencies in conjugated polymers. Chem Phys Lett 241, 89–96.
115. de Mello, J. C., Wittmann, H. F. and Friend, R. H. (1997) An improved experimental determination of external photoluminescence quantum efficiency. Adv Mater 9, 230–232.
116. He, L., Hattori, R. and Kanicki, J. (2000) Light output measurements of the organic light-emitting devices. Rev Sci Instrum 71, 2104–2107.
117. Rohwer, L. S. and Martin, J. E. (2005) Measuring the absolute quantum efficiency of luminescent materials. J Lumin 115, 77–90.
118. Madge, D., Brannon, J. H., Cremers T. L., Olmsted, III, J. (1979) Absolute luminescence yield of cresyl violet. A standard for the red. J Phys Chem 83, 696.
119. Chen, R. F. (1972) Measurements of absolute values in biochemical fluorescence spectroscopy. J Res Natl Bureau Stand 76A, 593–606.
120. Greenham, N. C., Friend, R. H. and Bradley, D. D.C. (1994) Angular dependence of the emission from a conjugated polymer light-emitting diode: Implications for efficiency calculations. Adv Mater 6, 491–494.
121. Porrès, L., Holland, A., Pålson, L.-O., Monkmann, A. P., Kemp, C. and Beeby, A. (2006) Absolute measurements of photoluminescence quantum yields of solutions using an integrating sphere. J Fluoresc 16, 267–272.
122. Pålsson, L.-O. and Monkman, A. P. (2002) Measurements of solid state photoluminescence quantum yields of films using a fluorimeter. Adv Mater 14, 757.
123. Froehlich, P. (1989) Understanding the sensitivity specification of spectrofluorometers. Int. Lab 19, 42–45.
124. This test, usually performed at an excitation wavelength of 350 nm and a detection wavelength of 397 nm, employs the Raman line of water to compare the long-term spectral sensitivity of a single fluorescence instrument or spectral sensitivities between instruments.
125. Gibeler, R., McGown, E., French, T. and Owicki, J. C. (2005) Performance validation of microplate fluorimeters. J Fluoresc 15, 363–375.
126. Parke, S., Watson, A. I. and Webb, R. S. (1970) Fluorescence decay times of divalent manganese in inorganic glasses. J Phys D: Appl Phys 3, 763–769.
127. Hoffmann, K., Resch-Genger, U. and Nitschke, R. (2005) Simple tool for the standardization of confocal spectral imaging systems. G.I.T. Imaging & Micros 3, 18–19.
128. Ray, K. G. and McCreery, R. L. (1997) Simplified calibration of instrument response function for Raman spectrometers based on luminescence intensity standards. Appl Spectrosc 51, 108–116.
129. Frost, K. J. and McCreery, R. L. (1998) Calibration of Raman spectrometer instrument response function with luminescence standards: An update. Appl Spectrosc 52, 1614–1618.

Membranes and Fluorescence Microscopy

Luis A. Bagatolli

Abstract Fluorescence spectroscopy-based techniques using conventional fluorimeters have been extensively applied since the late 1960s to study different aspects of membrane-related phenomena, i.e., mainly relating to lipid-lipid and lipid-protein (peptide) interactions. Even though fluorescence spectroscopy approaches provide very valuable structurally and dynamically related information on membranes, they generally produce mean parameters from data collected on bulk solutions of many vesicles and lack direct information on the spatial organization at the level of single membranes, a quality that can be provided by microscopy-related techniques. In this chapter, I will attempt to summarize representative examples concerning how microscopy (which provides information on membrane lateral organization by direct visualization) and spectroscopy techniques (which provides information about molecular interaction, order and microenvironment) can be combined to give a powerful new approach to study membrane-related phenomena. Additionally along this chapter, it will be discussed how membrane model systems can be further utilized to gain information about particular membrane-related process like protein(peptide)/membrane interactions.

Keywords Giant unilamellar vesicles · Membrane lateral structure · Membrane-peptide interaction · Polarity sensitive probes · Fluorescence microscopy

Introduction

Fluorescence spectroscopy-based techniques using conventional fluorimeters have been extensively applied since the late 1960s to study different aspects of membrane related phenomena, i.e., relating to lipid–lipid, lipid–protein (peptide), and lipid–DNA interactions. These types of studies encompass measurements of

L.A. Bagatolli (✉)
Membrane Biophysics and Biophotonics Group/MEMPHYS – Center for Biomembrane Physics, Department of Biochemistry and Molecular Biology, University of Southern Denmark, Campusvej 55 DK-5230, Odense M, Denmark
e-mail: bagatolli@memphys.sdu.dk

C.D. Geddes (ed.), *Reviews in Fluorescence 2007*, Reviews in Fluorescence 2007,
DOI 10.1007/978-0-387-88722-7_2,

fluorescence excitation and emission spectra, fluorescence time decays (lifetimes) and fluorescence polarization (or anisotropy) using of a large variety of fluorescent probes. The last also includes applications of Förster resonance energy transfer (FRET) or fluorescence quenching techniques [1–9]. In these types of studies, liposomes have had a significant role as model systems for cellular membranes. Liposome studies generally involve aqueous suspensions consisting of small unilamellar vesicles (SUVs, mean diameter 30–50 nm), large unilamellar vesicles (LUVs, mean diameter 100 nm) and multilamellar vesicles (MLVs) being the most popular model systems. The composition of these membranes can range from single lipid components to mixtures of lipids (synthetic or natural lipid extracts), both with and without membrane proteins, including in some cases closed vesicles obtained from native biological membranes. Even though the aforementioned fluorescence spectroscopy approaches provide very valuable structural and dynamical information on membranes, they generally produce mean parameters from data collected on bulk solutions of many vesicles and lack information on membrane organization at the level of a single vesicle, a quality that can be provided by microscopy related techniques. The lack of information regarding direct imaging of the (local) spatial organization of membranes can also be extended to other experiments performed with other spectroscopy and thermodynamics related techniques such as differential scanning calorimetry, IR spectroscopy, NMR, and X-ray diffraction to mention a few [10–16].

The first experiments reporting visualization of lipid domains in bilayers by means of fluorescence microscopy provided valuable new information (such as shape and size distribution of different lipid domains), not available before in the membrane field [17–19]. However, few reports explored the local fluorescence properties of the probes in any detail, using the classical fluorescent spectroscopy approaches mentioned above. The continued development of confocal microscopy, both one-photon and two-photon approaches, which has greatly increased the information available through imaging, has allowed for rapid advances in the biophotonics field. For example, fluorescence correlation spectroscopy (FCS), three dimensional particle tracking methods, including polarization, lifetime and emission spectral information based techniques can now be performed in the microscope environment [20–26]. At the present time, there are a number of laboratories actively advancing this concept, i.e., performing spectroscopy in a microscope for a variety of protocols and generating exciting results from studies ranging from cell physiology to the mechanics of polymer motion on surfaces.

The advantages of using a microscope as the optical arrangement are clear. The light collection efficiency of a well-designed microscope is greatly enhanced over other optical arrangements. In addition, the flexibility of present fluorescence microscopes creates for the spectroscopist a malleable optical compartment which can be designed and readily re-designed as needed. Of equal importance as the sensitivity and flexibility of a microscope is the addition of spectroscopy to the ability to collect spatially resolved information. With a properly designed system, it is possible at present to perform quantitative spectroscopic studies at each pixel and build an information image related to the sample at hand, be it a living cell, an extended surface polymer, or a membrane model system.

The main aim of this contribution is to elaborate and provide a concise review on the different approaches that use fluorescence spectroscopy tools in a microscope to explore membrane related phenomena. In other words, I will attempt to summarize representative examples concerning how microscopy (which provides information on membrane lateral organization by direct visualization) and spectroscopy techniques (which provides information about molecular interaction, order and microenvironment) can be combined to give a powerful new approach to study membrane related phenomena. Additionally, I will discuss how the membrane model system can be further utilized to gain information about particular membrane related process like protein(peptide)/membrane interactions.

Suitability of Different Model Membrane Systems for Fluorescence Microscopy Studies

The most popular membrane model systems (SUVs, LUVs, MLVs) may not be fully suitable to perform fluorescence microscopy experiments, particularly if the collection of structural details at the level of single (unilamellar) vesicles is the main purpose of the experiment. Even though a fraction of the population of MLV (micrometer size) can already be observed under a microscope (see Fig. 1A), SUVs and LUVs have sizes below the resolution of an optical microscope and the visualization of, for example, membrane lateral structure at the level of single vesicles cannot be obtained. It is interesting to mention, however, that some attempts to use small liposomes (LUVs) under a fluorescence microscope have been recently reported [27]. In these studies, determination of the full size distribution of different liposomes preparations (either extruded or non-extruded lipid vesicles) was successfully achieved. The last study opens a novel application involving membrane structures under the resolution limit of the fluorescent microscope with interesting potential applications using fluorescence to study membrane related phenomenon. By using small vesicles and the abilities of a fluorescence microscopy for instance, future ensemble-based assays investigating vesicle-size-related effects will benefit from the ability to perform accurate measurements in real time on a great number of single objects [27].

GUVs as Membrane Model Systems: Advantages and Disadvantages

From a historical perspective, the first documented report involving membranes and optical microscope related techniques comes from the mid-eighteenth century. Although the credit for discovery and characterization of unilamellar lipid vesicles (SUVs and LUVs) was given to Bangham and Horne during the early 1960s (using electron microscopy as an experimental technique) [28], reports about the colloid behavior of lecithin and other phospholipids can be found much earlier in the literature [29, 30]. In fact, hydration of dried extracts of lecithin from egg-yolk or brain and the subsequent formation of myelin figures have been documented in 1854 and

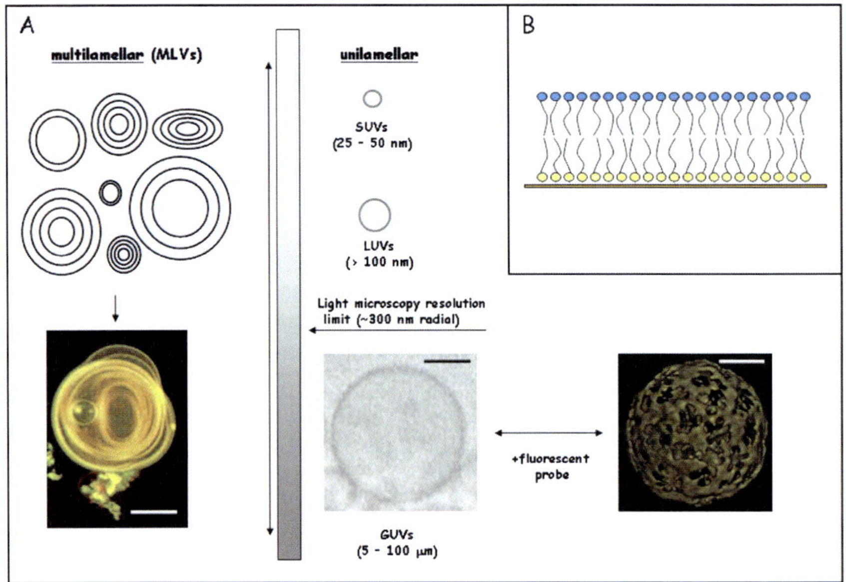

Fig. 1 Sketch of different membrane bilayer model systems. (**A**) A fraction of the population of MLVs (above 1 μm size) can be observed by fluorescence imaging (*left* fluorescent image, MLV composed of POPC, DiIC_{18} probe). Only GUVs can be observed by using light microscopy based techniques (notice that SUVs and LUVs have dimensions below the resolution limit of the microscope). Incorporation of fluorescent probes and application of confocal fluorescence microscopy techniques provide information about the 3D structure of the GUV including the lateral organization of the membrane (*right* fluorescent image, GUV composed of Ceramide/Sphingomyelin mixture, DiIC_{18} probe [118]). (**B**) Sketch of planar membrane bilayers on mica (atomically flat substrate). The different colors on the lipid polar head groups illustrate the possibility to generate compositionally asymmetric membranes (generally using Langmuir–Blodgett technique). The *bars* on the fluorescence microscopy images represent 10 μm

1867 by R. Virchow and C. Neubauer respectively [31, 32] using light microscopy. These membranes structures have dimensions of tens of micrometers and are the natural precursor of giant unilamellar vesicles (GUVs) when rehydration of dried lipid layers is performed. GUVs were first reported as alternative membrane model systems in 1969 by Reves and Dowben [33]. In recent years, GUVs have become very popular objects on which to perform fluorescence microscopy related experiments (see Fig. 1A). One of the reasons is that the dimensions of GUVs (mean diameter around ~25 μm) are higher than the intrinsic resolution limit of light microscopy related techniques (~250 nm radial), allowing observation of structural details in membrane organization practically above ~300 nm. Also the average size of GUVs is similar to that of the plasma membrane of a variety of cells. The last circumstance allows us to perform experiments at the level of single vesicles on the same size scale (curvature) as natural membrane systems (for example, cell plasma membrane). Since the experiments are performed at the level of single vesicles, heterogeneity in shape and size or the presence of multilamellar vesicles are ruled out.

One of the significant aspects in using giant vesicles as model systems is the ability to control the molecular composition of the membrane as well as the environmental conditions. For instance, studies of the lateral structure of membranes using giant vesicles as model systems and fluorescence microscopy as experimental technique were normally restricted to giant vesicles composed of single lipid species or mixtures with few components [17–19, 34–44]. However, as recently reported in the literature, it is also possible to form giant vesicles from natural lipid extracts [36, 45–47] and native membranes [45, 48–50]. Additionally, GUVs containing membrane proteins can also be generated [51–56] often involving an electroformation method using proteoliposomes as starting point. Curiously and even though the description of giant vesicles was done almost 40 years ago, it was not until 1999 that a wide application of fluorescence microscopy related techniques was practically tested in this model system. Seminal reports however about the use of GUV, fluorescence microscopy, and detection of lipid lateral heterogeneity were reported in the late 1980s from the group of M. Glaser [57].

Regarding the generation of these giant lipid structures, there is no general agreement about the experimental conditions required to obtain GUVs. This lack of consensus may be due to the fact that the mechanism underlying giant vesicle formation still remains obscure. As a consequence, there are many different methods described in the literature to prepare giant vesicles [33, 45, 52, 54, 58–61]. Although the aforementioned scenario may look complex, most of these methods are based on two main experimental protocols: the gentle hydration method, originally described by Reeves and Dowben [33], and the electroformation method introduced by Angelova and collaborators [59, 60]. Of these two experimental protocols, the electroformation method has the advantage that it provides the more homogeneous population of GUVs, with sizes between 5 and 100 μm in diameter. Additionally, the electroformation protocol requires less time compared to the gentle hydration method ($\sim$ 1–2 h vs 12–24 h, respectively) and provides a high yield of giant unilamellar vesicles ($\sim$95%) [62, 63].

Perhaps the main drawback of GUVs and their applications as model systems for biological membranes is the fact that low salt concentrations (generally below 10 mM NaCl) are normally required for their preparation using the above-mentioned protocols. This problem has recently been overcome by the introduction of a new electroformation protocol by Pott et al. [64]. This method, originally developed for single phospholipid species was successfully tested for preparing GUVs from compositionally complex lipid mixtures, including native biological membranes [49]. This protocol produces membrane objects that practically keep the composition and organization of biological membranes (i.e., a more complete model system respect to those composed only for lipids). This protocol offers another choice with respect to that reported by Baumgart et al. [50], in which cell blebbing is used to produce GUVs. This last method is limited to cell plasma membranes, precluding the use of membranes from internal cell organelles to generate GUVs.

Planar Membranes

Even though it is not the aim of this chapter to extensively review fluorescence microscopy applications in other model membranes other than GUVs, it is fair to mention that planar bilayer membranes are very interesting membrane models to perform fluorescence microscopy experiments (Fig. 1B). An interesting example regarding the combination of planar membranes and fluorescence microscopy is that recently reported by Simonsen [65], where the evolution of the lateral structure of planar membranes upon changes in molecular composition induced by $PLA_2$is explored. This last approach allows simultaneous correlation of membrane lateral pattern and enzyme kinetics. In fact many planar membranes systems were investigated (lipid–lipid but also lipid–protein interactions) using fluorescence microscopy as is nicely reviewed by Crane and Tamm [66].

Most of the fluorescence microscopy studies reported in planar membranes focus on measurements of fluorescence intensity of particular selection of probes with almost no practical combinations of fluorescence spectroscopic parameters (such as lifetime [67], polarization or emission spectra shift [68]) and microscopy. In fact, the use of probe partitioning into the different membrane regions as a criterion to assign lipid phases was questioned recently for this model system in experiments that combine atomic force microscopy (AFM) and fluorescence microscopy [69]. The last criterion was also stressed for GUV systems comparing and discussing the information obtained from various "partition-like" fluorescence probes with polarity sensitive fluorescent dyes [70] (see next section).

In summary, planar bilayer membrane systems are excellent candidates to apply fluorescence spectroscopy related studies [67] as well as fluorescence fluctuation analysis similar to those done in GUVs systems [24, 25]. In particular, this model system offers the possibility to easily generate asymmetric membranes, something that is possible, but practically difficult using GUVs [71]. Another advantage of planar membranes over free standing bilayers models (i.e., GUVs) is related to the geometrical features of the model system. For example determination of size and area analysis of different lipid domains in membranes displaying lateral phase separation is much simpler in planar membrane systems [72] than in GUVs (where quasi-spherical membrane structures are present) [73].

Fluorescent Probes and Lateral Structure of Biological Membranes

From the work reported using fluorescence spectroscopy (microscopy) and membrane systems, it is very obvious that a large number of options regarding lipid-like fluorescent probes are presently available (to have an idea one can refer to the catalog of Invitrogen/Molecular probes, for example). With the attempt to summarize the available information I decided to divide the lipid-like fluorescent probes into two main families. The last takes into account the partition properties among coexisting lipid phases and the particular mode of excitation under a fluorescence microscope. The first family contains those probes where excitation is viable by one photon excitation mode, particularly in the visible range. These probes are the most used in epi-fluorescence and confocal fluorescence microscopy, particularly given the availability of the equipment in common laboratories. Generally, these probes have a two-chain lipid-like structure of variable size and unsaturation (even though some fluorescent moieties can themselves be used without further chemical modifications, i.e., without attachment of alkyl chains). For this group of probes, the fluorescent moiety can be attached either to the lipid chain or to the polar head group (Fig. 2A). One of the common features of these fluorescent probes is the uneven

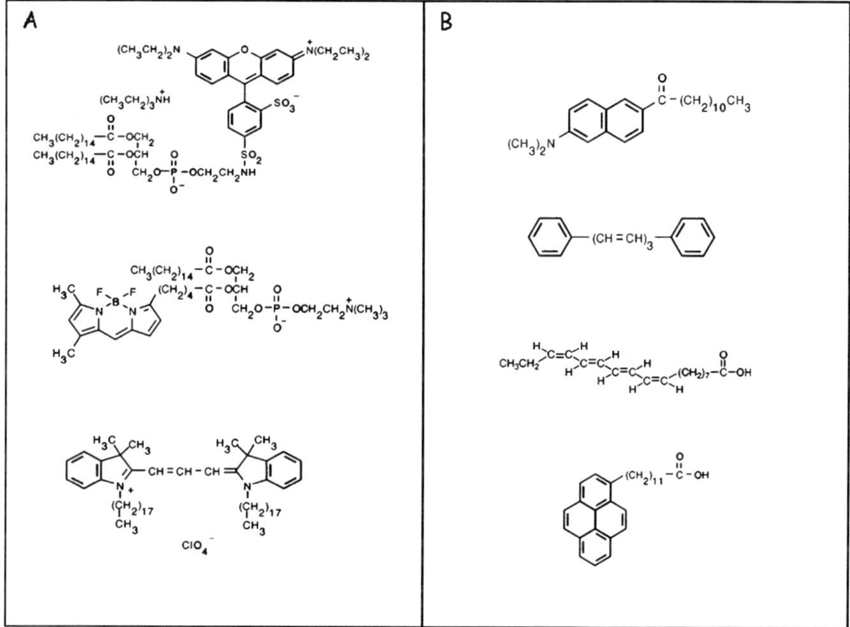

Fig. 2 Chemical structure of representative fluorescent probes: (**A**) Rhodamine-DPPE (*top*), Bodipy-PC (*center*), DiIC$_{18}$ (*bottom*). (**B**) From *top* to *bottom*: LAURDAN, DPH, Parinaric acid, C12 Pyrene

partition into coexisting membrane regions [17, 34, 38, 70, 74, 75]. Examples of these probes are amphiphilic derivatives of rhodamines, fluoresceins, dialkylcarbocyanine (DiI, DiO), dialkylaminostyryl (DiA), and coumarins, including bodipy, perylene, and naphtopyrene (see Fig. 2 for representative example). As was mentioned above, the partition of this family of fluorescence probes has been used as a criterion to assign lipid phases in GUVs displaying lipid phase coexistence [70]. Even thought an extensive characterization of the partition of several fluorescent dyes *on selected lipid mixtures* was reported [74], the aforementioned criterion to define the nature of a lipid phase is still risky. Changes in the partition properties of the probes on the different membrane regions are highly dependent on the chemical composition of the local membrane domain *and not on the phase state* [18, 34, 70]. A practical solution to strength the data interpretation when fluorescent images are used to assign lipid phases is the use of fluorescence correlation spectroscopy [17, 38]. With this last approach the diffusion of the probe can be related to the nature of the existing lipid phases. Alternative measurements of lifetime [76] or polarization under the microscope are also strong alternatives to explore which type of lipid phase is present.

As mentioned in the introduction section, the vast majority of the studies on the characterization of lipid membrane's lateral structure using fluorescence spectroscopy techniques (cuvette experiments) exploits other fluorescent parameters such as fluorescence lifetime, rotational correlation time (obtained by anisotropy measurements), and position of the emission spectral maximum. Several fluorescent probes, particularly those UV-excited fluorescent probes (such as Pyrene, DPH and their derivatives, parinaric acid, LAURDAN, PRODAN), are used in regular fluorescence cuvette experiments [7, 77–84]. These dyes represent a second family of membrane probes and have not been fully exploited yet using fluorescence microscopy related techniques. Two main reasons for the dearth of such studies are (i) the inaccessibility of these advanced microscopy techniques, i.e., lifetime microscopy or polarization fluorescence microscopy (expensive and specialized equipment is required along with significant user expertise) and (ii) it is practically difficult to perform fluorescence microscopy experiments (one photon excitation, i.e., epifluorescence and confocal) with UV-fading fluorescent probes, since the extent of photobleaching is high and is often technically difficult to obtain reliable fluorescence images. An alternative solution is the use of multiphoton microscopy experiments. In this way an extensive application was reported with the probe LAURDAN under multiphoton excitation microscopy [25, 70, 85, 86]. Perhaps one of the most remarkable features of these probes (particularly the ones that keep a fatty acid like structure, i.e., DPH, LAURDAN, parinaric acid) is the fact that they show an even distribution in membranes displaying phase coexistence. In this last situation correlation of the probe's spatial distribution and fluorescence parameters such as spectral shift, polarization, or lifetimes is ideal to relate the membrane region of interest with lipid phases. Other recently reported polarity sensitive probes to be fully exploited under fluorescence microscopy are 3-hydroxyflavone derivatives [87], C-LAURDAN [88], and di-4-ANEPPDHQ [89]. These probes show different emission spectra depending to the membrane's phase scenario and also show even distributions on membranes displaying lateral phase separation.

Lateral Structure of Compositionally Simple and Complex Membrane Model Systems

As mentioned above, it was not until the end of the 1990s when several papers appeared applying fluorescence microscopy related techniques (epifluorescence, confocal, and two-photon excitation fluorescence microscopy) to bilayer membrane systems. These reports showed for first time images of the temperature-dependent lateral structure of giant vesicles composed of different phospholipids, phospholipid binary mixtures, ternary lipid mixtures containing cholesterol, natural lipid extracts, and native membranes [17–19, 34, 36–38, 40–42, 45, 46, 85, 90, 91]. These papers presented a correlation between micron-size (visual) lipid domain structures and local lipid dynamics under different environmental conditions.

The experimental data involving fluorescence microscopy and GUVs have also been used to construct lipid phase diagrams for artificial lipid mixtures (i.e., phase diagrams that include visual information about membrane lateral structure) [38, 41, 75, 92]. If this type of experiments are performed, it is very important to corroborate the local phase state in each particular membrane region with different approaches beside probe partition as mentioned above (i.e., FCS, LAURDAN GP, FLIM) [17, 25, 38, 70, 76]. Importantly, if phase diagrams are constructed with these data, careful evaluation of thermodynamic equilibrium is required (something that is not really demonstrated in the work reported until now). In fact, some differences in particular regions of the cholesterol containing ternary mixtures (canonical "raft" mixtures) phase diagrams was reported in the literature when phase diagrams were obtained either from GUVs experiments or bulk techniques involving solutions of LUVs (fluorescence spectroscopy in cuvette) [41, 93]. These discrepancies may be related to the resolution of the fluorescence microscope [9, 75]; however it is interesting to notice that in the GUV experiments reported on no additional fluorescence spectroscopy data beside the partition of the lipophilic probe is used to construct the phase diagram [75]. By confirming the existence of phases by fluorescence parameters other than imaging fluorescence intensity and validating basic thermodynamic rules in GUVs systems, one can be assured that the information obtained is suitable to construct a phase diagram. By applying image analysis, for example, quantitative area information and further tests of thermodynamics rules can be performed, including experiments to evaluate the presence of domains below the resolution of the microscope (LAURDAN GP, see [85]). Quantification of the area contribution of the different lipid phases in GUVs displaying lipid phase coexistence is presently under evaluation in our laboratory [73].

The information obtained from model systems can be utilized to explore biological membranes with the possibility to display lipid mediated lateral heterogeneity. Examples are reports studying, for example, GUVs composed of native lung surfactant [45], stratum corneum (SC) skin lipids membranes [94], brush border membranes [48], and red blood cells [49]. Figure 3 shows a representative example of skin lipid mixtures (extracted from stratum corneum of human skin tissue; a complex mixture of ceramide, fatty acids, and cholesterol) and pulmonary surfactant from pig lungs (containing a complex mixture of DPPC, cholesterol, and unsaturated lipids including surfactant proteins). Both mixtures at the correspondent biologically relevant temperature show coexistence of different lipid phases (gel/gel and liquid ordered/liquid disordered-like phase coexistence for skin lipid mixtures and lung surfactant respectively; Fig. 3A and B, left panel). Perturbations of either composition (removing of cholesterol) and temperature in lung surfactant membranes (Fig. 3 B) or temperature and pH for human SC skin lipid mixtures (Fig. 3 A) generate significant changes in the lateral structure of these different membranes [45, 94]. This behavior can be

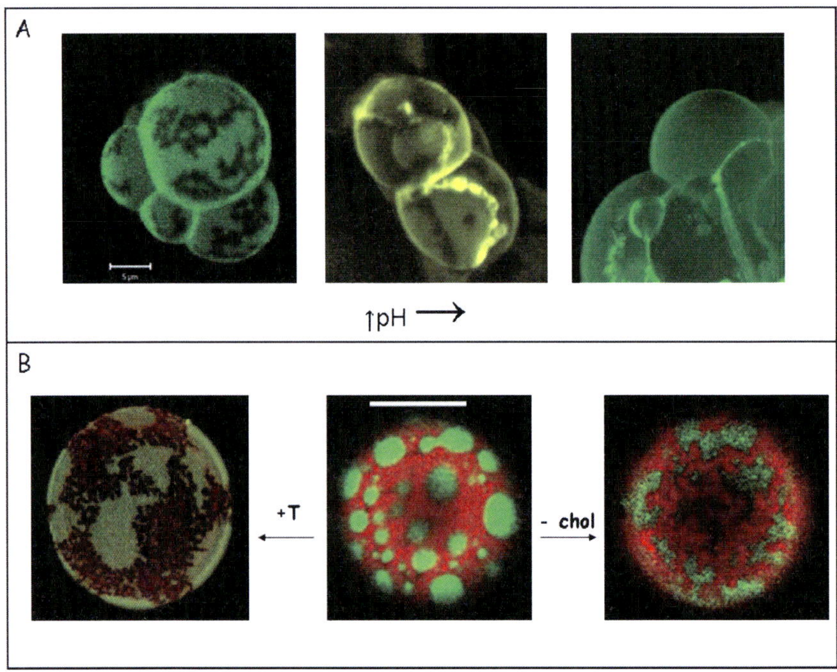

Fig. 3 Examples of giant lipid structures composed of compositionally complex membranes. (**A**) Effect of pH on giant lipid membranes composed of native lipid mixtures extracted from human skin stratum corneum membranes. The figure in the *left* show the particular lateral structure of this membrane (coexistence of two gel –like phases) at physiological conditions (pH 5, T = 32°C, DiIC$_{18}$ probe). Increasing the pH to 6 and 7 (center and right fluorescence microscopy image respectively) generates a change in the lateral organization of the lipids (for more details see [94]). (**B**) GUVs composed of pig native lung surfactant at physiological conditions (center fluorescence microscopy image showing presence of liquid ordered (*red*)/liquid disordered (*green*) -like phase coexistence; T = 37°C). An increase in the temperature cause a phase transition on the native material (*left* fluorescence microscopy image, T = 38°C). Extraction of cholesterol generates a lateral pattern that resembles coexistence of gel (*red*)/liquid disordered (*green*) phases (*right* fluorescence microscopy image). For additional details see [45]. The *bars* represent 15 μm

reproduced in model membranes using representative lipids. For example ternary mixtures composed of DOPC/DPPC/cholesterol have been utilized to mimic lung surfactant [45].

Peptide(Protein)–Lipid Interactions Studies in GUVs

As was emphasized above GUVs are micrometer size objects that allow one to perform fluorescence microscopy at the level of single vesicles. The additional visual information extracted from these experiments can be fully exploited to interrogate the effect of proteins and peptides on the structure of membranes. In this section representative examples will be summarized with particular emphasis in the utilized experimental strategy.

Peptide–Membrane Interactions

A wide range of peptides and toxins are harmful to cells by altering the hydrophobic-hydrophilic seal of biological membranes. Generally this last effect is related to the composition of the membrane and the chemical structure of the peptide [95–99]. In fact the membrane damage can be exerted by different mechanisms: (i) *membrane lysis (or membrane solubilization)*: peptides that remain tightly bound to the membrane interface and promote bilayer damage via solubilization of membrane components (so-called detergent-like or carpet-like mechanism) [100–103] and (ii) *pore formation*: peptides that acquire a particular conformation across the lipid bilayer and consequently form size-defined permeating structures named as pores but do not alter the long-range interactions between membrane components [96–98, 100, 101, 104–108].

A common approach to study the aforementioned membrane perturbation is to observe leakage of fluorescent molecules entrapped inside liposome solutions. However, this fluorescent cuvette experiment lacks visual information, a feature that can be obtained using GUVs and fluorescence microscopy. The use of GUVs and fluorescence microscopy not only allows the direct observation of the leakage phenomenon caused by the interaction of the peptide with the membrane (pore formation or membrane lysis, see Fig. 4), but also can be potentially used to examine simultaneously lateral structure of the membrane and peptide distribution in the plane or across the membrane (if the peptide is fluorescently labeled). Few

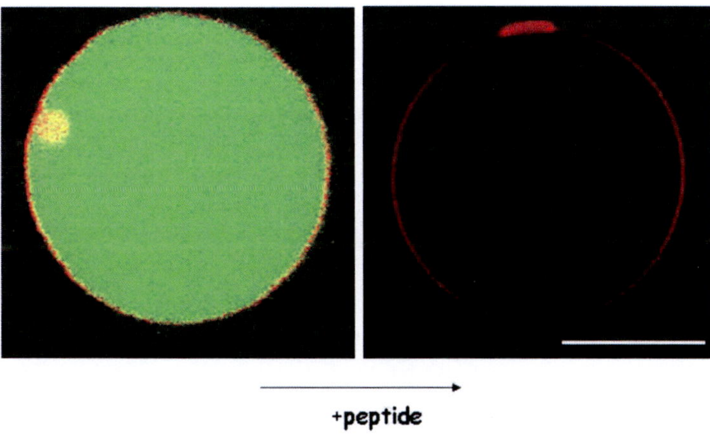

Fig. 4 GUV composed of POPC labeled with DiIC8 (membrane, *red*) filled with carboxyfluorescein (*green*). Upon peptide addition (mellitin in this case) a leakage of the fluorophore is observed due to formation of pores (see [104, 105]). The *bar* represents 20 μm

applications were reported in the literature using slight variations of the aforementioned approach [102, 104, 105, 107, 109, 110]. For example, in those applications reported by Ambroggio et al, the lytic mechanism of different peptides (Alzheimer β-amyloid peptide 1–42, Citropin, Aurein, and Maculatin) has been studied [104, 105]. In these experiments GUVs are loaded with different-sized water soluble fluorophores including a membrane reporter (lipid-like fluorescent probe, Fig. 4) [104, 105]. For example, when GUVs are exposed to Maculatin peptides, a low molecular weight probe entrapped in the GUVs (Alexa^{633}-maleimide in this case) first leaked out without significant leakage of a high molecular weight probe (Alexa^{488}-dextran, 10 kDa molecular weight). Additionally it can be simultaneously observed that there is not any evident change in the overall three-dimensional structure and integrity of the GUVs during the leakage event (using the membrane probe DiIC$_{18}$, red) [104, 105]. The interpretation of the above results is that the Maculatin peptide generates pore-like structures in the membrane allowing the differential leakage of the soluble fluorescent probes from the inner side of the vesicles. A similar effect is observed for the well-known pore-forming peptide melittin and the Alzheimer β-amyloid peptide 1–42 [104, 105]. On the other hand, a different behavior is achieved by the small antibiotic peptides Citropin and Aurein at the same concentration as the one used for the Maculatin experiments. In the case of Citropin and Aurein concomitant leakage of both dyes and membrane destruction is observed when GUVs are exposed to the peptides. Consequently, quantitative information regarding leakage kinetics (fluorescence intensity vs. time) and pore size (leakage of entrapped fluorophores of different molecular weight) can be simultaneously obtained in this type of experiment using GUVs and fluorescence microscopy.

A similar approach was used to observe the interaction of the cell penetrating peptide pep-1 using GUVs composed of POPC or POPC/POPG (4:1 molar) [102]. In this case the GUVs were loaded with three different sized soluble fluorescent markers, Alexa Fluor^{633}-maleimide (1300 MW), Alexa Fluor^{488}-Dextran (3000 MW), and Alexa Fluor^{546}-Dextran (10000 MW). In these experiments it was observed that pep-1 does not alter the bilayer integrity and does not provoke changes in membrane permeability at peptide to lipid ratios below 1. However, above this ratio liposomes suffer an absolute destruction with the simultaneous leakage of the three dyes, ending in deformed lipid aggregates. This experiment gives an idea of the cytotoxic mechanism of pep-1 when it is present at high concentrations [102]. Similar experimental strategies (but using a single water soluble fluorophore) were used by Tamba and Yamazaki [109] to observe the effect of magainin 2 on GUVs composed of PC/PG mixtures and Hasper et al. [107] to explore the bactericidal mechanism of lantibiotic peptides in lipid II containing GUVs.

In summary, these "leakage" experiments at the level of single vesicles allowing simultaneous determination of membrane-destabilization mechanisms mediated by different membrane-active peptides including estimation of pore size (if pore formation is present), spatial localization of peptides (if fluorescent labeled peptides are available), and membrane lateral structure upon peptide addition. The last approach allows more complete information than those based on fluorescent cuvette measurements where visual information about the microscopic scenario is not available.

Protein–Membrane Interactions

The majority of the experiments reported using GUVs containing membrane proteins focus on studying the spatial correlation between certain membrane proteins and cholesterol enriched domains (liquid ordered phase). These experiments particularly involve the use of lipid-like probes and fluorescently labeled proteins in the GUVs. Imaging of the fluorescence intensity distribution in the GUV membrane and/or diffusion measurements (using FCS) were reported in the literature in order to ascertain the distribution of membrane proteins into particular lipid phases [45, 48, 53, 56]. Another type of fluorescence microscopy experiments involving GUVs is related to the potential interaction of proteins with the outer leaflet of the GUVs bilayer. Generally the effects reported in these experiments are related to changes in membrane lateral structure or eventually changes in the 3D structure of the GUVs [111–113]. Other reports that explored the aforementioned effects are related to proteins that cause lipid hydrolysis, i.e., lipids that change the membrane composition (lipases for examples) [114, 115].

Future Directions

A large amount of information relating to various membrane-related phenomena were obtained in the last 10 years using GUVs and fluorescence microscopy. However, development of more sophisticated membrane model systems is still required to achieve more realistic information concerning biological membranes. For instance, the potential effect of the cytoskeleton on membrane lateral structure has not been yet deeply explored. However, some papers reported polymerization of actin inside giant vesicles [116] and modulation on the lateral structure of the membranes displaying phase coexistence upon protein polymerization [117]. Last but not least the local dynamics of domain formation was recently described by Celli et al. [25]. These authors used a combination of imaging and fluctuation techniques to investigate the temporal evolution of gel phase domains at the onset of phase separation. These types of studies are extremely valuable and likely to be extended to cholesterol containing ternary mixtures. The full and concerted characterization of structural and dynamical aspects of lateral phase separation phenomenon in model membranes will establish the basis to evaluate if related processes happen in biological membranes. The last will also help to ascertain when lipid mediated phenomenon (or liquid ordered phases) are really relevant in particular biological membranes.

Acknowledgments The author wants to thank Dr. David Jameson for the critical reading of the manuscript. Research in the laboratory of L. A. B. is funded by grants from Forskningsrådet for Natur og Univers (FNU, Denmark), Forskningsrådet for Sundhed og Sygdom (FSS) and the Danish National Research Foundation (which supports MEMPHYS-Center for Biomembrane Physics).

References

1. V. Borenstain, Y. Barenholz, Characterization of liposomes and other lipid assemblies by multiprobe fluorescence polarization. *Chem Phys Lipids* **64** (1–3), 117–127 (1993).
2. L.M. Loura, A. Fedorov, M. Prieto, Exclusion of a cholesterol analog from the cholesterol-rich phase in model membranes. *Biochim Biophys Acta* **1511** (2), 236–243 (2001).
3. L.M. Loura, A. Fedorov, M. Prieto, Fluid–fluid membrane microheterogeneity: a fluorescence resonance energy transfer study. *Biophys J* **80** (2), 776–788 (2001).
4. M.D. Yeager, G.W. Feigenson, Fluorescence quenching in model membranes: phospholipid acyl chain distributions around small fluorophores. *Biochemistry* **29** (18), 4380–4392 (1990).
5. J.R. Silvius, I.R. Nabi, Fluorescence-quenching and resonance energy transfer studies of lipid microdomains in model and biological membranes. *Mol Membr Biol* **23** (1), 5–16 (2006).
6. C. Reyes Mateo, A. Ulises Acuna, J.C. Brochon, Liquid-crystalline phases of cholesterol/lipid bilayers as revealed by the fluorescence of trans-parinaric acid. *Biophys J* **68** (3), 978–987 (1995).
7. T. Parasassi, G. De Stasio, A. d'Ubaldo, E. Gratton, Phase fluctuation in phospholipid membranes revealed by Laurdan fluorescence. *Biophys J* **57** (6), 1179–1186 (1990).
8. E. Perochon, A. Lopez, J.F. Tocanne, Polarity of lipid bilayers. A fluorescence investigation. *Biochemistry* **31** (33), 7672–7682 (1992).
9. R.F. de Almeida, L.M. Loura, A. Fedorov, M. Prieto, Lipid rafts have different sizes depending on membrane composition: a time-resolved fluorescence resonance energy transfer study. *J Mol Biol* **346** (4), 1109–1120 (2005).
10. K. Arnold, A. Losche, K. Gawrisch, 31p-NMR investigations of phase separation in phosphatidylcholine/phosphatidylethanolamine mixtures. *Biochim Biophys Acta* **645** (1), 143–148 (1981).
11. A. Filippov, G. Oradd, G. Lindblom, Domain formation in model membranes studied by pulsed-field gradient-NMR: the role of lipid polyunsaturation. *Biophys J* **93** (9), 3182–3190 (2007).
12. R.N. Lewis, R.N. McElhaney, Fourier transform infrared spectroscopy in the study of lipid phase transitions in model and biological membranes: practical considerations. *Methods Mol Biol* **400**, 207–226 (2007).
13. S. Mabrey, J.M. Sturtevant, Investigation of phase transitions of lipids and lipid mixtures by sensitivity differential scanning calorimetry. *Proc Natl Acad Sci USA* **73** (11), 3862–3866 (1976).
14. P.W. van Dijck, A.J. Kaper, H.A. Oonk, J. de Gier, Miscibility properties of binary phosphatidylcholine mixtures. A calorimetric study. *Biochim Biophys Acta* **470** (1), 58–69 (1977).
15. A. Blume, R.J. Wittebort, S.K. Das Gupta, R.G. Griffin, Phase equilibria, molecular conformation, and dynamics in phosphatidylcholine/phosphatidylethanolamine bilayers. *Biochemistry* **21** (24), 6243–6253 (1982).
16. M. Caffrey, F.S. Hing, A temperature gradient method for lipid phase diagram construction using time-resolved x-ray diffraction. *Biophys J* **51** (1), 37–46 (1987).
17. J. Korlach, P. Schwille, W.W. Webb, G.W. Feigenson, Characterization of lipid bilayer phases by confocal microscopy and fluorescence correlation spectroscopy. *Proc Natl Acad Sci USA* **96** (15), 8461–8466 (1999).
18. L.A. Bagatolli, E. Gratton, Two photon fluorescence microscopy of coexisting lipid domains in giant unilamellar vesicles of binary phospholipid mixtures. *Biophys J* **78** (1), 290–305 (2000).
19. L.A. Bagatolli, E. Gratton, Two-photon fluorescence microscopy observation of shape changes at the phase transition in phospholipid giant unilamellar vesicles. *Biophys J* **77** (4), 2090–2101 (1999).

20. S. Breusegem, M. Levi, N. Barry, Fluorescence correlation spectroscopy and fluorescence lifetime imaging microscopy. *Nephron Exp Nephrol* **103**(2), e41–e49. (2005).
21. Q.S. Hanley, K.A. Lidke, R. Heintzmann, D.J. Arndt-Jovin, T.M. Jovin, Fluorescence lifetime imaging in an optically sectioning programmable array microscope (PAM). *Cytometry A* **67** (2), 112–118 (2005).
22. S.A. Sanchez, E. Gratton, Lipid– protein interactions revealed by two-photon microscopy and fluorescence correlation spectroscopy. *Acc Chem Res* **38** (6), 469–477 (2005).
23. Y. Chen, B.C. Lagerholm, B. Yang, K. Jacobson, Methods to measure the lateral diffusion of membrane lipids and proteins. *Methods* **39** (2), 147–153 (2006).
24. J. Ries, P. Schwille, New concepts for fluorescence correlation spectroscopy on membranes. *Phys Chem Chem Phys* **10** (24), 3487–3497 (2008).
25. A. Celli, S. Beretta, E. Gratton, Phase fluctuations on the micron–submicron scale in GUVs composed of a binary lipid mixture. *Biophys J* **94** (1), 104–116 (2008).
26. K.A. Lidke, B. Rieger, D.S. Lidke, T.M. Jovin, The role of photon statistics in fluorescence anisotropy imaging. *IEEE Trans Image Process* **14** (9), 1237–1245 (2005).
27. A.H. Kunding, M.W. Mortensen, S.M. Christensen, D. Stamou, A Fluorescence-based technique to construct size distributions from single object measurements, application to the extrusion of lipid vesicles. *Biophys J* **95** (3), 1176–1188 (2008).
28. A.D. Bangham, R.W. Horne, Negative staining of phospholipids and their structural modification by surface-active agents as observed in the electron microscope. *J Mol Biol* **8**, 660–668 (1964).
29. S. Segota, D. Tezak, Spontaneous formation of vesicles. *Adv Colloid Interface Sci* **121** (1–3), 51–75 (2006).
30. D.D. Lasic, Giant Vesicles: A Historical Introduction, in: P.L. Luisi, P. Walde (Eds.), Giant Vesicles, Wiley, New York, (2000), pp. 11–24.
31. R. Virchow, Ueber das ausgebreitete Vorkommen einer dem Nervenmark analogen Substanz in den thierischen Geweben. *Virchows Archiv* **6** (4), 562–572 (1854).
32. C. Neubauer, Ueber das Myelin. *Zeitschrift für Analytische Chemie* **6**, 189–195 (1867).
33. J.P. Reeves, R.M. Dowben, Formation and properties of thin-walled phospholipid vesicles. *J Cell Physiol* **73** (1), 49–60 (1969).
34. L.A. Bagatolli, E. Gratton, A correlation between lipid domain shape and binary phospholipid mixture composition in free standing bilayers: A two-photon fluorescence microscopy study. *Biophys J* **79** (1), 434–447 (2000).
35. T. Baumgart, S.T. Hess, W.W. Webb, Imaging coexisting fluid domains in biomembrane models coupling curvature and line tension. *Nature* **425** (6960), 821–824 (2003).
36. C. Dietrich, L.A. Bagatolli, Z.N. Volovyk, N.L. Thompson, M. Levi, K. Jacobson, E. Gratton, Lipid rafts reconstituted in model membranes. *Biophys J* **80** (3), 1417–1428 (2001).
37. G.W. Feigenson, J.T. Buboltz, Ternary phase diagram of dipalmitoyl-PC/dilauroyl-PC/cholesterol. nanoscopic domain formation driven by cholesterol. *Biophys J* **80** (6), 2775–2788 (2001).
38. N. Kahya, D. Scherfeld, K. Bacia, B. Poolman, P. Schwille, Probing lipid mobility of raft-exhibiting model membranes by fluorescence correlation spectroscopy. *J Biol Chem* **278** (30), 28109–28115 (2003).
39. N. Kahya, D. Scherfeld, P. Schwille, Differential lipid packing abilities and dynamics in giant unilamellar vesicles composed of short-chain saturated glycerol-phospholipids, sphingomyelin and cholesterol. *Chem Phys Lipids* **135** (2), 169–180 (2005).
40. S.L. Veatch, S.L. Keller, Organization in lipid membranes containing cholesterol. *Phys Rev Lett* **89** (26), 268101 (2002).
41. S.L. Veatch, S.L. Keller, Separation of liquid phases in giant vesicles of ternary mixtures of phospholipids and cholesterol. *Biophys J* **85** (5), 3074–3083 (2003).
42. S.L. Veatch, S.L. Keller, Miscibility phase diagrams of giant vesicles containing sphingomyelin. *Phys Rev Lett* **94** (14), 148101 (2005).

43. S.L. Veatch, I.V. Polozov, K. Gawrisch, S.L. Keller, Liquid domains in vesicles investigated by NMR and fluorescence microscopy. *Biophys J* **86** (5), 2910–2922 (2004).
44. J. Sot, L.A. Bagatolli, F.M. Goñi, A. Alonso, Detergent-resistant, ceramide-enriched domains in sphingomyelin/ceramide bilayers. *Biophys. J.* **90**, 903–914 (2006).
45. J. Bernardino de la Serna, J. Perez-Gil, A.C. Simonsen, L.A. Bagatolli, Cholesterol rules: direct observation of the coexistence of two fluid phases in native pulmonary surfactant membranes at physiological temperatures. *J Biol Chem* **279** (39), 40715–40722 (2004).
46. K. Nag, J.S. Pao, R.R. Harbottle, F. Possmayer, N.O. Petersen, L.A. Bagatolli, Segregation of saturated chain lipids in pulmonary surfactant films and bilayers. *Biophys J* **82** (4), 2041–2051 (2002).
47. L.A. Bagatolli, E. Gratton, T.K. Khan, P.L. Chong, Two-photon fluorescence microscopy studies of bipolar tetraether giant liposomes from thermoacidophilic archaebacteria Sulfolobus acidocaldarius. *Biophys J* **79** (1), 416–425 (2000).
48. Q. Ruan, M.A. Cheng, M. Levi, E. Gratton, W.W. Mantulin, Spatial-temporal studies of membrane dynamics: scanning fluorescence correlation spectroscopy (SFCS). *Biophys J* **87** (2), 1260–1267 (2004).
49. L.R. Montes, A. Alonso, F.M. Goni, L.A. Bagatolli, Giant unilamellar vesicles electroformed from native membranes and organic lipid mixtures under physiological conditions. *Biophys J* **93** (10), 3548–3554 (2007).
50. T. Baumgart, A.T. Hammond, P. Sengupta, S.T. Hess, D.A. Holowka, B.A. Baird, W.W. Webb, Large-scale fluid/fluid phase separation of proteins and lipids in giant plasma membrane vesicles. *Proc Natl Acad Sci USA* **104** (9), 3165–3170 (2007).
51. K. Bacia, C.G. Schuette, N. Kahya, R. Jahn, P. Schwille, SNAREs prefer liquid-disordered over "raft" (liquid-ordered) domains when reconstituted into giant unilamellar vesicles. *J Biol Chem* **279** (36), 37951–37955 (2004).
52. P. Girard, J. Pecreaux, G. Lenoir, P. Falson, J.L. Rigaud, P. Bassereau, A new method for the reconstitution of membrane proteins into giant unilamellar vesicles. *Biophys J* **87** (1), 419–429 (2004).
53. N. Kahya, D.A. Brown, P. Schwille, Raft partitioning and dynamic behavior of human placental alkaline phosphatase in giant unilamellar vesicles. *Biochemistry* **44** (20), 7479–7489 (2005).
54. N. Kahya, E.I. Pecheur, W.P. de Boeij, D.A. Wiersma, D. Hoekstra, Reconstitution of membrane proteins into giant unilamellar vesicles via peptide-induced fusion. *Biophys J* **81** (3), 1464–1474 (2001).
55. G. Koster, M. VanDuijn, B. Hofs, M. Dogterom, Membrane tube formation from giant vesicles by dynamic association of motor proteins. *Proc Natl Acad Sci USA* **100** (26), 15583–15588 (2003).
56. N. Kahya, Targeting membrane proteins to liquid-ordered phases: molecular self-organization explored by fluorescence correlation spectroscopy. *Chem Phys Lipids* **141** (1–2), 158–168 (2006).
57. D.M. Haverstick, M. Glaser, Visualization of domain formation in the inner and outer leaflets of a phospholipid bilayer. *J Cell Biol* **106** (6), 1885–1892 (1988).
58. K. Akashi, H. Miyata, H. Itoh, K. Kinosita, Jr., Preparation of giant liposomes in physiological conditions and their characterization under an optical microscope. *Biophys J* **71** (6), 3242–3250 (1996).
59. M.I. Angelova, D.S. Dimitrov, Liposome electroformation. *Faraday Discuss Chem Soc* **81**, 303–311. (1986).
60. M.I. Angelova, S. Soléau, P. Meléard, J.F. Faucon, P. Bothorel, Preparation of giant vesicles by external AC fields. Kinetics and application. *Progr Colloid Polym Sci* **89**, 127–131 (1992).
61. A. Moscho, O. Orwar, D.T. Chiu, B.P. Modi, R.N. Zare, Rapid preparation of giant unilamellar vesicles. *Proc Natl Acad Sci USA* **93** (21), 11443–11447 (1996).

62. L.A. Bagatolli, T. Parasassi, E. Gratton, Giant phospholipid vesicles: comparison among the whole lipid sample characteristics using different preparation methods: a two photon fluorescence microscopy study. *Chem Phys Lipids* **105** (2), 135–147 (2000).
63. N. Düzgünes, L.A. Bagatolli, P. Meers, Y.K. Oh, R.M. Straubinger, Fluorescence Methods in Liposome Research, in: V. Weissig, V. Torchilin (Eds.), Liposomes: A Practical Approach (2nd Edition), Oxford University Press, Oxford, 2003, pp. 105–147.
64. T. Pott, H. Bouvrais, P. Meleard, Giant unilamellar vesicle formation under physiologically relevant conditions. *Chem Phys Lipids* **154** (2), 115–119 (2008).
65. A.C. Simonsen, Activation of phospholipase A2 by ternary model membranes. *Biophys J* **94** (10), 3966–3975 (2008).
66. J.M. Crane, L.K. Tamm, Fluorescence microscopy to study domains in supported lipid bilayers. *Methods Mol Biol* **400**, 481–488 (2007).
67. A. Benda, V. Fagul'ova, A. Deyneka, J. Enderlein, M. Hof, Fluorescence lifetime correlation spectroscopy combined with lifetime tuning: new perspectives in supported phospholipid bilayer research. *Langmuir* **22** (23), 9580–9585 (2006).
68. J. Brewer, U. Bernchou, L.A. Bagatolli, 391-Pos. Generalized polarization and fluorescence lifetime imaging analyses of laurdan labeled supported lipid bilayers. *Biophys. J.* **94**, 391 (2008).
69. J.E. Shaw, R.F. Epand, R.M. Epand, Z. Li, R. Bittman, C.M. Yip, Correlated fluorescence-atomic force microscopy of membrane domains: structure of fluorescence probes determines lipid localization. *Biophys J* **90** (6), 2170–2178 (2006).
70. L.A. Bagatolli, To see or not to see: lateral organization of biological membranes and fluorescence microscopy. *Biochim Biophys Acta* **1758** (10), 1541–1556 (2006).
71. S. Pautot, B.J. Frisken, D.A. Weitz, Engineering asymmetric vesicles. *Proc Natl Acad Sci USA* **100** (19), 10718–10721 (2003).
72. M.H. Jensen, E.J. Morris, A.C. Simonsen, Domain shapes, coarsening, and random patterns in ternary membranes. *Langmuir* **23** (15), 8135–8141 (2007).
73. M. Fidorra, S. Hartel, A. Garcia, J. Ipsen, L.A. Bagatolli, 1198-Pos. Do giant unilamellar vesicles composed of binary lipid mixtures obey the lever rule?: A quantitative microscopy imaging approach. *Biophys. J.* **94**, 1198 (2008).
74. T. Baumgart, G. Hunt, E.R. Farkas, W.W. Webb, G.W. Feigenson, Fluorescence probe partitioning between Lo/Ld phases in lipid membranes. *Biochim Biophys Acta* **1768** (9), 2182–2194 (2007).
75. S.L. Veatch, S.L. Keller, Seeing spots: complex phase behavior in simple membranes. *Biochim Biophys Acta* **1746** (3), 172–185 (2005).
76. R.F. de Almeida, J. Borst, A. Fedorov, M. Prieto, A.J. Visser, Complexity of lipid domains and rafts in giant unilamellar vesicles revealed by combining imaging and microscopic and macroscopic time-resolved fluorescence. *Biophys J* **93** (2), 539–553 (2007).
77. B.R. Lentz, Y. Barenholz, T.E. Thompson, Fluorescence depolarization studies of phase transitions and fluidity in phospholipid bilayers. 2. Two component phosphatidylcholine liposomes. *Biochemistry* **15** (20), 4529–4537 (1976).
78. B.R. Lentz, Y. Barenholz, T.E. Thompson, Fluorescence depolarization studies of phase transitions and fluidity in phospholipid bilayers. 1. Single component phosphatidylcholine liposomes. *Biochemistry* **15** (20), 4521–4528 (1976).
79. H.A. Garda, A.M. Bernasconi, R.R. Brenner, Possible compensation of structural and viscotropic properties in hepatic microsomes and erythrocyte membranes of rats with essential fatty acid deficiency. *J Lipid Res* **35** (8), 1367–1377 (1994).
80. M.E. Jones, B.R. Lentz, Phospholipid lateral organization in synthetic membranes as monitored by pyrene-labeled phospholipids: effects of temperature and prothrombin fragment 1 binding. *Biochemistry* **25** (3), 567–574 (1986).
81. T. Parasassi, M. Di Stefano, M. Loiero, G. Ravagnan, E. Gratton, Influence of cholesterol on phospholipid bilayers phase domains as detected by Laurdan fluorescence. *Biophys J* **66** (1), 120–132 (1994).

82. E.K. Krasnowska, E. Gratton, T. Parasassi, Prodan as a membrane surface fluorescence probe: partitioning between water and phospholipid phases. *Biophys J* **74** (4), 1984–1993 (1998).
83. C.R. Mateo, M.P. Lillo, J. Gonzalez-Rodriguez, A.U. Acuna, Lateral heterogeneity in human platelet plasma membrane and lipids from the time-resolved fluorescence of trans-parinaric acid. *Eur Biophys J* **20** (1), 53–59 (1991).
84. M. Velez, M.P. Lillo, A.U. Acuna, J. Gonzalez-Rodriguez, Cholesterol effect on the physical state of lipid multibilayers from the platelet plasma membrane by time-resolved fluorescence. *Biochim Biophys Acta* **1235** (2), 343–350 (1995).
85. T. Parasassi, E. Gratton, W.M. Yu, P. Wilson, M. Levi, Two-photon fluorescence microscopy of laurdan generalized polarization domains in model and natural membranes. *Biophys J* **72** (6), 2413–2429 (1997).
86. K. Gaus, E. Gratton, E.P. Kable, A.S. Jones, I. Gelissen, L. Kritharides, W. Jessup, Visualizing lipid structure and raft domains in living cells with two-photon microscopy. *Proc Natl Acad Sci USA* **100** (26), 15554–15559 (2003).
87. G. M'Baye, Y. Mely, G. Duportail, A.S. Klymchenko, Liquid ordered and gel phases of lipid bilayers: fluorescent probes reveal close fluidity but different hydration. *Biophys J* **95** (3), 1217–1225 (2008).
88. H.M. Kim, H.J. Choo, S.Y. Jung, Y.G. Ko, W.H. Park, S.J. Jeon, C.H. Kim, T. Joo, B.R. Cho, A two-photon fluorescent probe for lipid raft imaging: C-laurdan. *Chembiochem* **8** (5), 553–559 (2007).
89. L. Jin, A.C. Millard, J.P. Wuskell, X. Dong, D. Wu, H.A. Clark, L.M. Loew, Characterization and application of a new optical probe for membrane lipid domains. *Biophys J* **90** (7), 2563–2575 (2006).
90. L. Bagatolli, E. Gratton, T.K. Khan, P.L. Chong, Two-photon fluorescence microscopy studies of bipolar tetraether giant liposomes from thermoacidophilic archaebacteria Sulfolobus acidocaldarius. *Biophys J* **79** (1), 416–425 (2000).
91. M. Fidorra, L. Duelund, C. Leidy, A.C. Simonsen, L.A. Bagatolli, Absence of fluid-ordered/fluid-disordered phase coexistence in ceramide/POPC mixtures containing cholesterol. *Biophys J* **90** (12), 4437–4451 (2006).
92. G.W. Feigenson, Phase boundaries and biological membranes. *Annu Rev Biophys Biomol Struct* **36**, 63–77 (2007).
93. R.F. de Almeida, A. Fedorov, M. Prieto, Sphingomyelin/phosphatidylcholine/cholesterol phase diagram: boundaries and composition of lipid rafts. *Biophys J* **85** (4), 2406–2416 (2003).
94. I. Plasencia, L. Norlen, L.A. Bagatolli, Direct visualization of lipid domains in human skin stratum corneum's lipid membranes: effect of pH and temperature. *Biophys J* **93** (9), 3142–3155 (2007).
95. K. Matsuzaki, Magainins as paradigm for the mode of action of pore forming polypeptides. *Biochim Biophys Acta* **1376** (3), 391–400 (1998).
96. T. Hara, Y. Mitani, K. Tanaka, N. Uematsu, A. Takakura, T. Tachi, H. Kodama, M. Kondo, H. Mori, A. Otaka, F. Nobutaka, K. Matsuzaki, Heterodimer formation between the antimicrobial peptides magainin 2 and PGLa in lipid bilayers: a cross-linking study. *Biochemistry* **40** (41), 12395–12399 (2001).
97. F.Y. Chen, M.T. Lee, H.W. Huang, Evidence for membrane thinning effect as the mechanism for peptide-induced pore formation. *Biophys J* **84** (6), 3751–3758 (2003).
98. M. Zasloff, Antimicrobial peptides of multicellular organisms. *Nature* **415** (6870), 389–395 (2002).
99. R.I. Lehrer, Primate defensins. *Nat Rev Microbiol* **2** (9), 727–738 (2004).
100. N. Papo, Y. Shai, Exploring peptide membrane interaction using surface plasmon resonance: differentiation between pore formation versus membrane disruption by lytic peptides. *Biochemistry* **42** (2), 458–466 (2003).

101. Y. Shai, Mechanism of the binding, insertion and destabilization of phospholipid bilayer membranes by alpha-helical antimicrobial and cell non-selective membrane-lytic peptides. *Biochim Biophys Acta* **1462** (1–2), 55–70 (1999).
102. S.T. Henriques, A. Quintas, L.A. Bagatolli, F. Homble, M.A. Castanho, Energy-independent translocation of cell-penetrating peptides occurs without formation of pores. A biophysical study with pep-1. *Mol Membr Biol* **24** (4), 282–293 (2007).
103. R. Mani, J.J. Buffy, A.J. Waring, R.I. Lehrer, M. Hong, Solid-state NMR investigation of the selective disruption of lipid membranes by protegrin-1. *Biochemistry* **43** (43), 13839–13848 (2004).
104. E.E. Ambroggio, D.H. Kim, F. Separovic, C.J. Barrow, K.J. Barnham, L.A. Bagatolli, G.D. Fidelio, Surface behavior and lipid interaction of Alzheimer beta-amyloid peptide 1–42: a membrane-disrupting peptide. *Biophys J* **88** (4), 2706–2713 (2005).
105. E.E. Ambroggio, F. Separovic, J.H. Bowie, G.D. Fidelio, L.A. Bagatolli, Direct visualization of membrane leakage induced by the antibiotic peptides: maculatin, citropin, and aurein. *Biophys J* **89** (3), 1874–1881 (2005).
106. M.P. Boland, F. Separovic, Membrane interactions of antimicrobial peptides from Australian tree frogs. *Biochim Biophys Acta* **1758** (9), 1178–1183 (2006).
107. H.E. Hasper, N.E. Kramer, J.L. Smith, J.D. Hillman, C. Zachariah, O.P. Kuipers, B. de Kruijff, E. Breukink, An alternative bactericidal mechanism of action for lantibiotic peptides that target lipid II. *Science* **313** (5793), 1636–1637 (2006).
108. H.W. Huang, F.Y. Chen, M.T. Lee, Molecular mechanism of Peptide-induced pores in membranes. *Phys Rev Lett* **92** (19), 198304 (2004).
109. Y. Tamba, M. Yamazaki, Single giant unilamellar vesicle method reveals effect of antimicrobial peptide magainin 2 on membrane permeability. *Biochemistry* **44** (48), 15823–15833 (2005).
110. P.E. Thoren, D. Persson, E.K. Esbjorner, M. Goksor, P. Lincoln, B. Norden, Membrane binding and translocation of cell-penetrating peptides. *Biochemistry* **43** (12), 3471–3489 (2004).
111. P.D. Moens, L.A. Bagatolli, Profilin binding to sub-micellar concentrations of phosphatidylinositol (4,5) bisphosphate and phosphatidylinositol (3,4,5) trisphosphate. *Biochim Biophys Acta* **1768** (3), 439–449 (2007).
112. L.A. Bagatolli, D.D. Binns, D.M. Jameson, J.P. Albanesi, Activation of dynamin II by POPC in giant unilamellar vesicles: a two-photon fluorescence microscopy study. *J Protein Chem* **21** (6), 383–391 (2002).
113. C. Arnulphi, S.A. Sanchez, M.A. Tricerri, E. Gratton, A. Jonas, Interaction of human apolipoprotein A-I with model membranes exhibiting lipid domains. *Biophys J* **89** (1), 285–295 (2005).
114. S.A. Sanchez, L.A. Bagatolli, E. Gratton, T.L. Hazlett, A two-photon view of an enzyme at work: crotalus atrox venom PLA2 interaction with single-lipid and mixed-lipid giant unilamellar vesicles. *Biophys J* **82** (4), 2232–2243 (2002).
115. J.M. Holopainen, M.I. Angelova, T. Soderlund, P.K. Kinnunen, Macroscopic consequences of the action of phospholipase C on giant unilamellar liposomes. *Biophys J* **83** (2), 932–943 (2002).
116. H. Miyata, K. Ohki, G. Marriot, S. Nishiyama, K. Akashi, K. Kinosita Jr, Cell Deformation Mechanics Studied with Actin-Containing Giant Vesicles, a Cell Mimicking System, in: P.L. Luisi, P. Walde (Eds.), Giant Vesicles, Wiley, New York, 2000, pp. 319–334.
117. A.P. Liu, D.A. Fletcher, Actin polymerization serves as a membrane domain switch in model lipid bilayers. *Biophys J* **91** (11), 4064–4070 (2006).
118. J. Sot, L.A. Bagatolli, F.M. Goni, A. Alonso, Detergent-resistant, ceramide-enriched domains in sphingomyelin/ceramide bilayers. *Biophys J* **90** (3), 903–914 (2006).

Electronic Energy Transport and Fluorescence Spectroscopy for Structural Insights into Proteins, Regular Protein Aggregates and Lipid Systems

Therese Mikaelsson, Radek Šachl, and Lennart B.-Å. Johansson

Abstract The present review aims at surveying recent theoretical development and applications of electronic energy transport between chromophoric molecules (i.e. donors and acceptors) in various protein and lipid systems. Reversible, partly reversible, and irreversible energy transport within pairs of interacting chromophoric molecules are considered. Also energy migration/transfer within ensembles of many donor and acceptor molecules is discussed. An extended Förster theory of interacting pairs is summarised, which brings the analyses of data to the same level of molecular description as in ESR and NMR spectroscopy. Recent applications of energy transfer/migration on protein systems concern their structure, folding, and their formation of non-covalent protein polymers. The latter systems are of particular interest in e.g. the study of amyloid formation and the molecular functioning of muscles. The energy transfer/migration processes have also been utilised to study the spatial distribution of lipid molecules, which is of interest in the study of biological membranes and their functioning, e.g. the presumed formation of so-called rafts.

Keywords Electronic energy transfer/migration · Donor–acceptor energy transfer · (Partial) donor–donor energy migration · Homotransfer · Protein polymerization · Lipid membranes · Protein folding

Abbreviations

A	acceptor of electronic energy
BD	Brownian dynamics
CTBX	cholera toxin B-subunit
$\chi(t)$	the time-dependent excitation probability
D	donor of electronic energy
$D(t)$	experimental difference curve created from fluorescence depolarisation data

L.B.-Å. Johansson (✉)
Departments of Chemistry, Biophysical Chemistry, Umeå University, S-901 87 Umeå, Sweden
e-mail: lennart.johansson@chem.umu.se

C.D. Geddes (ed.), *Reviews in Fluorescence 2007*, Reviews in Fluorescence 2007,
DOI 10.1007/978-0-387-88722-7_3,

$\boldsymbol{D_j}$	director frame for the *j*th donor
DAET	donor–acceptor energy transfer
DDEM	donor–donor energy migration
EFT	extended Förster theory
EM	energy migration
(F)RET	Förster resonance energy transfer
$F_{\mathrm{D}}(t)$	fluorescence relaxation of a donor in absence of acceptors
ϕ_{c}	the rotational correlation time
$G^{\mathrm{s}}(t)$	the excitation probability of the initially excited donor
κ	the angular dependence of dipole–dipole coupling
L	the Liouville operator
L	laboratory frame
M	molecule fixed frame
MD	molecular dynamics
MC	Monte Carlo
N	the aggregation number
SLE	stochastic Liouville equation
SME	stochastic master equation
τ_{D}	the fluorescence lifetime of the donor
PDDEM	partial donor–donor energy migration
pDNA	plasmid DNA
R	the distance between the centres of mass of the donor groups
R	a coordinate system fixed in a protein
R_0	the Förster radius
r_0	the fundamental anisotropy
S_{j}	second rank order parameter
TCSPC	time-correlated single-photon counting
ω	the rate of electronic energy transport
$\Omega = (\alpha, \beta, \gamma)$	the Eulerian angles of orientation

Introduction

A diverse of methods based on fluorescence and NMR spectroscopy are frequently applied to study the functioning, structure and dynamics of biomacromolcules, such as proteins and nucleic acids, as well as in studies of biological and model membranes. Methods based on X-ray diffraction [1, 2] and NMR spectroscopy [3, 4] are of central importance for the determination of macromolecular structures, and these can often provide three-dimensional structures with atomic resolution. While the former method relies on the preparation of crystals of high quality, the NMR methods may suffer from sufficient spectral resolution for larger molecules. Furthermore NMR is a rather insensitive technique that typically requires protein concentrations in the mM range, which are often associated with an unwanted protein aggregation. However, intense work has been and still is devoted to the development of complementary methods based on electronic energy transfer and fluorescence spectroscopy. These methods may become important, especially in studies that concern the structure of complex macromolecular assemblies. Most of these applications are based on monitoring the irreversible energy transfer from an excited donor to an acceptor molecule [5–8], while hitherto electronic energy migration between fluorescent molecules of the same kind, so-called donor–donor energy migration (DDEM) or homotransfer [9], is used much less frequently.

Distances within macromolecules can be estimated from studies of the electronic energy transfer between extrinsic or intrinsic chromophoric groups localised in macromolecules. The range of distances and the resolution from such experiments is different from that obtained by X-ray and NMR methods. While the X-ray and NMR methods measure short distances, typically representing nearest neighbouring atoms, the fluorescence techniques provide information about longer distances (10–100 Å), which are comparable to the size of proteins. The latter range of distances is also representative for intermacromolecular distances within supramacromolecular structures.

NMR spectroscopy [10–13] and X-ray diffraction [14–16] have long been applied to obtain information about structure and dynamics of lipid membrane systems. For a long time fluorescence quenching has also been extensively applied in various investigations, e.g. for the determination of micellar size [17, 18], the size of the building units of cubic liquid crystalline phases [19, 20], as well as to examine the characteristic properties of micro-heterogeneous surfactant and lipid systems [21]. More recently fluorescence experiments have been used to monitor the heterogeneous lipid distributions in giant lipid vesicles [22]. These distributions are sometimes called rafts, and they might be of biological interest.

In quite many applications, donor–acceptor pairs are used to estimate distances in macromolecules and supra-macromolecular structures. This process is here referred to as donor–acceptor energy transfer (DAET). In the literature this process is also named fluorescence resonance energy transfer (FRET). DAET/FRET has been applied in an immense number of studies and publications, and surveys of related papers are found in books describing fluorescence spectroscopy [6, 23] and in specialised editions on electronic energy transport [5, 8]. When DAET and extrinsic probes are used to explore macromolecules (e.g. proteins or nucleic acids), a major practical difficulty is to achieve specific attachment of one donor and one acceptor group within the *same* macromolecule. This problem is circumvented by using two chemically identical fluorophores in the labelling procedure [24].

Energy transfer among identical fluorescent groups usually can be considered reversible. Hence, the electronic energy jumps back and forth within a pair; moreover it can migrate among several molecules within an ensemble. For this reason the concept *energy migration* is used to distinguish from the case of irreversible electronic energy transfer. In what follows, the abbreviation DDEM is used for the process of donor–donor energy migration. Although the problem with specific labelling is largely solved, the use of DDEM introduces other methodological questions to solve, which cause the need of fluorescence depolarisation experiments to monitor the energy migration process. The depolarisation experiments are influenced by the reorienting motions of the fluorophores and the energy migration process, which itself depends on these motions. Therefore, the analysis at a molecular scale becomes rather complex. However, these questions can be sorted out and solved by adopting an extended Förster theory (EFT) [25, 26], which is presented in this review.

The need to use fluorescent groups, which *irrespective* of their localisation in a macromolecular structure (e.g. a protein) exhibit very similar photophysics, limits the applicability of DDEM. Most fluorescent molecules do not fulfil this criterion, whereby the number of useful probes is considerably reduced. Nevertheless, it is possible to make use of the interaction between two chemically identical, but photophysically non-identical donor groups. The energy migration rate within such a DD-pair will influence the photophysics observed, and it is then possible to show how the fluorescence relaxation depends on the rate of energy migration [9, 27]. To distinguish from DDEM and DAET, this case is referred to as partial donor–donor energy migration (PDDEM), because one deals with donor molecules for which the excitation probability is partially reversible. The theoretical treatment of PDDEM is very similar to that of partly reversible DAET [28]. Strictly speaking, PPDEM can occur in any DD-pair, for which each D in the absence of energy migration exhibits non-exponential photophysics [29]. This statement is valid even if the two interacting D-molecules have an identical non-exponential decay [30].

Several biological functions and diseases are connected with proteins which exhibit an inherent ability [31] to form various non-covalent protein polymers. For instance, the α,β,γ-crystallines form rigid three-dimensional protein structures, which are of utmost importance for the properties of the human lens [32]. Several diseases are related to non-covalent protein polymers, e.g. the formation of amyloids [33, 34] and prions [35, 36]. The cytolytic toxins [37] constitute another interesting class of proteins that form regular proteins aggregates [38, 39], which are thought to create pores in membranes. The X-ray and NMR-methods have the highest potential for exploring such structures at an atomic level. While the challenge of preparing crystal for X-ray should be easier with regularly aggregating proteins, this is in practise not found, where rather mixtures of crystalline structures are obtained. Furthermore the interpretation of NMR data becomes very complicated [40]. This motivates the development of methods based on fluorescence DAET and DDEM for the study of structures and functioning of non-covalent protein polymers. Recently such a method has been presented [41] and applied to investigate the structure of filamentous actin [42].

A general challenge with fluorescence and electronic energy transfer data concerns the molecular interpretation. A theoretical molecular description involves several parameters, which are usually difficult to unambiguously quantify. In this review, recent theoretical developments are presented together with a new approach to the qualitative improvement of data analysis. This review also highlights some recent DDEM- and DAET-applications on biomacromolecular systems. In particular, the emphasis is on studies of protein and lipid systems.

Theoretical Development

Applications of electronic energy transfer/migration deal with either pairs or ensembles of interacting chromophores. From a theoretical point of view the former case accounts for the interaction within two particles, while the latter involves a many-particle system with energy transfer/migration among spatially and orientationally distributed molecules. The latter complexity can be handled by combining Monte Carlo (MC) and Brownian dynamics (BD) simulations [41]. Each of these two classes of systems is separately discussed in the following subsections.

Electronic Energy Migration/Transfer Within a Pair

Electronic energy transport between two chromophoric molecules can be divided into three separate cases. Each case is dictated by the chemical and photophysical properties of the interacting chromophores. Most often the energy transfer between a donor and an acceptor group is studied, i.e. between chemically and photophysically different molecules. This DAET process is schematically illustrated by the reaction

$$\mathrm{D}^* + \mathrm{A} \longrightarrow \mathrm{D} + \mathrm{A}^*. \tag{1}$$

The electronic energy is here transferred irreversibly to the acceptor at a rate (ω_{DA}). Thereafter the excited acceptor A^* either emits a photon or relaxes to its electronic ground state via non-radiative routes. For DA pairs an EFT has recently been derived by solving the stochastic master equation (SME) of energy transfer. The SME, which accounts for the reorientational and the spatial dynamics, has been derived from the stochastic Liouville equation (SLE). The solution of the SME provides the following expression of the time-dependent excitation probability $\{\chi(t)\}$ of the donor group [43, 44]:

$$\chi(t) = \left\langle \exp\left(-\int_0^t \omega(t')\mathrm{d}t' \right) \right\rangle F_{\mathrm{D}}(t). \tag{2}$$

In Eq. (2), $F_{\mathrm{D}}(t)$ stands for the fluorescence relaxation of the donor in the absence of an acceptor. The expression within the bracket (<...>) is a stochastic average over the rate of electronic energy transfer between a donor and an acceptor group. The stochastic time-dependent rate of electronic energy transfer is given by

$$\begin{aligned} \omega(t) &= \Lambda\kappa^2(t), \\ \Lambda &= \frac{3}{2\tau_{\mathrm{D}}}\left(\frac{R_0}{R}\right)^6, \\ \kappa(t) &= \hat{\mu}_{\mathrm{A}}(t)\cdot\hat{\mu}_{\mathrm{B}}(t) - 3(\hat{\mu}_{\mathrm{A}}(t)\cdot\hat{R})(\hat{\mu}_{\mathrm{A}}(t)\cdot\hat{R}). \end{aligned} \tag{3}$$

Here the κ^2-function, $\hat{R}$, Λ and τ_D denotes the stochastic orientational dependence of the dipole-dipole coupling, a unit vector of the distance vector $\vec{R}(= R \cdot \hat{R})$, the coupling strength and the fluorescence lifetime of the donor in absence of an acceptor, respectively. The angles defining the κ^2-function are indicated in Fig. 1. The restricted local reorienting motions of the interacting D and A groups, with respect to the coordinate systems denoted by $\mathbf{D}_A$ and $\mathbf{D}_B$, are described by the transformation angles $\Omega_{M_A D_A}$ and $\Omega_{M_B D_B}$, respectively. The mutual orientation of the director frames $\mathbf{D}_A$ and $\mathbf{D}_B$ relative to a common $\mathbf{R}$-frame is described by the angles $\Omega_{D_j R}$, which are referred to as the *configuration angles*. Notice that in the short-time limit, Eq. (2) and the EFT are transferred into the well-known theoretical expression given by Förster's theory [45], i.e. the fluorescence decay of the donor is

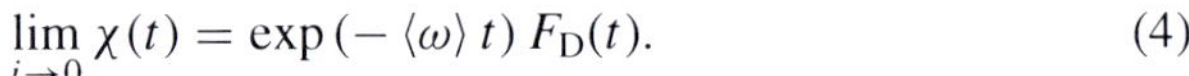

$$\lim_{i \to 0} \chi(t) = \exp(-\langle\omega\rangle t)\, F_D(t). \tag{4}$$

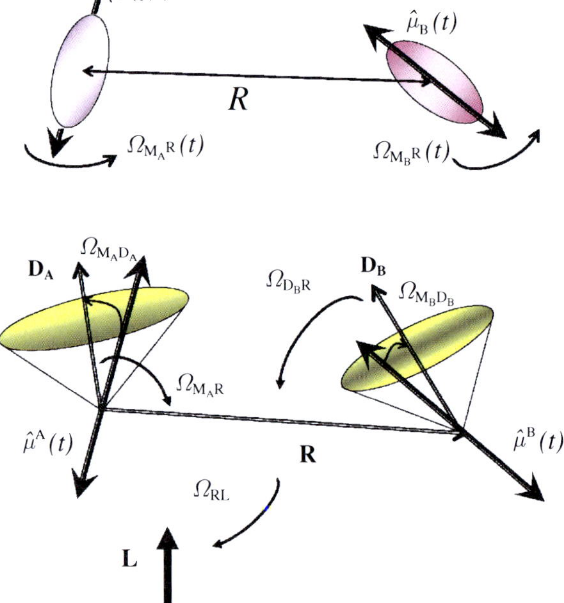

Fig. 1 The *upper* schematic shows two interacting chromophores A and B. These are either two donors, or one donor and one acceptor. A and B are thought to be covalently linked to a rigid macromolecule at a fixed distance (R), but they can undergo local reorienting motions in their binding sites. The local reorientations of the transition dipoles ($\hat{\mu}_A$and$\hat{\mu}_B$) are described by the Eulerian angles ($\Omega_{M_j D_j}, j =$ A or B) relative to the local coordinate systems $\mathbf{D}_A$ and $\mathbf{D}_B$. The different orientation transformations that relate the angular dependences (Ω_{IJ}) of the macromolecule, the A and B groups, as well as the energy transport to a laboratory coordinate system (**L**) are illustrated in the *lower* part of the figure. The Z-axis of the **R** frame connects the centres of mass of A and B

In Eq. (4) the exponent ($\langle\omega\rangle$) corresponds to the *dynamic* average of the transfer rate between the donor and the acceptor groups. In the numerous applications of Förster's theory, the fluorescence of a donor is monitored in the presence of an acceptor, while both groups are attached to a macromolecule. In the analyses of data it is frequently assumed that the two groups are isotropically oriented, and that the reorienting motions are fast as compared to the rate of energy transfer, i.e. in the dynamic limit. In practise these assumptions are usually not fulfilled. In fact, the fluorescence relaxation and the reorienting motions typically occur on the same timescale, i.e. in the ns range. Some resulting consequences of these approximations are illustrated in Fig. 2. Here the D and A groups interact over a fixed distance, which is taken to be $R = 0.8 \cdot R_0$. The donor's lifetime in the absence of A is 10 ns, and both groups undergo rotational motions with a rotational correlation time of $\phi_c = 5$ ns. Using the EFT, the time-correlated single-photon counting (TCSPC) data have been created to mimic the corresponding to the D-fluorescence relaxation experiment. According to Eq. (4), the initial decay of the fluorescence relaxation equals the decay expected according to the FT. It is obvious that the FT initially coincides with the EFT (cf. Fig. 2), but the two curves gradually diverge over time. Unlike Förster's theory, the EFT predicts a multi-exponential decay. As a consequence, significant errors in the distance calculations are expected when Förster's theory is used in analyses of experimental data. For isotropic orientation $\langle\kappa^2\rangle = 2/3$, the errors can be minimised by determining the initial mono-exponential part of the decay. This means that the decay is fitted to data over the range for which Eq. (4) provides a good statistical fit. Thus, from the value of $\langle\omega\rangle$ obtained and $\langle\omega\rangle = \langle\kappa^2\rangle\Lambda$, an accurate value of the distance can be calculated.

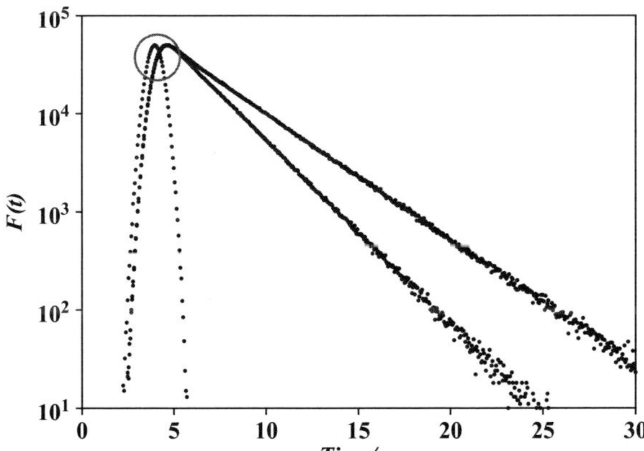

Fig. 2 Generated fluorescence relaxation data that mimic TCSPC experiments for DA-pairs of which the two groups are separated at a fixed distance $R = 0.8R_0$. The lifetime of the donor $\tau_D =$ 10.0 ns. The D and A groups are reorienting in an isotropic potential, i.e. the molecular order is isotropic. The photophysics' relaxation $\{F(t)\}$ of the D as predicted by the EFT (*upper trace*) and the FT. The ring indicates the time range where the FT agrees with the EFT

A second interesting case of energy transport is the reversible migration between two donors that exhibit the same mono-exponential fluorescence decay. This refers to the DDEM process, which is schematically illustrated by the reaction

$$D_A{}^* + D_B \rightleftarrows D_A + D_B{}^*. \tag{5}$$

It is obvious that the forward and backward rates of energy migration are equal ($\omega_{AB} = \omega_{BA}$), which implies that the total excitation probability is conserved. Thus, the total excitation probability and the observed fluorescence decay become invariant to the energy migration process. The SME of energy transport for a DD pair can be derived from the stochastic Liouville equation, in a similar manner as was used for DAET. The solution of this SME provides an expression for the time-dependent excitation probability of the primary (p) and secondary (s) excited donor [46]. These excitation probabilities are given by

$$\begin{aligned}
\chi^{p}(t) &= \frac{1}{2}\left\langle\left[1+\exp\left(-2\int_0^t \omega(t')\mathrm{d}t'\right)\right]\right\rangle F_D(t) \equiv \tilde{\chi}^{p}(t)F_D(t), \\
\chi^{s}(t) &= \frac{1}{2}\left\langle\left[1-\exp\left(-2\int_0^t \omega(t')\mathrm{d}t'\right)\right]\right\rangle F_D(t) \equiv \tilde{\chi}^{s}(t)F_D(t), \\
\tilde{\chi}^{s}(t) &= 1-\tilde{\chi}^{p}(t).
\end{aligned} \tag{6}$$

The last equation shows that the excitation due to energy migration is conserved, whereby the total excitation probability, i.e. the observable $F_D(t) = \chi^{p}(t) + \chi^{s}(t)$, becomes insensitive to the DDEM process. However, information about the DDEM process can be obtained from fluorescence depolarisation experiments. From time-resolved depolarisation experiments the time-resolved fluorescence anisotropy can be constructed [6, 23]. For an ensemble of fluorescent molecules the anisotropy corresponds to an orientational correlation function of second rank according to

$$r(t) = r_0 \left\langle P_2\left[\hat{\mu}(0)\cdot\hat{\mu}(t)\right]\right\rangle \equiv r_0\rho(t). \tag{7}$$

In Eq. (7) r_0, $\langle\ldots\rangle$ and P_2 denote the fundamental anisotropy [47], the orientational average over the ensemble and the second Legendre polynomial, respectively. The orientation of the electronic transition dipole moment at the times of excitation ($t = 0$) and emission ($t = t$) are denoted by the unit vectors $\hat{\mu}(0)$ and $\hat{\mu}(t)$. In the following we consider two interacting donor groups (D_A and D_B), which are covalently attached to a macromolecule. Here the macromolecule is assumed to undergo global rotational diffusion like a spherical particle with a characteristic rotational correlation time denoted $\phi_{c,glob}$. Because of energy migration between D_A and D_B, the experimental fluorescence anisotropy $\{r(t)\}$ is composed of the following contributions [26]:

$$r(t) = \frac{r(0)}{2}\left[\rho_{AA}(t) + \rho_{BB}(t) + \rho_{AB}(t) + \rho_{BA}(t)\right]\exp(-t/\phi_{c,glob}) \tag{8a}$$

The initial excitation probability of D_A and D_B is equal (= 1/2). The fluorescence anisotropy of each donor in the absence of DDEM is described by $\rho_{ii}(t)$ (i = A or B). The anisotropy contribution due to DDEM from D_i to D_j is given by $\rho_{ij}(t)$. The ρ-terms in Eq. (8a) are given by

$$\rho_{ii}(t) = \left\langle P_2\left[\hat{\mu}_i(0)\cdot\hat{\mu}_i(t)\right]\tilde{\chi}^{\mathrm{P}}(t)\right\rangle \quad i = \mathrm{A, B}, \tag{8b}$$

$$\rho_{ij}(t) = \left\langle P_2\left[\hat{\mu}_i(0)\cdot\hat{\mu}_j(t)\right]\left(1-\tilde{\chi}^{\mathrm{P}}(t)\right)\right\rangle, \quad i, j = \mathrm{A, B}; (\mathrm{A} \neq \mathrm{B}). \tag{8c}$$

The third case concerns the energy transport between two chemically identical fluorophores which exhibit different photophysics. This partly reversible process is illustrated by the following reaction:

$$\mathrm{D_A}^* + \mathrm{D_B} \rightleftarrows \mathrm{D_A} + \mathrm{D_B}^*. \tag{9}$$

For this case the time-dependent excitation probabilities of the donors are different, even if for equal rates $\omega_{AB} = \omega_{BA}$. Thus, the PDDEM process represents the intermediate case of DAET and DDEM. An EFT of PDDEM has also been developed. Starting from the stochastic Liouville equation, it is possible to derive the following stochastic master equation [48]:

$$\begin{bmatrix} \dot{\chi}_A(t) \\ \dot{\chi}_B(t) \end{bmatrix} = \begin{bmatrix} -1/\tau_A - \omega(t) & \omega(t) \\ \omega(t) & -1/\tau_B - \omega(t) \end{bmatrix} \begin{bmatrix} \chi_A(t) \\ \chi_B(t) \end{bmatrix}. \tag{10}$$

This SME is relevant for two chemically identical donor molecules which exhibit the same absorption and fluorescence spectra, while their fluorescence lifetimes differ. The latter might be due to e.g. different chemical environments or different exposures of a quencher. For the numerical solution of Eq. (10), it is convenient to adopt the following transformations:

$$\begin{bmatrix} \xi_A(t) = e^{t/\tau_A}\chi_A(t) \\ \xi_B(t) = e^{t/\tau_B}\chi_B(t) \end{bmatrix}, \tag{11}$$

whereby the integrated form of Eq. (10) can be written as

$$\begin{bmatrix} \xi_A(t) \\ \xi_B(t) \end{bmatrix} = \exp\left\{\int_0^t 2\omega(s)\mathbf{P}(s)\mathrm{d}s\right\} \begin{bmatrix} \xi_A(0) \\ \xi_B(0) \end{bmatrix}. \tag{12}$$

Here $\mathbf{P}(s)$ stands for the two-dimensional matrix:

$$\mathbf{P}(t) = \begin{bmatrix} -1/2 & (1/2)\exp(\tau_B^{-1} - \tau_A^{-1})t \\ (1/2)\exp(\tau_A^{-1} - \tau_B^{-1})t & -1/2 \end{bmatrix}. \tag{13}$$

In the studies of PDDEM it is worth noticing that fluorescence lifetime and depolarisation experiments contain information about the excitation probabilities of the donors. Again, we consider two interacting donor groups (A and B), which are attached specifically to e.g. a macromolecule. The orientation of the pairs averaged over the ensemble studied is assumed isotropic. For a coupled AB pair, the observed theoretical fluorescence decay of the photophysics $\{s(t)\}$ is given by

$$s(t) = \frac{1}{2} < \chi_{\mathrm{A}}^{\mathrm{p}}(t) + \chi_{\mathrm{A}}^{\mathrm{s}}(t) + \chi_{\mathrm{B}}^{\mathrm{p}}(t) + \chi_{\mathrm{B}}^{\mathrm{s}}(t) > . \tag{14}$$

For example, $\chi_{\mathrm{A}}^{\mathrm{p}}(t)$ and $\chi_{\mathrm{A}}^{\mathrm{s}}(t)$ denote the probability that donor A is primarily (p) and secondarily (s) excited by donor B and thereafter emits a photon at the time t, respectively. Notice that the observed fluorescence decay is not invariant to the energy migration process as is the case in DDEM. Therefore, a rather complex expression for the fluorescence anisotropy is obtained:

$$r(t) = \frac{r_0 < \rho_{\mathrm{AA}}(t)\chi_{\mathrm{A}}^{\mathrm{p}}(t) + \rho_{\mathrm{AB}}(t)\chi_{\mathrm{B}}^{\mathrm{s}}(t) + \rho_{\mathrm{BB}}(t)\chi_{\mathrm{B}}^{\mathrm{p}}(t) + \rho_{\mathrm{BA}}(t)\chi_{\mathrm{A}}^{\mathrm{s}}(t) > \exp\left(-t/\Phi_{\mathrm{prot}}\right)}{< \chi_{\mathrm{A}}^{\mathrm{p}}(t) + \chi_{\mathrm{B}}^{\mathrm{s}}(t) + \chi_{\mathrm{B}}^{\mathrm{p}}(t) + \chi_{\mathrm{A}}^{\mathrm{s}}(t) >}, \tag{15a}$$

where

$$\rho_{ij}(t) = P_2\left[\hat{\mu}_i(0) \cdot \hat{\mu}_j(t)\right] \quad i, j \in \mathrm{A, B}. \tag{15b}$$

In the limit of DDEM, i.e. for identical fluorescence decays $\{= F_{\mathrm{D}}(t)\}$ of donors A and B, one obtains that [46]

$$\chi_i^{\mathrm{p}}(t) = \tilde{\chi}_i^{\mathrm{p}}(t)F_{\mathrm{D}}(t) = \left\{1 - \tilde{\chi}_i^{\mathrm{s}}(t)\right\} F_{\mathrm{D}}(t) \quad i \in \mathrm{A, B}. \tag{16}$$

Here the excitation probabilities due to energy migration are denoted by $\tilde{\chi}_i(t)$. The following expression of fluorescence anisotropy is then obtained by inserting Eq. (16) into Eq. (15b):

$$\begin{aligned} r(t) = r_0 < &\rho_{\mathrm{AA}}(t)\tilde{\chi}_{\mathrm{A}}^{\mathrm{p}}(t) + \rho_{\mathrm{AB}}(t)\left\{1 - \tilde{\chi}_{\mathrm{A}}^{\mathrm{p}}(t)\right\} + \rho_{\mathrm{BB}}(t)\tilde{\chi}_{\mathrm{B}}^{\mathrm{p}}(t) + \rho_{\mathrm{BA}}(t)\left\{1 - \tilde{\chi}_{\mathrm{B}}^{\mathrm{p}}(t)\right\} \\ &> \exp\left(-t/\Phi_{\mathrm{prot}}\right). \end{aligned} \tag{17}$$

Thus, by using Eq. (15a) it follows that the anisotropy decay is invariant to the photophysics decay as is expected for DDEM.

The different angular transformations needed to evaluate the orientation correlations functions $\rho_{ij}(t)$ are illustrated in Fig. 1. The two correlation functions for which $i{=}j$ are given by

$$\rho_{ii}(t) = \sum_{m_i=-2}^{2} \left\langle D_{m_i,0}^{(2)}(\Omega_{\mathrm{M_iD_i}}^{0}) D_{-m_i,0}^{(2)}(\Omega_{\mathrm{M_iD_i}})\right\rangle (-1)^{m_i} \tag{18}$$

while for $i \neq j$, the corresponding functions read

$$\rho_{ij}(t) = \sum_{q,q',m} D^{(2)}_{m,q}(\Omega_{\mathrm{D_iR}}) D^{(2)}_{-m,q'}(\Omega_{\mathrm{D_jR}}) \left\langle D^{(2)}_{q,0}(\Omega^0_{\mathrm{M_iD_i}}) D^{(2)}_{q',0}(\Omega_{\mathrm{M_jD_j}}) \right\rangle (-1)^m. \quad (19)$$

The second rank Wigner rotational matrix elements [49], $D^{(2)}_{p,q}(\Omega)\,(\Omega \equiv \alpha, \beta, \gamma)$, depend on the Euler angles $\Omega^0_{\mathrm{M_iD_i}}$ and $\Omega_{\mathrm{M_iD_i}}$ which transform the transition dipole to a fixed and immobilised director frame at the times $t = 0$ and $t = t$. The orientations of the two director frames relative to a common **R**-frame are given by the configuration angles $\Omega_{\mathrm{D_iR}}$(cf. Fig. 1).

In depolarisation experiments the fluorescence intensities $F_{||}(t)$ and $F_{\perp}(t)$ are collected with the emission polariser set parallel ($||$) and perpendicular ($\perp$) relative to the excitation polariser. In terms of contribution from the different correlation functions $(\rho_{ij}(t))$ and excitation probabilities $\left(\chi^k_m(t),\ k = p \text{ or } s \text{ and } m \in \text{ A, B}\right)$ these observables are given by

$$\begin{aligned} F_{||}(t) &= C\left\{\sum_{i=\mathrm{A,B}} \left\langle (1 + 2\rho_{ii}(t))\, \chi^{\mathrm{p}}_i(t)\right\rangle + \sum_{i \neq j = \mathrm{A,B}} \left\langle \left(1 + 2\rho_{ij}(t)\right) \chi^{\mathrm{s}}_i(t)\right\rangle\right\} \\ F_{\perp}(t) &= C\left\{\sum_{i=\mathrm{A,B}} \left\langle (1 - \rho_{ii}(t))\, \chi^{\mathrm{p}}_i(t)\right\rangle + \sum_{i \neq j = \mathrm{A,B}} \left\langle \left(1 - \rho_{ij}(t)\right) \chi^{\mathrm{s}}_i(t)\right\rangle\right\} \end{aligned} . \quad (20)$$

DDEM in Regular Polymer Structures

The theoretical description of reversible energy transport among many donors in an ensemble or structure is extremely complex. In the most general case, the rates of energy migration are determined by the reorientation and spatial dynamics, as well as the spatial and orientational distributions of the donors. Unfortunately, there exists no analytical expression theory that relates the fluorescence depolarisation experiment or the anisotropy ($r(t)$) to the above-mentioned properties. Fortunately, these properties can be taken into account by using BD and MC simulations, as described previously [41]. For describing the local reorientations of the donor bound to a macromolecule, e.g. a protein, BD simulations can be implemented by using suitable orienting potentials, usually a Maier–Saupe [50–52] or a cone potential [53, 54]. The parameters of modelling potentials can be adjusted by fitting BD simulations to the experimental orientation correlation functions, which are obtained for the donors in absence of energy migration. To mimic the DDEM among labelled monomers within a regular structure (cf. Fig. 3), MC simulations are used. In experiments/simulations this structure is preferably prepared/examined for different mixtures of D-labelled and unlabelled monomers, which then provide independent observations/conditions. In the MC simulations, the spatial positions of all the

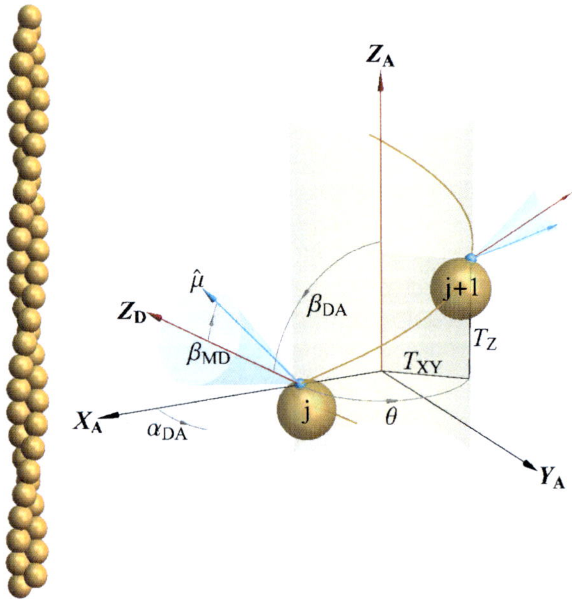

Fig. 3 Schematic showing the coordinate systems used to describe the protein position in regular structures forming helical, linear and ring-shaped aggregates. The Z_A axis coincides with the C_∞-axis of the aggregate and T_{xy} denotes the distance from this axis to the position of a fluorescent group. The translational and rotational transformations between nearest protein neighbours are θ and T_z, respectively. The fluorophore undergoes local reorienting motions about an effective symmetry axis Z_D that is transformed to the aggregate fixed frame by $\Omega_{DA} = (\alpha_{DA}, \beta_{DA})$. The electric transition dipole $\vec{\mu}$ is transformed to the D-frame by the angles $\Omega_{MD} = (\alpha_{MD}, \beta_{MD})$

donors as well as the neighbours within a certain cut-off distance need to be defined. The BD trajectories are calculated over a long period of time, referred to as T_∞. The total migration rate is calculated from

$$\Omega(t) = \sum_{j=-n}^{n} \omega_{0j}(t), \tag{21}$$

where $\omega_{0j}(t)$ is the rate of energy migration (cf. Eq. 3) from the 0th to the jth donor. Because the donors undergo reorienting motions on the time scale of migration, κ_{0j}^2 is time-dependent. The coordinates of $\hat{\mu}_j$ with respect to the aggregate fixed frame are given by $\hat{\mu}_j(t) = \vec{A}_{\mu_j}(t) - \vec{O}(T_{XY}, T_Z, \theta, j)$. Here $\vec{A}_{\mu_j}(t)$ denotes the vector directed from the origin of (X_A, Y_A, Z_A) to the point described by $\hat{\mu}_j$ and $\vec{O}(T_{XY}, T_Z, \theta, j)$ describes the position of the centre of mass of the jth donor (cf. Fig. 3).

The first question to answer in the MC simulation is the time (τ_{EM}) *when* an energy migration event takes place. A time interval Δt is chosen, which is smaller than the characteristic time for the variation of $\Omega(t)$ caused by reorienting motions.

Hence, $\Omega(t)$ is approximated to be constant within the time interval Δt. Now a random number is generated from a uniform distribution $\eta \in (0, 1]$, and τ_{EM} is calculated according to

$$\tau_{\mathrm{EM}} = -\frac{1}{\Omega(t)} \ln \eta \, . \tag{22}$$

The obtained value is only accepted if $\eta < \Delta t$, which ensures that the assumption of a constant $\Omega(t)$ is correct. If $\eta > \Delta t$ one steps forward a time unit Δt and calculates $\Omega(t + \Delta t)$. Using this Ω-value, a new random number is generated, and the procedure is repeated until a value of $\eta < \Delta t$ is reached.

The second decision concerns *where* the energy migrates, i.e. to which D group among the labelled monomers. From the above calculations one knows the time (T) of the energy migration event. By using BD simulations we can therefore account for the reorienting motions of all donors within the cut-off distance $\vec{A}_{\mu_j}(T)$, which enables calculation of $\kappa_{0j}^2(T)$, $\omega_{0j}(T)$, and $\Omega(T)$. The simulation of the orientational trajectories is performed for all particles in the time window $T \in [0, T_\infty]$ within the cut-off distance and is only repeated for the new fluorophore when moving along the aggregate. To select the next excited donor, energy migration rates are normalised and sorted in decreasing order according to

$$|\omega_{0j}(T)| = \frac{\omega_{0j}(T)}{\Omega(T)}. \tag{23}$$

Here a random number is generated from a uniform distribution $\eta \in (0, 1]$ and the ith donor is selected, for which

$$\eta \in \left(\sum_{j=1}^{j=i-1} \omega_{0j}(T), \sum_{j=1}^{j=i} \omega_{0j}(T) \right]. \tag{24}$$

The calculations above (Eqs. 21, 22, 23 and 24) account for the local anisotropic motions of the donors groups, i.e. energy migration under dynamic conditions. For energy migration in the static limit the scheme also holds, whereas the time-dependence of ω_{0j} and Ω is no longer relevant. The time-dependent fluorescence anisotropy is calculated for times $[T - \tau, T)$ as outlined above until $T \geq T_\infty$. The procedure is repeated many times to form the final ensemble average ($= \langle \ldots \rangle$):

$$r(t) = r_0 \sum_{j=-n}^{j=n} \left\langle p_j(t) P_2 \left(\hat{\mu}_0(0) \cdot \hat{\mu}_j(t) \right) \right\rangle . \tag{25}$$

For aggregates effectively being infinitely long, relative to the distance of energy migration, the value of n must be adjusted to probe sufficiently many monomers. When the jth donor is excited the probability, $p_j(t)$, is set to 1, if not it is set to 0.

To exemplify the influence of aggregate symmetry on the time-resolved fluorescence depolarisation, data have been generated for a configuration (α_{DA}, β_{DA}) and a local order parameter ($= \langle D_{00}^{(2)}(\beta_{MD})\rangle \equiv S$) of the donor group, as well as for different labelling efficiencies [41]. Three principal aggregate symmetries were considered, namely the helical, linear and circular one. The influence of dynamics was also examined. The data show (cf. Fig. 4) that the energy migration in the static limit is faster in linear and circular aggregates, as compared to the helical geometry. More evident is the large difference between the $r(t)$-decays in the static limit, at

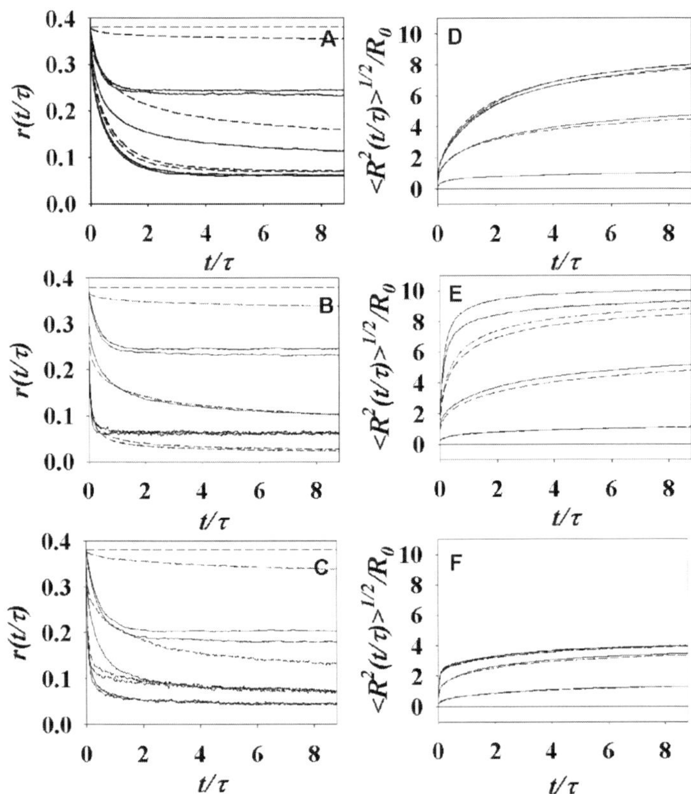

Fig. 4 The time-dependent anisotropy decay $\{r(t)\}$ and the normalised mean square displacement of the excitation $\{\ \langle R^2(t)\rangle^{1/2}/R_0\ \}$. The unit on the x-axis is reduced time, i.e. t/τ where τ is the fluorescence lifetime. The data refers to the following aggregate geometries: helical (**A**, **D**), linear (**B**, **E**) and circular (**C**, **F**). The *dashed lines* correspond to DDEM in the static limit (for local uniaxial anisotropic D-distribution), while the remaining $r(t)$-decays (*solid lines*) account for Brownian motions within the corresponding uniaxial Maier–Saupe potentials. The rotational correlation time is 0.86 in units of reduced time. In graphs (**A**, **B**) and (**C**) the residual anisotropy $\{r(t_\infty)\}$ decreases, while the mean square displacement (**D**, **E**, **F**) increases with increasing fraction of labelling. The labelling fractions (f) start by a very low value ($f \approx 10^{-4}$) and are increased to 5, 50, 95 and 100 %. The local order parameter $\left\langle D_{00}^{(2)}(\beta_{MD})\right\rangle = 0.80$

low degrees of D-labelling (f), and in the presence of reorienting motion of the donor groups. As can be expected this difference decreases with increasing f-values, so that the $r(t)$-decays become very similar when approaching 100% labelling. The DDEM simulation algorithm used also provides information about how far (in terms of the root mean-square displacement) the initial excitation migrates within an aggregate. In units of the average number of Förster radii, one finds that the migration in the linear aggregate is much faster and takes place over a larger distance, as compared to the helical aggregate. This is compatible with the difference in dimensionality and a similar behaviour was previously observed [55]. Because the energy displacement in a ring is directly related to its radius, a comparison is less straightforward between the circular geometry on one hand and the helical and linear geometries on the other hand.

Electronic Energy Migration/Transfer in Model Membranes

Quite often lipid molecules are labelled with fluorescent groups which are used to probe the membrane in different lateral positions. A realistic scenario is illustrated in Fig. 5, where the fluorescent label is covalently attached to the lipid head-group. If the labels are donor groups, DDEM occurs within and between the bilayer leaflets, i.e. as intra- and interlayer migration, respectively. This is also the case if one studies mixtures of donor- and acceptor-labelled lipids. To model the fluorescence relaxation and depolarisation of the donor in DAET and DDEM experiments, one needs to calculate the probability ($= G^{s}(t)$) that the initially excited donor is still excited at a time t later. Models for the probabilities of the intra- and interlayer processes have previously been derived [56]:

$$\ln G^{s}_{\text{intra}}(t) = -C_2 \frac{1}{\lambda^{1/3}} \Gamma\left(\frac{2}{3}\right)\left(\frac{t}{\tau}\right)^{1/3}, \tag{26}$$

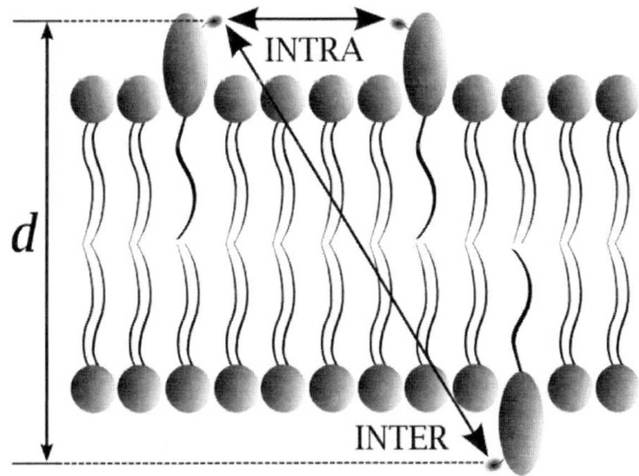

Fig. 5 Energy transfer/migration occurs either between fluorophore labels in one leaf of the lipid bilayer, i.e. an INTRA layer process, or between fluorophores located in different layers of bilayer, i.e. an INTER layer process. The energy transport is intra as well as inter between layers separated by the distance d

$$\ln G^{\mathrm{s}}_{\mathrm{inter}}(\mu) = -\frac{C_2}{3\lambda\nu^2}\left(\frac{2}{3}\mu\right)^{1/3}\int_0^{\frac{2\mu}{3}}\{1-\exp(-s)\}\,s^{-4/3}\mathrm{d}s\,. \tag{27}$$

Here C_2 denotes the reduced concentration, which stands for the number of electronically interacting molecules within the area of a circle determined by the Förster radius, R_0. The reduced concentration can be calculated from $C_2 = \pi\rho R_0^2$ with knowledge of the surface acceptor density (ρ). The parameter λ is a number equal to 1 or 2 for DAET and DDEM, respectively. In Eq. (27) $\mu = (3\lambda/2)\,t\nu^6\tau^{-1}$, $s = \cos^6\theta_r$and $\nu = R_0 d^{-1}$. Here d denotes the distance between the monolayers, which is approximately equal to the thickness of a lipid bilayer (cf. Fig. 5). The angle θ_r is between the surface normal and a line connecting the centres of mass of each interacting dipole. For the energy transfer/migration within and between the two-dimensional planes, the electronic transition dipoles are approximated to be either isotropic or in-plane oriented [56]. The Eqs. (26) and (27) were derived for *isotropically* oriented transition dipole moments.

The total excitation probability, $G^{\mathrm{s}}(t)$, that accounts for both intra- and interlayer energy transfer/migration is given by the joint probability

$$G^{\mathrm{s}}(t) = G^{\mathrm{s}}_{\mathrm{intra}}(t)\prod_j G^{\mathrm{s}}_{j,\mathrm{inter}}(t)\,. \tag{28}$$

In Eq. (28), the multiplication accounts for the general case of energy transfer/migration between several planes of interacting donors/acceptors. In the case when there is no intralayer energy transport $G^{s}_{\mathrm{intra}}(t) = 1$.

For a donor–acceptor system the fluorescence relaxation of the donor is given by

$$F(t) = G^{\mathrm{s}}(t)\,F_{\mathrm{D}}(t). \tag{29}$$

To monitor the rate of energy migration among donors, one needs to measure the fluorescence depolarisation, which is conveniently expressed by the time-resolved anisotropy, *r*(*t*), according to [57]

$$r(t) = r_0\left\{\rho(t) - S^2\right\}G^{\mathrm{s}}(t) + r_0 S^2. \tag{30}$$

Here S describes the order of transition dipole S with respect to the bilayer normal. The reorientation of the excited donors is described by

$$\rho(t) = \sum a_j\,\exp(-t/\phi_j), \tag{31}$$

where ϕ_j are the rotational correlation times which describe the local motions of the donors in the lipid bilayer.

DAET/DDEM in Micelles

When considering electronic energy transfer/migration within vesicles, it is usually not needed to account for the membrane curvature, since the vesicle radii (> 500 Å) are much larger than R_0. However, this approximation is usually not valid for micellar systems, whose radii rarely exceeds 150 Å (see for example reference [58]). In order to account for the micelle curvature, the Eq. (26) above must be slightly modified. When donor and acceptors are distributed on the surface of a sphere (at the core–shell interface of micelles) and all donors occupy equivalent positions one obtains that $G^{\mathrm{s}}(t)$) of the donor with respect to DAET [59] is given by

$$\ln G^{\mathrm{s}}(t) = -2\pi C_{\mathrm{A}} \int_0^{2R_{\mathrm{s}}} r\,\{1 - \exp[-\omega(r)t]\}\,\mathrm{d}r, \tag{32}$$

where C_{A} is the acceptor surface density, R_{s} the radius of a sphere, r the distance between the donor and the acceptor and finally $\omega(r)$ stands for the rate of energy transfer of the donor–acceptor pair separated by distance r. This equation can be transferred into a more convenient form [60]:

$$\ln G^{\mathrm{s}}(t) = -\,C_2\Gamma\left(\frac{2}{3}\right)\left(\frac{t}{\tau_{\mathrm{D}}}\right)^{1/3}\left[1 + 0.0231\left(\frac{R_0}{R_{\mathrm{s}}}\right)^4\left(\frac{t}{\tau_{\mathrm{D}}}\right)^{2/3} - 7.21 \times 10^{-5}\left(\frac{R_0}{R_{\mathrm{s}}}\right)^{10}\left(\frac{t}{\tau_{\mathrm{D}}}\right)^{5/3} + \cdots\right]. \tag{33}$$

For $R_{\mathrm{s}} > R_0$, the time-dependent terms in the bracket ([]) become negligible and the equation transfers into Eq. (26), which accounts for homogenously distributed donors and acceptors in two dimensions (i.e. $\lambda = 1$).

From a physical point of view it is reasonable to assume that the donors and the acceptors are spatially distributed in a core-shell interface of thickness Δ (cf. Fig. 6), since the core-shell thickness usually cannot be neglected. A more general model that accounts for these circumstances has been derived [60, 61]. The fluorescence decay is then given by

$$F_{\mathrm{D}}(t) = \exp\left(\frac{-t}{\tau_{\mathrm{D}}}\right) \int_{V_{\mathrm{mic}}} P(r_{\mathrm{D}})G^{\mathrm{s}}(t, r_{\mathrm{D}})r_{\mathrm{D}}^2\mathrm{d}r_{\mathrm{D}}, \tag{34a}$$

$$G^{\mathrm{s}}(t, r_{\mathrm{D}}) = \exp\left(\frac{-2\pi}{r_{\mathrm{D}}}\int_{R_e}^{\infty}\{1 - \exp[-\omega(r)t]\}\left\{\int_{|r_{\mathrm{D}}-r|}^{r_{\mathrm{D}}+r} n_{\mathrm{A}}P(r_{\mathrm{A}})r_{\mathrm{A}}\,\mathrm{d}r_{\mathrm{A}}\right\} r\,\mathrm{d}r\right), \tag{34b}$$

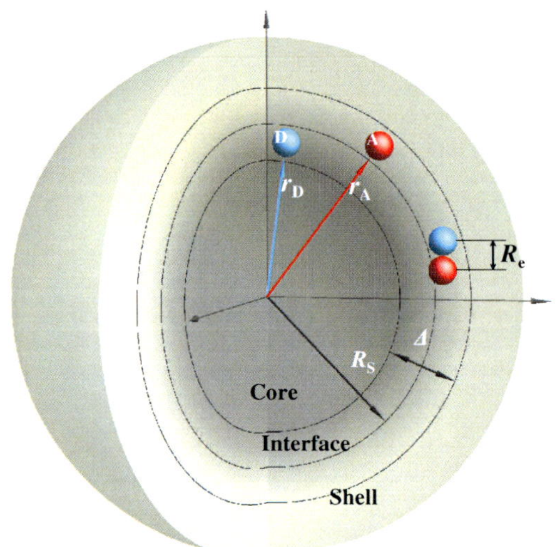

Fig. 6 A schematic of a spherical micelle, which indicates the core radius (R_s), the interface thickness (Δ), as well as the shell region. Donor and acceptor molecules are solubilised in the interface at the distances r_D or r_A, respectively. The encounter radius (R_e) refers to the shortest effective distance between the donor and acceptor molecules

where V_{mic}, $P(r)$, R_e, and n_A stands for the volume of the micelle, the distribution of donors and acceptors within its interface, the minimal distance between D and A, and the total number of acceptors within the micelle, respectively (cf. Fig. 6). Now imagine that the donors and acceptors are attached between the hydrophobic and hydrophilic block in the place of block junctions. Then, according to Farinha et al. [62], the distribution of fluorescent probes can be replaced by the distribution of block junctions derived by Helfand and Tagami [63]. The distribution has a shape of inverse hyperbolic cosine function, is less peaked and has broader wings as compared to a Gaussian distribution

$$P(r) = A\,[\cosh\{2(r - R_s)/\Delta\}]^{-1}\,, \tag{35}$$

where Δ is assigned to the interface thickness and A is a normalisation constant.

Applications of DAET and DDEM

Intra- and Intermolecular Distances in Proteins

Few scientists have hitherto used energy migration to determine distances between two chemically identical fluorophores in a protein molecule. Important reasons are the difficulties in obtaining well-defined labelling positions of the interacting chromophores and the complexity in analysing experimental data, which was discussed in the section "Theoretical Development". The energy migration may take place between intrinsic fluorophores, e.g. Trp residues [64–66], or between extrinsic probes [67–69] such as e.g. BODIPY-derivatives [70–72]. Applications of energy migration have been used to study protein unfolding [67], active sites [65] and conformational structures and motions [73, 74]. In these studies more or less qualitative methods were used for analysing the data. Only recently, the EFT was used for a quantitative analysis of fluorescence depolarisation experiments [26].

Heyn and coworkers [75] have convincingly combined DDEM and DAET experiments to study the dimerisation and inter-chromophore distance in the Cph1 Phytochrome from *Synechocystis*. The chromophore analogue phycoerythrobilin (PEB) was used as the donor group, while either the natural chromophore phycocyanobilin (PCB) or PEB was used as the acceptor in DAET and DDEM studies, respectively. Since each Cph1 monomer has a single binding site for the chromophore, i.e. for the PEB or PCB groups, energy migration/transfer occurs within the Cph1 dimer. Time-resolved fluorescence experiments were used to study the PEB/PCB dimers. The average lifetime of PEB was found to decrease linearly with the fraction of occupied PCB sites, which is also expected because the influence of energy transfer increases linearly with the population of DA pairs. For the PEB/PCB dimers, the anisotropy of PEB did not depend on the rate of energy transfer, which is expected if the donor is immobilised on its timescale of fluorescence. The fluorescence depolarisation experiments on PEB/PEB dimers revealed a gradual decrease of the anisotropy decay with increasing concentration of DD pairs, while the lifetime was constant. This clearly shows that DDEM is present within the Cph1 dimers. In the determination of the distance between the PEB-PEB chromophores, the uncertainty of κ^2 was considered. The allowed κ^2-value angles were determined in an elegant manner by combining the anisotropies obtained in the presence of DDEM and knowledge about the twofold symmetry of the dimer.

Recently, Heyn and coworkers [76] have also examined the energy transfer from a single tryptophan to 4-hydroxy-cinnamoyl in a photoactive yellow protein. The aim was to monitor changes of the chromophore structure during the photocycle. For each of the structures corresponding to the two isomerisation states P and I_2, the configuration angles were estimated from the X-ray structure. It was then possible to calculate the κ^2-value and thus predict the rate of energy transfer, which was found to be in excellent agreement with the measured values. The values of the obtained fluorescence anisotropy were compatible with immobilised chromophores. For the isomerisation of the P to the I_2 structure, the κ^2 value decreased by a factor of 10, whereas the spectral overlap increased by a factor of 2.3. These effects were confirmed from time-resolved fluorescence experiments, which show that the rate of energy transfer is much faster in the P isomer, as compared to the I_2 isomer.

Tryptophan and BODIPY {N-(4,4-difluoro-5,7-dimethyl-4-bora-3a,4a-diaza-sindacene-3-yl)methyl iodoacetamide} is a suitable donor–acceptor pair for intraprotein distance measurements, applicable to the study of protein folding. This is demonstrated in studies of electronic energy transfer between Trp and BODIPY located protein, S6, obtained from *Thermus thermophilus* [77]. Revealing the mechanisms that dictate protein stability is of large relevance for instance to enable design of temperature tolerant enzymes with high enzymatic activity over a large temperature interval. In an effort to identify mechanisms that dictate protein stability, the pair Trp–BODIPY has been used in a comparative study of the folding thermodynamics and kinetics of the ribosomal protein S16, which was isolated from a mesophilic and hyperthermophilic bacterium [78].

Non-covalent Protein Polymers

Proteins may form oligomeric and large aggregates, which constitute regular structures commonly referred to as protein polymers. Using a chemist's terminology, it would be more correct to name the structures non-covalent protein polymers. For instance, transthyretin forms tetramers and mutants of transthyretin form structures, which are associated with the human amyloid disease, *familial amyloidic polyneuropathy* [31, 79]. A common feature of the amyloid diseases is the formation of crystalline non-covalent protein polymers, so-called amyloid fibrils. These are, depending on the kind of disease, found as protein deposits in different extra cellular spaces of tissues or in a few cases as intracellular inclusions.

The structure of filamentous actin (F-actin) is another non-covalent polymer, resulting from the spontaneous polymerisation of actin molecules in the presence of salt, and characterised by a fast and a slow growing end where actin subunits add and dissociate in a constant flux balanced by a stable concentration of unpolymerised G-actin. The helically twisted structure of F-actin is of methodological interest in this context since it incorporates special cases of linear and ring-shaped structures [41]. The characteristic parameters of the non-covalent polymer (cf. Fig. 3) constitute the translational distance (T_z), the rotation (θ) of each neighbour protein molecule, and the radial distance between the helical axis and the position of the donor (T_{xy}). Recently, the structure of F-actin was studied by performing DDEM experiments and analysing the data by means of Monte Carlo and Brownian dynamics simulations, combined with a fitting procedure, which is based on a genetic algorithm [42]. The method enables prediction of the distribution of a repeating position for the centre of mass of the donor molecule without any *quantitative* preliminary information of the polymer; the only requirement is that the structure is regular. It was applied to studies of BODIPY labelled filamentous actin. The samples studied were composed of different fractions of labelled actin molecules. The obtained structural parameters T_z and θ were compatible with previously published data [80].

Yeow and Clayton [81] have studied the enumeration of oligomerisation of membrane proteins in living cells by using DDEM and fluorescence depolarisation experiments. In systems exhibiting oligomerisation, the steady-state anisotropy obeys a characteristic feature when plotted as a function of labelling efficiency. The paper also presents a model for interpreting anisotropy data obtained in studies of dilute solutions of oligomers and oligomerisation distributions.

Tubulin is another well-known protein that forms the building unit of microtubules (see e.g. reference [82]). Acuna and coworkers [83] have studied the interaction of the anticancer drug Taxol with microtubules.

Micelles

Micelles are built up by a hydrophobic core and shell-forming hydrophilic part and have been studied for decades. The motivations are several, e.g. micelles appear useful in applications to environmental questions (removal of organic pollutants) or in pharmaceutical applications as drug releasing units. Although the micellar structure seems simple, micelles may exhibit quite a complex response to physicochemical properties such as the ionic strength, the solvent polarity, temperature, pH [84, 85]. Different techniques have been implemented to elucidate their unique behaviour, ranging from static and dynamic light and neutron scattering, various fluorescence techniques, NMR, AFM, TEM [84, 85], and more recently also electrophoresis [86]. In the following focus will be on the implementation of DAET/DDEM to the study of micellar structures.

Block copolymer micelles. The models of the fluorescence donor decay, which are briefly described in the theoretical section, can be used to obtain information about the aggregation numbers, (N), the core radius of a micelle and the interfacial thickness. For cases when the higher terms can be neglected in Eq. (33), a similar expression to Eq. (26) has been published [87]. This equation contains the parameter $\xi = 0.339 f_A\ N(R_0/R_s)^2$, which can be determined from experiments. However, neither N nor R_s can be determined unless the mole fraction of acceptor (f_A) is known [87]. A more extended model [60, 61] accounts also for the finite interfacial thickness {cf. Eq. 34 and 35}. Farinha et al. have successfully applied this model in studies of a PI-PMMA diblock copolymer dissolved in acetonitrile [62]. For this system the interfacial thickness was estimated to be 9 ± 1 Å. The above model is valid only when donors and acceptors are distributed within the interface. If this cannot be assumed other models are needed. Jones et al. [88] have applied such a model for the case of acceptors being solubilised at the interface, while donors are distributed above the surface in a shell. More than 30 years ago Tachiya [17] developed another model which has been frequently used to determine aggregation numbers. It is then assumed that fluorescent probes residing in a micelle are quenched by quenchers present in the bulk phase. However, the model, although not yet used, should also be applicable to the quenching of donors by acceptors which are incorporated into micelles.

In addition to the above mentioned parameters of micelles, it is also possible to use DAET for determining the critical micelle concentration (cmc), as was shown by Liu et al. [89]. In this work polystyrene-*block*-poly(2-hydroxyethyl methacrylate) was labelled by either fluorene or pyrene for which a sudden increase of the DAET efficiency was observed upon micelle formation. DAET enables the estimation of cmc, as well as qualitatively monitor the monomer-micelle equilibrium itself. In such studies two solutions containing micelles are mixed, after being doped with a low fraction of the donor and acceptor-labelled monomers, respectively. Upon mixing, the monomers exchange between micelles and the donor intensity decreases over time due to DAET [90].

Towards biological micellar systems. Several research groups have applied DAET to show that the plasmid DNA (pDNA) condenses in confined environments of nanoparticles. These can be of pharmacological interest as potential drug/gene delivery devices [91, 92]. To investigate this idea, Itaka et al. [91] used pDNA labelled with both fluorescein (donor) and X-rhodamine (acceptor) and followed changes in the donor spectra upon solubilising pDNA in micelles. Since the donor intensity drastically decreased upon binding, it has been suggested that pDNA comprise a more packed (condensed) state in the nano-environment. Itaka et al. went even further and used this finding to explore the stability of nanoparticles in serum, which could serve as potential gene (DNA) delivery devices. They solubilised pDNA inside a variety of nanoparticles and monitored repeatedly donor spectra in the course of time. While the donor intensity did not change substantially within several hours for pDNA incorporated into poly(ethylene glycol)-poly(L-lysine) micelles, this process was significantly faster for pDNA solubilised in lipoplex particles. The latter is a consequence of the fast degradation of lipoplex particles in serum. Therefore, DAET enables one to observe directly the stability of gene-delivery devices in serum.

Another study that focuses on conformational changes monitored by DAET has been carried out by Kim et al. [93]. They investigated the conformational switching of a β-amyloid peptide, which is known to be connected with Alzheimer decease. For this purpose, they used the β-amyloid peptide labelled by amine-reactive TMR (donor) and thiol-reactive DABMI, which enabled estimating the DA distances during the aggregation process. The observed changes in distances with increasing peptide concentration were assigned to monomeric and random coiled amyloid peptides forming micelles, followed by the formation of monomers with β-sheet conformation.

Sarkar et al. [94] studied the folding of heart cytochrome *c* using a buffer in the absence and presence of urea, and in reverse CTAB micelles. In these experiments a dansyl was covalently attached to the surface of the cytochrome. The dansyl group then acts as a donor to the intrinsic heme group. The structural perturbations observed were compatible with circular dichroism and dynamic light scattering experiments. Recently, Taniguchi et al. [95] have employed a DAET-based method to investigate light-induced structural changes of phoborhodopsin/transducer complex in *n*-dodecyl β-D-maltoside micelles. Then the protein complex was labelled at different positions by a DA pair. The absorption spectrum of this pair was not overlapping with that of the protein complex. Hereby it was possible to follow the dynamic change of the protein complex upon excitation. By monitoring the donor intensity, the structural changes decayed at a rate of $1.1 \pm 0.2\,s^{-1}$.

Electronic Energy Transfer in Membranes

Searching for rafts. It has been proposed that the cell membrane may contain liquid-ordered domains, usually called rafts. It is then assumed that these coexist with the liquid-crystalline/liquid-disordered phase of the biomembrane [96]. Since rafts are thought to be involved in various biological functions, such as endo- and exocytosis, cell–cell signalling and protein sorting [97, 98], any evidence for their existence is of outmost interest. The difference in spatial resolution of the techniques available for exploring membranes is very large. Fluorescence microscopy and AFM reveal domains on the micrometer scale [99], while NMR methods are sensitive to domains on the sub-micrometer scale [100]. Moreover DAET experiments are able to detect heterogeneities as small as few nanometers. Hence, DAET can become an important method in the search for lipid rafts.

To fully utilise the potential of the DAET process, it is important to choose suitable fluorescent probes needed, which partition differently in crystalline and liquid-ordered phases. If both partners of a DA pair incorporate preferentially in the same phase, the energy transfer rate will increase upon the formation of phase-separated domains. On the other hand, if the probes have different affinity to the phases, energy transfer efficiency decreases as a consequence of donors being separated from acceptors. As a result it is important to also account for the Förster radii and for possible dimensions of the presumed lipid rafts. As a rule of thumb, the domain sizes should be comparable to the Förster radius of the DA pair.

The search for lipid rafts were initiated by Pedersen et al. [101], who successfully observed domain formation in one model lipid membrane composed of 1,2-dihexadecanoyl-*sn*-glycero-3-phosphocholine and at temperatures close to the main gel to liquid crystal transition. A donor–acceptor pair derivative of phosphoethanolamine were used, which show different affinity to the coexisting gel/liquid phases. The formation of domains was indicated by the decrease in the donor fluorescence intensity. Later this experimental approach has been extended to the study of more complex systems with an effort to also examine living cells [102, 103].

A slightly different approach based on DAET/DDEM was chosen to prove the intrinsic self-aggregation property of G_{M1} gangliosides in DOPC vesicles [104]. For this purpose, a variety of BODIPY-labelled G_{M1} were synthesised, which could act as DD or DA pairs. The time-resolved fluorescence depolarisation/relaxation data were analysed by means of models, which are very similar to those given by Eqs. (26), (27), (28), (29), (30) and (31). The model applied is strictly valid for a homogenous distribution of fluorescent probes in membranes. It was found, however, that the recovered reduced concentrations were considerably higher than the values expected for a homogeneous distribution. Therefore the models were modified and synthetic data were generated to mimic various non-uniform distributions of the probes. The reduced concentrations obtained from the reanalyses of these data were also found significantly higher than the expected ones when using the idealised model. Taken together, this strongly suggests that the G_{M1} lipids show an intrinsic self-aggregation property in lipid bilayers of DOPC.

The determination of raft-sizes from fluorescence decay curves of DAET experiments is a challenging task. Almeida and Loura et al. [105, 106] have estimated an upper limit of the domain sizes. The following model assumed

$$F_{\mathrm{DA}}(t) = x_1 \tilde{G}^{\mathrm{s}}_{\mathrm{intra},1}(t)\tilde{G}^{\mathrm{s}}_{\mathrm{inter},1}(t)F_{\mathrm{D1}}(t) + x_2 \tilde{G}^{\mathrm{s}}_{\mathrm{intra},2}(t)\tilde{G}^{\mathrm{s}}_{\mathrm{inter},2}(t)F_{\mathrm{D2}}(t). \tag{36}$$

This means that the donor decay in the presence of acceptor $F_{\mathrm{DA}}(t)$ is expressed as the sum of two contributions, the first one originating from the liquid-ordered and the other one from the liquid-disordered phase. In Eq. (36), x_j and $F_{Dj}(t)$ denotes the mole fraction and the fluorescence decay of the donor in the jth phase, respectively. $\tilde{G}^{s}_{\mathrm{intra},j}(t)$ and $\tilde{G}^{s}_{\mathrm{inter},j}(t)$ are extended probabilities of $G^{s}_{\mathrm{intra}}(t)$ and $G^{s}_{\mathrm{inter}}(t)$, which account for the excluded volume of the donor. One should notice that Eq. (36) only accounts for energy transfer between a donor and an acceptor occupying the same phase. Thus, the domains studied need to be much larger than R_0. The analyses of the donor decays yield the partition coefficient between both phases of the acceptors. This coefficient can also be obtained in an independent experiment (e.g. from the variation of the lifetime weighted quantum yield or from steady state experiments). From a comparison of the partition coefficient provided by the model, and that obtained independently, the upper limits of domain sizes can be obtained by means of MC simulations [105]. It is found that:

i) The two partition coefficients coincide for domain radii, corresponding to at least 15–20R_0.
ii) The recovered partition coefficient approaches unity if the domain sizes are close to R_0. This is because all acceptors are then distributed randomly in both phases with respect to the donors.
iii) For a domain size of $\sim 9R_0$, the recovered partition coefficient for the acceptor probe is still closer to unity than the input value, which means that the transfer across the interface is not negligible.

Towles et al. [107] have presented another DAET-approach to the determination of domain sizes. In this work the above presented model was extended. In brief a constant density of acceptors within a shell around a donor is assumed. The contribution of this shell to the overall decay was estimated and summoned up over the whole space. An analytical expression for the donor decay was derived for heterogeneously distributed acceptors. Numerical calculations were used to estimate the average probability of finding a domain at a distance from the point where the initially excited donor is located. The donor decay is expressed as

$$F_{\mathrm{DA}}(t) = F_{\mathrm{D}}(t)\prod_{j=1}^{\infty}\frac{\left[\tilde{G}^{s}_{\mathrm{intra}}(t)\tilde{G}^{s}_{\mathrm{inter}}(t)\right]_{R_{\mathrm{ex}}=\eta j}}{\left[\tilde{G}^{s}_{\mathrm{intra}}(t)\tilde{G}^{s}_{\mathrm{inter}}(t)\right]_{R_{\mathrm{ex}}=\eta j+1}}, \tag{37}$$

where R_{ex} is the radius of the excluded zone (shell), η is the thickness of the shell and the survival probability $\tilde{G}^{s}_{j}(t)$ (j = intra, inter) depends on the distribution of acceptors surrounding the donors. From a comparison of the analytical model with MC simulations it was concluded that the model predicts domain sizes within an accuracy of $\sim$20%, for domain sizes $< 4R_0$. In addition to the above mentioned models, another one also exists that enables an estimation of domain sizes from fluorescence steady-state experiments. Using this model heterogeneities both in binary and ternary systems have been reported [108].

Interaction membrane-peptide/membrane-protein. Lipids play a crucial role in protein/peptide assembling behaviour. Interestingly, some helical peptides that by themselves do not assemble actually form peptide–peptide assemblies in the presence of lipids bilayer [109]. Sparr et al. [110] have studied self-association of transmembrane α-helices in model membranes by means of DAET and the formation of pyrene excimers. Either the C or N terminus of the peptide was labelled by pyrene. The efficiency of excimer formation was compared for C–C, C–N and N–N pyrene-labelled peptides. Supporting information has been obtained by DAET between the intrinsic tryptophan group and pyrene-labelled helices. The self-association of α-helices that did not contain any specific recognition motive occurred almost exclusively between antiparallel helices. Relatively high peptide/lipid ratios (at least 1/25) were needed to obtain association in DOPC vesicles. Self-assembly was promoted by an increase as well as a decrease of the bilayer thickness because of the increased mismatch [111] between the bilayer thickness and the hydrophobic length of the peptide. It was therefore concluded that the lipids influence the association of α-helices.

The specific interactions between rhodopsine and the G-protein in phospholipid vesicles have been studied by DAET [112]. For this purpose, these molecules were labelled with coumarin or pyrene, which form an efficient DA pair. Since the G-protein can bind to the membrane by non-specific interactions, additional experiments were carried for estimating this fraction. In the first experiment non-hydrolysable analogues of GTP were added which promote dissociation of the rhodopsine-G-protein complex. In further experiments excess of unlabeled G-protein was added, which competes for the binding sites in rhodopsine. It was possible to observe specific binding of the G-protein, even in the presence of non-specific interactions with a membrane.

In a complex DAET study Loura et al. have studied the interaction of the K6 W peptide with lipid bilayers composed of dipalmitoyl-phosphatidylcholine and dipalmitoyl-phosphatidylserine [113]. These model membranes were examined at two temperatures (45 or 60°C) and for different peptide concentrations. The DA-pair used was DPH-labelled phosphocholine (donor) and NBD-labelled phosphocholine (acceptor). The donor group is preferably located in the hydrophobic part, while the acceptor group is likely localised in the interface region of the membrane. Therefore energy transfer occurs between a plane of randomly spaced donors and two more planes containing the acceptors located in the opposite interfaces. A model accounting for this case was successfully applied only in the presence of the peptide at the lower temperatures. The data obtained for the higher temperature are compatible with heterogeneous lipid bilayers (containing domains). This was concluded from the donor decay in the presence of acceptor, which could be described by assuming two infinitively separated phases (see above). Although addition of peptide did not cause any substantial perturbations to the heterogeneous membrane at the lower temperature, the DAET efficiency changed drastically for the previously homogenous membrane. Interestingly, no improvement in describing the donor decay was obtained by a model which accounts for the two-phase separation. Because the peptide is positively charged and the lipid bilayer is negative, the formation of multi-bilayers was suggested, where stacking is promoted by the peptides. The data were also analysed with a model that describes DAET between two or more bilayers from which a repeat distance of 57 Å was obtained.

A very similar approach that was used to prove the self-aggregation property of G_{M1} (see above) can also be applied in the study of peptide/protein aggregation in membranes. Several years ago this was demonstrated by Bogen et al. [114] in studies of the WALP16 peptide labelled with rhodamine 101. The obtained reduced concentration for the transmembrane peptide WALP16 was about 10 times higher than that expected for randomly distributed peptides.

Pore formation. The action of some antibacterial molecules is ascribed to their presumed ability to form pores in the membrane. Although several investigators have addressed this question [115, 116], a minority of them make use of the DAET/DDEM techniques. A possible explanation is the lack of fluorescent compounds that would preferentially reside inside pores. Rather, other methods have been used, such as the measurements of leakage of fluorescent probes from the inside of vesicles [116] or fluorescence microscopy [117]. A real challenge would be to estimate the pore size by means of the DDEM methods.

Viero et al. [118] have published an interesting DAET study on the pore formation. They observed the oligomerisation of two different proteins HlgA/HlgB in a model membrane. Site specific mutagenesis was used to replace certain residues of HlgA and HlgB by cysteine. These Cys groups were labelled with ALEXA-488 (donor) and ALEXA-546 (acceptor). The mutated residues were deliberately selected on the right or left interface of the protein surfaces. The reason for the choice was a proposal that the wall of the pore is build up by rings consisting of alternating HlgA/HlgB proteins. From an analysis of the energy transfer efficiency for all possible DA pairs (such as $\mathrm{HlgA_{right}}$/$\mathrm{HlgB_{right}}$, $\mathrm{HlgA_{left}}$/$\mathrm{HlgB_{right}}$, etc.), it was shown that heterogeneous interfaces HlgA/HlgB or HlgB/HlgA have a much higher affinity than corresponding HlgA/HlgA, HlgB/HlgB pairs. In another study of pore formation Silva et al. [119] labelled an amino group of nystatin polyene molecule by NBD. The intention was to prove the ability of the nystatin polyene molecule to aggregate and form pores in lipid membranes by means of DDEM. Surprisingly, they could not confirm the previous results obtained by O'Neill et al. [120], who observed pore formation by photo-bleaching measurements. The authors have suggested [119] that the function of the NBD-labelled amino group is crucial for the action of nystatin.

Membrane–DNA interactions. There are surprisingly few fluorescence spectroscopic studies of lipid–DNA interactions. However, such methods have been applied to examine the encapsulation efficiency into lipoplex particles (cf. the above section about micelles and a paper by Madeira et al. [121] for further references). Madeira et al. [121] were among the first to apply DAET/DDEM in quantitative studies. DNA/lipid complexes were examined by DAET by using DNA, labelled by either BOBO-1 (donor) or ethidium bromide (acceptor). The charged membrane was doped with either DPH-PC (donor) or BODIPY-PC (acceptor). It was confirmed that the DNA molecules intercalate between membranes and orient with their long axis parallel to the bilayer surface. In the theoretical modelling it was therefore assumed that the donors and acceptors are randomly distributed in different parallel planes. Since there was an indispensable fraction of unbound DNA, the authors modified the theoretical model, depending on whether the DNA served as a donor or acceptor. From the analyses of the fluorescence decays, the distance between membrane-bound DNA and the donors/acceptors which were attached to the PC chain was obtained, as well as the encapsulation efficiencies.

Approaching studies of the living cell. Up to now this review has dealt with model membranes, which are more or less relevant for mimicking the properties of biological membranes. Nevertheless this approach is useful, especially for testing complex models aimed for quantitative studies. In studies of really complex systems, such as living cells, it is usually very difficult to determine absolute values of energy transfer efficiencies, and a more qualitative approach is wanted. For instance, the aim might be to study the formation of various associates (such as protein dimers, lipid domains, etc.) in living cells. It is then necessary to find out whether the donors and acceptors involved in the formation of studied aggregates are randomly distributed in the membrane, or whether they form aggregates and/or partition into the domain structure. For exploring this, the energy transfer efficiency is usually monitored as a function of the acceptor surface density and the molar donor/acceptor ratio. For donors and acceptors forming aggregates, the DAET efficiency depends on the donor/acceptor ratio, and it is independent of the surface density of acceptors. On the other hand, for randomly distributed probes, the DAET efficiency increases with increasing surface density of the acceptors. This methodology has been applied, for example by Herrick-Davis et al. [122], who studied the formation of homo-dimers of serotonin-5HT$_{2C}$ in the membrane of living cells. In addition to DAET, bioluminescence and energy transfer was used in studies which do not require an external excitation by for instance a laser.

The above outlined concept has also been frequently used to prove the presumed existence of rafts in living cells. However, this proposal is still controversial, as is evident from the review by Silvius et al. [123]. For example, Kenworthy et al. [124] monitored energy transfer between the donor- and acceptor-labelled cholera toxin B-subunit (CTBX), between donor- and acceptor-labelled GPI-anchored proteins and CTBX, as well as between CTBX and GPI-anchored proteins in living cells. It was found that the DAET efficiency correlates with the acceptor surface density, and it approaches zero at low surface densities. This is indeed consistent with randomly distributed probes within cell membranes and consequently with no or a very low formation of rafts. Sharma et al. [125] have used GPI-anchored fluorescent proteins (GFP or mYFP) and the GPI-anchored folate receptor to monitor the raft formation by means of both DDEM and DAET. The analysis of the depolarisation data suggests that the fractions of rafts are 20–40%. Their size is estimated to be less than 5 nm in diameter and they comprise 4–5 GPI molecules. Notice however that the DAET studies by Kenworthy et al. [124] did not reveal any formation of rafts. A possible explanation might be that the DDEM method in the present case is more sensitive to small heterogeneities, which consist of only ca. 4–5 molecules. Sengupta et al. [126] have recently investigated inhomogeneities in plasma membranes by labelling lipids with carbocyanine derivatives of different acyl chain length. These studies provide consistent results from DAET as well as DDEM experiments, which are compatible with the existence of nanometer sized domains.

Acknowledgments This work was financially supported by the Swedish Research Council. We are grateful to Mr. Oleg Opanasyuk for preparing Figs. 3 and 6.

Reference

1. R.E. Dickerson, X-Ray Analysis and Protein Structure, Academic Press, New York, (1964).
2. T.L. Blundell, L.N. Johnson, Protein Crystallography, Academic Press, London, (1976).
3. G.M. Clore, A.M. Gronenborn, Topics in Molecular and Structural Biology, The Macmillan Press Ltd, London, (1993).
4. K. Wütrich, NMR of Proteins and Nucleic Acids, Wiley, New York, (1986).
5. B.W. Van der Meer, G. Coker III, S.-Y.S. Chen, Resonance Energy Transfer: Theory and Data, VCH Publishers, Inc., New York, (1994).
6. B. Valeur, Molecular Fluorescence. Principles and Applications, Wiley-VCH, (2002).
7. J.R. Lakowicz, Principles of Fluorescence Spectroscopy, 2nd ed., Kluwer Academic/Plenum Publishers, New York, (1999).
8. N.L. Vekshin, Energy Transfer in Macromolecules, SPIE Press, Bellingham, WA, (1997).
9. S.Kalinin, L.B.-Å. Johansson, Utility and considerations of donor-donor energy migration as a fluorescence method for exploring protein structure-function, J. Fluoresc. 14 (2004) 681–691.
10. J.H. Davis, Deuterium nuclear magnetic resonance and relaxation in partially ordered systems, Adv. Magn. Reson. 13 (1989) 195–223.
11. G. Lindblom, G. Orädd, NMR studies of translational diffusion in lyotropic liquid crystals and lipid membranes, Progr. Nucl. Magn. Reson. Spectrosc. 26 (1994) 483–516.
12. J.H. Davis, M. Auger, Static and magic angle spinning NMR of membrane peptides and proteins, Prog. NMR Spectrosc. 35 (1999) 1–84.
13. G. Orädd, G. Lindblom, Lateral diffusion studied by pulsed field gradient NMR oriented lipid membranes, Magn. Reson. Chem. 42 (2004) 123–131.
14. F. Österberg, L. Rilfors, Å. Wieslander, G. Lindblom, S.M. Gruner, Lipid extracts from membranes of Acholeplasma laidlawii. A grown with different fatty acids have a nearly constant spontaneous curvature, Biochim. Biophys. Acta 1257 (1989) 18–24.
15. M.W. Tate, E.F. Eikenberry, D.C. Turner, E. Shyamsunder, S.M. Gruner, Nonbilayer phases of membrane lipids, Chem. Phys. Lipids 57 (1991) 147–164.
16. G.C. Shearman, O. Ces, R.H. Templer, J.M. Seddon, Inverse lyotropic phases of lipids and membrane curvature, J. Phys. Condens. Matter 18 (2006) 1105–1124.
17. M. Tachiya, Application of a generating function to reaction kinetics in micelles. kinetics of quenching of lumuniscence probes in micelles, Chem. Phys. Lett. 33 (1975) 289–292.
18. M. Almgren, J.-E. Löfroth, Effects of polydispersity on fluorescence quenching in micelles, J. Chem. Phys. 76 (1982) 2734–2743.
19. O. Söderman, L.B.-Å. Johansson, The cubic phase (I1) in the dodecyltrimethylammonium chloride/water system. A fluorescence quenching study, J. Phys. Chem. 91 (1987) 5275–5278.
20. O. Söderman, L.B.-Å. Johansson, Aggregate size and surfactant/hydrocarbon diffusion in the cubic phase of sodium octanoate/hydrocarbon/water, J. Coll. Interface Sci. 179 (1996) 570.
21. M. Almgren, Diffusion-influenced deactivation processes in the study of surfactant aggregates, Adv. Colloid Interface Sci. 41 (1992) 9–32.

22. L.A. Bagatolli, Direct observation of lipid domains in free standing bilayers: from simple to complex lipid mixtures, Chem. Phys. Lipids 122 (2003) 137–145.
23. J.R. Lakowicz, Principles of fluorescence spectroscopy, 3rd ed., Springer, Singapore, (2006).
24. J. Karolin, M. Fa, M. Wilczynska, T. Ny, L.B.-Å. Johansson, Donor–donor energy migration (DDEM) for determining intramolecular distances in proteins: I. Application of a model to the latent plasminogen activator inhibitor-1 (PAI-1), Biophys. J. 74 (1998) 11–21.
25. L.B.-Å. Johansson, F. Bergström, P. Edman, I.V. Grechishnikova, J.G. Molotkovsky, Electronic energy migration and molecular rotation within bichromophoric macromolecules. Part 1. Test of a model using bis(9-anthrylmethylphosphonate) bisteroid, J. Chem. Soc. Faraday Trans. 92 (1996) 1563–1567.
26. M. Isaksson, P. Hägglöf, P. Håkansson, T. Ny, L.B.-Å. Johansson, Extended Förster Theory for determining intraprotein distances: 2. An accurate analysis of fluorescence depolarisation experiments, Phys. Chem. Chem .Phys. 9 (2007) 3914–3922.
27. S.V. Kalinin, J.G. Molotkovsky, L.B.-Å. Johansson, Partial donor–donor energy migration (PDDEM) as a fluorescence spectroscopic tool of measuring distances in biomacromolecules, spectrochim. Acta Part A 58 (2002) 1087–1097.
28. P. Woolley, K.G. Steinhauser, B. Epe, Forster-type energy-transfer – simultaneous forward and reverse transfer between unlike fluorophores, Biophy. Chem. 26 (1987) 367–374.
29. M. Isaksson, S. Kalinin, S. Lobov, S. Wang, T. Ny, L.B.-Å. Johansson, Distance measurements in proteins by fluorescence using partial donor–donor energy migration (PDDEM), Phys. Chem. Chem. Phys. 6 (2004) 3001–3008.
30. S. Kalinin, L.B.-Å. Johansson, Energy migration and transfer rates are invariant to modelling the fluorescence relaxation by discrete and continuous distributions of lifetimes, J. Phys. Chem. B 108 (2004) 3092–3097.
31. C.M. Dobson, Protein folding and misfolding, Nature 426 (2003) 884–890.
32. C.A. Paterson, N.A. Delamere, The Lens, 9th ed., Mosby-Year Book, St. Louis, MO, (1992).
33. M. Malisauskas, V. Zamotin, J. Jass, W. Noppe, C.M. Dobson, L.A. Morozova-Roche, Amyloid protofilaments from the calcium-binding protein equine lysozyme: formation of ring and linear structures depends on pH and metal ion concentration, J. Molec. Biol. 330 (2003) 879–890.
34. M. Malisauskas, J. Ostman, A. Darinskas, V. Zamotin, E. Liutkevicius, E. Lundgren, L.A. Morozova-Roche, Does the cytotoxic effect of transient amyloid oligomers from common equine lysozyme in vitro imply innate amyloid toxicity? J. Biol. Chem. 280 (2005) 6269–6275.
35. A. Aguzzi, M. Polymenidou, Mammalian prion biology: one century of evolving concepts, Cell 116 (2004) 313–327.
36. F. Houston, W. Goldmann, A. Chong, M. Jeffrey, L. Gonzalez, J. Foster, D. Parnham, N. Hunter, Prion diseases: BSE in sheep bred for resistance to infection, Nature 423 (2003) 498–498.
37. S.N. Wai, M. Westermark, J. Oscarsson, J. Jass, E. Maier, R. Benz, B.E. Uhlin, Characterisation of dominantly negative mutant ClyA cytotoxin proteins in *Echerichia coli*, J, Bacteriol. 185 (2003) 5491–5499.
38. P. Stanley, V. Koronakis, C. Hughes, Acylation of *Escherichia coli* hemolysin: a unique protein lipidation mechanism underlaying toxin function, Microbiol. Mol. Biol. Rev. 62 (1998) 309–333.
39. L. Abrami, M. Fivaz, F.G. van der Goot, Adventures of a pore-forming toxin at the target cell surface, Trends Microbiol. 8 (2000) 168–172.
40. A. Olofsson, A.E. Sauer-Eriksson, A. Öhman, The solvent protection of Alzheimer amyloid-b-(1– 42) fibrils as determined by solution NMR spectroscopy, J. Biol. Chem. 281 (2006) 477–483.
41. D. Marushchak, L.B.-Å. Johansson, On the quantitative treatment of donor-donor energy migration in regularly aggregated proteins, J. Fluoresc. 15 (2005) 797–804.

42. D. Marushchak, S. Grenklo, T. Johansson, R. Karlsson, L.B.-Å. Johansson, Fluorescence depolarisation studies of filamentous actin analysed with a genetic algorithm, Biophys. J. 93 (2007) 3291–3299.
43. P.-O. Westlund, H. Wennerström, Electronic energy transfer in liquids. The effect of molecular dynamics, J. Chem. Phys. 99 (1993) 6583–6589.
44. M. Isaksson, N. Norlin, P.-O. Westlund, L.B.-Å. Johansson, On the quantitative molecular analysis of electronic energy transfer within donor-acceptor Pairs, Phys. Chem. Chem. Phys. 9 (2007) 1941–1951.
45. T. Förster, Zwischenmolekulare Energiewanderung und Fluorescenz, Ann. Phys. 2 (1948) 55–75.
46. L.B.-Å. Johansson, P. Edman, P.-O. Westlund, Energy migration and rotational motion within bichromophoric molecules. II. A derivation of the fluorescence anisotropy, J. Chem. Phys. 105 (1996) 10896–10904.
47. A. Jablonski, Influence of torosional vibrations of luminescent molecules on the fundamental polarisation of photoluminiscence of solutions, Acta Phys. Polon. 10 (1950) 33–36.
48. N. Norlin, P. Håkansson, P.-O. Westlund, L.B.-Å. Johansson, Extended Förster theory of partial donor-donor energy migration (PPDEM): the κ^2-dynamics and fluorophore reorientation, Phys. Chem. Chem. Phys. 10 (2008) 6962–6970.
49. D.M. Brink, G.R. Satchler, Angular momentum, Clarendon Press, Oxford, (1993).
50. M. Doi, Molecular-dynamics and rheological properties of concentrated solutions of rod-like polymers in isotropic and liquid-crystalline phases, J. Polymer. Sci. Poly. Phys. Ed. 19 (1981) 229.
51. W. Maier, A. Saupe, Z. Naturforsch. A 14 (1959) 882–889.
52. H.C. Öttinger, Stochastic Processes in Polymeric Fluids, Eine einfache molekulär-statistische theorie der nematischen kristallinflüssigen phase 1, Springer, Berlin, Heidelberg, New York, (1996).
53. I. Fedchenia, P.-O. Westlund, U. Cegrell, Brownian dynamic simulation of restricted molecular diffusion the symmetric and deformed cone model, Mol. Sim. 11 (1993) 373.
54. W.H. Press, S.A. Teukolsky, B.P. Flannery, Numerical Recipes in C, Cambridge, Cambridge, (1992).
55. L.B.-Å. Johansson, S. Engström, M. Lindberg, Electronic energy transfer in anisotropic systems. III. Monte Carlo simulations of energy migration in membranes, J. Chem. Phys. 96 (1992) 3844–3856.
56. J. Baumann, MD. Fayer, Excitation transfer in disordered two-dimensional and anisotropic three dimensional systems: effects of spatial geometry on the time-resolved observables, J. Chem. Phys. 85 (1986) 4087–4108.
57. B. Medhage, E. Mukhtar, B. Kalman, L.B.-Å. Johansson, J.G. Molotkovsky, Electronic energy transfer in anisotropic systems. 5. Rhodamine-lipid derivatives in model membranes, J. Chem. Soc. Faraday Trans. 88 (1992) 2845–2851.
58. R. Xu, M.A. Winnik, G. Riess, B. Chu, M.D. Croucher, Micellization of polystyrene-poly(ethylene oxide) block copolymers in water. 5. A test of the star and mean-field models, Macromolecules 25 (1992) 644–652.
59. P. Levitz, J.M. Drake, J. Klafter, Critical evaluation of the application of direct energy transfer in probing the morphology of porous solids, J. Chem. Phys. 89 (1988) 5224–5236.
60. A. Yekta, M.A. Winnik, J.P.S. Farinha, J.M.G. Martinho, Dipole–dipole electronic energy transfer. Fluorescence decay functions for arbitrary distributions of donors and acceptors. II. Systems with spherical symmetry, J. Phys. Chem. A 101 (1997) 1787–1792.
61. J.P.S. Farinha, J.M.G. Martinho, S. Kawaguchi, A. Yekta, M.A. Winnik, Latex Film Formation probed by nonradiative energy transfer: effect of grafted and free poly(ethylene oxide) on a poly(*n*-butyl methacrylate) Latex, J. Phys. Chem. 100 (1996) 12552–12558.
62. J.P.S. Farinha, K. Schillén, M.A. Winnik, Interfaces in self-assembling diblock copolymer systems: characterization of poly(isoprene-b-methyl methacrylate) micelles in acetonitrile, J. Phys. Chem. B 103 (1999) 2487–2495.

63. E. Helfand, Y. Tagami, Theory of the interface between immiscible polymers. II, J. Chem. Phys. 56 (1972) 1592–3601.
64. A.H.C. de Oliveira, J.R. Giglo, S.H. Andrião-Escarso, R.J. Ward, The effect of resonance energy homotransfer on the intrinsic tryptophan fluorescence emission of the bothopstoxin-I dimer, Biochem. Biophys. Res. Comm. 284 (2001) 1011–1015.
65. M. Kyoung, S.Y. Kim, H.-Y. Seok, I.-S. Park, M. Lee, Probing the caspase-3 active site by fluorescence lifetime measurements, Biochem. Biophys. Acta 1598 (2002) 74–79.
66. P.D.J. Moens, M.K. Helms, D.M. Jameson, Detection of tryptophan to tryptophan energy transfer in proteins, Protein J. 23 (2004) 79–83.
67. X. Duan, Z. Zhao, J. Ye, H. Ma, A. Xia, G. Yang, C.-C. Wang, Donor-donor energy-migration measurements of dimeric DsbC labeled at its N-terminal amines with fluorescent probes: a study of protein unfolding, Angew. Chemie Internat. Ed. 43 (2004) 4216–4219.
68. M. Fa, F. Bergström, J. Karolin, L.B.-Å. Johansson, T. Ny, Conformational studies of plasminogen activator inhibitor type 1 by fluorescence spectroscopy: Analysis of the reactive centre of inhibitory-, substrate- and reactive centre cleaved forms, Eur. J. Biochem. 267 (2000) 3729–3734.
69. P. Zou, K. Surendhran, H.S. Mchaourab, Distance measurements by flourescence energy homotransfer: evaluation in T4 Lysozyme and correlation with dipolar coupling between spin labels, Biophys. J. Biophys. Lett. 92 (2007) L27–L29.
70. F. Bergström, I. Mikhalyov, P. Hägglöf, R. Wortmann, T. Ny, L.B.-Å. Johansson, Dimers of dipyrrometheneboron difluoride (BODIPY) with light spectroscopic applications in chemistry and biology, J. Am. Chem. Soc. 124 (2002) 196–204.
71. J. Karolin, L.B.-Å. Johansson, L. Strandberg, T. Ny, Fluorescence and absorption spectroscopic properties of dipyrrometheneboron difluoride (BODIPY) derivatives in liquids, lipid membranes, and proteins, J. Am. Chem. Soc. 116 (1994) 7801–7806.
72. D. Marushchak, S. Kalinin, I. Mikhalyov, N. Gretskaya, L.B.-Å. Johansson, Pyrromethene dyes (BODIPY) can form ground state homo and hetero dimers: photophysics and spectral properties, Spectrochim. Acta, A 65 (2006) 113–122.
73. M. Wilczynska, M. Fa, J. Karolin, P.-I. Ohlsson, L.B.-Å. Johansson, T. Ny, Structural insights into serpin-protease complexes reveal the inhibitory mechanism of serpins, Nat. Struct. Biol. 4 (1997) 354–356.
74. M. Wilczynska, S. Lobov, P.-I. Ohlsson, T. Ny, A redox-sensitive loop regulates plasminogen activator inhibitor type-2 (PAI-2) polymerization, EMBO J. 22 (2003) 1753–1761.
75. H. Otto, T. Lamparter, B. Borucki, J. Hughes, M.P. Heyn, Dimerization and inter-chromophore distance of Cph1 phytochrome from synechocystis, as monitored by fluorescence homo and hetero energy transfer, Biochemistry 42 (2003) 5885–5895.
76. H. Otto, D. Hoersch, T. Meyer, M. Cusanovich, M.P. Heyn, Time-resolved single tryptophan fluorescence in photoactive yellow protein monitors changes in the chromophore structure during the photocycle via energy transfer, Biochemistry 44 (2005) 16804–16816.
77. M. Olofsson, S. Kalinin, M. Oliveberg, L.B.-Å. Johansson, Tryptophan – BODIPY: a versatile donor-acceptor pair for probing generic changes of intraprotein distances, Phys. Chem. Chem. Phys. 8 (2006) 3130–3140.
78. M. Wallgren, J. Ådén, O. Pylypenko, T. Mikaelsson, L.B.-Å. Johansson, A. Rak, M. Wolf-Watz, Extreme temperature tolerance of a hyperthermophilic protein coupled to residual structure in the unfolded state, J. Mol. Biol. 379 (2008) 845–858.
79. J.-C. Rochet, P.T. Lansbury, Jr, Amyloid fibrillogenesis: themes and variations, Curr. Opin. Struct. Biol. 10 (2000) 60–68.
80. E. Egelman, The structure of F-actin, J. Muscle Res. Cell Motil. 6 (1985) 129–151.
81. E.K.L. Yeow, A.H.A. Clayton, Enumeration of oligomerization states of membrane proteins in living cells by Homo-FRET spectroscopy and microscopy: theory and application, Biophys. J. 92 (2007) 3098–3104.

82. J. Howard, A.A. Hyman, Dynamics and mechanics of the microtubule plus end, Nature 422 (2003) 753–758.
83. M.P. Lillo, O. Canadas, R.E. Dale, A.U. Acuna, Location and properties of the taxol binding center in microtubules: a picosecond laser study with fluorescent taxoids, Biochemistry 41 (2002) 12436–12449.
84. P. Brocca, L. Cantù, M. Cortia, E.D. Favero, A. Raudino, Collective phenomena in confined micellar systems of gangliosides, Physica A 304 (2002) 177–190.
85. C. Giacomelli, L.L. Men, R. Borsali, Phosphorylcholine-based pH-responsive diblock copolymer micelles as drug delivery vehicles: light scattering, electron microscopy, and fluorescence experiments, Biomacromolecules 7 (2006) 817–828.
86. M. Milnera, M. Štěpánek, I. Zusková, K. Procházka, Experimental study of the electrophoretic mobility and effective electric charge of polystyrene-block-poly(methacrylic acid) micelles in aqueous media, Inter. J. Polym. Anal. Char. 12 (2007) 23–33.
87. K. Schillén, A. Yekta, S. Ni, M.A. Winnik, Characterization by fluorescence energy transfer of the core of polyisoprene-poly(methylmethacrylate) diblock copolymer micelles. Strong segregation in acetonitrile, Macromolecules 31 (1998) 210–212.
88. G.M. Jones, C. Wofsy, C. Aurell, L.A. Sklar, Analysis of vertical fluorescence resonance energy transfer from the surface of a small-diameter sphere, Biophys. J. 76 (1999) 517–527.
89. G. Liu, C.K. Smith, N. Hu, J. Tao, Formation and properties of polystyrene-block-poly(2-hydroxyethyl methacrylate) micelles, Macromolecules 29 (1996) 220–227.
90. K. Procházka, B. Bednář, E. Mukhtar, P. Svoboda, J. Trněná, M. Almgren, Nonradiative energy transfer in block copolymer micelles, J. Phys. Chem. 95 (1990) 4563–4568.
91. K. Itaka, A. Harada, K. Nakamura, H. Kawaguchi, K. Kataoka, Evaluation by fluorescence resonance energy transfer of the stability of nonviral gene delivery vectors under physiological conditions, Biomacromolecules 3 (2002) 841–845.
92. A.K. Shaw, R. Sarkar, S.K. Pal, Direct observation of DNA condensation in a nano-cage by using a molecular ruler, Chem. Phys. Lett. 408 (2005) 366–370.
93. J. Kim, M. Lee, Observation of multi-step conformation switching in b-amyloid peptide aggregation by fluorescence resonance energy transfer, Biochem. Biophys. Res. Com. 316 (2004) 393–397.
94. R. Sarkar, A.K. Shaw, S.S. Narayanan, F. Dias, A. Monkman, S.K. Pal, Direct observation of protein folding in nanoenvironments using a molecular ruler, Biophys. Chem. 123 (2006) 40–48.
95. Y. Taniguchi, T. Ikehara, N. Kamo, H. Yamasaki, Y. Toyoshima, Dynamics of light-induced conformational changes of the phoborhodopsin/transducer complex formed in the *n*-dodecyl b-D-maltoside micelle, Biochemistry 46 (2007) 5349–5357.
96. S.L. Veatch, S.L. Keller, Seeing spots: complex phase behavior in simple membranes, Biochim. Biophys. Acta 1746 (2005) 172–185.
97. K. Simons, E. Ikonen, Functional rafts in cell membranes, Nature 387 (1997) 569–572.
98. S.R. Mayor, M. Rao, Rafts: scale-dependent, active lipid organization at the cell surface, Traffic 5 (2004) 231–240.
99. J.E. Shaw, R.F. Epand, R.M. Epand, Z. Li, R. Bittman, C.Y. Yip, Correlated fluorescence-atomic force microscopy of membrane domains: structure of fluorescence probes determines lipid localization, Biophys. J. 90 (2006) 2170–2178.
100. A. Filippov, G. Orädd, G. Lindblom, Lipid lateral diffusion in ordered and disordered phases in raft mixtures, Biophys. J. 86 (2004) 891–896.
101. S. Pedersen, K. Jørgensen, T.R. Baekmark, O.G. Mouritsen, Indirect evidence for lipid-domain formation in the transition region of phospholipid bilayers by two-probe fluorescence energy transfer, Biophys. J. 71 (1996) 554–560.
102. I. Fedchina, P.-O. Westlund, Influence of molecular reorientation on electronic energy transfer between a pair of mobile chromophores: The stochastic Liouville equation combined with Brownian dynamic simulation techniques, Phys. Rev. E 50 (1994) 555–565.

103. C. Leidy, W.F. Wolkers, K. Jørgensen, O.G. Mouritsen, J.H. Crowe, Lateral organization and domain formation in a two-component lipid membrane system, Biophys. J. 80 (2001) 1819–1828.
104. D. Marushchak, N. Gretskaya, I. Mikhalyov, L.B.-Å. Johansson, Self-aggregation – an intrinsic property of GM1 in lipid bilayers, Mol. Membr. Biol. 24 (2007) 102–112.
105. R.F.M. de Almeida, L.M.S. Loura, A. Fedorov, M. Prieto, Lipid rafts have different sizes depending on membrane composition: a time-resolved fluorescence resonance energy transfer study, J. Mol. Biol. 346 (2005) 1109–1120.
106. L.M.S. Loura, A. Fedorov, M. Prieto, Fluid–fluid membrane microheterogeneity: a fluorescence resonance energy transfer study, Biophys. J. 80 (2001) 776–788.
107. K.B. Towles, A.C. Brown, S.P. Wrenn, N. Dan, Effect of membrane microheterogeneity and domain size on fluorescence resonance energy transfer, Biophys. J. 93 (2007) 655–667.
108. A.C. Brown, K.B. Towles, S.P. Wrenn, Measuring raft size as a function of membrane composition in PC-based systems: Part 1 – binary systems, Langmuir 23 (2007) 11180–11187.
109. S. Mall, R. Broadbridge, R.P. Sharma, J.M. East, A.G. Lee, Self-association of model transmembrane a-helices is modulated by lipid structure, Biochemistry 40(41), (2001) 12379–12386.
110. E. Sparr, W.L. Ash, P.V. Nazarov, D.T.S. Rijkers, M.A. Hemminga, D.P. Tieleman, J.A. Killian, self-association of transmembrane α-helices in model membranes, J. Biol. Chem. 280 (2005) 39324–39331.
111. M. Bloom, E. Evans, O.G. Moritsen, Physical properties of the fluid lipid-bilayer component of cell membranes: a perspective, Q. Rev. Biophys. 24 (1991) 293–297.
112. H. Borochov-Neori, M. Montal, Rhodopsin-G-protein interactions monitored by resonance energy transfer, Biochemistry 28 (1989) 1711–1178.
113. L.M.S. Loura, A. Coutinho, A. Silva, A. Fedorov, M. Prieto, Structural effects of a basic peptide on the organization of dipalmitoylphosphatidylcholine/dipalmitoylphosphatidylserine membranes: a fluorescent resonance energy transfer study, J. Phys. Chem. B 110 (2006) 8130–8141.
114. S.-T. Bogen, G. de Korte-Kool, G. Lindblom, L.B.-Å. Johansson, Aggregation of an α-helical transmembrane peptide in lipid phases, studied by time resolved fluorescence spectroscopy, J. Phys. Chem. B 103 (1999) 8344–8352.
115. H.E. Hasper, B. de Kruijff, E. Breukink, Assembly and stability of nisin-Lipid II pores, Biochemistry 43 (2004) 11567–11575.
116. Y. Tamba, M. Yamazaki, Single giant unilamellar vesicle method reveals effect of antimicrobial peptide magainin 2 on membrane permeability, Biochemistry 44 (2005) 15823–15833.
117. A. Miszta, R. Macháň, A. Benda, A.J. Ouellette, W.T. Hermens, M. Hof, Combination of ellipsometry, laser scanning microscopy and Z-scan fluorescence correlation spectroscopy elucidating interaction of cryptdin-4 with supported phospholipid bilayers, J. Pept. Sci. 14 (2008) 503–509.
118. G. Viero, R. Cunaccia, G. Prévost, S. Werner, H. Monteil, D. Keller, O. Joubert, G. Menestrina, M.D. Serra, Homologous versus heterologous interactions in the bicomponent staphylococcal γ -haemolysin pore, Biochem. J. 394 (2006) 217–225.
119. L.C. Silva, A. Coutinho, A. Fedorov, M. Prieto, Conformation and self-assembly of a nystatin nitrobenzoxadiazole derivative in lipid membranes, Biochim. Biophys. Acta 1617 (2003) 69–79.
120. L.J. O′Neill, J.G. Miller, N.O. Petersen, Evidence for nystatin micelles in L-cell membranes from fluorescence photobleaching measurements of diffusion, Biochemistry 25 (1986) 177–181.
121. C. Madeira, L.M.S. Loura, M.R. Aires-Barros, A. Fedorov, M. Prietoy, Characterization of DNA/lipid complexes by fluorescence resonance energy transfer, Biophys. J. 85 (2003) 3106–3119.

122. K. Herrick-Davis, E. Grinde, J.E. Mazurkiewicz, Biochemical and biophysical characterization of serotonin 5-HT2C receptor homodimers on the plasma membrane of living cells, Biochemistry 43 (2004) 13963–13971.
123. J.R. Silvius, I.R. Nabi, Fluorescence-quenching and resonance energy transfer studies of lipid microdomains in model and biological membranes, Mol. Membr. Biol. 23 (2006) 5–16.
124. A.K. Kenworthy, N. Petranova, M. Edidin, High-resolution FRET microscopy of cholera toxin B-subunit and GPI-anchored proteins in cell plasma membranes, Mol. Biol. Cell 11 (2000) 1645–1655.
125. P. Sharma, R. Varma, R.C. Sarasij, G.K. Ira, G. Krishnamoorthy, M. Rao, S. Mayor, nanoscale organization of multiple GPI-anchored proteins in living cell membranes, Cell 116 (2004) 577–589.
126. P. Sengupta, D. Holowka, B. Baird, Fluorescence resonance energy transfer between lipid probes detects nanoscopic heterogeneity in the plasma membrane of live cells, Biophys. J. 92 (2007) 3564–3574.

Spectra FRET: A Fluorescence Resonance Energy Transfer Method in Live Cells

Ekaterina A. Bykova and Jie Zheng

Abstract The technique of Fluorescence Resonance Energy Transfer (FRET) has been extensively used in optical microscopy for the study of bio-molecular interactions. As most transient and weak intermolecular interactions only occur in intact cells under physiological conditions, it is essential to be able to record FRET from live cells. This chapter introduces a new spectra FRET approach that has recently been developed and tested. It provides the spatial resolution for separation of membrane and intracellular signals. Its spectrum-based procedure is easy to carry out and is powerful in separating donor and acceptor emissions as well as detecting the presence of other background light. Moreover, spectra FRET allows for a fine correction for common contaminations in FRET experiments, such as cross-talk and bleed-through, therefore contributing to the reduction of errors in FRET analysis.

Keywords Microscopy · Spectroscopy · Live cell imaging · Protein-protein interaction

Introduction

Why Do We Use FRET?

Routinely monitoring real-time biological processes in live cells has been a biologist's dream since the cellular structure of live organisms was revealed. Technology developments in optical microscopy and achievements in molecular biology made it possible to light up particular biological molecules within a cell with fluorescently-tagged immunoglobulins. Fluorescence microscopy greatly enhanced the contrast of the cellular structures and the resolution of optical microscopy. Later discoveries

E.A. Bykova (✉)
Department of Physiology and Membrane Biology, University of California, One Shields Avenue, Davis, CA 95616, USA
e-mail: eabykova@ucdavis.edu

C.D. Geddes (ed.), *Reviews in Fluorescence 2007*, Reviews in Fluorescence 2007,
DOI 10.1007/978-0-387-88722-7_4,

of fluorescent proteins further improved our ability to study sub-cellular events. Genetic labeling of target proteins with fluorescent proteins allows the investigator to conveniently follow protein synthesis, trafficking, and interactions in both expression systems and native cells. In general, the resolution of optical microscopy is limited to about one-half of the wavelength of the observing light (hundreds of nanometers). Based on this resolution, many cellular processes and functions have been successfully studied with fluorescence microscopy. However, it has become increasingly clear that simply knowing the sub-cellular location at optical resolution is often insufficient. For example, to demonstrate whether two proteins interact one needs to establish that they are within proximity much closer than the optical resolution limit. Electron microscopy possesses a desirable resolution. Nonetheless, it is restricted to fixed cells and therefore lacks a dynamic component of intermolecular interactions. Similarly, biochemical and immunochemical methods have low temporal resolutions. In addition, these methods often miss weak and transient interactions. Fluorescence resonance energy transfer (FRET) is a technique that empowers the investigator with the ability to detect molecular proximity within a distance range of 10–100 Å. It dramatically extends the resolution of fluorescence microscopy and thus allows the detection of dynamic intermolecular interactions in live cells. FRET has been widely used in biomedical research as a molecular ruler.

Fluorescence Resonance Energy Transfer

Fluorescence (Föster) resonance energy transfer refers to the physical phenomenon occurring between two fluorescent molecules when one excited molecule (donor) transfers energy to another nearby molecule (acceptor) through non-radiative dipole–dipole coupling. The efficiency of energy transfer between donor and acceptor molecules depends on the distance separating the fluorophores:

$$E = \frac{R_0^6}{R_0^6 + R^6}, \tag{1}$$

where R is the distance between the two fluorophores and R_0 refers to the distance at which energy transfer efficiency is 50%, a characteristic parameter for a given pair of fluorescent molecules.

When the FRET phenomenon is used for detecting intermolecular interactions or estimating molecular proximity, molecules of interests are labeled with a FRET pair – the donor fluorophore and the acceptor fluorophore. In order to achieve sufficient energy transfer, there should be a large overlap between the donor emission spectrum and the acceptor excitation spectrum (Fig. 1a). Upon donor excitation, the donor fluorophore can release the acquired energy by either emitting light or transferring the energy to an acceptor molecule if the acceptor is in its vicinity. When energy transfer occurs, the acceptor fluorophore emits light while the donor emission is reduced.

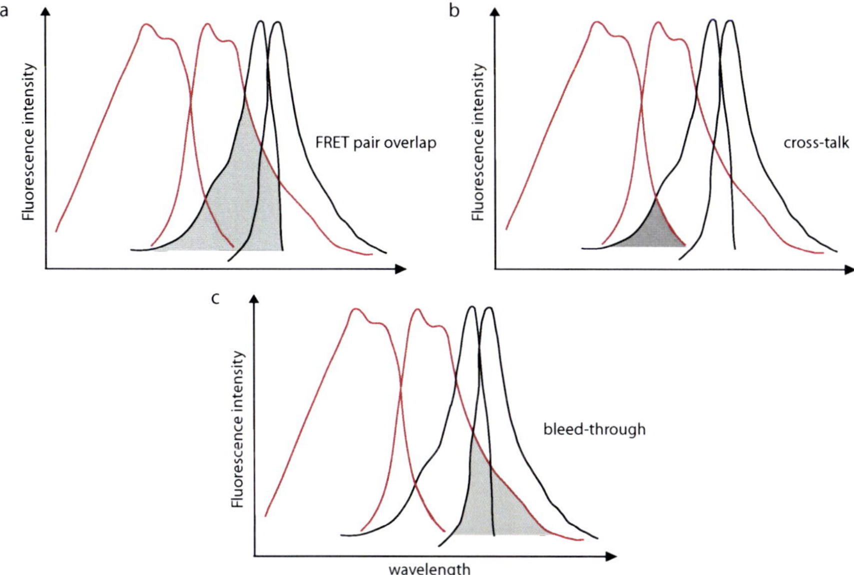

Fig. 1 Absorption and emission spectra of Cerulean (donor) in *red* and eYFP (acceptor) in *black*. *Grey* area shows (**a**) overlap between Cerulean emission spectra and eYFP absorption, (**b**) absorption spectra overlap, which leads to cross-talk, and (**c**) emission spectra overlap, which leads to bleed-through

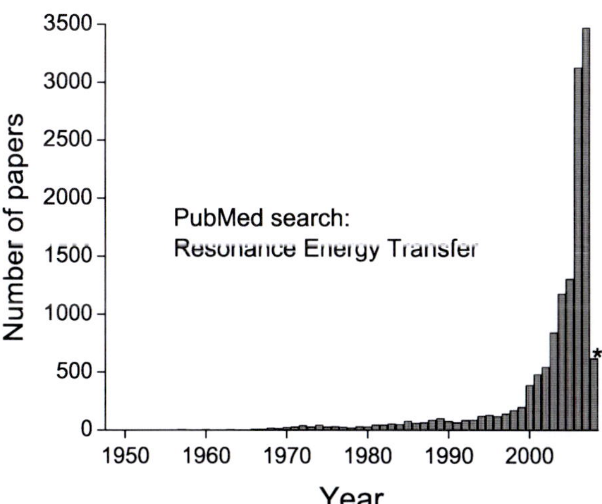

Fig. 2 Diagram illustrating the number of published papers over the years after the theory of resonance energy transfer was first proposed. $*$ indicates an incomplete number for the current year (2008)

The theory of resonance energy transfer was originally proposed by Theodor Föster in 1948 [1]. Its introduction to biomedical research was partly attributed to a classical experiment conducted in 1967 by Lubert Stryer and Richard Haugland [2]. In the experiment, poly-L-proline oligomers, which formed rigid structures with predictable lengths, were used as spacers to separate small fluorophores dansyl and naphthyl that formed a FRET pair. The efficiency of energy transfer was inversely proportional to the sixth power of the distance between dansyl and naphthyl, ranging from 100 to 16% for distances between 12 Å and 46 Å. Nonetheless, FRET was not widely used in biomedical research until the development of green fluorescent protein (GFP) mutants [3]. These mutant fluorescent proteins greatly facilitated the application of FRET in live cells to elucidate interactions between biological macromolecules. As a result, the number of studies utilizing FRET in biological systems has been enormously increasing over the last decade (Fig. 2). FRET has become a routine technique for many laboratories.

Methods to Measure FRET

A key to successful application of FRET methods is reliable quantification of the efficiency of energy transfer. FRET efficiency can be quantified by many methods depending on investigator preferences, sample specificity or system restrictions, FRET pairs, and equipment used.

Methods based on detecting emission intensity changes were commonly used in measuring FRET efficiency. *Sensitized emission* (acceptor enhanced emission) [4, 5] and *acceptor photobleaching* approaches [6–8] utilize changes in the emission intensity upon FRET. *Sensitized emission* is the most straightforward approach which measures increased acceptor emission caused by energy coupling. *Acceptor photobleaching* technique requires an additional step – photobleaching. When acceptor molecules are destroyed by excessive radiation of acceptor-specific excitation light, donor fluorophore emission increases due to the absence of FRET. Another group of techniques, including *Fluorescence Lifetime Imaging Microscopy* (FLIM) [9–12] and *polarized anisotropy imaging* [13–15], utilizes change in fluorescence lifetime of fluorophores due to FRET. Basic principal behind *FLIM* lies in measuring fluorescence lifetime of a donor molecule. When FRET occurs, donor molecule fluorescence lifetime decreases due to donor energy been transferred to an acceptor. *Polarized anisotropy imaging* technique measures FRET efficiency based on a change in fluorescence polarization of the donor or acceptor molecules. For instance, when the donor fluorescence lifetime is shortened due to FRET, its polarization is increased.

All FRET quantification methods have certain advantages and limitations. Intensity-based methods, in spite of their conceptual simplicity and easiness, have to deal with contaminations coming from overlaps of donor and acceptor excitation and emission spectra (Fig. 1b and c). Partial overlap of the excitation spectra of the donor and acceptor fluorophores (Fig.1b) causes direct excitation of the acceptor molecule while exciting the donor, which leads to overestimation of FRET efficiency. This type of contamination is often called "cross-talk". Another problem, so-called "bleed-through", comes from the emission spectra overlap (Fig. 1c). In this case, donor emission contaminates acceptor emission, again resulting in overestimation of FRET efficiency. Standard measurements have to be properly performed to eliminate these contaminations when calculating FRET efficiency. Techniques based on fluorescence lifetime measurements do not have to deal with spectral contaminations because these are kinetic parameters of the fluorophores, not intensity parameters. The FLIM approach relays on specialized, expensive equipments required for lifetime measurements. Anisotropy imaging does not require sophisticated equipments. However, it is primarily used for qualitative detection of the presence of FRET but not for quantifying FRET efficiency. Furthermore, complications common to any FRET technique include the variable donor to acceptor expression level in live cells and a mixture of fluorophore populations when random association between donor-labeled and acceptor-labeled molecules takes place. Depending on the system one works with, the expected FRET estimate will be different even if the underlying true FRET efficiency, determined by the distance between the donor and acceptor in the interacting complexes, is the same. In order to correctly calculate FRET efficiency, a model accounting for different donor–acceptor levels in live cells and a mixture of fluorophore populations is often needed, as demonstrated in our example.

Spectral Imaging: Spectra FRET

The *spectra FRET* approach is a kind of *sensitized emission* methods in that it is based on quantification of the increased acceptor emission upon FRET. The difference between spectra FRET and many other intensity measurement-based methods is in the modality of the collected data. Instead of fluorescence intensity values at peak emission wavelength ranges, the row data from the spectra FRET method are emission spectra of the donor and acceptor fluorophores. Expansion into the wavelength dimension provides a number of advantages. Let us review the spectra FRET approach in detail.

Instrumentation

An advantage of the spectra FRET technique over high-tech FRET detection methods lies in the equipment it requires. A regular fluorescence microscope can be used with slight and inexpensive modifications. The extra gear necessary for spectra FRET quantification includes an appropriate set of filters for a given FRET pair, a spectrograph to separate donor and acceptor emissions by wavelength, and a CCD camera to capture spectroscopic images. (Except for the spectrograph, the other parts are normally used also in other intensity-based FRET methods.) A schematic FRET recording system is shown in Fig. 3. A mercury lamp or a laser generator is used as the source of excitation light. Sample excitation time is controlled by a shutter. Light beam coming through the shutter goes through an appropriate filter (for donor or acceptor excitation) and hits a sample. Fluorescence emission from the sample is directed into an adjustable spectrograph input slit. The fluorescence signal

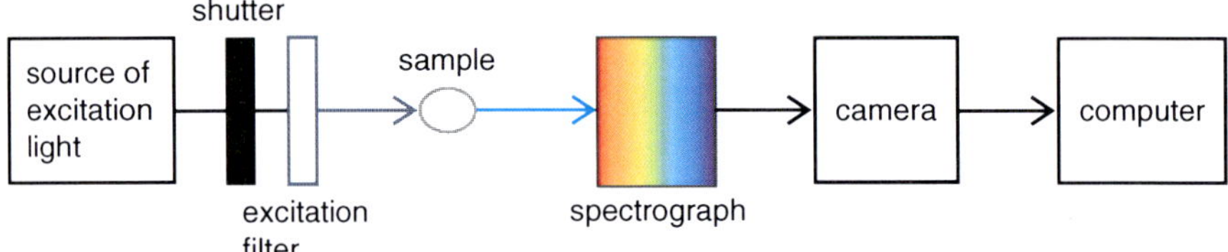

Fig. 3 Diagram illustrating the spectra FRET set-up

from a thin section of a cell, covered by the vertical slit, hits a grating mirror inside the spectrograph, resulting in a horizontal shift of the emission light. The amount of shift depends on the wavelength. A spectroscopic image, with the wavelength along the x-axis and the cell position along the y-axis, is recorded by the camera. The data from the camera are saved to a computer. Spectra are reconstructed from spectroscopic images by scanning along the x-axis at the positions of the cell membrane.

Data Acquisition

To record FRET from live cells, target proteins are labeled with a FRET pair by genetically attaching fluorescent proteins, by modification of engineered cysteines in the target proteins with sulfhydryl-reactive fluorophores, or by other available site-specific labeling techniques. A bright fluorescent cell is brought to the middle of the microscope's field of view (Fig. 4a). The position and width of the spectrograph input slit (white rectangle in Fig. 4a) is adjusted, letting only emission light from the selected section of the cell into the spectrograph. In order to quantify FRET efficiency, two spectroscopic images of the same cell are acquired. The first image is taken with excitation light that specifically activates the acceptor fluorophore (Fig. 4a, b). The second image is taken with a donor-specific excitation (Fig. 4c, d). In the spectroscopic images shown in Fig. 4, the sharp bright lines (indicated by arrows) correspond to fluorescence signals coming from the plasma membrane, while the bright bands between membrane fluorescence signals are from cytoplasmic structures. Fluorescence intensity readouts along horizontal lines along each bright line constitute the emission spectrum (Fig.4 k–o). By moving along the vertical axis of a spectroscopic image (which corresponds to the position of the cell) signals from different parts of the cell can be analyzed. In this way, spectroscopic imaging allows easy separation of fluorescence signals from different cellular structures.

Spectroscopic imaging also provides additional checks for quality of the acquired data. For instance, contaminating auto-fluorescence from cytoplasmic organelles will distort the emission spectra. This can be conveniently spotted when the spectra are constructed, or later on when RatioA is calculated (see below). In addition, poor focus adjustment or low signal intensity will result in blurred spectroscopic images with not well-defined membrane signal. These can be easily identified and discarded during off-line FRET quantification to avoid possible contaminations from surrounding regions. Therefore, the spectral approach to FRET efficiency

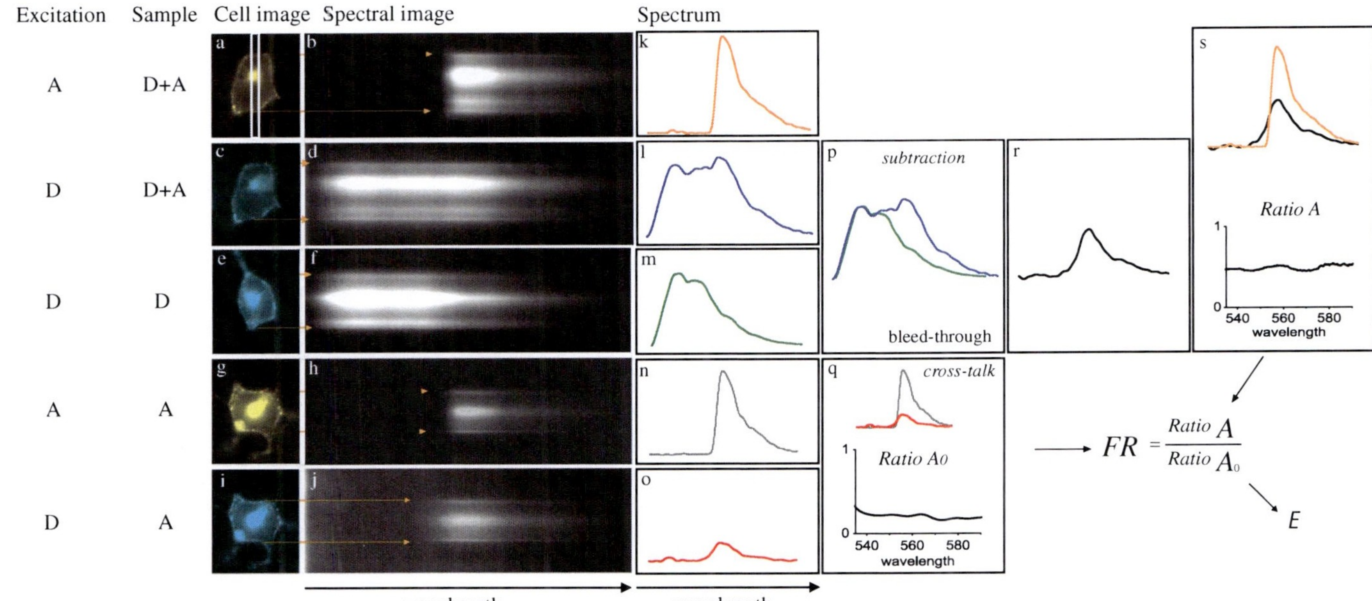

Fig. 4 Spectra FRET data acquisition and analysis. Fluorescence images of cells (**a**, **c**, **e**, **g**, **i**), their corresponding spectral images (**b**, **d**, **f**, **h**, **j**), and the emission spectra from cellular membrane (**k**, **l**, **m**, **n**, **o**). The fluorophore(s) present in the cell and the excitation light are labeled on the *left*. (**p–s**) Procedures for processing emission spectra to get FRET efficiency estimates

measurement provides the spatial resolution for easy separation of membrane and cytoplasmic signals. This is especially important when studying membrane proteins in cells over-expressing exogenous proteins, as most of the fluorescently labeled proteins synthesized by the cell are trapped in the cytosol.

Data Analysis

Besides the two spectroscopic images of a cell carrying both donor- and acceptor-labeled proteins, three additional standard spectroscopic images are required to accurately calculate FRET efficiency. The standard spectral images do not have to be taken for every single experiment for a given pair of molecules of interest if the same FRET recording system and same FRET pair are used. These additional images are necessary to correct for bleed-through and cross-talk contaminations.

Bleed-Through

To form a good FRET pair, two fluorophores should have extensive overlap between the donor fluorophore emission spectrum and the acceptor excitation spectrum (Fig. 1a). Such overlap rarely occurs without overlap between the excitation spectra of the pair of fluorophore and between the emission spectra (Fig. 1b, c), as the excitation and emission spectra of most fluorophores are rather broad and the Stokes shift is relatively small. Overlap between donor and acceptor emission spectra results in acceptor emission being contaminated with donor emission (bleed-through) (Fig. 1c). If not properly corrected, the donor emission intensity will be added to the acceptor emission, which leads to an overestimated FRET efficiency.

In order to correct bleed-through, a spectral image from a cell carrying only donor fluorophores has to be taken (Fig. 4e, f). From this image, a standard donor emission spectrum is constructed (Fig. 4m). Bleed-through is corrected by subtracting a scaled donor-only spectrum, $F_{\mathrm{D}}^{\mathrm{d}}$, from a spectrum originated from a sample carrying both donor and acceptor molecules that is excited with a donor-specific excitation, $F_{\mathrm{DA}}^{\mathrm{d}}$ (Fig. 4p). The difference (Fig. 4r) contains the acceptor emission due to FRET, $F_{\mathrm{a}}^{\mathrm{FRET}}$, plus a direct acceptor excitation component, $F_{\mathrm{a}}^{\mathrm{direct}}$, which will be dealt with below under the section "cross-talk":

$$F_{\mathrm{DA}}^{\mathrm{d}} - F_{\mathrm{D}}^{\mathrm{d}} = F_{\mathrm{a}}^{\mathrm{FRET}} + F_{\mathrm{a}}^{\mathrm{direct}} . \tag{2}$$

In order to calculate FRET efficiency, a ratio between the calculated difference $F_{\mathrm{a}}^{\mathrm{FRET}} + F_{\mathrm{a}}^{\mathrm{direct}}$ and an acceptor spectrum from a sample with both fluorophores, $F_{\mathrm{DA}}^{\mathrm{a}}$, is taken (Fig. 4s):

$$\frac{F_{\mathrm{a}}^{\mathrm{FRET}} + F_{\mathrm{a}}^{\mathrm{direct}}}{F_{\mathrm{DA}}^{\mathrm{a}}} = \text{ratio } A . \tag{3}$$

As both the denominator and the numerator of Eq. (3) are from the same sample, Ratio A is concentration-independent. This makes it convenient to compare between cells with different expression levels.

Cross-Talk

Cross-talk is the direct excitation of acceptor fluorophores when exciting at the donor-specific wavelength. It is the result of donor and acceptor excitation spectra overlap (Fig. 1b). In other words, acceptor fluorophores emit light upon donor excitation as if there is FRET. This problem, when not properly solved, will result in overestimation of FRET efficiency or apparent FRET efficiency even when molecules are not within the distance for energy transfer to occur.

To account for cross-talk, a ratio between spectra from a sample carrying only acceptor fluorophores (Fig. 4n, o) is taken:

$$\frac{F_{\mathrm{A}}^{\mathrm{d}}}{F_{\mathrm{A}}^{\mathrm{a}}} = \text{ratio } A_0, \tag{4}$$

where $F_{\mathrm{A}}^{\mathrm{d}}$ is fluorescence coming from the acceptor sample excited at the donor-specific wavelength (Fig. 4o) and $F_{\mathrm{A}}^{\mathrm{a}}$ is fluorescence coming from the same sample excited at the acceptor-specific wavelength (Fig. 4n). Again, the ratio is unitless and concentration-independent.

Ratio A_0 (Fig. 4q) reflects the degree of direct excitation of the acceptor fluorophore by the donor excitation. It is equivalent to the $\frac{F_{\mathrm{a}}^{\mathrm{direct}}}{F_{\mathrm{DA}}^{\mathrm{a}}}$ component in Ratio A (Eq. 3). To eliminate cross-talk a FRET ratio (FR) between Ratio A and Ratio A_0 is calculated:

$$FR = \frac{Ratio\ A}{Ratio\ A_0} = \frac{F_a^{FRET}}{F_a^{direct}} + 1. \tag{5}$$

As can be seen in Eq. (5), FR is proportional to FRET efficiency, E:

$$E = \frac{e_{\mathrm{A}}}{e_{\mathrm{D}}}(FR - 1) \tag{6}$$

where e_{A} and e_{D} are acceptor and donor extinction coefficients at the donor excitation wavelength, respectively.

Spectra FRET in Live Cells

Measuring FRET efficiency in live cells is advantageous in many ways comparing to purified systems. With live cells, it becomes possible to monitor cellular processes in a more relevant environment where natural interactions between biological molecules take place. It is also possible to detect weak and transient interactions or events which only happen during normal cellular functioning. FRET measurements in live cells, however, is more complex than in purified systems. Here three problems commonly occurring to FRET measurements from live cells are discussed.

These complications can be severe and, when not properly treated, can corrupt experimental results and lead to incorrect conclusions [16].

Mixed Fluorophores Populations

The problem of mixed fluorophore populations arises under a number of circumstances (Fig. 5). For example, if the interaction between target molecules is transient, free unpaired fluorophores will co-exist with paired fluorophores (Fig. 5c). Even for permanent interactions, it is possible that the induced expression of fluorescently labeled interacting molecules is not at a stoichiometrical level, leaving the excess fluorophores being unpaired. If interacting molecular complexes are homomeric, associations between like fluorophores (donor–donor or acceptor–acceptor, which does not give FRET) are found together with donor–acceptor complexes (which does give FRET) (Fig. 5b). Free fluorophores and like fluorophores complexes do not contribute to FRET but still contribute to recorded fluorescence intensities, therefore causing underestimation of FRET efficiency. In these cases, the FRET efficiency calculated from such samples will not reflect the true transfer efficiency between donor–acceptor pairs but rather an apparent FRET efficiency that depends additionally on the composition of the fluorophore population. In the following case, we examine the mixed fluorophore population problem with a simple system of permanent homomeric dimers. For simplicity reason, we also assume that no monomers exist in the system. There are many examples of homodimers complexes in biology, for example, voltage-dependent proton channel Hv1 [17, 18].

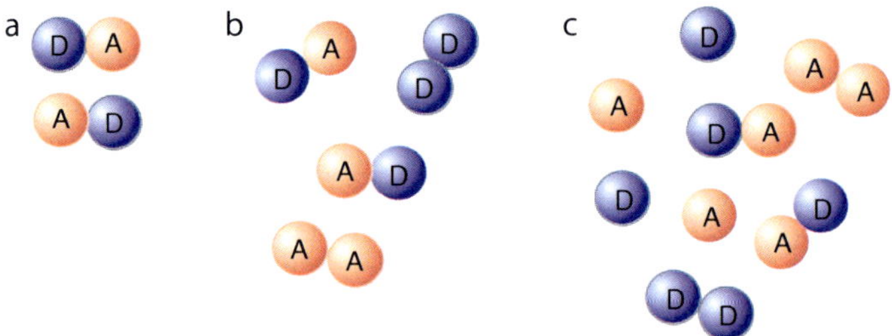

Fig. 5 Mixed fluorophore populations. (**a**) A homogenous D–A dimer population. (**b**) Heterogenous populations of fluorophore pairs. (**c**) Heterogenous populations of dimers and monomers. *A* and *D* stand for acceptor and donor fluorophore labeled molecules, respectively

If dimer formation between the donor-labeled molecule, D, and the acceptor-labeled molecule, A, is random and free monomers do not exist, the system contains three populations: D–D, D–A, and A–A, as shown in Fig. 5b. The binomial distribution can be used to describe the probabilities of forming these populations (Table 1):

Table 1 Probabilities of the formation of different fluorophore populations and their contributions to fluorescence intensities.

Fluorophores populations	D–D	D–A, A–D	A–A
Probability of forming each population	P_D^2	$2P_D\,P_A$	P_A^2
Total donor intensity	$2P_D^2\,S_D$	$2P_D\,P_A\,(1-E)\,S_D$	0
Total acceptor intensity	0	$2P_D\,P_A\,(1+E\,\frac{e_D}{e_A})\,S_A$	$2P_A^2\,S_A$

D, a molecule of interest labeled with a donor fluorophore; A, a molecule labeled with an acceptor; P_D and P_A, probability of a donor-labeled and acceptor-labeled molecule to participate in a dimer formation, respectively; S_D and S_A, donor and acceptor coefficients reflecting the system and the fluorophores used; e_D and e_A, donor and acceptor extinction coefficients, respectively; E, FRET efficiency.

If the donor and acceptor expression levels are known, P_D and P_A can be determined. Donor and acceptor intensities can be calculated based on probabilities of the formation of each fluorophore population, as shown in Table 1. The apparent FRET efficiency, calculated from the enhanced emission of the acceptor, is equal to

$$E_{\text{app}} = E\frac{2P_D P_A \frac{e_D}{e_A} S_A}{2P_D P_A S_A + 2P_A^2 S_A} \cdot \frac{e_A}{e_D}. \quad (7)$$

Variable Donor/Acceptor Expression Level

Another pitfall in measuring FRET from live cells is the variable expression level of transfected constructs. As each cell randomly picks up DNA/RNA constructs, expression levels differ from cell to cell throughout a plate. It can be seen from Eq. (7) that the donor-to-acceptor ratio (P_D/P_A) does affect FRET efficiency. Variability of the expression level in culture cells will result in a different FRET efficiency calculated from cells with different donor-to-acceptor ratios. In order to obtain the true FRET, which is independent of the donor-to-acceptor ratio, variable expression levels have to be accounted for when estimating FRET efficiency. Again let us examine the simple case of homodimers.

If r is the concentration ratio between the donor and acceptor fluorophores expressed in a cell, then the probabilities of a donor and an acceptor molecule to be incorporated into a dimer are

$$P_D = \frac{D}{D+A} = \frac{r}{1+r}, \quad (8)$$

$$P_A = 1 - P_D = \frac{1}{1+r}. \quad (9)$$

When the above equations are combined with Eq. (7), FRET efficiency can be calculated in spite of variable expression levels of donor and acceptor and mixed fluorophore populations:

$$E_{\text{app}} = E\frac{r}{1+r}. \tag{10}$$

In the live cell experiment, the concentration ratio r is generally not measurable. However, it can be estimated from the measurable donor and acceptor fluorescence intensities, F_{D} and F_{A}, respectively, based on Table 1:

$$\frac{F_{\text{D}}}{F_{\text{A}}} = \frac{1+r-E}{1+\frac{1}{r}+E\frac{e_{\text{D}}}{e_{\text{A}}}}\frac{S_{\text{D}}}{S_{\text{A}}}. \tag{11}$$

Combination of Eqs. (10) and (11) eliminates the unknown factor r, allowing the true FRET efficiency to be calculated based on measurable parameters – apparent energy transfer efficiency and the donor-to-acceptor intensity ratio.

An example of a data set from the electric ray chloride channel ClC-0 fitted with the discussed model is shown on Fig. 6. ClC-0 exists as a homodimer in the plasma membrane of the electric organ electrocytes [19]. In the example experiment, CLC-0 subunits are labeled with a donor, cerulean, or an acceptor, eYFP, and are co-expressed in HEK 293 cells. Each symbol in the figure represents the apparent FRET efficiency value measured from a cell using spectra FRET. The apparent FRET efficiency approaches the true efficiency only at very high donor-to-acceptor fluorescence intensity ratios (Fig. 6). This is expected from Eq. (10), in which $r/(1+r)$ approaches 1 when r is large. The convergence of the apparent FRET efficiency to its true value also makes physical sense, as at high donor-to-acceptor ratios most acceptor molecules are paired with a donor fluorophore. As the energy transfer efficiency is measured in this case from the enhanced emission of the

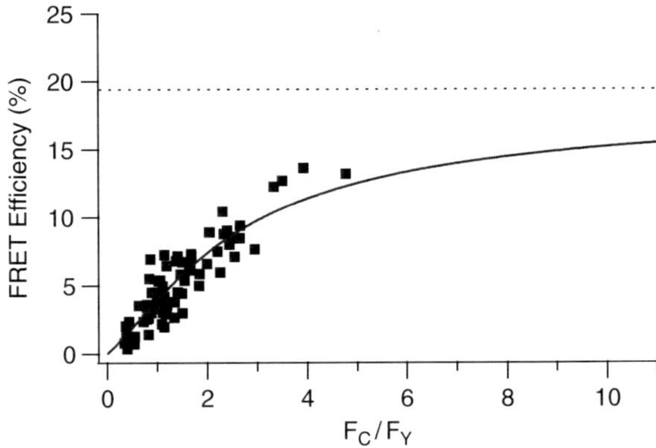

Fig. 6 FRET efficiency measured from the chloride channel ClC-0 labeled with Cerulean and eYFP. The *curve* represents model fit. *Dotted line* shows plateau which reflects the true FRET efficiency. Modified with permission from Bykova et al. [25]

acceptor, at very high donor-to-acceptor ratios virtually all the acceptors contribute to FRET. The recorded efficiency values would be close to the true FRET efficiency.

The example in Fig. 6 highlights the potentially large error resulted from a not-so-uncommon practice of FRET measurements. If one attempts to estimate the true FRET efficiency from cells having near equal donor and acceptor intensities ($F_C/F_Y = 1$), the apparent FRET efficiency can be much lower than the true value. Should we, then, select cells with very high donor-to-acceptor ratios and estimate FRET only from those cells? Unfortunately, high donor-to-acceptor ratios generally correspond to low acceptor intensities (due to a limited camera detection window), which are more likely affected by potential contaminations and are hard to accurately quantify due to the low signal-to-noise ratio. More amicable FRET data are collected at close donor-to-acceptor intensities. But as discussed above, in these cells only half of the fluorophores are correctly paired and the apparent efficiency is much smaller than the true efficiency. Therefore, the most reliable way to estimate FRET efficiency is to collect data from cells with various donor-to-acceptor intensities and fit the data with an appropriate model that accounts for mixed fluorophore populations and variable expression levels.

The simple model discussed above only works for two molecules forming a dimer with a very low dissociation constant (no free monomers). When working with more complicated interactions new models have to be built based on new assumptions. For example, in a tetrameric complex there will be multiple pathways through which energy transfer can occur [20].

Intramolecular FRET is a "lucky" exception that does not require a model to calculate FRET efficiency. When a single molecule is labeled stoichiometrically with a FRET pair to monitor conformational rearrangements of this molecule, every donor fluorophore has an acceptor fluorophore and no other forms of fluorophore complexes are expected (Fig. 5a). In this case calculation of FRET efficiency is straightforward and one can simply follow the steps discussed in the "data analysis" section. Indeed, a vast number of assays are based on energy transfer between two fluorophores attached to one target molecule [21–24].

Background Fluorescence

Specificity of labeling target molecules has always been an issue when working with live cells. Fluorescent proteins offer a good opportunity to avoid nonspecific labeling. Because target proteins are genetically linked with the fluorescent proteins, they are exclusively tagged with appropriate fluorescent proteins while other proteins are not. The size of fluorescent proteins is one major disadvantage. When fine movements of a target molecule is the subject of an experiment, fluorescent proteins would not be suitable for revealing this event due to their size and the load they apply. Small cysteine-reactive fluorophores can be used instead of fluorescent proteins for these FRET experiments. The relatively small size of many commercially available fluorophores is beneficial for fine FRET measurements. Specificity of labeling is based on the specific reaction between the sylfhydryl group of a cysteine and the reactive group of the fluorescent molecule. While it is often possible to introduce a cysteine to a protein without significantly changing its function, preventing unwanted labeling of endogenous cysteines in the target protein as well as other native proteins is a major challenge. Specificity of labeling remains to be a hot topic in the protein structure field and requires further advancements.

Conclusions

Every technique intended to measure FRET has its limitations – whether it relays on specialized equipment for taking measurements or extensive corrections and standards are essential for reaching the correct answer. Successful FRET analysis often relies on choosing the right technique and carrying out the necessary corrections. The spectroscopy-based spectra FRET approach offers a good alternative to measure FRET in live cells. Equipment requirements are minimal and inexpensive; a regular fluorescence microscope can be easily upgraded to become a FRET recording system. Standard measurements are still required to calculate FRET efficiency but spectral data have some advantages over intensity values. It allows fine corrections for contaminations due to spectral overlaps; it can be selectively applied to plasma membrane or intracellular structures; it contains controls over data quality through direct monitoring of the emission spectrum (which is characteristic for each fluorophore) instead of intensity values; it provides information about changes in local environment around a fluorophore (for instance, change in pH or hydrophobicity), which is essential since local environment can affect fluorescence intensity and be misinterpreted as change in FRET efficiency. Overall, the technique has a potential to become a common lab approach to record FRET from live cells.

References

1. L. Stryer, Fluorescence energy transfer as a spectroscopic ruler, *Annu Rev Biochem* 47: 819–46 (1978).
2. L. Stryer and R.P. Haugland, Energy transfer: a spectroscopic ruler, *Proc Natl Acad Sci USA* 58(2): 719–26 (1967).
3. R.Y. Tsien, The green fluorescent protein, *Annu Rev Biochem* 67: 509–44 (1998).
4. G.W. Gordon, G. Berry, X.H. Liang, B. Levine and B. Herman, Quantitative fluorescence resonance energy transfer measurements using fluorescence microscopy, *Biophys J* 74(5): 2702–13 (1998).
5. M.G. Erickson, B.A. Alseikhan, B.Z. Peterson and D.T. Yue, Preassociation of calmodulin with voltage-gated Ca(2+) channels revealed by FRET in single living cells, *Neuron* 31(6): 973–85 (2001).
6. A.K. Kenworthy, Imaging protein–protein interactions using fluorescence resonance energy transfer microscopy, *Methods* 24(3): 289–96 (2001).
7. F.S. Wouters and P.I. Bastiaens, Imaging protein–protein interactions by fluorescence resonance energy transfer (FRET) microscopy, *Curr Protoc Protein Sci* 19: Unit19 5 (2001).
8. A. Miyawaki and R.Y. Tsien, Monitoring protein conformations and interactions by fluorescence resonance energy transfer between mutants of green fluorescent protein, *Methods Enzymol* 327: 472–500 (2000).

9. F. Festy, S.M. Ameer-Beg, T. Ng and K. Suhling, Imaging proteins in vivo using fluorescence lifetime microscopy, *Mol Biosyst* 3(6): 381–91 (2007).
10. M. Peter and S.M. Ameer-Beg, Imaging molecular interactions by multiphoton FLIM, *Biol Cell* 96(3): 231–6 (2004).
11. D. Elson, J. Requejo-Isidro, I. Munro, F. Reavell, J. Siegel, K. Suhling, P. Tadrous, R. Benninger, P. Lanigan, J. McGinty, C. Talbot, B. Treanor, S. Webb, A. Sandison, A. Wallace, D. Davis, J. Lever, M. Neil, D. Phillips, G. Stamp, and P. French, Time-domain fluorescence lifetime imaging applied to biological tissue, *Photochem Photobiol Sci* 3(8): 795–801 (2004).
12. K. Suhling, P.M. French and D. Phillips, Time-resolved fluorescence microscopy, *Photochem Photobiol Sci* 4(1): 13–22 (2005).
13. M. Cohen-Kashi, Y. Namer and M. Deutsch, Fluorescence resonance energy transfer imaging via fluorescence polarization measurement, *J Biomed Opt* 11(3): 34015 (2006).
14. D.W. Piston and M.A. Rizzo, FRET by fluorescence polarization microscopy, *Methods Cell Biol* 85: 415–30 (2008).
15. M.A. Rizzo and D.W. Piston, High-contrast imaging of fluorescent protein FRET by fluorescence polarization microscopy, *Biophys J* 88(2): L14–6 (2005).
16. C.L. Takanishi, E.A. Bykova, W. Cheng and J. Zheng, GFP-based FRET analysis in live cells, *Brain Res* 1091(1): 132–9 (2006).
17. S.Y. Lee, J.A. Letts and R. Mackinnon, Dimeric subunit stoichiometry of the human voltage-dependent proton channel Hv1, *Proc Natl Acad Sci USA* 105(22): 7692–7695 (2008).
18. F. Tombola, M.H. Ulbrich and E.Y. Isacoff, The voltage-gated proton channel Hv1 has two pores, each controlled by one voltage sensor, *Neuron* 58(4): 546–56 (2008).
19. C. Miller and M.M. White, Dimeric structure of single chloride channels from Torpedo electroplax, *Proc Natl Acad Sci USA* 81(9): 2772–5 (1984).
20. W. Cheng, F. Yang, C.L. Takanishi and J. Zheng, Thermosensitive TRPV channel subunits coassemble into heteromeric channels with intermediate conductance and gating properties, *J Gen Physiol* 129(3): 191–207 (2007).
21. J. Zhang and M.D. Allen, FRET-based biosensors for protein kinases: illuminating the kinome, *Mol Biosyst* 3(11): 759–65 (2007).
22. A. Miyawaki, J. Llopis, R. Heim, J.M. McCaffery, J.A. Adams, M. Ikura and R.Y. Tsien, Fluorescent indicators for Ca2+ based on green fluorescent proteins and calmodulin, *Nature* 388(6645): 882–7 (1997).
23. I.L. Medintz, Recent progress in developing FRET-based intracellular sensors for the detection of small molecule nutrients and ligands, *Trends Biotechnol* 24(12): 539–42 (2006).
24. L. Hodgson, O. Pertz and K.M. Hahn, Design and optimization of genetically encoded fluorescent biosensors: GTPase biosensors, *Methods Cell Biol* 85: 63–81 (2008).
25. E.A. Bykova, X.D. Zhang, T.Y. Chen and J. Zheng, Large movement in the C terminus of CLC-0 chloride channel during slow gating, *Nat Struct Mol Biol* 13(12): 1115–9 (2006).

Boronic Acid Based Modular Fluorescent Saccharide Sensors

John S. Fossey and Tony D. James

Abstract The ability to monitor analytes within physiological, environmental and industrial scenarios is of prime importance. Since recognition events occur on a molecular level, gathering and processing this information pose a fundamental challenge. Therefore, robust chemical molecular sensors with the capacity to detect chosen molecules selectively and signal this presence continue to attract considerable attention. Real-time monitoring of saccharides is of particular interest, in aqueous systems such as D-glucose in blood. The covalent coupling interaction between boronic acids and saccharides has been exploited with some success to monitor the presence of such saccharides.

Keywords Boronic acid · Fluorescent · Sensor · Saccharide · Glucose

Introduction

The challenge of recognising a molecular analyte with a synthetically prepared receptor has inspired many supramolecular chemists. Since inception, research in this area has been instrumental in elucidating the mechanisms of many biological events encompassing recognition and catalysis [1]. The significance of this work was underlined with the award of the Nobel Prize in Chemistry to Cram, Lehn and Peterson in 1987 "for their development and use of molecules with structure-specific interactions of high selectivity" [2]. Since then the diversity of compounds studied within the remit of supramolecular chemistry has significantly grown. Of particular interest are molecular sensors, single molecules which can both recognise and signal analyte presence in real time [3, 4].

Molecular recognition underpins all sensor chemistry; the process involves the interaction between two substances. Importantly, recognition is not just defined as a binding event but requires an element of selectivity between the guest and the host. Optimal selectivity occurs between compounds with carefully matched electronic,

T.D. James (✉)
Department of Chemistry, University of Bath, Bath, BA2 7AY, UK
e-mail: t.d.james@bath.ac.uk

C.D. Geddes (ed.), *Reviews in Fluorescence 2007*, Reviews in Fluorescence 2007,
DOI 10.1007/978-0-387-88722-7_5,

geometric and polar elements. For synthetic receptors the potential exists to engineer receptors for a chosen analyte through selective structural design and complementary functional groups. The power of this concept is illustrated within Nature, where biological systems have evolved with exquisite binding sites, sequestering guest molecules with high selectivity.

A channel of communication must be established between the receptor and the "outside world", thus a receptor becomes a sensor. For a sensor to function it must permit selective binding between host and guest and report these binding events by generating a signal. By performing these fundamental tasks sensors have the potential to relay information concerning the presence and location of important species in a potentially quantifiable manner, acting as a conduit for information about events occurring at the molecular level and our own.

Chemical sensors may be categorised as either biosensors, or synthetic sensors. Biosensors make use of existing biological units for recognition. Many physiologically important analytes already have corresponding biological receptors with intrinsically selectivity, if these receptors can be coupled to a signal transducer a biosensor can be developed [5].

Synthetic sensors incorporate a synthetically manufactured element for recognition. Whilst biomimetic receptors have been prepared with synthetic receptors mimicking the active sites of naturally occurring biological molecules, synthetic receptors can be designed entirely from first principles.

The development of coherent strategies for the selective binding of analyte molecules, by rational design of synthetic receptors, remains one of chemistry's most aspirational goals. Research conducted to this end is driven by a fundamental curiosity and the need to monitor compounds of industrial, environmental and biological importance.

We have exploited the interactions of boronic acids and diols, since the primary interaction of a boronic acid with a diol is covalent and involves the reversible and rapid formation of a cyclic boronate ester. An array of hydroxyl groups presented by saccharides provide an ideal architecture for these interactions and has led to the development of boronic acid based sensors for saccharides [6–23] (Scheme 1).

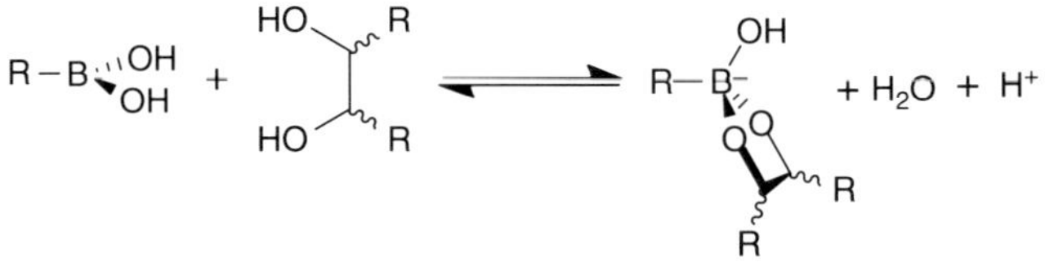

Scheme 1 The rapid and reversible formation of a cyclic boronate ester

Many synthetic receptors developed for neutral guests have relied on non-covalent interactions, such as hydrogen bonding, for recognition, yet in aqueous systems neutral guests may become heavily solvated. Whilst biological systems are able to expel water from their binding pockets to sequester analytes, using non-covalent interactions, synthetic monomeric receptors have not yet been designed where hydrogen bonding has been able to compete with water (solvent) for low concentrations of monosaccharides [24]. However, it should be noted that progress has been made in this area; Davis reported a hydrogen bonding receptor capable of binding D-glucose in water with a weak but significant stability constant [25]. More recently Davis has been able to develop a synthetic lectin analog for biomimetic disaccharide recognition which works very well in water [26]. The interaction of boronic acids with saccharides and diols is well reported; we have exploited it in a range of applications as diverse as NMR shift reagents [27–31], functional polymers for electrophoresis [32] and molecular capsules derived from the facile boronic acid – diol interactions [33, 34].

The most popular class of the boronic acid based sensors utilise an amine group proximal to boron coupled to a fluorescence output. The Lewis acid–Lewis base interaction between the boronic acid and the tertiary amine has a dual role. Firstly, it enables molecular recognition at neutral pH. Secondly, it can be used to signal binding by modulating the intensity of fluorescence emissions.

Fluorescent Sensors

Optical signals convey information through space'; fluorescent sensors can be used in dynamic systems, such as living tissue, and relay information remotely. Submillisecond response times are usual, allowing information to be communicated in essentially real-time. If targeted correctly, fluorophores can be located with subnanometre accuracy, in effect permitting real-space monitoring [35]. Fluorescence also demonstrates exceptionally high sensitivity; under controlled conditions detection of responses from single fluorescent molecules [36–38] and in the case of fluorescent sensors, from single guest molecules [39].

As fluorescent sensors are capable of reporting a wealth of physical information at low concentrations (micromolar concentrations are typical), they can operate with the minimum disruption to the system being investigated. From an analytical perspective these characteristics are attractive and commercially the sparing quantities of compound required can offset synthetic costs.

Fluorescent sensors can be found in many recent analytical advances, such as the continuous monitoring systems developed by immobilising fluorescent sensors onto fibre optic sensing arrays [40] or the live imaging of analytes within cells through confocal microscopy [41]. Commercially fluorescent sensors include clinical tools such as the blood gas analyzers that are now commonplace within hospital high-dependency wards and ambulances allowing point-of-care diagnostic monitoring [42–45] or the glucose responsive contact lenses currently being pioneered by the Lakowicz research group [46]. These examples underline the robust and adaptable nature of fluorescent sensors, which in turn permit rapid and accurate analyte detection by portable devices.

The use of boronic acids in the development of fluorescent sensors for saccharides is a comparatively new field. Following the first report by Czarnik in 1992 [47], D-glucose selectivity was achieved in 1994 by Fossey and Shinkai [48]. A year later this was followed up by enantioselective saccharide recognition [49]. The intervening years have seen the field grow to the point where hundreds of publications now report on boronic acid – saccharide recognition [19–23] (Scheme 2).

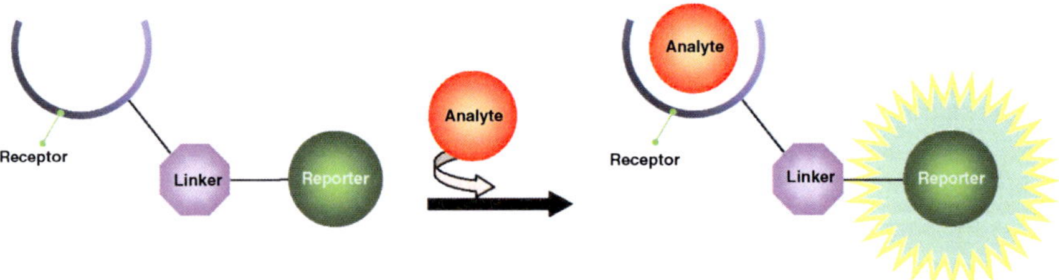

Scheme 2 The complementary interaction between a guest analyte and a host binding pocket, illustrated here by a *red–orange* guest analyte and *blue* host, allows selective binding to occur between two elements. Linking a unit capable of reporting this binding event, converts the receptor into a sensor. In this figure an optical "off–on" response is depicted from an appended fluorophore, illustrated in *green*

For fluorescent sensors design is often discussed in relation to a receptor and a fluorophore separated by a spacer so as to match the Donor-Bridge-Acceptor motif. For saccharide recognition via boronic acid complexation, the interaction between *o*-methylphenylboronic acids (Lewis acidic) and proximal tertiary amines (Lewis basic) has been exploited.

Whilst elucidating the precise nature of the amino base–boronic acid (N$\cdots$B) interaction has been debated, it is clear that an interaction exists which provides two distinct advantages [22, 23, 50, 51].

First Wulff proposed that the interaction between a boronic acid and proximal amine lowers the pK_a of the boronic acid [52] allowing binding to occur at neutral pH, useful in some biological scenarios.

Second is the contraction of the O–B–O bond angle upon complexation with a saccharide and the associated increase in acidity at the boron centre. The increase in acidity of the already Lewis acidic boron enhances the N$\cdots$B interaction which in turn influences the fluorescence. The reduction in pK_a at boron on saccharide binding has the net effect of modulating fluorescence intensity.

HO B OH N

1

The first fluorescent sensor for saccharides to employ the *N*-methyl-*o*-(aminomethyl)phenylboronic acid fragment was reported by Fossey et al. in 1994, sensor **1** [53, 54]. A large increase in fluorescence on addition of saccharide was observed, over a broad pH range. The monoboronic acid **1** displayed the same trend in selectivity for saccharide binding preference as that reported by Lorand and Edwards for phenylboronic acid 35 years earlier, a trend which appears to be inherent to all monoboronic acid receptors of this type. Qualitatively, binding constants (K) with monoboronic acids increase as follows: D-glucose < D-galactose < D-fructose [55].

Modular Fluorescent Sensors

In the design of boronic acid based sensor systems it has been established that two receptor units are required if saccharide selectivity is to be achieved [54]. Retaining the same dual boronic acid recognition units throughout, a modular system in which the linker and fluorophore units of these sensors could be modified independently needed to be developed. Altering the dimensions of the binding pocket permit the emission wavelength of the systems to be altered and provided conclusive data on the role of individual variables by tuning them in a controlled manner. Producing a range of modular sensors in order to derive quantitative trends must employ a generic scaffold and adhered to throughout, if a meaningful comparison is to be obtained across the series.

Whilst sensors developed around an anthracene core unit have proved to be selective for saccharides such as D-glucose [48,54], the rigid core unit which acts triply as scaffold, linker and fluorophore limits the modifications that can be made to any one part of the system without influencing the sensor as a whole.

Compound **2** represents a generic template on which to develop saccharide selective sensors. The design includes two boronic acid groups required for selectivity but allows the separation between them to be varied by altering the linker. It also permits the fluorophore to be varied independently and using only one fluorophore overcomes the problems that may arise from excimer emission, insolubility, excessive hydrophobicity and steric crowding at the binding pocket.

Modular PET sensors $\mathbf{3}_{(n=3)}$–$\mathbf{8}_{(n=8)}$ contained two phenylboronic acid groups, a pyrene fluorophore and a variable linker. The linker was varied from trimethylene $\mathbf{3}_{(n=3)}$ to octamethylene $\mathbf{8}_{(n=8)}$ [56, 57].

The fluorescence titrations of $\mathbf{3}_{(n=3)}$–$\mathbf{8}_{(n=8)}$ and $\mathbf{9}_{(\text{pyrene})}$ (1.0×10^{-7} mol dm^{-3}) with various saccharides was carried out in a pH 8.21 aqueous methanolic buffer solution [58]. The fluorescence intensity of sensors $\mathbf{3}_{(n=3)}$–$\mathbf{8}_{(n=8)}$ and $\mathbf{9}_{(\text{pyrene})}$ increased with increasing saccharide concentration. The observed stability constants (K_{obs}) of PET sensors $\mathbf{3}_{(n=3)}$–$\mathbf{8}_{(n=8)}$ and $\mathbf{9}_{(\text{pyrene})}$ were calculated by fitting the emission intensity at 397 nm versus saccharide concentration and are given in Table 1.

To help visualise the trends in the observed stability constants (K_{obs}) documented in Table 1, the observed stability constants (K_{obs}) of the diboronic acid sensors $\mathbf{3}_{(n=3)}$–$\mathbf{8}_{(n=8)}$ are reported in Fig. 1 divided by (i.e. relative to) the observed stability constants (K_{obs}) of their equivalent monoboronic acid analogue $\mathbf{9}_{(\text{pyrene})}$. In most cases, the observed stability constants (K_{obs}) with diboronic acid sensors $\mathbf{3}_{(n=3)}$–$\mathbf{8}_{(n=8)}$ are higher than for the monoboronic acid sensor $\mathbf{9}_{(\text{pyrene})}$.

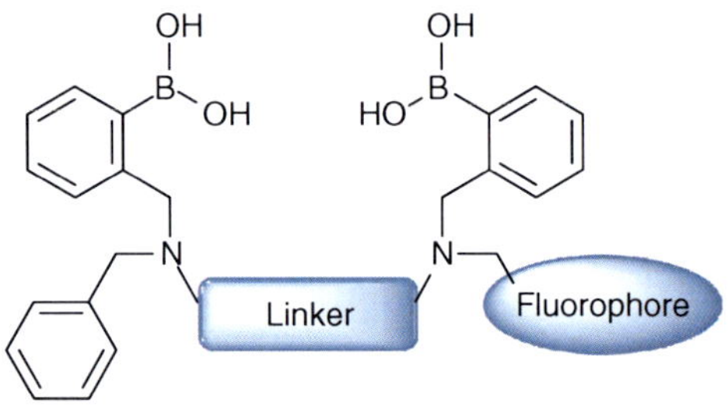

2

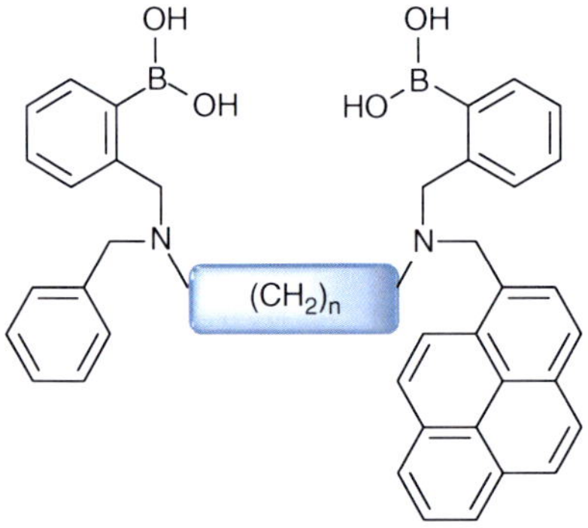

$\mathbf{3}_{(n=3)}$, $\mathbf{4}_{(n=4)}$, $\mathbf{5}_{(n=5)}$, $\mathbf{6}_{(n=6)}$, $\mathbf{7}_{(n=7)}$, $\mathbf{8}_{(n=8)}$ $\mathbf{9}_{(pyrene)}$

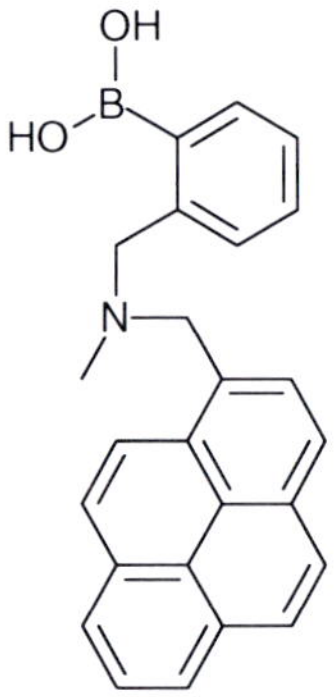

It is known that D-glucose and D-galactose will bind to diboronic acids readily using two sets of diols, thus forming stable, cyclic 1:1 complexes. The allosteric binding of the two boronic acid groups is clearly illustrated by the relative difference between the observed stability constants (K_{obs}) of the equivalent di- and monoboronic acid compounds. The observed stability constants (K_{obs}) of the diboronic acid sensors are up to ~22 times larger than with their monoboronic acid counterparts, see Fig. 1.

The observed stability constants (K_{obs}) for the diboronic acid sensors with D-fructose and D-mannose are up to twice as strong as with the monoboronic acid sensor $\mathbf{9}_{(pyrene)}$. Each D-fructose and D-mannose molecule will only bind to one boronic acid unit through one diol. This allows complexes to form with overall 2:1 (saccharide/sensor) stoichiometry. The relative values of ca. 2 are indicative of two independent saccharide binding events on each sensor, with no concomitant increase in stability derived from cooperative binding.

The highest observed stability constants (K_{obs}) for D-glucose within these systems was obtained by sensor $\mathbf{6}_{(n=6)}$. The flexible six carbon linker provided the optimal selectivity for D-glucose over other saccharides.

Curiously there is an inversion in the selectivity displayed by these systems upon moving from a six to a seven carbon linker. From Fig. 1 it is evident that the trimethylene-linked $\mathbf{3}_{(n=3)}$ shows little specificity between D-glucose and

Table 1 Quantum yield (qFM) values for compounds $\mathbf{3}_{(n=3)}$–$\mathbf{8}_{(n=8)}$ and $\mathbf{9}_{(\text{pyrene})}$ in the absence of saccharides. Observed stability constant (K_{obs}) (determination of coefficient, r^2) and fluorescence enhancements for compounds $\mathbf{3}_{(n=3)}$–$\mathbf{8}_{(n=8)}$ and $\mathbf{9}_{(\text{pyrene})}$ with monosaccharides. The system with the highest observed stability constant (K_{obs}) is highlighted in bold

		D-glucose		D-galactose	
Sensor	qFM	K_{obs}/dm^3 mol^{-1}	Fluorescence enhancement	K_{obs}/dm^3 mol^{-1}	Fluorescence enhancement
$\mathbf{3}_{(n=3)}$	0.16	103 ± 3 (1.00)	3.9	119 ± 5 (1.00)	3.5
$\mathbf{4}_{(n=4)}$	0.16	295 ± 11 (1.00)	3.3	222 ± 17 (1.00)	3.7
$\mathbf{5}_{(n=5)}$	0.20	333 ± 27 (1.00)	3.4	177 ± 15 (1.00)	3.0
$\mathbf{6}_{(n=6)}$	**0.24**	**962 ± 70 (0.99)**	**2.8**	657 ± 39 (1.00)	3.1
$\mathbf{7}_{(n=7)}$	0.16	336 ± 30 (0.98)	3.0	542 ± 41 (0.99)	2.9
$\mathbf{8}_{(n=8)}$	0.19	368 ± 21 (1.00)	2.3	562 ± 56 (0.99)	2.3
$\mathbf{9}_{(\text{pyrene})}$	0.17	44 ± 3 (1.00)	4.5	51 ± 2 (1.00)	4.2
		D-fructose		D-mannose	
Sensor	qFM	K_{obs}/dm^3 mol^{-1}	Fluorescence enhancement	K_{obs}/dm^3 mol^{-1}	Fluorescence enhancement
$\mathbf{3}_{(n=3)}$	0.16	95 ± 9 (0.99)	3.6	45 ± 4 (1.00)	2.7
$\mathbf{4}_{(n=4)}$	0.16	266 ± 28 (0.99)	4.2	39 ± 1 (1.00)	3.4
$\mathbf{5}_{(n=5)}$	0.20	433 ± 19 (1.00)	3.4	48 ± 2 (1.00)	3.0
$\mathbf{6}_{(n=6)}$	0.24	784 ± 44 (1.00)	3.2	74 ± 3 (1.00)	2.8
$\mathbf{7}_{(n=7)}$	0.16	722 ± 37 (1.00)	3.3	70 ± 5 (1.00)	2.7
$\mathbf{8}_{(n=8)}$	0.19	594 ± 56 (0.99)	2.3	82 ± 3 (1.00)	2.2
$\mathbf{9}_{(\text{pyrene})}$	0.17	395 ± 11 (1.00)	3.6	36 ± 1 (1.00)	3.7

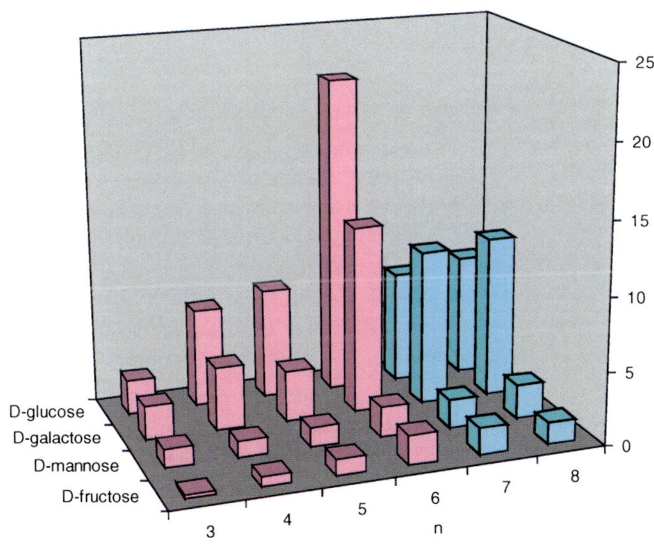

Fig. 1 Observed stability constants (K_{obs}) of $\mathbf{3}_{(n=3)}$–$\mathbf{8}_{(n=8)}$ divided by the observed stability constants ($\underline{K}_{obs}$) of $\mathbf{9}_{(\text{pyrene})}$, to yield relative values with saccharides

D-galactose. Increasing the size of the binding pocket, tetramethylene $\mathbf{4}_{(n=4)}$ through to hexamethylene $\mathbf{6}_{(n=6)}$, induces a clear selectivity for D-glucose, with $\mathbf{6}_{(n=6)}$ providing the strongest binding. However, there is an inversion in this selectivity on increasing the linker length to heptamethylene $\mathbf{7}_{(n=7)}$ and octamethylene $\mathbf{8}_{(n=8)}$, with the enlarged binding pocket being D-galactose selective.

The difference in configuration of these diastereomers was initially thought to be the origin of the selectivity differences [57]. For example the 1,2- and 4,6-diols of D-glucose are oriented in the same direction (down), but in D-galactose (the 4-epimer of D-glucose), the 1,2-diol is down and the 4,6-diol is up.

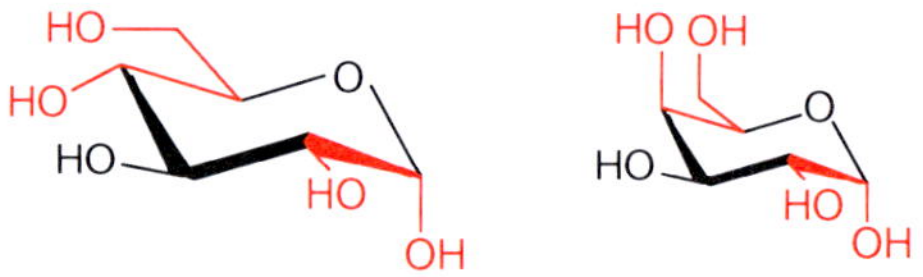

glucose galactose

Consideration of the saccharidic forms where a syn-periplanar arrangement of the anomeric hydroxyl pair can be attained [59]. Generally this requires formation of the furanose form of the saccharide. However, *in silico* analyses has revealed that in the case of D-galactose the α-D-furanose form of the saccharide is not the only species that can be considered with a syn-periplanar alignment of the anomeric hydroxyl pair [59].

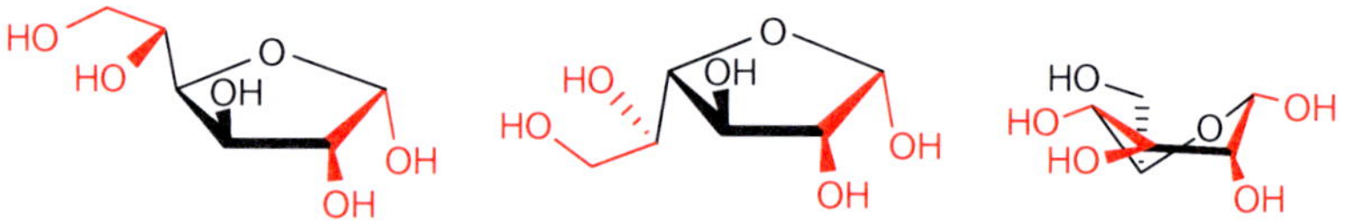

α-D-glucofuranose α-D-galactofuranose α-D-galactopyranose (twist boat)

On complexation of a boronic acid to the diol in the 1,2-position of α-D-galactopyranose a reorientation of the saccharide can afford an acetal ring with a boat or twisted boat conformation. This reorientation permits a second boronic acid group to bind with the diol in the 3,4-position producing an energetically stable complex with the pyranose form of the saccharide [60].

In light of this data it seems plausible that the preference of D-galactose for a slightly larger binding pocket than D-glucose can be ascribed to the stability of the preferred ring form. D-Galactose can generate complexes with boronic acids in its pyranose form, whereas D-glucose will prefer the smaller furanose ring form.

The approach to use modular or core scaffolds for saccharide recognition has also been championed by Hall [61, 62], Wang [63–66] and Singaram [67–80].

Having developed the basic saccharide selective scaffold our current aim is to develop synthetic routes through which we can rapidly access libraries of sensors and then screen for saccharide selectivity. Towards that end we have recently developed a convenient fluorescent boronic acid unit "click fluor" assembled via

Huisgen [3+2] cycloadditions [81], which are ideally suited to a modular synthetic approach towards sensors [82]. The so-called "*click reaction*" [83, 84] forms an aromatic 1,2,3-triazole ring following the addition of an azide to a terminal alkyne has the potential to create a fluorescent sensor from non-fluorescent constituent parts. Fluorescent boronic acids have also been prepared using Huisgen [3+2] cycloadditions by Smietana and Vasseur who employed the Seyferth–Gilbert procedure in a one pot process to generate alkynes from boronic acid aldehydes which were then coupled in a Huisgen [3+2] cycloaddition with fluorophore azides [85]. Wang has prepared acetylene substituted boronic acids as potential [3+2] cycloaddition units [86] and suggested the use as units for *click reactions*. Wang has also recently used [3+2] cycloaddition reactions to prepare a boronic-acid-labelled thymidine triphosphate for incorporation into DNA [87].

In our preliminary investigations fluorescent sensor **10** was prepared employing a copper (I) catalysed azide-alkyne [3+2] cycloaddition developed by Ham [88] enabling the synthesis of 1,2,3-triazole ring as predominantly the 1,4-regioisomer. At this point during the synthesis X-ray quality crystals of the pinacol protected intermediate **11** were obtained. The target sensor **10** was then obtained via a two-step deprotection of the pinacol ester [89, 90].

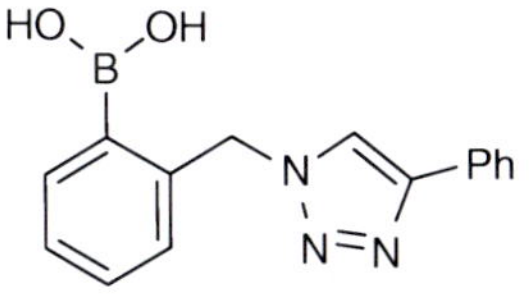

HO B OH N N N Ph

10

O2 O1 B1 N1 N2 N3

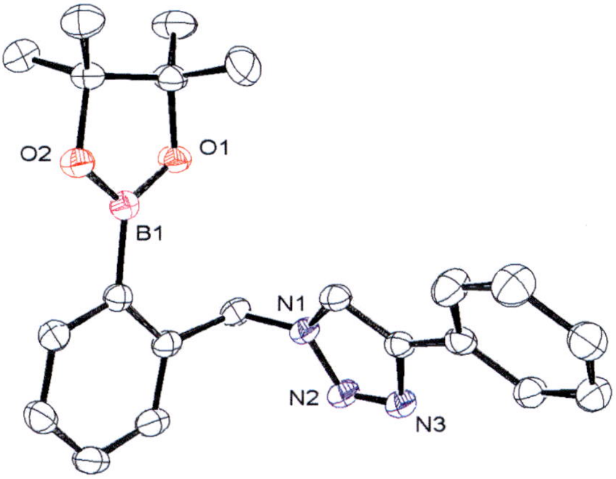

11

The fluorescence titrations of **10** (1.67×10^{-6} mol dm^{-3}) with different saccharides were carried out in a pH 8.21 buffer (52.1 wt% methanol in water with KCl, 0.01000 mol dm^{-3}; KH_2PO_4, 0.002752 mol dm^{-3}; Na_2HPO_4, 0.002757 mol dm^{-3}). The fluorescence spectra of **10** in the presence of D-fructose (0–2.2 mol dm^{-3}) are shown in Fig. 2.

The fluorescence enhancements (I/I_0) obtained for **10** on the addition of D-fructose, D-galactose and D-mannose are 27, 20 and 16 fold, respectively. We believe that these large fluorescence enhancements can be attributed to fluorescence recovery of the 1,2,3-triazole fluorophore [91–93]. In the absence of saccharides the normal fluorescence of the LE (Locally Excited) state of the 1,2,3-triazole donor of

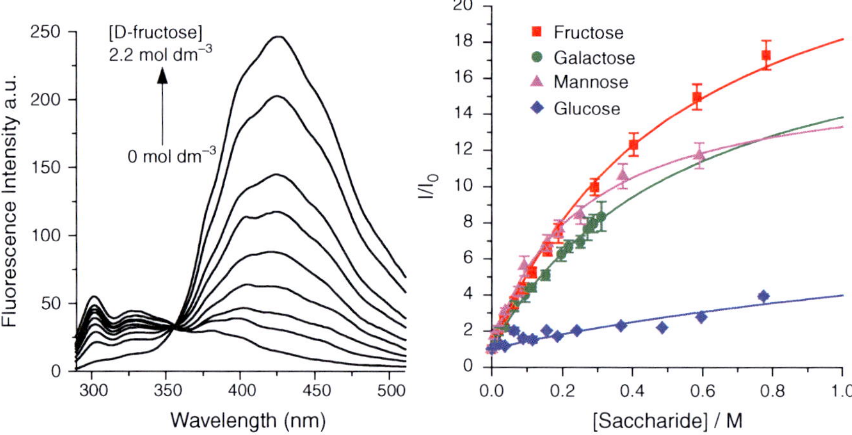

Fig. 2 Fluorescence spectra of **10** in the presence of increasing concentrations of D-fructose, with the peak at 300 nm decreasing in intensity and the peak at 430 nm increasing in intensity, resulting in an isoemissive point at 360 nm (*left*) and binding curves for ■ D-fructose, ◇ D-glucose, • D-galactose and D-mannose (*right*)

sensor **10** is quenched by energy transfer to the neutral phenylboronic acid acceptor, weakly Lewis acidic boron centre. When saccharides are added, a negatively charged boronate anion is formed due to the enhancement of the Lewis acidity of the boron centre on saccharide binding. Under these conditions, energy transfer from the 1,2,3-triazole donor becomes unfavourable and fluorescence is recovered.

We coined the phrase "*click-fluor*" to describe the generation of a fluorophore from non-fluorescent constituent parts via a so-called "*click reaction*". In this case, the triazole ring forms an integral part of the fluorophore, i.e. a new property is imparted on a molecule conceived by a "*click reaction*". Using our simple "*click-fluor*" unit we believe that it will become possible to develop fluorescent modular sensor arrays for rapid screening of target saccharides. The two most attractive aspects of "*click-fluor*" are that a fluorophore is generated upon triazole formation and the wide availability of acetylene units.

Although they have yet to be developed into a fluorescent sensor systems we envision the use of molecular self-assembled systems as an important step for the next generation of fluorescent sensors. In particular collaborative work with the Kubo group has resulted in ion pair-driven heterodimeric capsule [33]. The system consists of cyclotricatechylene and a boronic aciD-appended hexahomotrioxacalix[3]arene (Scheme 3). The two components do not interact with each other until Et_4NAcO is added to the solution. On addition of Et_4NAcO quantitative formation a capsule by boronate esterification is observed. The self-assembly process is a direct result of anion directed boronate ester formation and the Et_4N^+ template. Reversible boronate esterification also allowed for selective control of capsule formation as a function of pH (Scheme 3).

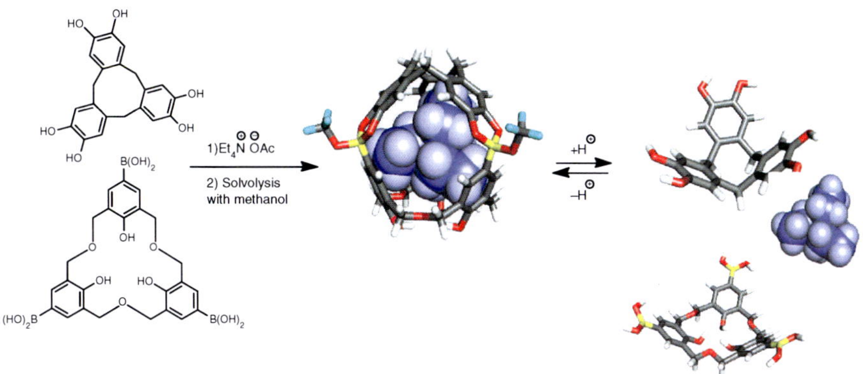

Scheme 3 Anion driven capsule formation

Conclusions

The development of consistent and logical strategies for the selective binding of saccharides, by rational design of synthetic receptors, is one of chemistry's most intriguing and rewarding goals. We believe that the best way forward in the development of new modular sensors will be to employ some basic design criteria but also to use synthetic aids such as reliable so-called "*click-reactions*" and physical aids such as "self-assembly". Boronic acid molecular sensors have proven their worth in the development of glucose selective sensors and receptors; the next challenge will be to develop sensors for the diverse array of other biologically significant saccharides.

References

1. G. R. Desiraju, Chemistry Beyond the Molecule, *Nature*, 412, 397–400, 2001.
2. B. G. Malmström, *Nobel Lectures in Chemistry (1991–1995)*, World Scientific: Singapore, 1997.
3. A. P. de Silva, H. Q. N. Gunaratne, T. Gunnlaugsson, A. J. M. Huxley, C. P. McCoy, J. T. Rademacher and T. E. Rice, Signaling Recognition Events with Fluorescent Sensors and Switches, *Chemical Reviews*, 97, 1515–1566, 1997.
4. A. W. Czarnik, *Fluorescent Chemosensors for Ion and Molecule Recognition*, American Chemical Society Books: Washington DC, 1993.
5. R. Jelinek and S. Kolusheva, Carbohydrate Biosensors, *Chemical Reviews*, 104, 5987–6015, 2004.
6. T. D. James, P. Linnane and S. Shinkai, Fluorescent Saccharide Receptors: A Sweet Solution to the Design, Assembly and Evaluation of Boronic Acid Derived Pet Sensors, *Chemical Communications*, 281–288, 1996.
7. T. D. James, K. R. A. S. Sandanayake and S. Shinkai, Saccharide Sensing with Molecular Receptors Based on Boronic Acid, *Angewandte Chemie-International Edition in English*, 35, 1911–1922, 1996.

8. M. Granda-Valdes, R. Badia, G. Pina-Luis and M. E. Diaz-Garcia, Photoinduced Electron Transfer Systems and Their Analytical Application in Chemical Sensing, *Quimica Analitica (Barcelona)*, 19, 38–53, 2000.
9. W. Wang, X. Gao and B. Wang, Boronic AciD-Based Sensors, *Current Organic Chemistry*, 6, 1285–1317, 2002.
10. T. D. James and S. Shinkai, Artificial Receptors as Chemosensors for Carbohydrates, *Topics in Current Chemistry*, 218, 159–200, 2002.
11. S. Striegler, Selective Carbohydrate Recognition by Synthetic Receptors in Aqueous Solution, *Current Organic Chemistry*, 7, 81–102, 2003.
12. H. Cao and M. D. Heagy, Fluorescent Chemosensors for Carbohydrates: A Decade's Worth of Bright Spies for Saccharides in Review, *Journal of Fluorescence*, 14, 569–584, 2004.
13. H. Fang, G. Kaur and B. Wang, Progress in Boronic AciD-Based Fluorescent Glucose Sensors, *Journal of Fluorescence*, 14, 481–489, 2004.
14. E. A. Moschou, B. V. Sharma, S. K. Deo and S. Daunert, Fluorescence Glucose Detection: Advances toward the Ideal in vivo Biosensor, *Journal of Fluorescence*, 14, 535–547, 2004.
15. S. Shinkai and M. Takeuchi, Molecular Design of Synthetic Receptors with Dynamic, Imprinting, and Allosteric Functions, *Biosensors & Bioelectronics*, 20, 1250–1259, 2004.
16. M. D. Phillips and T. D. James, Boronic Acid Based Modular Fluorescent Sensors for Glucose, *Journal of Fluorescence*, 14, 549–559, 2004.
17. S. Shinkai and M. Takeuchi, Molecular Design of Synthetic Receptors with Dynamic, Imprinting, and Allosteric Functions, *Bulletin of the Chemical Society of Japan*, 78, 40–51, 2005.
18. J. Yan, H. Fang and B. Wang, Boronolectins and Fluorescent Boronolectins: An Examination of the Detailed Chemistry Issues Important for the Design, *Medicinal Research Reviews*, 25, 490–520, 2005.
19. A. P. Davis and T. D. James, In *Functional Synthetic Receptors*; Schrader, T., Hamilton, A., Eds.; Wiley-VCH: Weinheim, 2005, 45–109.
20. T. D. James, In *Boronic Acids*; Hall, D. G., Ed.; Wiley-VCH: Weinheim, 2005, 441–480.
21. T. D. James and S. Shinkai, In *Topics in Fluorescence Spectroscopy*; Geddes, C. D., Lakowicz., J. R., Eds.; Springer: New York, 2005; Vol. 10, 41–67.
22. T. D. James, M. D. Phillips and S. Shinkai, *Boronic Acids in Saccharide Recognition*, RSC: Cambridge, 2006.
23. T. D. James, Saccharide-Selective Boronic Acid Based Photoinduced Electron Transfer (PET) Fluorescent Sensors, *Topics in Current Chemistry*, 277, 107–152, 2007.
24. A. P. Davis and R. S. Wareham, Carbohydrate Recognition through Noncovalent Interactions: A Challenge for Biomimetic and Supramolecular Chemistry, *Angewandte Chemie-International Edition*, 38, 2978–2996, 1999.
25. E. Klein, M. P. Crump and A. P. Davis, Carbohydrate Recognition in Water by a Tricyclic Polyamide Receptor, *Angewandte Chemie-International Edition*, 44, 298–302, 2005.
26. Y. Ferrand, M. P. Crump and A. P. Davis, A Synthetic Lectin Analog for Biomimetic Disaccharide Recognition, *Science* (*Washington, DC, United States*), 318, 619–622, 2007.
27. A. M. Kelly, Y. Perez-Fuertes, S. Arimori, S. D. Bull and T. D. James, Simple Protocol for NMR Analysis of the Enantiomeric Purity of Diols, *Organic Letters*, 8, 1971–1974, 2006.
28. Y. Perez-Fuertes, A. M. Kelly, A. L. Johnson, S. Arimori, S. D. Bull and T. D. James, Simple Protocol for NMR Analysis of the Enantiomeric Purity of Primary Amines, *Organic Letters*, 8, 609–612, 2006.
29. A. M. Kelly, S. D. Bull and T. D. James, Simple Chiral Derivatisation Protocols for NMR Analysis of the Enantiopurity of 1,2-Diphenylethane-1,2-Diamine and N-Boccyclohexane-1,2-Diamine, *Tetrahedron-Asymmetry*, 19, 489–494, 2008.
30. A. M. Kelly, Y. Pérez-Fuertes, J. S. Fossey, S. L. Yeste, S. D. Bull and T. D. James, Simple Protocols for NMR Analysis of the Enantiomeric Purity of Chiral Diols, *Nature Protocols*, 3, 215–219, 2008.
31. Y. Pérez-Fuertes, A. M. Kelly, J. S. Fossey, M. E. Powell, S. D. Bull and T. D. James, Simple Protocols for NMR Analysis of the Enantiomeric Purity of Chiral Primary Amines, *Nature Protocols*, 3, 210–214, 2008.

32. T. R. Jackson, J. S. Springall, D. Rogallc, N. Masumoto, H. C. Li, F. D'Hooge, S. P. Perera, A. T. A. Jenkins, T. D. James, J. S. Fossey and J. M. H. van den Elsen, Boronate Affinity Saccharide Electrophoresis (BASE): A Novel Saccharide Analysis Tool, *Electrophoresis*, 29, 4185–4191, 2008.
33. K. Kataoka, T. D. James and Y. Kubo, Ion Pair-Driven Heterodimeric Capsule Based on Boronate Esterification: Construction and the Dynamic Behavior, *Journal of the American Chemical Society*, 129, 15126–15127, 2007.
34. N. Fujita, S. Shinkai and T. D. James, Boronic Acids in Molecular Self-Assembly, *Chemistry - An Asian Journal*, 3, 1076–1091, 2008.
35. R. A. Bissell, A. P. de Silva, H. Q. N. Gunaratne, P. L. M. Lynch, G. E. M. Maguire and K. R. A. S. Sandanayake, Molecular Fluorescent Signaling with Fluor Spacer Receptor Systems – Approaches to Sensing and Switching Devices *via* Supramolecular Photophysics, *Chemical Society Reviews*, 21, 187–195, 1992.
36. M. Böhmer and J. Enderlein, Fluorescence Spectroscopy of Single Molecules under Ambient Conditions: Methodology and Technology, *ChemPhysChem*, 4, 793–808, 2003.
37. W. E. Moerner and D. P. Fromm, Methods of Single-Molecule Fluorescence Spectroscopy and Microscopy, *Review of Scientific Instruments*, 74, 3597–3619, 2003.
38. W. P. Ambrose, P. M. Goodwin, J. H. Jett, A. van Orden, J. H. Werner and R. A. Keller, Single Molecule Fluorescence Spectroscopy at Ambient Temperature, *Chemical Reviews*, 99, 2929–2956, 1999.
39. M. Sauer, Single-Molecule-Sensitive Fluorescent Sensors Based on Photoinduced Intramolecular Charge Transfer, *Angewandte Chemie-International Edition*, 42, 1790–1793, 2003.
40. J. R. Epstein and D. R. Walt, Fluorescence-Based Fibre Optic Arrays: A Universal Platform for Sensing, *Chemical Society Reviews*, 32, 203–214, 2003.
41. M. Fehr, S. Lalonde, D. W. Ehrhardt and W. B. Frommer, Live Imaging of Glucose Homeostasis in Nuclei of Cos-7 Cells, *Journal of Fluorescence*, 14, 603–609, 2004.
42. A. P. de Silva, H. Q. N. Gunaratne, T. Gunnlaugsson and M. Nieuwenhuizen, Fluorescent Switches with High Selectivity Towards Sodium Ions: Correlation of Ion-Induced Conformation Switching with Fluorescence Function, *Chemical Communications*, 1967–1968, 1996.
43. H. R. He, M. A. Mortellaro, M. J. P. Leiner, S. T. Young, R. J. Fraatz and J. K. Tusa, A Fluorescent Chemosensor for Sodium Based on Photoinduced Electron Transfer, *Analytical Chemistry*, 75, 549–555, 2003.
44. A. J. Tudos, G. A. J. Besselink and R. B. M. Schasfoort, Trends in Miniaturized Total Analysis Systems for Point-of-Care Testing in Clinical Chemistry, *Lab on a Chip*, 1, 83–95, 2001
45. H. Schlebusch, I. Paffenholz, R. Zerback and R. Leinberger, Analytical Performance of a Portable Critical Care Blood Gas Analyzer, *Clinica Chimica Acta*, 307, 107–112, 2001.
46. R. Badugu, J. R. Lakowicz and C. D. Geddes, Boronic Acid Fluorescent Sensors for Monosaccharide Signaling Based on the 6-Methoxyquinolinium Heterocyclic Nucleus: Progress toward Noninvasive and Continuous Glucose Monitoring, *Bioorganic and Medicinal Chemistry*, 13, 113–119, 2004.
47. J. Yoon and A. W. Czarnik, Fluorescent Chemosensors of Carbohydrates – A Means of Chemically Communicating the Binding of Polyols in Water Based on Chelation-Enhanced Quenching, *Journal Of the American Chemical Society*, 114, 5874–5875, 1992.
48. T. D. James, K. Sandanayake and S. Shinkai, A Glucose-Selective Molecular Fluorescence Sensor, *Angewandte Chemie-International Edition In English*, 33, 2207–2209, 1994.
49. T. D. James, K. R. A. S. Sandanayake and S. Shinkai, Chiral Discrimination of Monosaccharides Using a Fluorescent Molecular Sensor, *Nature*, 374, 345–347, 1995.
50. L. Zhu, S. H. Shabbir, M. Gray, V. M. Lynch, S. Sorey and E. V. Anslyn, A Structural Investigation of the N-B Interaction in an *o*-(*N,N*-Dialkylaminomethyl)Arylboronate System, *Journal of the American Chemical Society*, 128, 1222–1232, 2006.
51. L. I. Bosch, T. M. Fyles and T. D. James, Binary and Ternary Phenylboronic Acid Complexes with Saccharides and Lewis Bases, *Tetrahedron*, 60, 11175–11190, 2004.
52. G. Wulff, Selective Binding to Polymers *via* Covalent Bonds – the Construction of Chiral Cavities as Specific Receptor-Sites, *Pure and Applied Chemistry*, 54, 2093–2102, 1982.

53. T. D. James, K. R. A. S. Sandanayake and S. Shinkai, Novel Photoinduced Electron-Transfer Sensor for Saccharides Based on the Interaction of Boronic Acid and Amine, *Journal of the Chemical Society, Chemical Communications*, 477–478, 1994.
54. T. D. James, K. R. A. S. Sandanayake, R. Iguchi and S. Shinkai, Novel Saccharide-Photoinduced Electron-Transfer Sensors Based on the Interaction of Boronic Acid and Amine, *Journal of the American Chemical Society*, 117, 8982–8987, 1995.
55. J. P. Lorand and J. O. Edwards, Polyol Complexes and Structure of the Benzeneboronate Ion, *Journal of Organic Chemistry*, 24, 769–774, 1959.
56. S. Arimori, M. L. Bell, C. S. Oh, K. A. Frimat and T. D. James, Modular Fluorescence Sensors for Saccharides, *Chemical Communications (Cambridge, United Kingdom)*, 1836–1837, 2001.
57. S. Arimori, M. L. Bell, C. S. Oh, K. A. Frimat and T. D. James, Modular Fluorescence Sensors for Saccharides, *Journal of the Chemical Society, Perkin Transactions* 1, 803–808, 2002.
58. D. D. Perrin and B. Dempsey, *Buffers for Ph and Metal Ion Control*, Chapman & Hall: London, 1974.
59. M. P. Nicholls and P. K. C. Paul, Structures of Carbohydrate-Boronic Acid Complexes Determined by Nmr and Molecular Modeling in Aqueous Alkaline Media, *Organic and Biomolecular Chemistry*, 2, 1434–1441, 2004.
60. C. R. Cooper and T. D. James, Selective Fluorescence Signalling of Saccharides in Their Furanose Form, *Chemistry Letters*, 883–884, 1998.
61. D. G. Hall, Editor, Boronic Acids: Preparation and Applications in Organic Synthesis and Medicine, Wiley: New York, 2005.
62. D. Stones, S. Manku, X. Lu and D. G. Hall, Modular SoliD-Phase Synthetic Approach to Optimize Structural and Electronic Properties of Oligo-Boronic Acid Receptors and Sensors for the Aqueous Recognition of Oligosaccharides, *Chemistry – A European Journal*, 10, 92–100, 2004.
63. W. Yang, H. Fan, X. Gao, S. Gao, V. V. R. Karnati, W. Ni, W. B. Hooks, J. Carson, B. Weston and B. Wang, The First Fluorescent Diboronic Acid Sensor Specific for Hepatocellular Carcinoma Cells Expressing Sialyl Lewis X, *Chemistry and Biology*, 11, 439–448, 2004.
64. W. Yang, S. Gao, X. Gao, V. V. R. Karnati, W. Ni, B. Wang, W. B. Hooks, J. Carson and B. Weston, Diboronic Acids as Fluorescent Probes for Cells Expressing Sialyl Lewis X, *Bioorganic and Medicinal Chemistry Letters*, 12, 2175–2177, 2002.
65. V. V. Karnati, X. Gao, S. Gao, W. Yang, W. Ni, S. Sankar and B. Wang, A Glucose-Selective Fluorescence Sensor Based on Boronic AciD-Diol Recognition, *Bioorganic and Medicinal Chemistry Letters*, 12, 3373–3377, 2002.
66. G. Kaur, H. Fang, X. Gao, H. Li and B. Wang, Substituent Effect on Anthracene-Based Bisboronic Acid Glucose Sensors, *Tetrahedron*, 62, 2583–2589, 2006.
67. J. N. Camara, J. T. Suri, F. E. Cappuccio, R. A. Wessling and B. Singaram, Boronic Acid Substituted Viologen Based Optical Sugar Sensors: Modulated Quenching with Viologen as a Method for Monosaccharide Detection, *Tetrahedron Letters*, 43, 1139–1141, 2002.
68. J. T. Suri, D. B. Cordes, F. E. Cappuccio, R. A. Wessling and B. Singaram, Monosaccharide Detection with 4,7-Phenanthrolinium Salts: Charge-Induced Fluorescence Sensing, *Langmuir*, 19, 5145–5152, 2003.
69. J. T. Suri, D. B. Cordes, F. E. Cappuccio, R. A. Wessling and B. Singaram, Continuous Glucose Sensing with a Fluorescent Thin-Film Hydrogel, *Angewandte Chemie-International Edition*, 42, 5857–5859, 2003.
70. F. E. Cappuccio, J. T. Suri, D. B. Cordes, R. A. Wessling and B. Singaram, Evaluation of Pyranine Derivatives in Boronic Acid Based Saccharide Sensing: Significance of Charge Interaction Between Dye and Quencher in Solution and Hydrogel, *Journal of Fluorescence*, 14, 521–533, 2004.
71. D. B. Cordes, S. Gamsey, Z. Sharrett, A. Miller, P. Thoniyot, R. A. Wessling and B. Singaram, The Interaction of Boronic AciD-Substituted Viologens with Pyranine: The Effects

of Quencher Charge on Fluorescence Quenching and Glucose Response, *Langmuir*, 21, 6540–6547, 2005.

72. D. B. Cordes, A. Miller, S. Gamsey, Z. Sharrett, P. Thoniyot, R. Wessling and B. Singaram, Optical Glucose Detection across the Visible Spectrum Using Anionic Fluorescent Dyes and a Viologen Quencher in a Two-Component Saccharide Sensing System, *Organic and Biomolecular Chemistry*, 3, 1708–1713, 2005.
73. D. B. Cordes, S. Gamsey and B. Singaram, Fluorescent Quantum Dots with Boronic Acid Substituted Viologens to Sense Glucose in Aqueous Solution, *Angewandte Chemie-International Edition*, 45, 3829–3832, 2006.
74. S. Gamsey, N. A. Baxter, Z. Sharrett, D. B. Cordes, M. M. Olmstead, R. A. Wessling and B. Singaram, The Effect of Boronic AciD-Positioning in an Optical Glucose-Sensing Ensemble, *Tetrahedron*, 62, 6321–6331, 2006.
75. S. Gamsey, J. T. Suri, R. A. Wessling and B. Singaram, Continuous Glucose Detection Using Boronic AciD-Substituted Viologens in Fluorescent Hydrogels: Linker Effects and Extension to Fiber Optics, *Langmuir*, 22, 9067–9074, 2006.
76. P. Thoniyot, F. E. Cappuccio, S. Gamsey, D. B. Cordes, R. A. Wessling and B. Singaram, Continuous Glucose Sensing with Fluorescent Thin-Film Hydrogels. 2. Fiber Optic Sensor Fabrication and in Vitro Testing, *Diabetes Technology and Therapeutics*, 8, 279–287, 2006.
77. D. B. Cordes, A. Miller, S. Gamsey and B. Singaram, Simultaneous Use of Multiple Fluorescent Reporter Dyes for Glucose Sensing in Aqueous Solution, *Analytical and Bioanalytical Chemistry*, 387, 2767–2773, 2007.
78. S. Gamsey, A. Miller, M. M. Olmstead, C. M. Beavers, L. C. Hirayama, S. Pradhan, R. A. Wessling and B. Singaram, Boronic AciD-Based Bipyridinium Salts as Tunable Receptors for Monosaccharides and a-Hydroxycarboxylates, *Journal of the American Chemical Society*, 129, 1278–1286, 2007.
79. Z. Sharrett, S. Gamsey, J. Fat, D. Cunningham-Bryant, R. A. Wessling and B. Singaram, The Effect of Boronic Acid Acidity on Performance of Viologen-Based Boronic Acids in a Two-Component Optical Glucose-Sensing System, *Tetrahedron Letters*, 48, 5125–5129, 2007.
80. Z. Sharrett, S. Gamsey, P. Levine, D. Cunningham-Bryant, B. Vilozny, A. Schiller, R. A. Wessling and B. Singaram, Boronic AciD-Appended Bis-Viologens as a New Family of Viologen Quenchers for Glucose Sensing, *Tetrahedron Letters*, 49, 300–304, 2008.
81. D. K. Scrafton, J. E. Taylor, M. F. Mahon, J. S. Fossey and T. D. James, "Click-Fluors": Modular Fluorescent Saccharide Sensors Based on a 1,2,3-Triazole Ring, *Journal of Organic Chemistry*, 73, 2871–2874, 2008.
82. L. Zhu and E. V. Anslyn, Signal Amplification by Allosteric Catalysis, *Angewandte Chemie-International Edition*, 45, 1190–1196, 2006.
83. M. G. Finn, H. C. Kolb, V. V. Fokin and K. B. Sharpless, Click Chemistry – Definition and Aims, *Progress in Chemistry*, 20, 1–4, 2008.
84. H. C. Kolb, M. G. Finn and K. B. Sharpless, Click Chemistry: Diverse Chemical Function from a Few Good Reactions, *Angewandte Chemie-International Edition*, 40, 2004–2021, 2001.
85. D. Luvino, C. Amalric, M. Smietana and J.-J. Vasseur, Sequential Seyferth-Gilbert/Cuaac Reactions: Application to the One-Pot Synthesis of Triazoles from Aldehydes, *Synlett*, 3037–3041, 2007.
86. S.-L. Zheng, S. Reid, N. Lin and B. Wang, Microwave-Assisted Synthesis of Ethynylarylboronates for the Construction of Boronic AciD-Based Fluorescent Sensors for Carbohydrates, *Tetrahedron Letters*, 47, 2331–2335, 2006.
87. N. Lin, J. Yan, Z. Huang, C. Altier, M. Li, N. Carrasco, M. Suyemoto, L. Johnston, S. Wang, Q. Wang, H. Fang, J. Caton-Williams and B. Wang, Design and Synthesis of Boronic-AciD-Labeled Thymidine Triphosphate for Incorporation into DNA, *Nucleic Acids Research*, 35, 1222–1229, 2007.
88. G. A. Molander and J. Ham, Synthesis of Functionalized Organotrifluoroborates *via* the 1,3-Dipolar Cycloaddition of Azides, *Organic Letters*, 8, 2767–2770, 2006.

89. A. K. L. Yuen and C. A. Hutton, Deprotection of Pinacolyl Boronate Esters *via* Hydrolysis of Intermediate Potassium Trifluoroborates, *Tetrahedron Letters*, 46, 7899–7903, 2005.
90. F. D'Hooge, D. Rogalle, M. J. Thatcher, S. P. Perera, J. M. H. van den Elsen, A. T. A. Jenkins, T. D. James and J. S. Fossey, Polymerisation Resistant Synthesis of Methacrylamido Phenylboronic Acids, *Polymer*, 49, 3362–3365, 2008.
91. S. Arimori, L. I. Bosch, C. J. Ward and T. D. James, Fluorescent Internal Charge Transfer (ICT) Saccharide Sensor, *Tetrahedron Letters*, 42, 4553–4555, 2001.
92. S. Arimori, L. I. Bosch, C. J. Ward and T. D. James, A D-glucose Selective Fluorescent Internal Charge Transfer (ICT) Sensor, *Tetrahedron Letters*, 43, 911–913, 2002.
93. L. I. Bosch, M. F. Mahon and T. D. James, The B-N Bond Controls the Balance between Locally Excited (LE) and Twisted Internal Charge Transfer (TICT) States Observed for Aniline Based Fluorescent Saccharide Sensors, *Tetrahedron Letters*, 45, 2859–2862, 2004.

Fluorescence Solvent Relaxation in Cationic Membranes

Agnieszka Olżyńska, Piotr Jurkiewicz, and Martin Hof

Introduction

The interest in positively charged lipid membranes stems from biology, where cationic lipids are used for delivery of nucleic acids into host cells to achieve transgene expression or gene regulation – techniques that are essential for genomics and proteomics. Electrostatic interaction of polyanionic DNA or RNA with cationic lipids leads to the formation of the so-called lipoplex, which functions as a transfection vector [1–4]. Complexation with lipids facilitates the cell uptake of genetic material, in particular crossing the hydrophobic barrier of cellular membrane. The efficiency of lipofection is still lower than transfection with viral vectors, but this is the safety and biocompatibility that drew attention to the lipofection in medicine. The long-term potential of gene therapy for the treatment of both inherited and acquired disorders is the motivation for many studies on transfection agents. The focus in lipid vector development is on protecting genetic material from degradation in the bloodstream of a patient (e.g., stealth liposomes), selectively targeting vector to the cells of interest, increasing transfection efficiency, and maintaining stability and repeatability of the formulation. All these require a thorough knowledge about the lipoplex structure and its interactions with physiological environment. Although it is already more than 20 years since cationic lipids were first used in DNA delivery in tissue cultures [5], still little is known about the possible structures of the lipoplexes and how the structure is determined by the lipid composition and physicochemical parameters. There are likely other factors besides electrostatics [6] that should be considered to better control the association process. The lipids that are used to form a lipoplex are usually mixtures of at least two lipid species in a form of liposomes (unilamellar or multilamellar). Moreover, the lipid surface is commonly altered by modifying the positively charged lipid compound [1, 7, 8] or grafting the surface with a polymer [9–11]. No quantitative measure of such alterations exists,

A. Olżyńska (✉)
J. Heyrovský Institute of Physical Chemistry of the ASCR, v. v. i., Dolejškova 3,
182 23 Prague 8, Czech Republic
e-mail: agnieszka.olzynska@jh-inst.cas

C.D. Geddes (ed.), *Reviews in Fluorescence 2007*, Reviews in Fluorescence 2007,
DOI 10.1007/978-0-387-88722-7_6, © Springer Science+Business Media, LLC 2009

therefore the approach is usually reduced to trial and error. It is reasonable to study the structure and properties of the mixed cationic lipid membrane itself before characterizing their interactions with nucleic acids and the structure and function of the resulting lipoplex. Basic studies on positively charged lipid membranes are relatively new, since the main interest was, for a long time, on neutral and negatively charged lipid bilayers, which are omnipresent in natural membranes, in contrast to the positive ones, which are rare in nature. Cationic lipids used in transfection protocols are synthetic products like, synthesized in 1988, DOTAP [12].

All lipid membranes are dynamic complex systems governed mainly by weak interactions between molecules of water and amphiphilic lipids. The self-organized structure of these aggregates depends on lipid composition, presence of ions or other molecules, pH, temperature, etc. The properties of the hydrophobic interior of lipid bilayer are relatively easy to determine experimentally, but the lipid–water interface, where the polar headgroups of the lipids are located, is less explored [13, 14]. This is an unfortunate situation, since many biologically relevant processes, including interactions with nucleic acids, take place in this region. The headgroup region is characterized by large gradients of all relevant physicochemical parameters, providing a complex environment, which can be modified by both the alterations of lipid bilayer composition and properties of the aqueous phase [15]. This picture is further complicated by the dynamic character of the membrane, the components of which are free to undergo a variety of motions due to thermal agitation as well as macro-scale membrane movements that affect the bilayer and adjacent water layers [16–18]. Still relatively little is known about hydration and dynamics of lipid bilayer. The limited availability of experimental data suppress the development of lipid based technologies [19, 20].

Fluorescence methods, which are very popular in natural and model membrane studies, are capable of measuring numerous important parameters of lipid bilayer, such as local pH, concentrations of surface adsorbed compound, permeability, lateral and rotational diffusion of lipids, and others [13, 21–26]. Solvent relaxation technique (SR), based on simple time-resolved fluorescence measurements, when applied to lipid bilayers, is a unique tool that allows characterization of the bilayer hydration and mobility, the factors that are responsible for the state of the membrane and the way it interacts with its surroundings [27]. Among other techniques employed to study those parameters (e.g., NMR [28, 29], fluorescence [30, 31], X-ray, and neutron diffraction [32, 33]) the solvent relaxation is unique in its ability to measure fully hydrated free-standing liquid crystalline lipid bilayers, which are the most biologically relevant membrane models [34–36]. In this review, we would like to demonstrate the potential of the solvent relaxation technique and summarize our studies on mixed cationic membranes.

Fluorescence Solvent Relaxation Technique

The methodology used to study the relaxation of hydrated lipid bilayers has already been described in detail in a few papers [27, 34, 36, 37]. Here, we will briefly outline the principle and most important aspects of the method.

The main interest is to measure the kinetics of the emission spectrum red-shift, which is the result of the relaxation of the polar environment of the dye as a response to the rapid change in the dipole moment of a dye upon electronic excitation. To do this, one needs to use a fluorescent dye that is characterized by sufficiently large change of its dipole moment. Also, the fluorescence lifetime of the chromophore should be long enough to capture the relaxation process of interest; this can be of importance in particularly viscous systems. The problem of a proper SR probe choice is discussed in the next paragraph.

The SR measurements consist of recording fluorescent decays at a series of emission wavelengths spanning the emission spectrum of the dye (usually 10–20 curves). The decays are usually fitted to multi-exponential functions, using the iterative reconvolution procedure. Then, the time-resolved emission spectra (TRES) are reconstructed [38] and fitted by an appropriate function (mainly log-normal) to determine their position, $\nu(t)$ and width, FWHM. In order to characterize the kinetics of the spectral shift, the spectral response function (or correlation function), defined as

$$C(t) = \frac{\nu(t) - \nu(\infty)}{\nu(0) - \nu(\infty)} = \frac{\nu(t) - \nu(\infty)}{\Delta\nu}, \tag{1}$$

is calculated. The $\nu(0)$ is the position of the so-called time-zero spectrum – the hypothetical fluorescence emission spectrum of the dye, which has vibrationally relaxed before any solvent motions. This spectrum cannot be measured directly in most systems due to the finite temporal resolution of the apparatus. Instead, it can be estimated using absorption and steady-state emission spectra measured in a non-polar solvent [39]. A comparison between the positions of the estimated time-zero spectrum and the one reconstructed from the measured decays gives the percentage of the solvation process that is faster than the time resolution of the instrumentation used [40, 41].

A quantitative description of the solvent relaxation process comprises two parameters. The first one, $\Delta\nu$, represents the overall emission shift, which was shown to be proportional to the polarity of the dye environment [38]. In a phospholipid/water system, polarity is mainly attributed to water molecules; therefore, $\Delta\nu$ reflects the extent of hydration. The second parameter is the relaxation time, τ_r, which describes the mobility of the polar environment of the dye. For the neat solvents it is attributed to their viscosity [38]. In the lipid bilayer at the level of glycerol, however, water hydrating the membrane is fully bound to the phospholipid molecules. Therefore, the slow relaxation kinetics observed in membranes is attributed to the collective motions of hydrated lipid moieties rather than the motions of water molecules themselves [40, 41]. The mean solvent relaxation time is often estimated calculating the so-called integrated relaxation time:

$$\tau_r \equiv \int_0^\infty C(t)dt. \tag{2}$$

This approach does not require any a priori knowledge about the solvation process [38], in contrast to fitting the relaxation function by kinetic models [40, 42]. The intrinsic uncertainty for this parameter can be estimated from the width of the instrument response function; in our case ~ 20 ps [27].

The temporal behavior of the TRES width provides additional information on the extent of the observed solvent relaxation. The experiments carried out in phospholipid bilayers [37, 43] and also in the supercooled liquids [44] have shown that the FWHM passes a maximum during solvation process. This is in a good agreement with the idea of a non-uniform spatial distribution of solvent response times [44, 45]. It has been shown that in homogeneous systems of low molar mass molecules, the FWHM decays monotonically. In a spatially inhomogeneous systems, however (i.e., where the microenviroments of the fluorophores differ), the relaxation behavior is different, since the solvation shells of individual fluorophores respond with different rates to changes in the local electric field. The overall transient inhomogeneity increases significantly during the solvation and than decreases once the solvation finishes and the equilibrated excited state is reached, so the FWHM of TRES, which gives the measure of inhomogeneity of dyes microenvironments, passes a pronounced maximum. This allows checking whether the entire response or merely part of it was captured within the time-window of the experiment [27]

SR Probes and Their Location in Lipid Bilayers

It has been already mentioned that the solvent relaxation dyes should undergo a large change in their dipole moment upon excitation (large Stokes shift). In the lipid membranes there are, however, further requirements that should be imposed on fluorescent probes. First of all, the chromophore has to be incorporated into the lipid bilayer, which usually requires that the probe be at least partly hydrophobic. It is very important to know the exact location and position of the chromophore in the bilayer, and to assure that these are not changed during experiment. Finally, the size and the structure of the probe should disturb the bilayer as little as possible; from the same reason the quantum efficiency should be high enough to keep the dye to lipid ratio low. For the list of the SR probes that are most often used in our laboratory see [27].

Solvent relaxation in the lipid membranes is four decades slower than in the bulk water – the mean time constant in pure water is 0.3 ps. [46], whereas at the glycerol level of a fully hydrated phospholipid bilayer in the liquid crystalline state it is about 2 ns [40]. Such a large change in solvation kinetics on a distance of only 1 nm results in enormous gradients of solvation properties along the membrane normal. Knowledge about the precise location of a polarity-sensitive fluorescent dye is crucial to characterization of the relaxation properties. It is also very advantageous if one has a series of probes with the same chromophore located at different depths within the bilayer. Examples of such dyes include naphthalene derivatives Prodan, Laurdan, and Patman (Fig. 1), which are often used in our laboratory. Their fluorophore consist of the electron-donating dimethylamino group and the

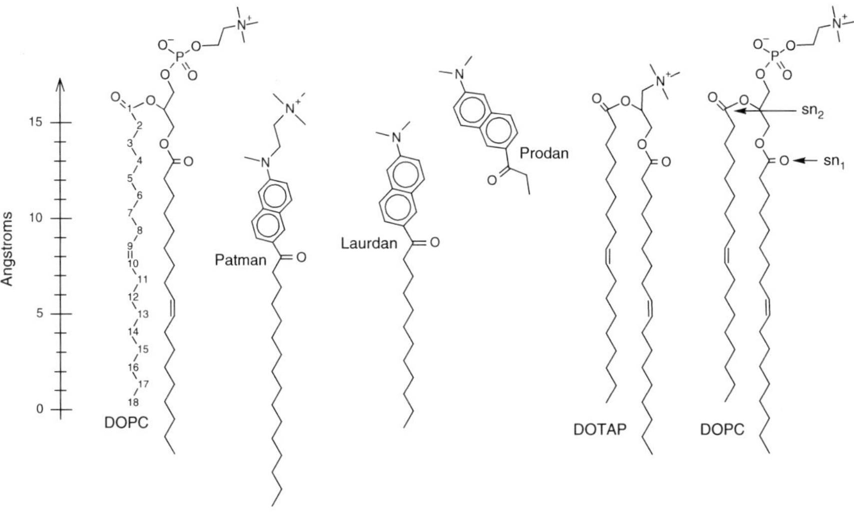

Fig. 1 Schematic diagram of the location of Patman, Laurdan, and Prodan in a membrane composed of dioleoylphosphatidylcholine (DOPC) and dioleoyltrimethylammoniumpropane (DOTAP). The positions of sn_1 and sn_2 carbonyls are indicated by *arrows*. The carbon atoms of the sn_2 chain are numbered. The scale on the *left* shows the approximate distance from the bilayer center. Upon the addition of DOTAP, the position of Patman changes and differs from the one shown here. See the text for details on the dye locations

electron-withdrawing carbonyl group of a different fatty acid residue, separated by a naphthalene ring. The naphthalene ring ensures a substantial increase of dipole moment upon electronic excitation.

Prodan [47], which contains a short propionyl group, is believed to be located in the outermost region of the interface [48]. The location is, however, poorly defined and, consequently, any measurements performed with Prodan should always be evaluated with care (see [49], for example). In addition, it was hypothesized that the two population of this probe that differ in their positions can be present in membrane: one perpendicular and one parallel to the membrane surface [48]. It is also the most water soluble dye from the set, hence some of the fluorescence signal can originate from the bulk which further complicates the interpretation of results. Nevertheless, Prodan is still a valuable membrane probe and can be successfully used in solvent relaxation studies [34, 49–51].

Laurdan and Patman molecules are modified so that their location within the membrane is more stable. Laurdan has a longer hydrocarbon chain (11 carbons) attached to the carbonyl group, which anchors the molecule in the hydrophobic part of the bilayer. It is commonly assumed that its fluorophore stays at the level of the glycerol backbone [52]. This assumption is a result of a comparison between its fluorescence properties under different experimental conditions [52–54].

Patman, in addition to a palmitoyl tail, possesses a trimethylammonium group, which orients the alkylamino end of the fluorophore toward the lipid–water interface

[55] and stabilizes the dye position in the bilayer due to interactions with charged phospholipid headgroups. The location of Patman is stable and unimodal. The NMR studies have shown that Patman is embedded deeper in the bilayer compared to Prodan, and that its movement is more restricted [35]. The positive charge of the ammonium group can be a serious drawback of this probe in some applications.

Easy and efficient tools to experimentally determine chromophore location within the lipid membrane are fluorescence quenching techniques. One of them is the parallax method, which can be used to measure the mean distance of the fluorophore from the bilayer center [24, 56–58]. The method is based on the quenching with spin labeled or brominated lipids or fatty acids with defined but different location of the quenching moieties. If more than two quenchers are used, the distribution of the fluorophore locations can be obtained [58]. For the polarity sensitive probes located within the membrane interface a better measure of their distribution can be obtained with wavelength-dependent parallax quenching [58]. The parallax method was used to measure Laurdan and Patman locations in dioleoylphosphatidylcholine (DOPC) bilayer. The distance from the bilayer center and the mean location of Laurdan fluorophore was about 11.4 Å, regardless of the selected pair of quenchers [58]. The obtained value was comparable to 9.8 Å measured in the cell membrane of Torpedo marmorata enriched with nicotine acetylcholine receptors [59]. The distance measured for Patman was 10.4 Å from the bilayer center. The results obtained for Prodan for different pairs of quenchers were not consistent and suggested a broad distribution of this dye. The statistical character of the location of both Laurdan and Patman and also quenchers implies that all the obtained values are averages and that the actual values can fluctuate [56]. The determined position of the two dyes are indicated in the schematic lipid bilayer representation in Fig. 1, together with the hypothetical Prodan position found in the literature [52].

The difference in localization between the chromophores of Patman and Laurdan is 1 Å, which is close to the uncertainty of the parallax method, but agrees with the SR results obtained for these probes in many lipid systems. It must be kept in mind that this small 1 Å difference in position occurs in the bilayer region where a large gradient of water concentration is present [16]. This fact, together with the sensitivity of the solvent relaxation method, is responsible for the differences in the measured relaxation times and $\Delta\nu$ values.

The average positions of Laurdan and Patman presented above were determined for the emission maxima, i.e., 493 and 475 nm, respectively. A strong dependence of quenching efficiency on the emission wavelength was observed. The calculated distances of the dyes from the membrane center increased with the wavelength (Fig. 2)

An increase in the detection wavelength of the emission leads to the photoselection of the chromophores, which are in a more relaxed and/or more polar environment, since solvent relaxation becomes faster and the amount of water increases when moving towards the lipid/water interface [36]. Thus, the wavelength dependence of the quenching rate provides information on the heterogeneity of the chromophore locations along the bilayer normal. These two dyes differs in this respect, i.e., a stronger dependence of the calculated position on the emission wavelength is observed for Laurdan ($\sim$2 Å) than for Patman ($<$1 Å). This can be a result of

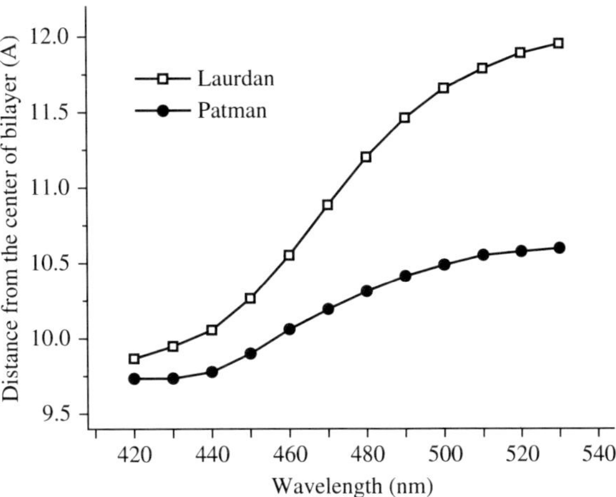

Fig. 2 Position of Laurdan (*open squares*) and Patman (*filled circles*) in terms of the distance from the center of bilayer, calculated for different emission wavelengths. The dye was excited at 373 nm. The measurements were performed at 20°C

the presence of the trimethylammonium group and longer hydrophobic chain in the molecule of Patman.

Once the location of a probe within lipid bilayer of interest is known, one can check if it is stable upon the changes in composition or physicochemical parameters. The easiest way to do this is to quench the fluorescence using a quencher that is water soluble and does not penetrate lipid membrane. One of such quenchers is acrylamide. Because of its neutral charge, it was selected to check whether the addition of cationic lipid DOTAP (dioleoyltrimethylammoniumpropane) to the DOPC lipid bilayer does not change the locations of the probes. The fluorescence intensity was being measured while the sample was titrated with the quencher. The data were analyzed in terms of the Stern–Volmer equation [58]. Only the quenching plots for Prodan departed slightly from linearity, which is yet another indication for a rather heterogeneous localization of this probe [48, 60]. The Stern–Volmer constants obtained for Laurdan and Patman are summarized in Table 1. An interesting result was the relocation of Patman, which moved out of the bilayer when the content of DOTAP was changed from 0 to 30 mol%. Further addition of DOTAP did not change the position of Patman any further. In the case of Laurdan there is

Table 1 The quenching of Patman and Laurdan embedded in DOPC/DOTAP LUVs with acrylamide

	Patman	Laurdan
[DOTAP] (mol%)	K_D (M^{-1})[a]	K_D (M^{-1})[a]
0	0.274 ± 0.003	0.340 ± 0.005
30	0.314 ± 0.003	0.347 ± 0.004
50	0.317 ± 0.004	0.350 ± 0.006
100	0.314 ± 0.006	0.369 ± 0.003

[a] K_D – Stern–Volmer constants calculated by fitting the data points to a straight line.

no significant change in its position. The increased accessibility of the fluorophore to acrylamide might also be explained by the increased penetration of the quencher into the interfacial region of the lipid bilayer. In such case, however, we would have observed the same effects for both dyes. Additionally the scenario of increased membrane penetration is unlikely in the view of the SR results, which indicate that the membrane becomes tighter when the amount of DOTAP rises to 30 mol%. In conclusion, the observed rise in the quenching efficiency for Patman is a consequence of its relocation upon the addition of DOTAP (up to 30 mol%) [58].

In addition to the quenching techniques, the analysis of the TRES width can give an indication whether the distribution of the probe fluorophore within the membrane is unimodal. We showed that if more than one homogenous population of the dye is present, or if other processes, besides solvation, take place, the FWHM profile clearly differs from the one normally observed [36, 40, 49].

SR in Cationic Membranes

Hydration and Mobility of the DOPC/DOTAP Liposome Membranes

Solvent relaxation measurements were performed for three fluorescent probes: Prodan, Laurdan and Patman embedded into mixed cationic membranes composed of cationic dioleoyltrimethylammoniumpropan (DOTAP) and zwitterionic dioleoylphosphatidylcholine (DOPC). Both the lipids contain the same hydrophobic parts consisting of two oleic chains, which assure their low transition temperature. The lipid bilayers were in the form of Large Unilamellar Vesicles with diameter of about 100 nm. Measurements were performed in 10 mM HEPES buffer with 100 mM NaCl at different temperatures. The fluorescence decays were measured for emission wavelengths of 400–540 nm, with 10 nm step (all three dyes share the same fluorophore, hence their emission spectra are similar). Examples of fluorescence decays of Laurdan in two different formulations are shown in Fig. 3.

The TRES reconstructed from the series of fluorescence decays (see Fig. 4 for examples) were fitted using log-normal function, and analyzed in terms of their positions and widths. Examples of TRES position and FWHM cures are given in Fig. 5. The gradual shift to longer wavelengths and a decrease of the intensity maxima are clearly visible in Fig. 4. The spectra became wider with time after excitation. The spectral difference between the two samples persists up to 8 ns, when the spectra superimpose. For liposomes formed from pure DOTAP, the spectral shift towards longer wavelengths occurs faster.

The Laurdan FWHM profile for liposomes formed from pure DOTAP is shifted to shorter times and larger wavenumbers, which indicates that the environment of Laurdan is more mobile and more heterogeneous, respectively (Fig. 5). The FWHM profiles for all three dyes in all investigated lipid systems showed clear maxima, providing evidence that solvent relaxation took place within the timescale of the

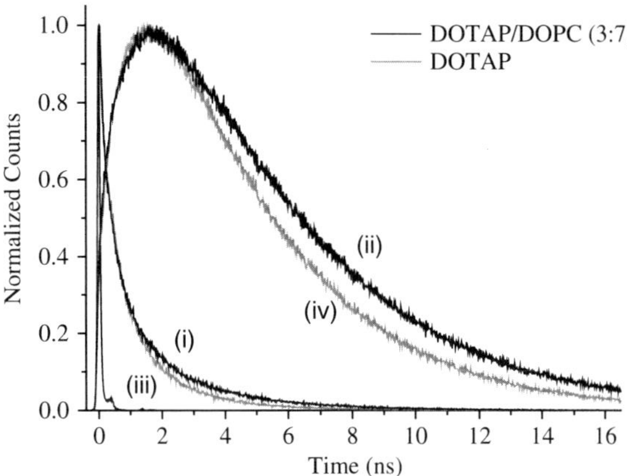

Fig. 3 Examples of fluorescence decays of Laurdan ($\lambda_{ex} = 373$ nm) at 10°C for 30 mol% of DOTAP in DOPC (*black lines*) at 400 nm *(i)* and 540 nm *(ii)* and pure DOTAP (*grey lines*) at 400 nm *(iii)* and 540 nm *(iv)*

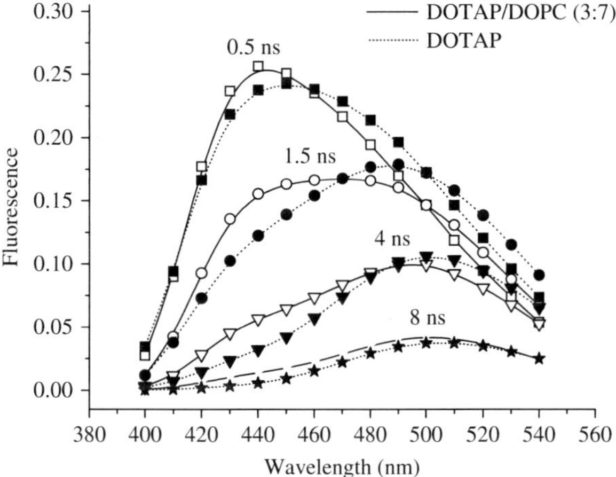

Fig. 4 Reconstructed time-resolved emission spectra (TRES) for Laurdan at 10°C for 30 mol% of DOTAP in DOPC (*empty symbols* and *solid lines*) and pure DOTAP (*filled symbols* and *dotted lines*), at 0.5 ns (*squares*), 1.5 ns (*circles*), 4 ns (*triangles*), and 8 ns (*stars*) after excitation

experiment and suggesting a unimodal distribution of the fluorophores. The $\nu(0)$ for the dyes used were estimated to be 23800 cm^{-1}.

Total emission shift, $\Delta\underline{\nu}$, calculated for Laurdan and Prodan was within the uncertainty of the method (50 cm^{-1}) (Fig. 6). For Patman, however, an increase of $\Delta\nu$ with increasing DOTAP content is observed. Typically, the larger the $\Delta\nu$ the more hydrated the membrane. The effect was, however, larger than any we had previously observed for liquid crystalline membranes. Correct interpretation of the SR results for Patman was possible only after the determination of the probe locations within the bilayer (see the previous chapter). The first part of the $\Delta\nu$ curve of Patman (in the range 0–30 mol% of DOTAP) can be attributed to the fact that

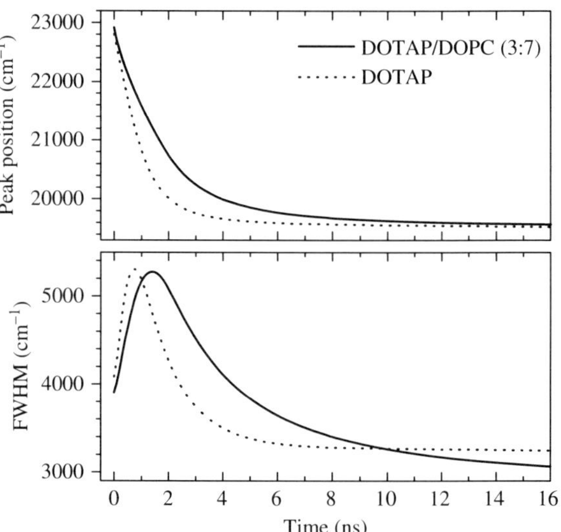

Fig. 5 The position of the maximum and full width at half maximum (FWHM) of TRES obtained for Laurdan at 10°C for 30 mol% of DOTAP in DOPC (*solid lines*) and pure DOTAP (*dotted lines*) as a function of time after excitation

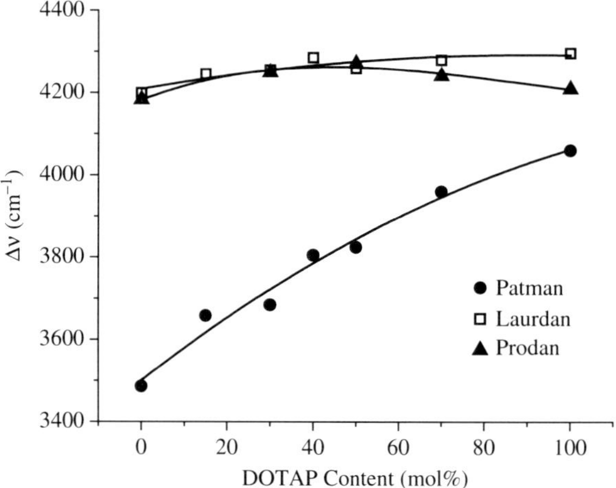

Fig. 6 Solvent relaxation $\Delta\nu$ parameter for Prodan (*filled triangles*), Laurdan (*open squares*), and Patman (*filled circles*) in DOPC/DOTAP LUVs as a function of DOTAP content. Measurements were performed at 10°C. The *solid lines* are polynomial fits and are shown only as a guide to the eye

the dye is moving outward from the bilayer where it probes more hydrated environment. For the fraction of DOTAP larger than 30 mol% the location of Patman does not change any more (see Table 1), thus the continuous increase of $\Delta\nu$ above 30 mol% of DOTAP indicates an increased hydration of the bilayer. This hydration increase is not observed for the two other dyes, which probe much more hydrated environment already in DOPC bilayer (Fig. 6). This means that water penetrates deeper into the membrane when the DOTAP content exceeds 30 mol%. Below this value we can only conclude that the hydration at the levels of Laurdan and Prodan does not change significantly.

The integrated solvent relaxation time (Equation 2) is a good measure of relaxation kinetics, provided the spectral response functions do not vary in shape, which was the case in the measured systems. The obtained values of τ_r are plotted in Fig. 7. Pronounced maxima at around 30 mol% of DOTAP in the lipid bilayer are reported by Laurdan and Prodan. The different behavior of Patman is once again attributed to its relocation. The kinetics of dielectric relaxation in the lipid membrane is often interpreted as the freedom of movement of water hydrating the membrane. At the level where the dyes we use are located water molecules are believed to be fully bound to membrane phospholipids. Therefore, the dynamics of the system as a whole should be considered [41], rather than the slow dynamics of individual water molecules, as it was previously postulated for subnanosecond dynamics observed at protein surfaces [61, 62]. We believe that this is the dynamics of the whole hydrated lipid moieties that we observe and we discuss the results, herein, in a general way, referring to bilayer mobility and rigidity [35]. The relaxation time constant depends nonmonotonically on the concentration of cationic lipid. The addition of up to 30 mol% of DOTAP to the DOPC bilayer increases the rigidity of the hydrated functional groups in the vicinity of the fluorophores of Laurdan and Prodan, which is reflected by the elevated integrated relaxation time. Further addition of DOTAP does not increase τ_r any more, instead the solvation kinetics becomes faster once again, which means that the lipid bilayer is getting more mobile in the closest surrounding of the fluorophores (headgroup region). This nonmnotonic dependence of the membrane headgroup mobility is in good agreement with anomalous mixing of zwitterionic and cationic lipids observed previously in calorimetric studies, where similar maximum was observed in the phase diagram

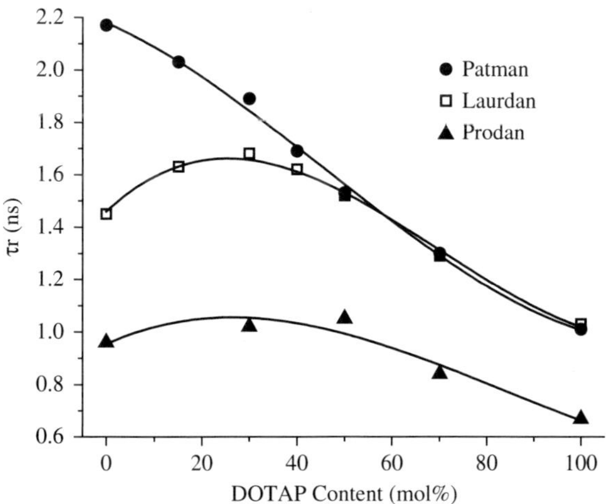

Fig. 7 Mean solvent relaxation time τ_r (defined by Eq. 2) for Prodan (*filled triangles*), Laurdan (*open squares*), and Patman (*filled circles*) in DOPC/DOTAP LUVs as a function of DOTAP content. Measurements were performed at 10°C. The *solid lines* are best polynomial fits and are shown only as a guide to the eye

of the mixture of dipalmitoylphosphatidylcholine and cationic dihexadecyldimethylammonium chloride at 35 mol% of cationic compound [63]. It is reasonable that the higher the temperature of the lipid system above its gel–liquid crystalline transition the more mobile the lipid molecules become. Consequently, at the certain temperature, at which all the systems studied are in their liquid crystalline state, the less mobile should be the system, the phase transition temperature of which is the highest. It holds true for our mixtures, which have higher phase transition temperatures than those of the individual pure components. But such relation does not have to be that straight forward and the SR kinetics does not always reflect the temperature offset above the phase transition. The reason for the nonideal mixing of zwitterionic and cationic lipids can be a rearrangement of the zwitterionic headgroups as a result of electrostatic interactions between both the compounds. The phosphatidylcholine headgroup is a dipole, with dipole moment of about 19 D [64], consisting of negatively charged phosphate group and positively charged quaternary ammonium group (in short P–N dipole). It was shown experimentally that in pure phosphatidylcholine bilayer the P–N vector lies almost parallel to the plane of the membrane [65]. Seelig et al., who measured the orientation of the P–N dipole using ^{31}P NMR, observed reorientation of this dipole of more than 30°, when a cationic amphiphile was added [66]. The authors attributed this reorientation to the electrostatic repulsion between the cationic headgroup and the N^+ end of the P–N dipole (see Fig. 1). More recent molecular dynamics simulations of the mixture of dimyristoylphosphatidylcholine (DMPC) and dimyristoyltrimethylammoniumpropane (DMTAP) showed the increase of the angle between the P–N dipole of DMPC and bilayer surfre DMPC to 60° in 50 mol% and more of DMTAP [67]. This reorientation resulted in decreased area per lipid for the DMPC/DMTAP (1:1) mixture. Above 50 mol% of DMTAP, the bilayer expanses again due to repulsion between the cationic headgroups. The curve of the area per lipid as a function of DMTAP content [67] was very similar to the τ_r curve we obtained for Laurdan (Fig. 7) except that the maximum, in our case, is at 30 mol% of the cationic lipid. To our best knowledge, our SR measurements gave the first experimental confirmation of these nonmonotonic changes in lipid packing in fully hydrated free standing mixed cationic membranes [58]. The discrepancy between the positions of the maxima can be the effect of different backbones of the lipids used (oleic versus myristic chains). This hypothesis motivated the SR studies on DMPC/DMTAP system [68].

DMPC/DMTAP Versus DOPC/DOTAP

SR measurements analogous to that presented above were performed for DMPC/DMTAP system, to study the influence of the hydrocarbon chains on the mobility and hydration of the headgroup region of these mixed cationic membranes [68]. The myristic chains are shorter than the oleic ones and fully saturated. To compare both systems only Laurdan data, for which the changes were the most pronounced, was used. The DMPC/DMTAP membranes were measured at 50° to keep approximately the same temperature offset above the main phase transition temperatures of both the systems. The $\Delta\nu$ and τ_r parameters are presented in Figs. 8 and 9, respectively; the results obtained previously for DOPC/DOTAP membrane are included for comparison.

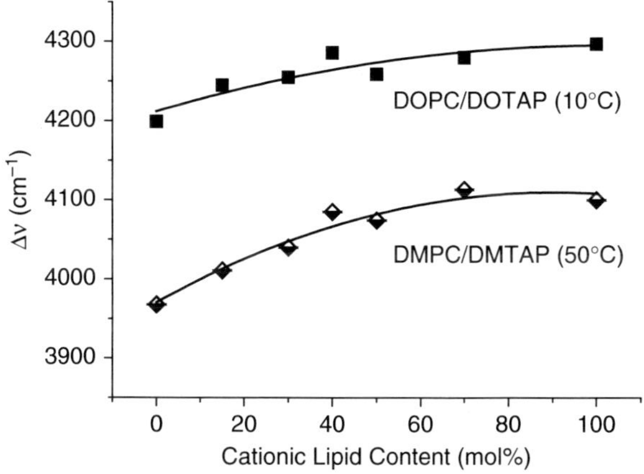

Fig. 8 Solvent relaxation $\Delta\nu$ parameter for Laurdan in DOPC/DOTAP LUVs (*squares*) and DMPC/DMTAP LUVs (*diamonds*) as a function of cationic lipid content. Measurements were performed at 10 and 50°C for DOPC/DOTAP and DMPC/DMTAP systems, respectively. The best polynomial fits (*solid lines*) are shown for eye guidance

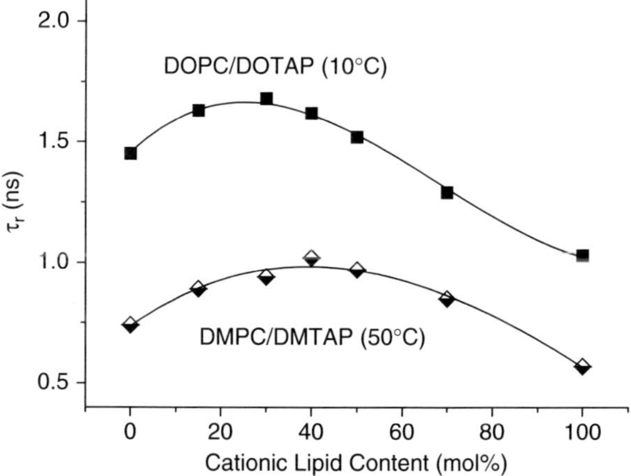

Fig. 9 Mean solvent relaxation time τ_r (defined by Eq. 2) for Laurdan in DOPC/DOTAP LUVs (*squares*) and DMPC/DMTAP LUVs (*diamonds*) as a function of cationic lipid content. Measurements were performed at 10 and 50°C for DOPC/DOTAP and DMPC/DMTAP bilayers, respectively. The results for DOPC/DOTAP are taken from Jurkiewicz et al. [58]. The best polynomial fits (*solid lines*) are shown for eye guidance

The increase of $\Delta\nu$ values with increasing TAP content was more pronounced for the DMPC/DMTAP system (Fig. 8). Although the overall change in DMPC/DMTAP hydration was not large, the bilayer hydrated gradually when percentage of DMTAP increased. The higher hydration of DOPC than DMPC is not a surprise as the double unsaturated backbones of DOPC create molecule with bulkier backbone, hence the larger area per lipid leaves more space for water at the headgroup region, in comparison with DMPC with saturated myristic chains.

The maximum of mean relaxation time is also present for DMPC/DMTAP membrane, although it is shifted to 45 mol% of DMTAP content (Fig. 9).This discrepancy is a straight forward consequence of the different geometries of DMPC and DOPC molecules. The membrane area per DMPC molecule is smaller and more dependent on its headgroup than on its hydrophobic part.

When the P–N dipoles raise, the DMPC bilayer compresses considerably, while the DOPC compressibility is restricted by its bulky hydrophobic part and finishes at lower cationic lipid content. It is worth noticing that the relative changes in the relaxation time are considerably bigger for DMPC/DMTAP bilayer (Fig. 9), i.e., τ_r at 45 mol% of DMTAP is 30% longer than for pure DMPC, while τ_r at 30 mol% of DOTAP is only 15% longer than the one for pure DOPC. The results obtained for DMPC/DMTAP membrane are in very good agreement with the previously mentioned molecular simulations of the same system, where the maximum compression was achieved at around 50 mol% of DMTAP [67]. It was shown that in some cationic membranes at high cationic lipid content the expansion of the membrane can cause the interdigitation of the bilayer leaflets [69]. In both the systems presented here no signs of interdigitation were observed.

The Effect of Temperature on SR Kinetics in DOPC/DOTAP Membrane

SR measurements of Laurdan in DOPC/DOTAP bilayers were performed at different temperatures (10, 15, 20, and 30°C). No changes in $\Delta\nu$ profiles were observed (data not shown), which means that in this range the temperature do not influence the extent of hydration of the headgroup region of this cationic bilayer. On the contrary, the relaxation kinetics strongly depends on the temperature of the system (Fig. 10). The temperature rise results in higher bilayer mobility (lower τ_r values) and a reduction of the maximum, which was most pronounced at 10°C, and gradually diminished at higher temperatures disappearing completely at 30°C. The electrostatic effects at higher temperatures are hindered by the elevated Brownian motions of all the elements of the system. As expected, this eliminates the compression effect for membranes formed from 0 to 30 mol% of DOTAP, whereas for higher DOTAP concentrations lateral repulsion is still present.

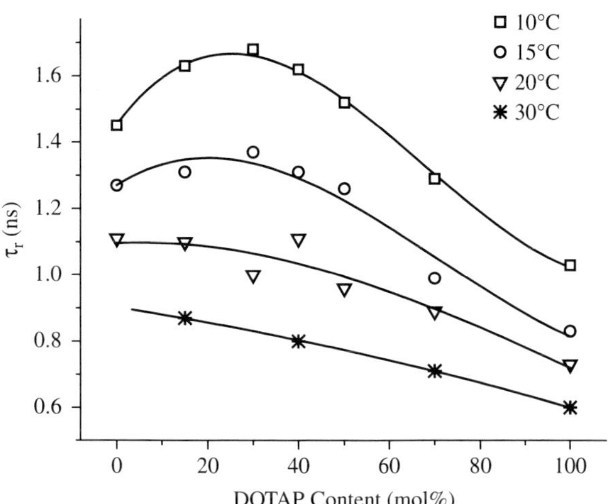

Fig. 10 Mean relaxation time τ_r for Laurdan in DOPC/DOTAP LUVs as a function of DOTAP content. Measurements were performed at 10°C (*squares*), 15°C (*circles*), 20°C (*triangles*), and 30°C (*asterisks*). The *solid lines* are best polynomial fits and are shown only as a guide to the eye

Conclusions

Solvent relaxation based on time-resolved measurements of fluorescent probes with defined locations within the lipid bilayer is a valuable and unique tool, suitable for monitoring the hydration and organization of fully hydrated biological and model membranes. The SR properties of DOPC/DOTAP cationic membranes as well as the locations of the fluorescent probes – Patman and Laurdan – were determined. Quenching with acrylamide provided a simple and reliable test for potential dye relocation, particularly important in studies of lipid membranes, in which large gradients of hydration properties are present. Wavelength-dependent parallax quenching, in turn, is capable of precisely defining the probing depth of the bilayer. Using a set of dyes that share the same spectroscopic properties, but differ in their location within the bilayer, allows the obtained SR parameters compare at different levels inside the membrane. The synthesis of other such sets of probes is still needed.

The observed nonmonotonic dependence of relaxation properties on DOTAP content in the DOPC bilayer supports a scenario of headgroup rearrangement in mixtures of cationic and zwitterionic lipids [66, 67, 69, 70]. This finding may have important implications for the development of synthetic genetic information carriers, e.g., it correlates with the interaction profile of cationic vesicles with oligonucleotides, for which maximum association was found at ~25% of DOTAP [71].

References

1. S. W. Hui, M. Langner, Y. L. Zhao, P. Ross, E. Hurley, and K. Chan, The role of helper lipids in cationic liposome-mediated gene transfer, *Biophys. J.* **71**, 590–599 (1996).
2. T. Kral, A. Benda, M. Hof, and M. Langner, Some aspects of DNA condensation observed by fluorescence correlation spectroscopy. In *Reviews in Fluorescence Vol. 2*, edited by C. D. Geddes and J. R. Lakowicz (Springer, New York, 2005), pp. 109–124.

3. T. Kral, M. Hof, P. Jurkiewicz, and M. Langner, Fluorescence correlation spectroscopy (FCS) as a tool to study DNA condensation with hexadecyltrimethylammonium bromide (HTAB), *Cell. Mol. Biol. Lett.* **7**, 203–211 (2002).
4. J. O. Radler, I. Koltover, T. Salditt, and C. R. Safinya, Structure of DNA-cationic liposome complexes: DNA intercalation in multilamellar membranes in distinct interhelical packing regimes, *Science* **275**(5301), 810–814 (1997).
5. P. L. Felgner, T. R. Gadek, M. Holm, R. Roman, H. W. Chan, M. Wenz, J. P. Northrop, G. M. Ringold, and M. Danielsen, Lipofection – a Highly Efficient, Lipid-Mediated DNA-Transfection Procedure, *Proc. Natl. Acad. Sci. USA* **84**(21), 7413–7417 (1987).
6. M. Langner, and K. Kubica, The electrostatics of lipid surfaces, *Chem. Phys. Lipids* **101**M(1), 3–35 (1999).
7. G. Byk, J. Sato, C. Mattler, M. Frederic, and D. Scherman, Novel non-viral vector for gene delivery: Synthesis of a second-generation library of mono-functionalized poly-(guanidinium)amines and their introduction into cationic lipids, *Biotechnol. Bioeng.* **61**, 81–87 (1998).
8. G. Byk and D. Scherman, Novel cationic lipids for gene delivery and gene therapy, *Exp. Opin. Ther. Patents* **8**, 1125–1141 (1998).
9. K. L. Hong, W. W. Zheng, A. Baker, and D. Papahadjopoulos, Stabilization of cationic liposome-plasmid DNA complexes by polyamines and poly(ethylene glycol)-phospholipid conjugates for efficient in vivo gene delivery, *FEBS Lett.* **400**(2), 233–237 (1997).
10. A. Martin-Herranz, A. Ahmad, H. M. Evans, K. Ewert, U. Schulze, and C. R. Safinya, Surface functionalized cationic lipid-DNA complexes for gene delivery: PEGylated lamellar complexes exhibit distinct DNA-DNA interaction regimes, *Biophys. J.* **86**(2), 1160–1168 (2004).
11. T. Borowik, K. Widerak, M. Ugorski, and M. Langer, Combined effect of surface electrostatic charge and poly(ethyl glycol) on the association of liposomes with colon carcinoma cells, *J. Liposome Res.* **15**(3–4), 199–213 (2005).
12. L. Stamatatos, R. Leventis, M. J. Zuckermann, and J. R. Silvius, Interactions of cationic lipid vesicles with negatively charged phospholipid-vesicles and biological-membranes, *Biochemistry* **27**(11), 3917–3925 (1988).
13. R. F. Epand, R. Kraayenhof, G. J. Sterk, H. W. W. F. Sang, and R. M. Epand, Fluorescent probes of membrane surface properties, *Biochim. Biophys. Acta* **1284**, 191–195 (1996).
14. R. D. Kaiser, and E. London, Location of diphenylhexatriene (DPH) and its derivatives within membranes: Comparition of different fluorescence quenching analysis of membrane depth, *Biochemistry* **37**, 8180–8190 (1998).
15. S. H. White, and M. C. Wiener, Determination of the structure of fluid lipid bilayer membranes. In *Permeability and stability of lipid bilayers*, edited by E. A. Disalvo and S. A. Simon (CRC Press, Boca Raton, 1995), pp. 1–19.
16. J. F. Nagle, and S. Tristram-Nagle, Structure of lipid bilayers, *Biochim. Biophys. Acta* **1469**, 159–195 (2000).
17. S. J. Marrink, and A. E. Mark, Effect of undulations on surface tension in simulated bilayers, *J. Phys. Chem. B* **105**, 6122–6127 (2001).
18. S. J. Marrink, D. P. Tieleman, A. R. v. Buuren, and H. J. C. Berendsen, Membrane and water: An interesting relationship, *Faraday Discuss.* **103**, 191–201 (1996).
19. S. J. Eastman, C. Siegel, J. Tousignant, A. E. Smith, S. H. Cheng, and R. K. Scheule, Biophysical characterization of cationic lipid: DNA complexes, *Biochim. Biophys. Acta* **1325**, 41–62 (1997).
20. M. Langner, The intracellular fate of non-viral DNA carriers, *Cell. Mol. Biol. Lett.* **5**, 295–313 (2000).
21. J. Gabrielska, S. Przestalski, A. Miszta, M. Soczynska-Kordala, and M. Langner, The effect of cholesterol on the absorption of phenyltin compounds onto phosphatidylcholine and sphingomyelin liposome membranes, *Appl. Organomet. Chem.* **18**, 9–14 (2004).

22. M. Langner, and H. Kleszczynska, Estimation of the organic compounds partition into phosphatidylcholine bilayers with pH sensitive fluorescence probe, *Cell. Mol. Biol. Lett.* **2**, 15–24 (1997).
23. R. Hutterer, F. W. Schneider, and M. Hof, Time-resolved emission spectra and anisotropy profiles for symmetric diacyl- and dietherphosphatidylcholines, *J. Fluorescence* **7**, 27–33 (1997).
24. A. S. Klymchenko, G. Duportail, A. P. Demchenko, and Y. Mely, Bimodal distribution and fluorescence response of environment-sensitive probes in lipid bilayers, *Biophys. J.* **86**(5), 2929–2941 (2004).
25. R. Hutterer, F. W. Schneider, W. T. Hermens, R. Wagenvoord, and M. Hof, Binding of prothrombin and its fragment 1 to phospholipid membranes studied by the solvent relaxation technique, *Biochim. Biophys. Acta* **1414**(1–2), 155–164 (1998).
26. A. Olzynska, M. Przybylo, J. Gabrielska, Z. Trela, S. Przestalski, and M. Langner, Di- and tri-phenyltin chlorides transfer across a model lipid bilayer, *Appl. Organomet. Chem.* **19**(10), 1073–1078 (2005).
27. P. Jurkiewicz, J. Sykora, A. Olzynska, J. Humplickova, and M. Hof, Solvent relaxation in phospholipid bilayers: Principles and recent applications, *J. Fluorescence* **15**(6), 883–894 (2005).
28. E. G. Finer, and A. Darke, Phospholipid hydration studied by deuteron magnetic-resonance spectroscopy, *Chem. Phys. Lipids* **12**(1), 1–16 (1974).
29. P. O. Westlund, Line shape analysis of NMR powder spectra of (H2O)-H-2 in lipid bilayer systems, *J. Phys. Chem. B* **104**(25), 6059–6064 (2000).
30. S. Mazeres, V. Schram, J. F. Tocanne, and A. Lopez, 7-Nitrobenz-2-oxa-1,3-diazole-4-yl-labeled phospholipids in lipid membranes: Differences in fluorescence behavior, *Biophys. J.* **71**(1), 327–335 (1996).
31. D. L. Bernik, D. Zubiri, E. Tymczyszyn, and E. A. Disalvo, Polarity and packing at the carbonyl and phosphate regions of lipid bilayers, *Langmuir* **17**(21), 6438–6442 (2001).
32. M. C. Rheinstadter, C. Ollinger, G. Fragneto, F. Demmel, and T. Salditt, Collective dynamics of lipid membranes studied by inelastic neutron scattering, *Phys. Rev. Lett.* **93**(10), (2004).
33. S. Tristram-Nagle, and J. F. Nagle, Lipid bilayers: Thermodynamics, structure, fluctuations, and interactions, *Chem. Phys. Lipids* **127**(1), 3–14 (2004).
34. R. Hutterer, A. B. J. Parusel, and M. Hof, Solvent relaxation of Prodan and Patman: A useful tool for the determination of polarity and rigidity changes in membranes, *J. Fluorescence* **8**(4), 389–393 (1998).
35. R. Hutterer, F. W. Schneider, H. Sprinz, and M. Hof, Binding and relaxation behaviour of Prodan and Patman in phospholipid vesicles: A fluorescence and H-1 NMR study, *Biophys. Chem.* **61**(2–3), 151–160 (1996).
36. J. Sykora, and M. Hof, Solvent relaxation in phospholipid bilayers: Physical understanding and biophysical applications, *Cell. Mol. Biol. Lett.* **7**(2), 259–261 (2002).
37. M. Hof, Solvent relaxation in biomembranes. In *Applied Fluorescence in Chemistry, Biology, and Medicine*, edited by W. Rettig, B. Strehmel and S. Schrader (Springer Verlag, Berlin, 1999), 439–456.
38. M. L. Horng, J. A. Gardecki, A. Papazyan, and M. Maroncelli, Subpicosecond measurements of polar solvation dynamics – coumarin-153 revisited, *J. Phys. Chem.* **99**(48), 17311–17337 (1995).
39. R. S. Fee, and M. Maroncelli, Estimating the time-zero spectrum in time-resolved emission measurements of solvation dynamics, *Chem. Phys.* **183**(2–3), 235–247 (1994).
40. J. Sykora, P. Kapusta, V. Fidler, and M. Hof, On what time scale does solvent relaxation in phospholipid bilayers happen? *Langmuir* **18**(3), 571–574 (2002).
41. L. Nilsson, and B. Halle, Molecular origin of time-dependent fluorescence shifts in proteins, *Proc. Natl. Acad. Sci. USA* **102**(39), 13867–13872 (2005).
42. J. Sykora, P. Jurkiewicz, R. M. Epand, R. Kraayenhof, M. Langner, and M. Hof, Influence of the curvature on the water structure in the headgroup region of phospholipid bilayer studied by the solvent relaxation technique, *Chem. Phys. Lipids* **135**(2), 213–221 (2005).

43. J. Sykora, V. Mudogo, R. Hutterer, M. Nepras, J. Vanerka, P. Kapusta, V. Fidler, and M. Hof, ABA-C-15: A new dye for probing solvent relaxation in phospholipid bilayers, *Langmuir* **18**(24), 9276–9282 (2002).
44. M. Yang, and R. Richert, Observation of heterogeneity in the nanosecond dynamics of a liquid, *J. Chem. Phys.* **115**(6), 2676–2680 (2001).
45. R. Richert, Spectral diffusion in liquids with fluctuating solvent responses: Dynamical heterogeneity and rate exchange, *J. Chem. Phys.* **115**(3), 1429–1434 (2001).
46. R. Jimenez, G. R. Fleming, P. V. Kumar, and M. Maroncelli, Femtosecond solvation dynamics of water, *Nature* **369**(6480), 471–473 (1994).
47. G. Weber, and F. J. Farris, Synthesis and spectral properties of a hydrophobic fluorescent-probe – 6-Propionyl-2-(Dimethylamino)Naphthalene, *Biochemistry* **18**(14), 3075–3078 (1979).
48. P. L. G. Chong, Effects of hydrostatic-pressure on the location of Prodan in lipid bilayers and cellular membranes, *Biochemistry* **27**(1), 399–404 (1988).
49. R. Hutterer and M. Hof, Probing ethanol-induced phospholipid phase transitions by the polarity sensitive fluorescence probes Prodan and Patman, *Z. Phys. Chem.* **216**, 333–346 (2002).
50. A. Olzynska, A. Zan, P. Jurkiewicz, J. Sykora, G. Grobner, M. Langner, and M. Hof, Molecular interpretation of fluorescence solvent relaxation of Patman and H-2 NMR experiments in phosphatidylcholine bilayers, *Chem. Phys. Lipids* **147**(2), 69–77 (2007).
51. K. Rieber, J. Sykora, A. Olzynska, R. Jelinek, G. Cevc, and M. Hof, The use of solvent relaxation technique to investigate headgroup hydration and protein binding of simple and mixed phosphatidylcholine/surfactant bilayer membranes, *Biochim. Biophys. Acta* **1768**(5), 1050–1058 (2007).
52. T. Parasassi, E. K. Krasnowska, L. Bagatolli, and E. Gratton, Laurdan and Prodan as polarity-sensitive fluorescent membrane probes, *J. Fluorescence* **8**(4), 365–373 (1998).
53. P. L. G. Chong and P. T. T. Wong, Interactions of Laurdan with phosphatidylcholine liposomes – a high-Pressure ftir study, *Biochim. Biophys. Acta* **1149**(2), 260–266 (1993).
54. M. Viard, J. Gallay, M. Vincent, and M. Paternostre, Origin of Laurdan sensitivity to the vesicle-to-micelle transition of phospholipid-octylglucoside system: A time-resolved fluorescence study, *Biophys. J.* **80**(1), 347–359 (2001).
55. J. R. Lakowicz, D. R. Bevan, B. P. Maliwal, H. Cherek, and A. Balter, Synthesis and characterization of a fluorescence probe of the phase-transition and dynamic properties of membranes, *Biochemistry* **22**(25), 5714–5722 (1983).
56. A. Chattopadhyay and E. London, Parallax method for direct measurement of membrane penetration depth utilizing fluorescence quenching by spin-labeled phospholipids, *Biochemistry* **26**(1), 39–45 (1987).
57. F. S. Abrams, and E. London, Extension of the parallax analysis of membrane penetration depth to the polar-region of model membranes – use of fluorescence quenching by a spin-label attached to the phospholipid polar headgroup, *Biochemistry* **32**(40), 10826–10831 (1993).
58. P. Jurkiewicz, A. Olzynska, M. Langner, and M. Hof, Headgroup hydration and mobility of DOTAP/DOPC bilayers: A fluorescence solvent relaxation study, *Langmuir* **22**(21), 8741–8749 (2006).
59. S. S. Antollini, and F. J. Barrantes, Disclosure of discrete sites for phospholipid and sterols at the protein–lipid interface in native acetylcholine receptor-rich membrane, *Biochemistry* **37**(47), 16653–16662 (1998).
60. C. D. Geddes, Optical halide sensing using fluorescence quenching: Theory, simulations and applications – a review, *Meas. Sci. Technol.* **12**(9), R53–R88 (2001).
61. K. Bhattacharyya, and B. Bagchi, Slow dynamics of constrained water in complex geometries, *J. Phys. Chem. A* **104**(46), 10603–10613 (2000).
62. S. K. Pal, and A. H. Zewail, Dynamics of water in biological recognition, *Chem. Rev.* **104**(4), 2099–2123 (2004).
63. J. R. Silvius, Anomalous mixing of zwitterionic and anionic phospholipids with double-chain cationic amphiphiles in lipid bilayers, *Biochim. Biophys. Acta* **1070**(1), 51–59 (1991).

64. J. C. W. Shepherd, and G. Buldt, Zwitterionic dipoles as a dielectric probe for investigating head group mobility in phospholipid membranes, *Biochim. Biophys. Acta* **514**(1), 83–94 (1978).
65. J. Seelig, P. M. Macdonald, and P. G. Scherer, Phospholipid head groups as sensors of electric charge in membranes, *Biochemistry* **26**(24), 7535–7541 (1987).
66. P. G. Scherer, and J. Seelig, Electric charge effects on phospholipid headgroups – phosphatidylcholine in mixtures with cationic and anionic amphiphiles, *Biochemistry* **28**(19), 7720–7728 (1989).
67. A. A. Gurtovenko, M. Patra, M. Karttunen, and I. Vattulainen, Cationic DMPC/DMTAP lipid bilayers: Molecular dynamics study, *Biophys. J.* **86**(6), 3461–3472 (2004).
68. A. Olzynska, P. Jurkiewicz, and M. Hof, Properties of mixed cationic membranes studied by fluorescence solvent relaxation, *J. Fluorescence* (in press) (2008).
69. S. J. Ryhanen, J. M. I. Alakoskela, and P. K. J. Kinnunen, Increasing surface charge density induces interdigitation in vesicles of cationic amphiphile and phosphatidylcholine, *Langmuir* **21**(13), 5707–5715 (2005).
70. L. F. Zhang, T. A. Spurlin, A. A. Gewirth, and S. Granick, Electrostatic stitching in gel-phase supported phospholipid bilayers, *J. Phys. Chem. B* **110**(1), 33–35 (2006).
71. P. Jurkiewicz, A. Okruszek, M. Hof, and M. Langner, Associating oligonucleotides with positively charged liposomes, *Cell. Molec. Biol. Lett.* **8**, 77–84 (2003).

Quantum Dot-Encoded Fluorescent Beads for Biodetection and Imaging

Jian Yang, Mark P. Sena, and Xiaohu Gao

Abstract The need to analyze increasingly large panels of biomolecules for biomedical research and clinical diagnostics has greatly stimulated the development of high-throughput screening technologies based on quantum dots. Herein, we describe recent advances in the synthesis of quantum-dot-encoded beads and control of their fluorescence signatures. Compared with commercial beads based on organic fluorophores, they exhibit significant advantages in terms of detection sensitivity, throughput, and photostability. We discuss bioapplications of quantum-dot-encoded beads in ultrasensitive detection and cellular imaging, review future challenges for their development, and highlight their most promising applications in bioanalytics.

Keywords Fluorescence · Encoding · Barcodes · Porous · Detection · Imaging · Quantum dots · Nanoparticle · Self assembly · Multiplexing

Recent advances in analytical chemistry and bioengineering have led to the development of a wide spectrum of enabling technologies for biomedical research. As current research in genomics and proteomics produces more sequence data, there is an increasing need for new technology platforms that can rapidly screen a large number of biomolecular targets such as nucleic acids, proteins, lipids, and enzymes. In this context, the development of miniaturized barcodes based on spectroscopic signatures (e.g., fluorescence, Raman spectrum, and infrared spectrum) of organic and inorganic materials has become an important strategy for multiplexed and sensitive detection. Although a large number of compounds are Raman or infrared active, their spectra are complicated, resulting in difficulties in decoding under highly multiplexed conditions. Similarly, fluorescence encoding techniques based on organic fluorophores suffer from relatively low coding capacity, photo-instability, and inability of simultaneous excitation of multiple colors. We and others have

X. Gao (✉)
Department of Bioengineering, University of Washington, Seattle, WA 98195, USA
e-mail: xgao@u.washington.edu

C.D. Geddes (ed.), *Reviews in Fluorescence 2007*, Reviews in Fluorescence 2007,
DOI 10.1007/978-0-387-88722-7_7,

recently addressed these problems by replacing organic fluorophores with semiconductor quantum dots (QDs). Compared to organic dyes, QDs possess many unique optical properties, such as symmetrical and narrow emission, broad absorption spectrum, large Stokes shift, large extinction coefficient, and high photostability, which make QDs the best fluorophore for optical barcoding.

In this chapter, we will discuss recent advances in QD-encoded microbeads and nanobeads, their optical properties, as well as their applications in ultrasensitive detection and cellular imaging. We will also provide our perspectives on future research directions.

QD-Encoded Microbeads

An early approach for incorporating QDs into microbeads is to first soak microbeads with chemical precursors and then initiate QD formation in situ [1–3]. This approach is straightforward; however, the resulting QDs generally have low quality in terms of particle monodispersity, quantum yield, and stability. In addition, it is nearly impossible to synthesize multicolor QDs inside the same microbeads for multiplexing. Here, we describe recent approaches that utilize high-quality QDs pre-made in hot organic solvents and assembled inside or on the surface of polymer scaffolds.

Porous Microbeads Doped with QDs

Optically encoded microbeads doped with QDs were first reported by Han, Gao, and Nie [4]. The barcoding was achieved by doping multicolor QDs at different intensity ratios. The use of 10 intensity levels and six colors can theoretically code for one million DNA or protein sequences. These QD-tagged microbeads were produced by first swelling polystyrene (PS) microbeads in the presence of 5–10% chloroform. Multicolor QDs suspended in butanol were then mixed with the swelled beads for doping. Because of the hydrophobic nature of polystyrene, partitioning of hydrophobic QDs into the microbeads over butanol is extremely high, leading to QD incorporation at precisely controlled doping levels (determined by the initial concentrations of microbeads and QDs). A limitation of using swelled polystyrene microbeads, however, is that the nanoparticle penetration depth is limited to a few hundred nanometers. Indeed, confocal microscopy studies conducted by Bradley et. al. have shown that the penetration depth of QDs into polystyrene microspheres correlates positively with the degree of swelling and negatively with the density of PS cross-linkers. In their study, QDs penetrated the bead core only when the microspheres were swollen to their maximum volume [5]. Unfortunately, harsh swelling conditions often induce PS microbead aggregation. As shown in Fig. 1, by combining QDs with mesoporous materials, we have solved this problem and have improved the fluorescence intensity and uniformity of barcode microbeads by two orders of magnitude [6,7]. Based on these highly porous microstructures, QDs

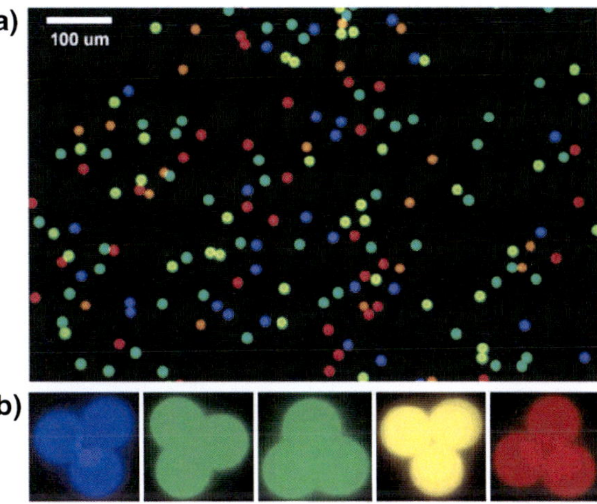

Fig. 1 True-color fluorescence images of mesoporous polystyrene beads (15.4 ± 0.2 μm diameter) doped with single-color quantum dots emitting light at 488 (*blue*), 520 (*green*), 550 (*yellow*), 580 (*orange*), or 610 nm (*red*). (**a**) Wide-field view of a large population of doped beads, prepared in batches and then mixed; and (**b**) detailed views of monochromatic bead clusters

can penetrate deeply inside the microbeads, revealed by fluorescence imaging and electron microscopy of 70 nm bead cross sections.

In contrast to the hydrophobic QDs discussed above, Nabiev's group incorporated water-soluble CdSe/ZnS core/shell nanocrystals into carboxylate-functionalized polystyrene microspheres with diameters of 0.3–1 μm [8]. These polymer beads displayed strong fluorescence comparable to commercial polymer microspheres doped with organic fluorophores. 3D confocal microscopy and time-resolved fluorescence measurements showed that more than 70% nanocrystals were stably incorporated into the polymer beads whereas the other 30% resided at or near the surface. Neither the position nor the width of the QD emission band was strongly affected by the encapsulating polymer. Based on fluorescence decay parameters, the highly fluorescent polymer beads exhibited excellent photostability in comparison to commercial, organic-fluorophore-doped polymer microspheres.

Stimuli-responsive hydrogel microspheres can also be used as matrices for the encapsulation of QDs [9]. Such materials allow for facile control of nanoparticle doping and release kinetics by altering pH or temperature. Wang and Gao demonstrated this concept by incorporating water-soluble CdTe QDs into a N-isopropylacrylamide and 4-vinylpyridine copolymer (PNIPVP) microspheres. PNIPVP microgels exhibited pH-dependent swelling behavior, which was utilized for control of QD loading and release from the gel matrix. When PNIPVP microgel particles were dispersed in a solution with a pH lower than the p*K*a of 4-vinylpyridine, the pyridine groups became fully protonated, resulting in internal charge repulsion and thus expansion of the hydrogel. At a pH of 3, an increase of the PNIPVP pore size enabled the homogeneous uptake of CdTe nanocrystals in controllable ratios. After incorporation, raising the pH effectively confined QDs within the hydrogel by reducing the PNIPVP pore size (Fig. 2). Within a pH range of 3–10, no leakage of QDs from the microgel was observed. Above a pH of 11,

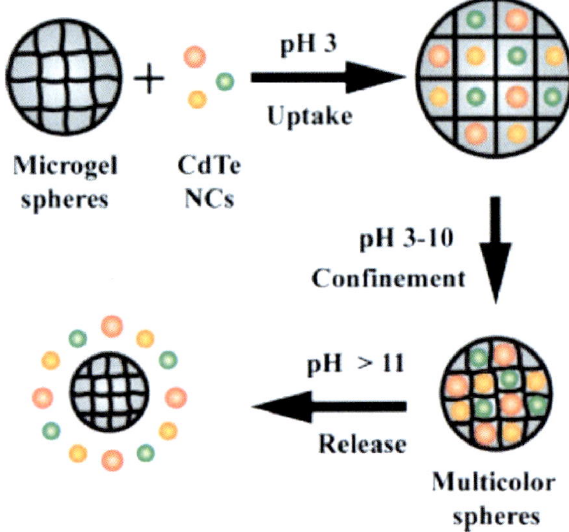

Fig. 2 Schematic illustration of the pH-controlled uploading and release of CdTe NCs into and out of PNIPVP spheres

however, CdTe nanocrystals were released from microspheres, due to complete collapse of the PNIPVP pores.

Sharing a similar mechanism, Wang and Gao also employed temperature-responsive poly(*N*-isopropylacrylamide) (PNIPAM) hydrogel microspheres for the incorporation of CdTe nanocrystals [10]. Below its lower critical solution temperature (LCST), PNIPAM is hydrophilic and swells with water. Above the LCST, the hydrogel collapses, expelling water and shrinking in size. CdTe nanocrystals modified by both thioglycerol and thioglycolic acid were incubated with PNIPAM microspheres at various temperatures. Stable incorporation into the hydrogel was driven by hydrogen bond formation between thioglycerol and PNIPAM's amide groups. It is important to note that without thioglycerol, QDs leached out of the hydrogel below the critical temperature. Thus, the authors deduced that through hydrogen bonding, QDs served as crosslinking agents—both increasing the effective LCST and decreasing the hydrodynamic radius of PNIPAM microspheres. Consistent fluorescence spectra taken before and after encapsulation indicated no QD aggregation. However, temperature-dependant fluorescence resonance energy transfer (FRET) was observed in microspheres containing a mixture of 2.5 and 3.5 nm QDs. Upon collapse of the hydrogel at increasing temperatures, the distance between neighboring QDs decreased to within the Forster radius and energy transfer from the 2.5 nm QDs to the 3.5 nm QDs was observed. This resulted in a reversible red shift in microsphere fluorescence (Fig. 3). In general, hydrogels serve as flexible platforms for a variety of biological applications. As demonstrated, environmentally sensitive composites show promise for use as sensors, or for controlled delivery of drug or nanoparticle payloads. Thus, hydrogels may be a useful class of scaffolds for the future development of multifunctional QD encoded beads.

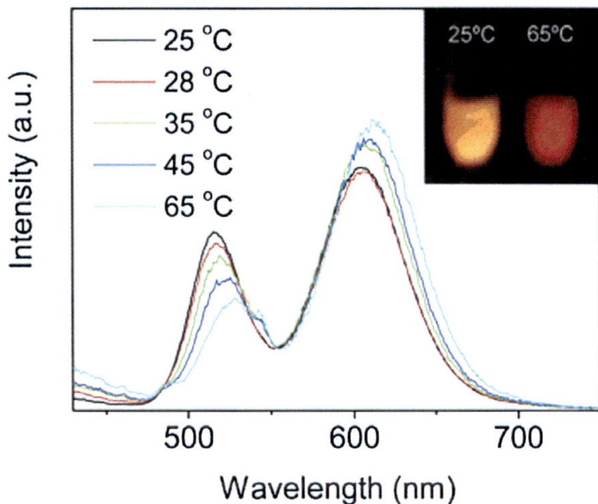

Fig. 3 Fluorescence spectra of PNIPAM spheres loaded with both 2.5 and 3.5 nm CdTe. (*Inset*) Fluorescent images of spheres under UV irradiation at 25 and 65°C, respectively

Layer-by-Layer Assembly

Layer-by-Layer assembly (LbL) is a surface-coating technique that involves the deposition of alternating, oppositely charged polyelectrolytes onto a surface or core particle. Using QDs with high surface charges, this technique can be employed for the preparation of fluorescent microspheres. The substantial features of the LbL technique are (1) it offers a facile and effective route to the incorporation of QDs without complicated chemical reactions; (2) it presents an excellent method to modulate QD loading and microsphere composition; and (3) it enables precise control of the radial distribution of QDs, as well as the size and topology of the microspheres.

Rogach and coworkers coated submicrometer-sized latex spheres with QDs for the engineering of photonic crystals based on LbL techniques [11]. Interestingly, they found that there was a strong electrostatic interaction between $-NH_3^+$ groups of poly(allylamine hydrochloride) (PAH) and –OH or –SH groups of 1-mercapto-2,3-propandiol and 1,2-dimercapto-3-propanol. TEM microscopy, EDX analysis, and confocal microscopy confirmed the successful adsorption of QDs and PAH onto latex spheres. Upon evaporation of the carrier solvent, these latex spheres self-assembled into an ordered array—the 3D architecture of sphere arrays exhibited face-centered cubic packing. The observed optical transmission spectrum of these arrays suggested that QDs significantly increase the refractive index contrast between latex spheres and interparticle voids, which dramatically affects the width and depth of the photonic crystal's stop band.

The same group also developed a method for encapsulating QDs into hollow polymer spheres [12]. In the report, PAH was first precipitated by citrate ions onto the surface of melamine formaldehyde (MF) microspheres 4.2 μm in diameter (Fig. 4b). The resulting core-shell microspheres were then coated with four alternating layers of polyelectrolytes (PAH and poly(styrene sulfonate), PSS). Next, the

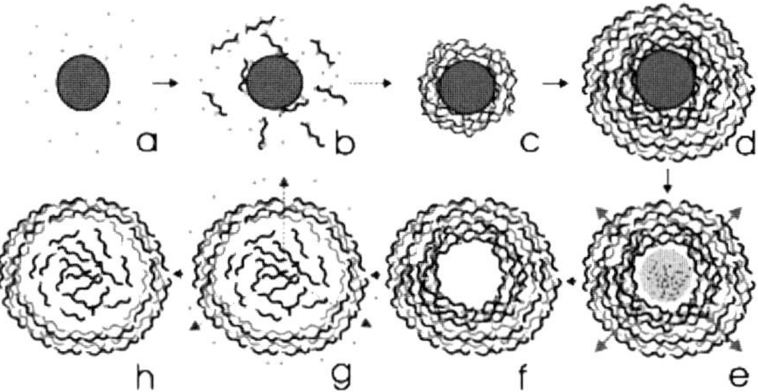

Fig. 4 Schematic illustration of the preparation of hollow polymer microcapsules. (**a–d**) LbL assembly of polyelectrolytes on the surface of a core particle. (**e**,**f**) Acidic removal of the core scaffold. (**g**,**h**) Ionic dissolution of the innermost layer resulting entrapped cationic PAH

MF core was dissolved in HCl and the citrate/PAH layer was disrupted by the addition of high ionic strength solution (Fig. 4e). Citrate ions were small enough to pass through polyelectrolyte multilayers, but positively charged PAH remained inside the microcapsules due to its high molecular weight. Upon mixing negatively charged CdTe nanocrystals with the hollow spheres in solution, QDs were electrostatically driven inside. These nanocrystals both adsorbed onto the inner wall of polymer spheres and combined with polyelectrolytes inside. The authors demonstrated that even after several rounds of centrifugation and washing, no leakage of nanocrystals into solution occurred. Interestingly, further studies on green and red fluorescent spheres showed that there was a nearly linear relationship between the green/red fluorescence intensity ratio and corresponding concentration ratio, indicating minimal energy transfer between different colors. This approach for QD encapsulation combines the advantages of facile QD doping after microparticle synthesis with the protective capacity of charge-stabilized multilayer polymer microcapsules.

LbL techniques for the preparation of QD-embedded microspheres have also been demonstrated by Caruso's group [13]. In their report, a three-layer PE film comprising PAH/PSS/PAH (PE_3) was first deposited onto $\sim$1 μm polystyrene beads in order to generate a smooth, positively charged surface. Negatively charged, thioglycolic acid-stabilized QDs were then electrostatically deposited on the microspheres. Several alternating layers of QDs and PE_3 were applied in order to generate the multilayer assembly. Finally, a PAH/PSS layer was applied, followed by adsorption of anti-IgG for use in biological assays. TEM images of the resulting microspheres revealed a smooth, aggregate-free surface (individual CdTe QDs were visible in high resolution mode). Based on the concentration of coated microspheres obtained from single particle light scattering (SPLS), and the known concentrations of QDs, the surface coverage of CdTe QDs was estimated as 20–30 mg/m^2, which is greater than the theoretical density from a close-packing model. The authors attributed this to infiltration of QDs into the PE thin film during deposition.

LbL assembly of polyelectrolytes onto core particles is a versatile approach to microsphere and microcapsule production as well as QD encapsulation. The use of various core materials, charged polymers, and layering schemes enables manipulation of the composition, luminescence, and biocompatibility of the resulting barcoded microparticles.

Polymerization/Silanization

Another notable approach to producing QD-encoded microbeads is the entrapment of QDs within the encapsulating matrix *during* microparticle synthesis. The key to success is the utilization of chemically reactive nanoparticle surface ligands. This technique enables covalent incorporation of QDs into polystyrene or silica beads. In contrast, when non-reactive surface ligands are used, the nanoparticle and polymer often phase separate during polymerization, leading to nanoparticle clustering on the surface of the resulting microspheres [14]. The polymerization approach offers significant stability over the diffusion-based or electrostatic-deposition-based methods described earlier.

As an example, O'Brien and coworkers employed suspension polymerization in order to produce QD-doped polystyrene microbeads [15]. Ligand exchange was first performed in order to replace hexadecylamine, a straight chain surfactant, with polymerizable ligands on the surface of QDs. Styrene polymerization in a QD suspension resulted in the generation of fluorescent microspheres with covalently incorporated QDs. Although the microspheres exhibited an emission peak similar to that of QDs, their sizes and brightness showed wide distribution. Later, the same group improved the size control in the case of CdS-doped spheres [16]. In this case, oleic acid was used as a QD capping ligand, which was believed to participate in the polymerization reaction through its reactive double bond. Laser scanning confocal microscopy confirmed that CdS QDs were distributed evenly within the microspheres. Although slight blue shifts in the fluorescence emission spectra occurred due to a refractive index change, aggregation (indicated by peak broadening and red-tailing) was not observed.

Professor Bawendi's group studied a similar route to producing polystyrene microspheres doped with oligomeric phosphine (OP)-functionalized CdSe/ZnCdS/ZnS QDs [17]. In order to meet the challenge of providing chemical compatibility between QDs and the encapsulating polymer, while simultaneously maintaining the fluorescent properties of QDs, cap-exchange was performed using OP ligands. These ligands were derivatized with methyl methacrylate (MMA) groups which bind strongly to QDs, confer QD solubility in polar solvents, and provide chemical stability to QDs upon self-polymerization and copolymerization with polystyrene. Detailed studies showed that the incorporation of QDs in polystyrene microspheres disturbed nucleation and growth, and led to the broadening of particle size distributions. Quantum efficiency also fell from 20 to 12% after encapsulation, indicating that despite protection by OP ligands, the optical properties of QDs are still sensitive to the local environment.

Gao's group employed a similar strategy but a different experiment design [18]. In this report, 3-mercaptopropionic acid (MPA)-modified CdTe QDs synthesized in aqueous solution were first transferred into a hydrophobic phase by didecyl-*p*-vinylbenzylmethylammonium chloride (DVMAC). Here, DVMAC acted not only as a surfactant and phase-transfer agent, but also as a chemical handle for polymerization due to its active vinyl group. Using octadecyl-*p*-vinylbenzyldimethylammonium chloride as an emulsifier, fluorescent polystyrene (PS) beads doped with CdTe QDs

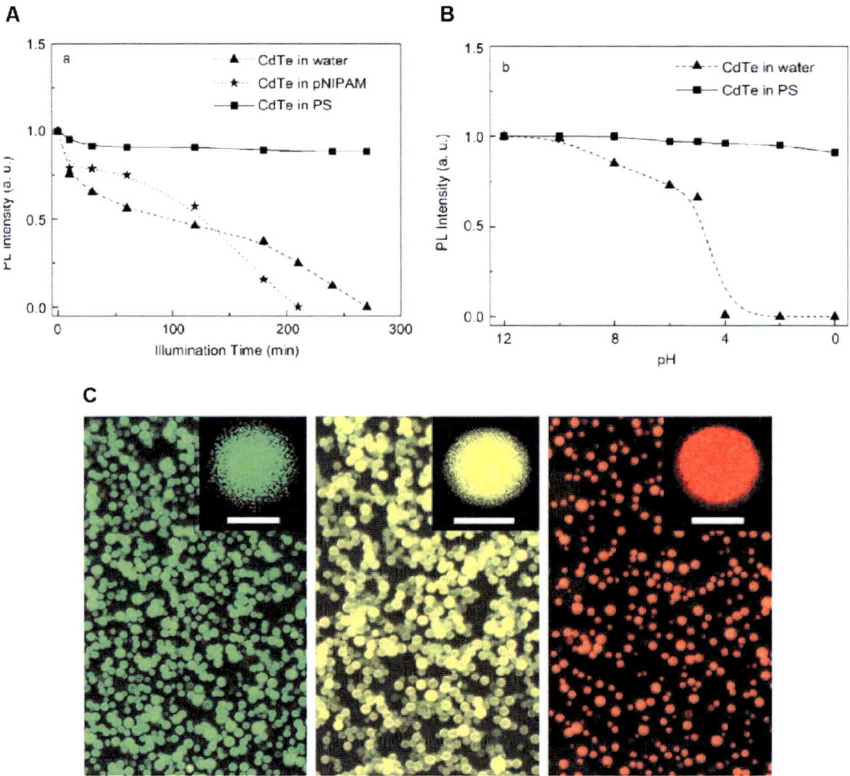

Fig. 5 (**a**) Photostability of fluorescent PS beads compared to QDs in water and within PNIPAM microspheres. (**b**) pH dependence of luminescence intensities of PS beads compared to CdTe nanocrystals in water. (**c**) Fluorescence images of PS beads loaded with *green* (*left*), *yellow* (*middle*), and *red* CdTe *dots* (*right*) and corresponding confocal fluorescence images (*insets*). The scale bars correspond to 2 μm

were prepared via a miniemulsion polymerization method. Despite size polydispersity and reduced photoluminescence efficiency, the fluorescent microspheres were observed to uptake QDs homogeneously and retain their emission spectra—key factors for optical encoding applications. In addition, protection from the external environment by the polymer matrix resulted in enhanced photostability and pH resistance (Fig. 5a,b). Dual-color beads were also synthesized using green and red CdTe QDs with emission bands at 546 and 634 nm, respectively (Fig. 5c). This demonstrated their efficacy for use in optical encoding.

Silanization has proven to be another useful means for producing QD-encoded microspheres. In this case, silica is the material for the encapsulating matrix rather than polystyrene. Bawendi's group developed a robust method to grow a QD-embedded silica shell on prefabricated monodisperse silica microspheres through hydrolysis of tetraethoxysilane (TEOS) [19]. In order to achieve high loading efficiency and to prevent phase separation of deposited silica and QDs, native trioctylphosphine oxide was substituted with 5-amino-1-pentanol (AP) and 3-aminopropyltrimethoxysilane (APS) on the QD surface. The hydroxyl group of AP conferred ethanol solubility, while APS provided chemical compatibility with the silica matrix through the formation of siloxane bonds. It is important to note that silanization using QDs functionalized with AP or APS alone resulted in nanocrystals aggregation. The microspheres produced by this method exhibited good luminescence with an emission profile almost identical to that of QDs alone.

QD-Encoded Nanobeads

The encoded microbeads discussed above are extremely bright and apt for multiplexing due to the large number of QDs embedded inside (thousands to millions per bead) and are the ideal platforms for suspension based biomolecular screening. However, because of the large size (typically 1–15 μm), these QD-doped microspheres are not suitable for applications such as gene, protein, and cell labeling. For these bioapplications, a number of technologies have been developed for the production of uniform and bright QD-encoded beads in the nanometer regime.

Silica-Based Nanobeads

Kotov and coworkers reported the preparation of "raisin bun" type silica nanobeads with multiple citrate-stabilized QDs doped inside [20]. QDs were modified with 3-mercaptopropyl trimethoxysilane (MPS) so that they could polymerize with siloxane compounds. This approach produced SiO_2 nanobeads from 40 to 80 nm, which is the size suitable for biomolecular labeling. Unfortunately, the absorption and emission profile was observed to shift and the quantum yield of the embedded QDs decreased significantly. Ying's group reported an improved method by combining reverse microemulsion with hydrophobic CdSe QDs [21]. QDs were coated with Igepal CO-520, an amphiphilic detergent, and confined within aqueous droplets of the microemulsion. Interestingly, the fluorescence intensity of the CdSe QDs increased after silica coating, similar to the result of capping CdSe dots with ZnS or CdS shells. In vitro cytotoxicity studies also revealed the non-toxic property of the silica nanobeads, although the long term (e.g., years) toxicity in animals is difficult to predict. Using this approach, the content of doped nanobeads is not limited to QDs alone. Multifunctional silica nanobeads simultaneously embedded with QDs and magnetic nanoparticles have also been produced [22]. Besides fluorescence, these silica beads also exhibited superparamagnetic behavior, with a coercivity value and blocking temperature comparable to those of the original γ-Fe2O3 particles. A problem observed for the dually functional nanobeads, however, was fluorescence attenuation by magnetic nanoparticles. This is not surprising because magnetic nanoparticles exhibit a broad absorption spectrum [23]. In addition, when multicolor QDs are used for doping, fluorescence resonance energy transfer can occur. One possible solution to this problem is to physically separate magnetic nanoparticles and QDs of different colors inside the nanobeads. In one example, Wang et al. used a modified LbL assembly approach to assemble layers of multicolor QD on the surface of a core containing magnetic nanoparticles [24].

An interesting silica-based nanostructure—mesoporous silica doped with magnetite nanoparticles and QDs—was recently reported by Hyeon and coworkers [25]. A detergent, cetyltri-methylammonium bromide (CTAB), served as a phase transfer agent to solubilize hydrophobic nanoparticles as well as a structural template for the formation of mesoporous silica spheres. The combination of nanoparticles and a porous bead structure could allow for applications in tracing studies and controlled drug delivery.

Polymer-QD Self-Assembly

The second important route for synthesis of nanoparticle doped nanobeads is based on polymer-nanoparticle self-assembly. Moffit's group carried highly systematic studies on CdS doped nanobeads [26]. In this approach, amphiphilic polystyrene-poly(acrylic acid) (PS-*b*-PAA) block copolymers were utilized as QD stabilizers and nanobead building blocks. Here, the PAA block bound to the Cd atoms on the QD surface, leaving the PS domains exposed to hydrophobic solution. Upon the dropwise addition of water (a poor solvent for hydrophobic QDs), CdS QDs aggregated in the presence of excess PS-*b*-PAA polymers (Fig. 6). The size of the nanobeads could be reproducibly controlled by the block copolymer concentration and the water addition rate.

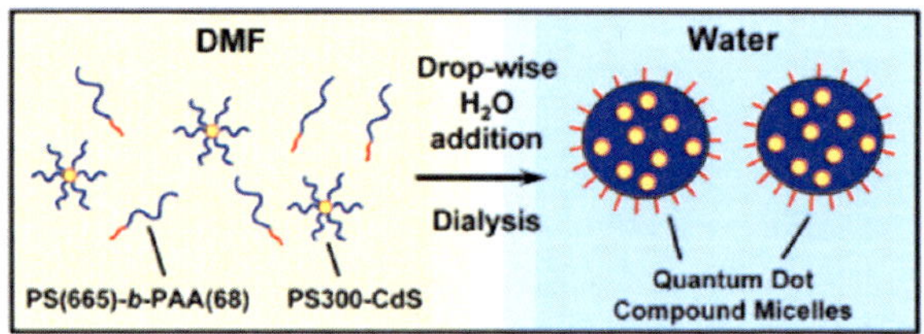

Fig. 6 Schematic of the formation of QD-encapsulating PS/PAA nanobeads by dropwise phase transfer

Recently, we reported a new approach for preparation of QD-tagged nanobeads based on epitaxial growth of nanoparticle-amphiphilic polymer complexes in homogeneous solution [27]. This next generation fluorescent probe is uniform in size, thousands of times brighter than single organic dyes, stable against photobleaching, and free of the 'blinking' effect—a phenomenon characteristic of single dye molecules and QDs [28,29]. The key feature was an amphiphilic alternating copolymer, poly(maleic anhydride-octadecene) (PMAO), which was capable of coating multicolor QDs via multivalent hydrophobic interactions. This effectively prevented QDs from forming irregular agglomerates upon nanobead formation, as well as from 'touching', which occurs when multiple QDs are sequestered within copolymer micelles. The QD-PMAO conjugates were initially highly soluble in tetrahydrofuran

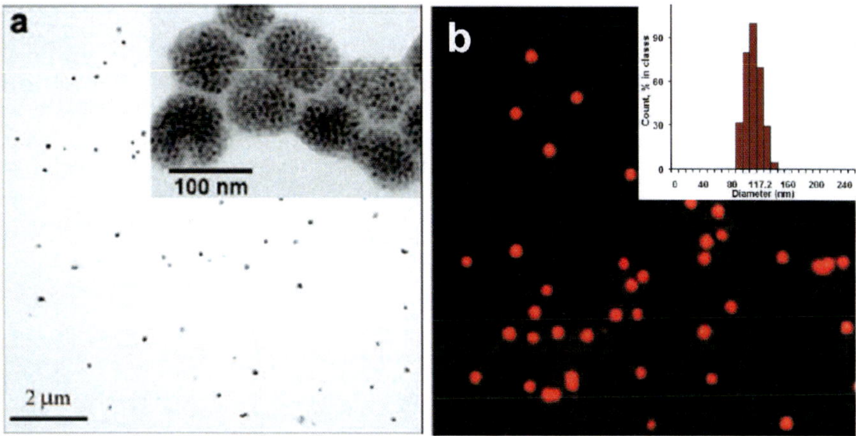

Fig. 7 (**a**) TEM and (**b**) fluorescence microscopy image of PMAO/QD nanobeads. (*Insets*) High resolution TEM and hydrodynamic size distribution by DLS (112 ± 18 nm).

(THF) forming aggregates in polar solvents such as dimethylformamide (DMF). However, by the slow addition of DMF into THF, a polarity gradient was created, which led to epitaxial growth of highly fluorescent nanobeads with narrow size dispersity. As seen in Fig. 7, the resultant beads were about 100 nm in diameter and approximately 200 times brighter than single QDs, which makes them suitable for molecular and cellular labeling.

Based on the same procedure, we have also demonstrated the preparation of multicolor QD doped nanobarcodes. Monodisperse QDs with fluorescence emission maxima at 520 nm (green) and 615 nm (red) were premixed at various fluorescence intensity ratios and used for nanobead synthesis. Using two intensities, three unique color barcodes were generated (green/red 1:2, 1:1, and 2:1). The nanobarcodes with ratios of 2:1, 1:1, and 1:2 appeared yellow, orange, and red by fluorescence imaging, which were confirmed by solution-based spectral measurements. Single bead spectroscopy using a hyperspectral imaging approach demonstrated that the fluorescence intensity ratios of individual nanobarcodes were remarkably robust, although the absolute intensities varied considerably from bead to bead. This investigation eliminated the possibility that the coded fluorescence spectra was representative of an ensemble-average rather than of an actual nanobarcode.

Self assembly of polymer-QD nanobeads enables a large number of QDs to be distributed evenly within the encapsulating matrix. With controllable color barcodes and narrow size distirbution, these beads represent a new class of nanoscale encoded probes.

In Vitro Biodetection and Imaging

Encoded micro/nanobeads are highly applicable in vitro for molecular diagnostics and biomedical research. Of the synthetic techniques described above, those that produce bright and easily modifiable beads are best suited for detection and labeling. In combination with single bead spectroscopy, encoded beads have great potential for highly multiplexed assays and biomolecular profiling. Meanwhile, bright nanobeads afford a high degree of detection sensitivity in assays and in biological contexts. Recent work has demonstrated the value for encoded beads in these fields.

In Vitro Screening Using Barcoded Microbeads

Using triple-color encoded microbeads, we have demonstrated a model oligonucleotide hybridization system for DNA detection [4]. Target DNA molecules were directly labeled with a fluorescent dye, and fluorescent microbeads were functionalized with complementary oligonucleotide probes. Optical spectroscopy at the single-bead level yielded both the coding and the target signals. The coding signals identified the DNA sequence, whereas the target signal indicated the presence and the abundance of that sequence. The assay involved hybridization of dye-labeled target sequences to complementary coded-bead probes (Fig. 8a). Analyte fluorescence signals were observed in the presence of complementary targets, whereas no analyte fluorescence was detected when control oligos (noncomplementary sequences) were used for hybridization (Fig. 8b). A limiting factor in target detection was the significant spectral overlap between the coding and reporter signals. To minimize this interference as well as enhance the detection sensitivity, target molecules could be labeled with single QDs rather than a fluorescent dye [30]. In our study, the barcode decoding was achieved using wavelength resolved spectroscopy. This decoding approach is sensitive and accurate, but slow, because the microbeads have to be measured individually. For high throughput analysis, this problem could be solved using flow cytometry, which can read thousands of fluorescent labeled cells and microparticles per second [7]. Due to high specificity and rapid reaction kinetics compared to planar chip systems, this type of assay is highly applicable for rapid, multiplexed screening, and sequencing.

In the first practical demonstration of QD-bead-based multiplexed analysis of patient samples, the Mahoney group used QD-encoded latex microbeads in conjunction with flow cytometry for high-throughput genotyping of single nucleotide polymorphisms (SNPs). They demonstrated the quantitation of 10 PCR-amplified SNPs using encoded beads containing three intensity levels of green and yellow (530 and 565 nm emission) QDs. Capture of the biotin-labeled amplicons was detected by 670 nm fluorescence emission of Streptavidin-phycoerythrin-Cy5. The authors demonstrated that their Qbead™ genotyping system was highly accurate, discerning 100% of the 286 genotypes tested, whereas direct sequencing discerned only 97.9%. It was also sensitive, requiring only 200 pg of initial genomic DNA. The Qbead™ system is less expensive and exhibited faster reaction kinetics than microarray technologies, which require 2D chip platforms.

Encoded microbeads are adaptable for use in gene expression and protein-protein assays. In combination with automated sample handling systems, this technology

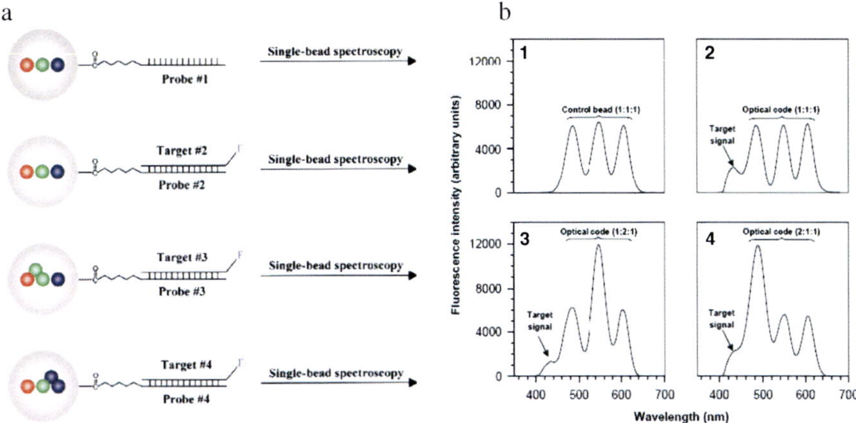

Fig. 8 (**a**) Schematic illustration of DNA hybridization assays using QD-tagged beads. Probe oligos (No. 1–4) were conjugated to the beads by crosslinking, and target oligos (No. 1–4) were detected with a *blue* fluorescent dye such as Cascade *Blue*. (**b**) Fluorescence spectra of the resulting bead-target conjugates (no target detected for the control probe no. 1)

has great potential for future use in high throughput screening – especially as the number of QD colors and intensity levels increases.

QD-Nanobarcodes as Fluorescence Reporters

We have recently studied the detection of protein targets using QD-embedded nanobeads as the reporter probe in an immunoassay [27]. QD-nanobeads were first functionalized with streptavidin using standard carbodiimide crosslinking chemistry; these replaced the organic dye or enzyme used in traditional sandwich immunoassays. For the detection of mouse IgG as a model target, goat anti-mouse IgG was adsorbed onto 96 well plates, followed by serially diluted target, then biotinylated rabbit anti-mouse IgG, and lastly, streptavidin-nanobeads (SA-NBs). Through the strong biotin-streptavidin interaction, SA-NBs bound to biotinylated antibodies. A low degree of nonspecific binding between SA-NBs and other proteins was observed. In comparison with the FITC labeled antibodies, the SA-NBs were 15 times more sensitive in detecting the mouse IgG molecules (Fig. 9a), likely due to the high brightness of the fluorescent nanobeads. We have further applied this technology to the detection of the tumor biomarker, prostate specific antigen (PSA) (Fig. 9b), and demonstrated that PSA was readily detected with sensitivity in the sub-nanomolar range (Fig. 9c). Further development and optimization of the SA-NBs and the sandwich immunoassay should significantly improve this detection limit. It is important to note that the encoding potential of nanobeads cannot be harnessed through the use of heterogeneous, surface-based assays due to the difficulty of optically interrogating single nanobarcodes. Homogeneous assays in conjunction

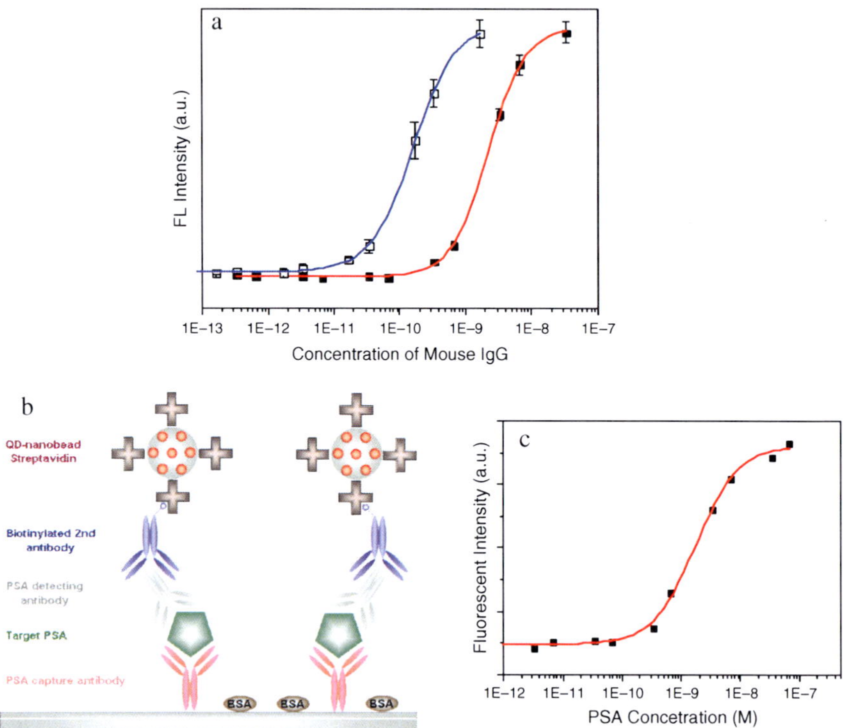

Fig. 9 (**a**) Mouse IgG detection sensitivity using QD-nanobeads (*blue*) in comparison to FITC (*red*). (**b**) Schematic illustration of the sandwich assay for PSA detection. (**c**) PSA detection sensitivity using QD-nanobeads

with flow cytometry are thus the most compatible systems for encoded QD-bead technology applied to protein detection.

Due to their small size, QD nanobeads are applicable for use in cellular and molecular imaging. Guo and coworkers delivered PNIPAM-coated QD-tagged nanoparticles into CHO cells [31]. Confocal fluorescence microscopy revealed the distribution of beads within cells, which suggested that the nanobeads were uptaken nonspecifically and resided mostly within the cytoplasm. Nabiev's group modified QD-nanobeads with anti-mouse IgG for detection of p-glycoprotein (a mediator of the multidrug resistance phenotype) overexpressed by MCF7r breast cells [8]. Fluorescence microscopy revealed the specific recognition of p-glycoprotein in different cell lines. The strong fluorescence and excellent binding selectivity of the QD nanobeads offer the possibilities for the detection of single biomolecules. Bawendi and Jain et al took this technology a step further by using QD-encoded silica beads for in vivo studies of mouse tumors [32]. In addition to performing extensive tumor imaging experiments using multicolor QDs in conjunction with multiphoton imaging, the investigators used variously sized QD-doped silica nanobeads to help guide the design of drug-delivering nanocarriers. In order to study the effect of nanocarrier size on efficacy for drug delivery, the ability of 100 and 500 nm diameter fluorescent nanobeads (blue and red, respectively) to extravasate tumor vasculature and permeate GFP-VEGF expressing perivascular cells was simultaneously assessed in vivo by multiphoton imaging. Because biological phenomena such as leakage of nanocarriers through tumor vasculaure into the surrounding tissue as well as uptake by the reticuloendothelial system are highly dependent on particle size, these studies were important for understanding the effect of nanocarrier size on biodistribution.

The high brightness of fluorescent nanobeads compared to single QDs allows for sensitive detection of biomolecules. Due to their small size compared to encoded microbeads, nanobeads are less prone to sterics-related problems during in vitro detection and cell or tissue labeling.

Perspectives

As we have described here, a number of strategies have been developed to incorporate QDs into nano- and micro-sized particles for the purpose of generating high-quality optical barcodes. Many techniques have significant advantages such as good size controllability, high capsulation efficiency, and strong fluorescence. However, an optimal QD-encoded particle possessing all of these characteristics has yet to be developed. Advances in the current tools and techniques that enable the production of high-quality QD-barcodes should eventually help push them to the market as precisely engineered reagents. With variable surface functionalities, such a product will be a valuable tool for use in many biodetection systems—much like the widely available suparaparamagnetic beads sold by numerous companies [33]. Further along the development path, we may witness QD-encoded particles being used for biological tracing, targeting, and drug delivery studies in vivo—although the challenges are complex and numerous.

As more precise control over QD loading is achieved, as well as over the optical properties of QDs themselves, beads with an unsurpassed number of optical codes will be generated. Because both QD size and composition can be used to tune fluorescence emission, beads possessing spectral signatures spanning both the visible and infrared spectrums will be possible. Based on 10 intensity levels, each QD color added to a library increases the number of possible optical barcodes by an order of magnitude. Such coding capacity makes QD-encapsulating beads candidates for security applications [34]. However, despite the power and versatility of QDs, a major challenge that will need to be addressed is maintaining their optical properties once incorporated into beads. Because oxidation, exposure to acidic conditions, or other forms of chemical degradation virtually destroy QDs, optimization of surface coatings that stabilize them within the encapsulating matrix and protect them from harsh environments will be critical.

Bioapplications of encoded bead technology present yet another challenge. Evolution has made it extremely difficult for foreign objects to persist in vivo. The reticuloendothelial system rapidly clears particles from the blood, while enzymes and acidic conditions within endosomes of the cell degrade anything that gets trapped inside. Even for in vitro diagnostic assays or cell imaging applications, biomolecules frequently adsorb onto particle surfaces, thus limiting the degree of specificity that coded-bead probes possess, and ultimately raising the detection limit due to enhanced background binding. To prevent this, surface coatings for barcode beads will have to be developed that prevent nonspecific adsorption of proteins.

Currently, the most applicable use for QD-based optical barcodes is in multi-analyte detection and biomolecular profiling. In combination with high-throughput single bead spectroscopy techniques such as flow cytometry, QD-encoded beads will serve as highly multiplexed platforms for a wide variety of in vitro diagnostic assays for both protein and nucleic acid detection. Flexible, bead-based detection platforms such as xMAP technology (Luminex®) that utilize up to 100 optical barcodes through the use of organic fluorophores are currently available (Fig. 10) [35]. Such systems are already being used for diagnostic applications such as molecular cancer profiling [36]. However, these are limited in their multiplexing capacity due to difficulties such as spectral overlap between coding dyes, a need for precise fluorophore excitation, and requirements for multiple excitation lasers. QD-encoded beads promise to resolve these issues due to their unique optical advantages. As mentioned, millions of optical codes can theoretically be produced, QDs are photostable (which resolves difficulties in fluorescence quantification), and most importantly for this application, multiple QD colors can be excited with a single, near-UV laser.

Overall, QD-encoded beads show tremendous promise as tools for molecular profiling, biological sensing, and imaging. The unique optical properties of quantum dots, along with the flexible surface and matrix chemistries, are making it possible to produce fluorescent beads with a variety of compositions, morphologies, and functionalities. Recent reductions in bead size from the micrometer to the nanometer regime (while maintaining unique spectral signatures) are allowing encoding applications to be extended to the cellular and molecular level. In the near future, both the quality and availability of QD-encoded beads will surely rise, especially as higher demands are made for multiplexed sensing and imaging technologies.

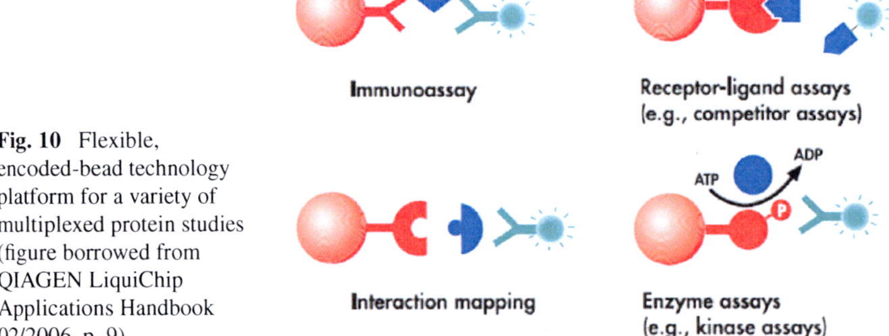

Fig. 10 Flexible, encoded-bead technology platform for a variety of multiplexed protein studies (figure borrowed from QIAGEN LiquiChip Applications Handbook 02/2006, p. 9)

Acknowledgments We thank the NIH, the NSF, the Seattle Foundation, the Department of Bioengineering at UW and the Department of Chemical Engineering at SCUT for generous financial support.

References

1. J. Zhang, S. Xu, and E. Kumacheva, Polymer microgels: reactors for semiconductor, metal, and magnetic nanoparticles, *J. Am. Chem. Soc.* **126**, 7908–7914 (2004).
2. J. Zhang, N. Coombs, E. Kumacheva, Y. Lin, and E. H. Sargent, A new approach to hybrid polymer-metal and polymer-semiconductor particles, *Adv. Mater.* **14**(23)1756–1759 (2002).
3. J. Zhang, N. Coombs, and E. Kumacheva, A new approach to hybrid nanocomposite materials with periodic structures, *J. Am. Chem. Soc.*, **124**(49), 14512–14513 (2002).
4. M. Han, X. Gao, J. Su, and S. Nie, Quantum-dot-tagged microbeads for multiplexed optical coding of biomolecules, *Nat. Biotech.* **19**(7), 631–635(2001).
5. M. Bradley, N. Bruno, and B. Vincent, Distribution of CdSe quantum dots within swollen polystyrene microgel particles using confocal microscopy, *Langmuir* **21**(7), 2750–2753 (2005).
6. X. Gao, and S. Nie, Doping mesoporous materials with multicolor quantum dots, *J. Phys. Chem. B* **107**(42), 11575–11578 (2003).
7. X. Gao, and S. M. Nie, Quantum dot-encoded mesoporous beads with high brightness and uniformity: rapid readout using flow cytometry, *Anal. Chem.* **76**(8), 2406–2410 (2004).
8. V. Stsiapura, A. Sukhanova, M. Artemyev, M. Pluot, J. H. M. Cohen, A. V. Baranov, V. Oleinikov, and I. Nabiev, Functionalized nanocrystal-tagged fluorescent polymer beads: synthesis, physicochemical characterization, and immunolabeling application, *Anal. Biochem.* **334**(2), 257–265 (2004).
9. M. Kuang, D. Wang, H. Bao, M. Y. Gao, H. Mohwald, and M. Jiang, Fabrication of multicolor-encoded microspheres by tagging semiconductor nanocrystals to hydrogel spheres, *Adv. Mater.* **17**(3), 267–250 (2005).
10. Y. Gong, M. Gao, D. Wang, and H. Mohwald, Incorporating fluorescent CdTe nanocrystals into a hydrogel via hydrogen bonding: toward fluorescent microspheres with temperature-responsive properties, *Chem. Mater.* **17**(10), 2648–2653 (2005).
11. A. Rogach, A. Susha, F. Caruso, G. Sukhorukov, A. Komowski, S. Kershaw, H. Mohwald, A. Eychmuller, and H. Weller, Nano- and microengineering: 3-D colloidal photonic crystals prepared from submicrometer-sized polystyrene latex spheres precoated with luminescent polyelectrolyte/nanocrystal shells , *Adv. Mater.* **12**(5), 333–337 (2000).
12. N. Gaponik, I. L. Radtchenko, G. B. Sukhorukov, H. Weller, and A. L. Rogach, Toward encoding combinatorial libraries: charge-driven microencapsulation of semiconductor nanocrystals luminescing in the visible and near IR, *Adv. Mater.* **14**(12), 879–882 (2002).
13. D. Wang, A. L. Rogach, and F. Caruso, Semiconductor quantum dot-labeled microsphere bioconjugates prepared by stepwise self-assembly, *Nano. Lett.* **2**(8), 857–861 (2002).
14. N. Joumaa, M. Lansalot, A. Theretz, A. Elaissari, A. Sukhanova, M. Artemyev, I. Nabiev, and J. H. M. Cohen, Synthesis of quantum dot-tagged submicrometer polystyrene particles by miniemulsion polymerization , *Langmuir* **22**(4), 1810–1816 (2006).
15. P. O'Brien, S. S. Cummins, D. Darcy, A. Dearden, O. Masala, N. L. Pickett, S. Ryley, and A. J. Sutherland, Quantum dot-labeled polymer beads by suspension polymerization, *Chem. Commun.* **3**(20), 2532–2533 (2003).
16. L. Yang, E. C. Y. Liu, N. Pickett, P. J. Skabara, S. S. Cummins, S. Ryley, A. J. Sutherland, and P. O'Brien, Synthesis and characterization of CdS quantum dots in polystyrene microbeads, *J. Mater. Chem.* **15**(12), 1238–1242 (2005).
17. W. Sheng, S. Kim, J. Lee, S. W. Kim, K. Jensen, and M. G. Bawendi, In-situ encapsulation of quantum dots into polymer microspheres, *Langmuir* **22**(8), 3782–3790 (2006).

18. Y. Yang, Z. Wen, Y. Dong, and M. Y. Gao, Incorporating CdTe nanocrystals into polystyrene microspheres: towards robust fluorescent beads, *Small* **7**(2), 898–901 (2006).
19. Y. Chan, J. P. Zimmer, M. Stroh, J. S. Steckel, R. K. Jain, and M. G. Bawendi, Incorporation of luminescent nanocrystals into monodisperse core-shell silica microspheres, *Adv. Mater.* **16**(23–24), 2092–2097 (2004).
20. A. L. Rogach, D. Nagesha, J W. Ostrander, M. Giersig and N. A. Kotov, "Raisin Bun"-type composite spheres of silica and semiconductor nanocrystals, *Chem. Mater.* **12**(9), 2676–2685 (2000).
21. S. T. Selvan, T. T. Tan and J. Y. Ying, Robust, non-cytotoxic, silica-coated CdSe quantum dots with efficient photoluminescence, *Adv. Mater.* **17**(13), 1620–1625 (2005).
22. D. K. Yi, S. T. Selvan, S. S. Lee, G. C. Papaefthymiou, D. Kundaliya, and J. Y. Ying, Silica-coated nanocomposites of magnetic nanoparticles and quantum dots, *J. Am. Chem. Soc.* **127**(14), 4990–4991 (2005).
23. R. F. Ziolo, E. P. Giannelis, B. A. Weinstein, M. P. Ohoro, B. N. Ganguly, V. Mehrotra, M. W. Russell, D. R. Huffman, Matrix-mediated synthesis of nanocrystalline γ-Fe2O3: a new optically transparent magnetic material, *Science* **257**, 219–223 (1992).
24. Q. Wang, Y. Liu, C. Lin and H. Yan, Layer-by-layer growth of superparamagnetic, fluorescent barcode nanospheres , *Nanotechnology* **18**(40), 405604 (2007).
25. J. Kim, J. E. Lee, J. Lee, J. H. Yu, B. C. Kim, K. An, Y. Hwang, C. H. Shin, J. Kim, and T. Hyeon, Magnetic fluorescent delivery vehicle using uniform mesoporous silica spheres embedded with monodisperse magnetic and semiconductor nanocrystals, *J. Am. Chem. Soc.* **128**(3), 688–689 (2006).
26. H. Yusuf. W. G. Kim, D. H. Lee, Y. Guo, and M. G. Moffitt, Size control of mesoscale aqueous assemblies of quantum dots and block copolymers, *Langmuir* **23**(2), 868–878 (2007).
27. J. Yang, S. R. Dave, and X. Gao, Quantum dot nanobarcodes: epitaxial assembly of nanoparticle-polymer complexes in homogeneous solution, *J. Am. Chem. Soc.* **130**(15), 5286–5292 (2008).
28. M. Nirmal, B. O. Dabbousi, M. G. Bawendi, J. J. Macklin, J. K. Trautman, T. D. Harris, and L. E. Brus, Fluorescence intermittency in single cadmium selenide nanocrystals, *Nature* **383**, 802–804 (1996).
29. S. A. Empedocles and M. G. Bawendi, Quantum-confined stark effect in single CdSe nanocrystallite quantum dots, *Science* **278**, 2114–2117 (1997).
30. P. S. Eastman, W. Ruan, M. Doctolero, R. Nuttall, G. de Feo, J. S. Park, J. S. F. Chu, P. Cooke, J. W. Gray, S. Li and F. Chen, QD nanobarcodes for multiplexed gene expression analysis, *Nano. Lett.* **6**(5), 1059–1064 (2006).
31. J. Guo, W. Yang, C. Wang, J. He, and J. Chen, Poly(N-isopropylacrylamide)-coated luminescent/magnetic silica microspheres: preparation, characterization, and biomedical applications, *Chem. Mater.* **18**(23), 5554–5562 (2006).
32. M. Stroh, J. P. Zimmer, D. G. Duga, T. S. Levchenko, K. S. Cohen, E. B. Brown, D. T. Scadden, V. P. Torchilin, M. G. Bawendi, D. Fukumura, and R. K. Jain, Quantum dots spectrally distinguish multiple species within the tumor milieu in vivo, *Nature Med.* **11**, 678–682 (2005).
33. Dynal® Products (online). Available: http://www.invitrogen.com/site/us/en/home/brands/Dynal.html (2008).
34. Evident Technologies *Security Inks* (online). Available: http://www.evidenttech.com/applications/security-inks.html (2008).
35. Luminex *xMAP Technology* (online). Available: http://www.luminexcorp.com/01_xMAPTechnology/index.html?&Collapse=115 (2008).
36. Z. A. Dehqanzada, C. E. Storrer, M. T. Hueman, R. J. Foley, K. A. Harris, Y. H. Jama, C. D. Shriver, S. Ponniah, and G. E. Peoples, Assessing serum cytokine profiles in breast cancer patients receiving a HER2/*neu* vaccine using Luminex® technology, *Oncol. Rep.* **17**: 687–694 (2007).

Study of Biological Assemblies by Ultrafast Fluorescence Spectroscopy

Sudip Kumar Mondal, Kalyanasis Sahu, and Kankan Bhattacharyya

Abstract Application of ultrafast time resolved fluorescence spectroscopy to the study of dynamics in biological assemblies is discussed. The recent results obtained using femtosecond time resolution and large scale computer simulations have significantly improved our understanding of the primary steps in solvation dynamics, FRET and other processes. Dynamics in many systems ranging from proteins (both in native and molten globule state) and DNA to micelles (tri-block co-polymer micelles and ionic liquid micelles) are discussed. Perhaps, the most interesting application is spatial resolution of solvation dynamics by variation of excitation wavelength.

Keywords Femtosecond fluorescence spectroscopy · Solvation · FRET · Proton/electron transfer · Excitation wavelength dependence · Protein · DNA Micelles

Introduction

The unique three-dimensional structure of a protein in the native or biologically active form arises from a delicate balance of hydrophobic and hydrophilic interactions. Similar interactions give rise to many other self-organized assemblies. Because of the fundamental role played by these assemblies in many biological and natural processes, there is a long standing interest to unravel dynamics in these assemblies. Most recently, ultrafast time resolved fluorescence spectroscopy in the picosecond and femtosecond time domain have yielded a good deal of new information on this issue. Progress in this field till 2004 has been summarized by us in a previous issue [1]. In this article, we survey the new results, in particular those published in the last two years (2005–2006).

K. Bhattacharyya (✉)
Department of Physical Chemistry, Indian Association for the Cultivation of Science, Jadavpur, Kolkata 700 032, India
e-mail: pckb@mahendra.iacs.res.in

C.D. Geddes (ed.), *Reviews in Fluorescence 2007*, Reviews in Fluorescence 2007,
DOI 10.1007/978-0-387-88722-7_8,

We will discuss mainly four different systems – cyclodextrins, micelles, proteins, and DNA. We will focus on recent results on ultrafast solvation dynamics, excited state proton transfer, anisotropy decay, fluorescence resonance energy transfer (FRET), and photo-induced electron transfer (PET) in these systems. For a general discussion on these phenomena, we refer to previous reviews and text books [1–6].

Cyclodextrins

A supramolecule consisting of a cyclodextrin (CD) cavity as a host and an organic fluorescent probe with several solvent molecules as guests (Fig. 1) is perhaps the most well-defined example of a molecule in a confined environment. Recently, several functionalized CDs (e.g., methyl, hydroxypropyl) have been developed which are much more soluble in water and more hydrophobic than the unsubstituted CDs [7–8]. The higher solubility of the alkyl CDs in water has been widely used in targeted delivery of drugs which are sparingly soluble in water [7]. The substituted CDs have also been utilized in preventing misfolding and aggregation of proteins

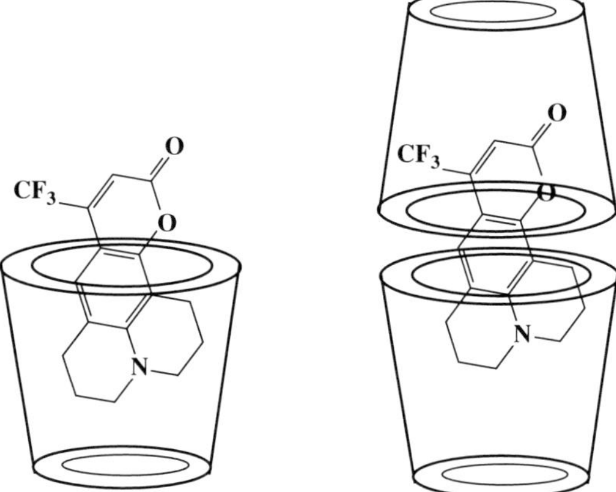

Fig. 1 1:1 and 1:2 complexes of C153 with methyl-β-cyclodextrin

by encapsulating the aromatic residues of a protein [8].

The stoichiometry of a guest-CD molecule may be easily obtained from fluorescence spectroscopy [9]. In an aqueous solution, both 1:1 and 1:2 (guest/CD) complexes may be present, simultaneously. If ϕ_0, ϕ_1, and ϕ_2 denote fluorescence quantum yields of the probe in free, 1:1 and 1:2 complexes, respectively and K_1 and K_2 denote the binding constants for 1:1 and 1:2 complexes, the observed emission quantum yield ϕ is given by [9–10]

$$\varphi = \frac{\varphi_0 + \varphi_1 K_1[\mathrm{CD}] + \varphi_2 K_2[\mathrm{CD}]^2}{1 + K_1[\mathrm{CD}] + K_2[\mathrm{CD}]^2}. \tag{1}$$

The values of ϕ_1, ϕ_2, K_1, and K_2 may be obtained by a non-linear least square fitting of ϕ_f against [CD] [10]. For a complex involving coumarin 153 (C153) and methylated β-CD, the values of K_1 for dimethyl β-CD and trimethyl β-CD are 220 M^{-1} and 1220 M^{-1}, respectively, while the values of K_2 are 3350 M^{-2} and 38500 M^{-2}, respectively [10]. From this, it is readily calculated that in an aqueous solution containing 130 mM dimethyl β-CD, 66% of the probe (C153) molecules are bound to dimethyl β-CD in the form of 1:2 complex, 33% as 1:1 complex and only 1% remains free. For 130 mM trimethyl β-CD, 80.3% of C153 are present as 1:2 complex, 19.6% as 1:1 complex, and only 0.1% remains in free form [10].

Fluorescence Anisotropy Decay in Cyclodextrin

The size of the 1:1 and 1:2 complex in an aqueous solution may be determined from the decay of fluorescence anisotropy of C153 [10]. The time constant of anisotropy decay (τ_R) is related to the hydrodynamic radius (r_h) as [2]

$$\tau_R = \frac{4\pi\eta r_h^3}{3kT}. \tag{2}$$

In this case, both the 1:1 and 1:2 complex contribute to the anisotropy decay, and hence the overall decay is bi-exponential. The faster decay component (1000–1150 ps) is ascribed to the smaller 1:1 complex and the slower one (2500–2700 ps) to the bigger 1:2 complex. The anisotropy decay data were fitted to a two exponential decay with amplitudes same as the relative contributions of the 1:1 and 1:2 complexes as obtained from steady state emission intensity measurements [10].

Using the measured viscosity of 130 mM dimethyl β-CD in water at 20°C (~1.7 mPa s) and the faster component of anisotropy decay, the hydrodynamic radius for the 1:1 complex is estimated to be 8.5 Å for 1:1 complex for both di- and trimethyl β-CD. This corresponds to a diameter of ~17 Å. This is larger than the reported height (10.9 Å) of dimethyl β-CD [10]. This suggests that in the 1:1 complex a part (~6 Å) of the probe is projected out of the cavity.

For the 1:2 complex (C153:CD), one may use the slower component of anisotropy decay (2700 ps for dimethyl β-CD and 2500 ps for trimethyl β-CD). From this, the hydrodynamic radius of the 1:2 complex of dimethyl β-CD and trimethyl β-CD are calculated to be 11.6±0.5 Å and 11.3±0.5 Å, respectively. This corresponds to a hydrodynamic diameter ($2r_h$) of ~22 Å which is roughly equal to the sum of height of two dimethyl β-CD (or trimethyl β-CD) cavities.

Douhal and coworkers [11–14] studied encapsulation of several drug molecules in cyclodextrins. They reported that the fluorescence lifetime of a drug, milrinone (MIR), increases from 65 ps in bulk water to 240–350 ps on confinement in cyclodextrins (α-, β-, γ-CD and dimethyl-β-CD) [13]. From anisotropy decay and PM3 calculations, they concluded that the drug is not fully encaged inside the cavity.

Most recently, several groups reported that in certain cases the cyclodextrin guest–host complexes display extremely slow anisotropy decay (time constant >20 ns) (Fig. 2) [15–17]. This has been attributed to the formation of linear nano-tube aggregates containing a large number (>50) cyclodextrins joined (stitched together) non-covalently by the guest molecules (Fig. 2).

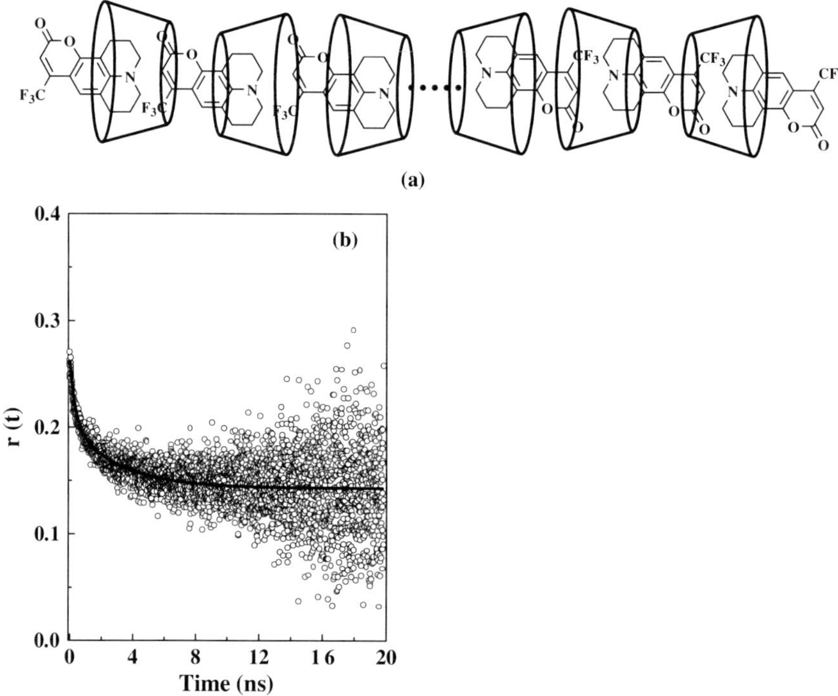

Fig. 2 (**a**) γ-Cyclodextrin nanotube aggregates with C153 [16]. (**b**) Anisotropy decay in γ-cyclodextrin nanoaggregates [16]. (Copyright © 2005 American Chemical Society)

Solvation Dynamics in Cyclodextrins

Several groups reported that solvation dynamics is markedly slowed down inside the cyclodextrin cavity [10–18,19–20]. Sen et al. studied the effect of methyl substitution on ultrafast solvation dynamics [10]. In trimethyl β-CD, all the three OH groups of the β-CD cavity are replaced by OMe groups. This causes substantial retardation of solvation dynamics with a fast component of 10 ps (24%) and two very slow components – 240 ps (45%) and 2450 ps (31%) [10]. In contrast, dimethyl β-CD contains only one OH group per glucose unit. It exhibits very fast dynamics with a major ultrafast component faster than 0.3 ps (40%), a fast component of 2.4 ps (24%), and two slow components –50 ps (18%) and 1450 ps (18%) (Fig. 3). It seems that the main requirement for ultrafast solvation is extended hydrogen bonded network between the water molecules inside the cavity with those outside the cavity [21]. In the case of trimethyl β-CD, there is no OH group at the rim of the cyclodextrin cavity and this leads to "loss of communication" between water molecules inside the cyclodextrin cavity and those outside.

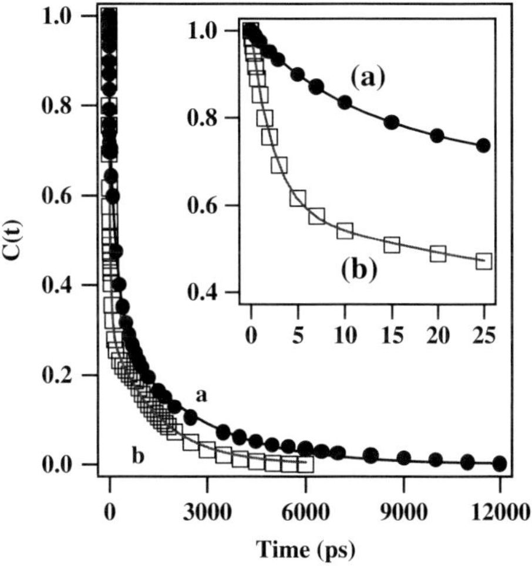

Fig. 3 Decay of solvent response function C(t) of C153 in (**a**) trimethyl-β-CD and (**b**) dimethyl-β-CD. The inset shows initial part of the decay [10]. (Copyright © 2005 American Chemical Society)

Excited State Proton Transfer in Cyclodextrins

Acidity of many molecules (e.g., pyranine) is markedly enhanced on electronic excitation. pK_a of pyranine in the ground state is 7.5 while that in the excited state (pK_a^*) decreases by 7 units to 0.5. In other words, acidity of pyranine is 10 million times stronger in the excited state compared to that in the ground state. As a result, in bulk water even at a neutral pH (~7) an excited pyranine molecule rapidly loses a proton in ~90 ps. Encapsulation of pyranine in the nanocavity of γ-cyclodextrin slows down the initial proton transfer step and accelerates the recombination of the geminate ion pair (proton and the anion) [21]. This leads to an overall increase in proton transfer time by about 15 times [22].

Micelles

Water soluble triblock copolymers have received a lot of recent attention because of their complex aggregation behavior and widespread industrial applications [23–28]. The symmetric (A-B-A) triblock copolymers composed of poly-(ethylene oxide) (PEO) and poly-(propylene oxide) (PPO) are denoted by $(PEO)_x$-$(PPO)_y$-$(PEO)_x$. For the copolymer Pluronic P123 (P123), $x = 20$ and $y = 70$. The PPO block is soluble in water below 288 K. The hydrophobicity of the PPO block above 288 K and consequent dehydration leads to the formation of a micelle consisting of the PEO-PPO-PEO triblock copolymers. According to SANS studies, such a micelle consists of a hydrophobic core presumably containing the PPO block and a hydrophilic corona of the PEO block (Fig. 4) [23–24]. Alexandridis et al. studied micellization of a series of PEO-PPO-PEO triblock copolymers using solubilization of a probe, diphenyl hexatriene (DPH) [25]. They reported that at 293 K, CMC of P123 in water is 0.18 wt% [25].

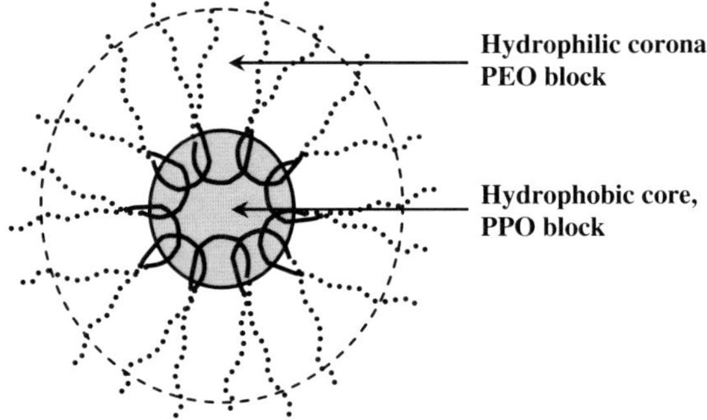

Fig. 4 Schematic representation of a P123 triblock copolymer micelle

Very recently, Castner and coworkers investigated temperature dependent micellization of aqueous PEO_{109}-PPO_{41}-PEO_{109} (Pluronic F88) [26–27]. They found that the fluorescence and excitation maxima of two coumarin dyes C480 and C153 and average lifetime of C153 undergo a dramatic change around the critical micellar temperature (CMT). Dutt showed that for PEO_{20}-PPO_{70}-PEO_{20} (Pluronic P123) the anisotropy decay becomes slower above CMT (289 K) when the micelles are formed [28]. More recently, Castner and coworkers studied phase transition and local friction (microviscosity in different regions) of F88 triblock polymer using fluorescence anisotropy decay [27].

Excitation Wavelength Dependence of Solvation Dynamics in Micelles and Lipids

Since the polarity of different regions of a triblock copolymer micelle is different, excitation of the fluorescent probes in different regions should lead to a variation of emission spectrum. In general, in such a heterogeneous system as the wavelength of excitation is increased, a more polar region is selected and the emission maximum shifts to the red. This phenomenon is known as Red Edge Excitation Shift (REES) [29–30]. Most recently, Sen et al. studied solvation dynamics in different regions of P123 micelle by varying excitation wavelength (λ_{ex}) [31]. The emission maximum of C480 exhibits a very large REES by 25 nm from 453 nm at $\lambda_{ex} = 375$ to 478 nm at $\lambda_{ex} = 435$ nm [31]. The solvation dynamics become faster with increase in the excitation wavelength (Fig. 5). This has been attributed to distribution of the probe in different locations ranging from – a fast bulk-like peripheral region (PEO-water interface) with solvation time $\leq$ 2 ps and a very slow (4500 ps), hydrophobic core region (PPO–PEO interface). With increase in λ_{ex}, contribution of the bulk-like ultrafast dynamics ($\leq$2 ps) increases from 7% at λ_{ex} = 375 nm to 78% at λ_{ex} = 425 nm with a concomitant decrease in the contribution of the core-like slow component (4500 ps) from 79% at λ_{ex} = 375 nm to 17% at λ_{ex} = 425 nm [31]. There is another component of 60 ps which is due to chain dynamics. Contribution of this component decreases from 14% at λ_{ex} = 375 nm to 5% at λ_{ex} = 425 nm [31].

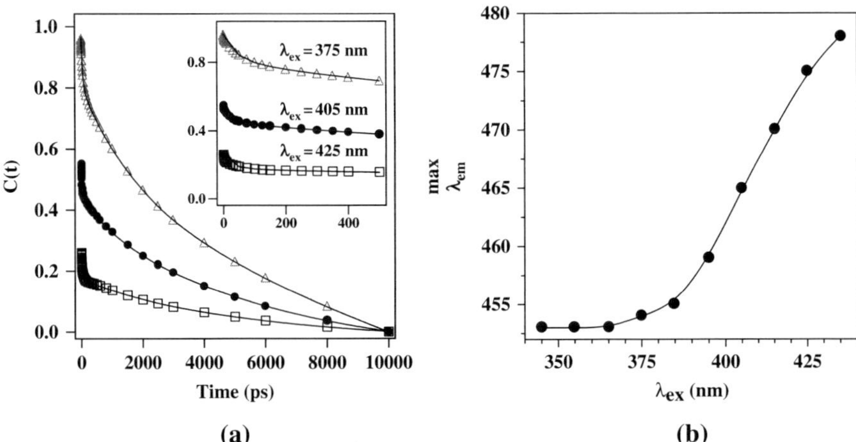

Fig. 5 λ_{ex} dependence of (**a**) solvation dynamics and (**b**) emission maxima of C480 in a P123 triblock copolymer micelle. (Reused with permission from Sen et al. [31] Copyright 2006, American Institute of Physics)

A lipid vesicle is a very heterogeneous system consisting of a highly hydrophobic bilayer membrane of surfactants which encloses a polar water pool. Sen et al. studied excitation wavelength (λ_{ex}) dependence in DMPC vesicles using C480 as a fluorescence probe [32–33]. They observed that at all λ_{ex} (390–430 nm) solvation dynamics is described by an ultrafast component of <0.3 ps, a fast component of 1.5 ps and two slow components of 250 ps and 2000 ps [33]. The relative contribution of the ultrafast components (<0.3 and 1.5 ps) increases with increase in λ_{ex} from 48% at λ_{ex} = 390 nm to 100% at λ_{ex} = 430 nm. Thus, at 430 nm there is no slow component. They ascribed the ultrafast dynamics to the very polar and mobile bulk-like region, deep inside the water pool and the slow dynamics (250 ps and 2000 ps) to a restricted and less polar interfacial region inside the bi-layer membrane [33].

Solvation Dynamics in Micelles: Worm Like, in Ormosils and in Ionic Liquids

On addition of certain electrolytes (e.g., sodium salicylate), the shape of a micelle changes from spherical shape to a worm like form. This results principally from binding of the counter ions to the micelle, and thereby decreasing electrostatic repulsion between the head groups. A worm-like micelle displays viscoelastic property. Sen et al. studied a worm like cetyltrimethylammonium bromide (CTAB) micelles formed on addition of Na-salicylate [34]. They observed that solvation dynamics in a worm like CTAB micelle is about 4 times slower than that in an ordinary spherical CTAB micelles [34].

A sol–gel matrix doped with an organic additive is called an ormosil. Sahu et al. investigated solvation dynamics in a CTAB doped sol–gel matrix [35]. They detected a fast component of 120 ps (25%) and a slow component of 7200 ps (75%), with an average solvation time, $<\tau_s> = 5430$ ps [35]. The fast component (120 ps) of the solvation dynamics in the ormosil is identical to the major component of solvation dynamics in an undoped matrix. They ascribed the ultraslow component of solvation (7200 ps) to the adsorption of CTAB micellar aggregates on the silica surface. The movement of the water molecules is highly restricted at the silica–CTAB interface because of the formation of hydrogen bonds of the water molecules with the silica surface and the polar head groups of CTAB [35]. Sen et al. reported solvation dynamics of DCM inside a DPPC entrapped sol–gel matrix [36]. The sol–gel matrixes are usually prepared by hydrolysis of orthosilicates generating alcohol as a by-product. The main advantage of a sol–gel matrix based on sodium silicate is that they may be prepared at a neutral pH without generation of alcohol [36]. Hence, DPPC vesicle retains its structure inside the sol–gel matrix. Solvation dynamics inside sol–gel entrapped DPPC vesicle is retarded ~2.5 times compared to that in DPPC liposomes in bulk water. The slow solvation dynamics is attributed to the restricted movement of the water molecules trapped in between the lipid vesicles and the sol–gel matrix [36].

Usually, micelles are formed in a polar solvent (e.g., water, methanol, and acetonitrile). Most recently, many groups reported formation of micelles [37] and reverse micelles [38] in ionic liquids (IL). Petrich and coworkers studied solvation dynamics of coumarin 153 in amphiphilic ionic liquids (1-cetyl-3-vinylimidazolium bromide and 1-cetyl-3-vinylimidazolium bis-[(trifluoromethyl)sulfonyl]imide) and their corresponding micelles in water [39]. They found that the imidazolium moiety is responsible for the majority of the solvation [39]. Sarkar and coworkers studied solvation dynamics in neat [bmim]-[PF_6] and in [bmim][PF_6]-Brij micelles [40]. They found that in pure [bmim][PF_6] the average solvation time is ~3 ns with a fast component of 0.80 ns (80%) and a slow component of 12.5 ns (20%). In the presence of 250 mM Brij-35, the average solvation time increases to 5.45 ns with a fast component of 0.98 ns (70%) and a slow component of 15.90 ns (30%) [40]. They ascribed the 2–3 fold increase of solvation time in the micelle in IL to the very high viscosity of neat IL. The magnitude of retardation of solvation dynamics on confinement in a micelle is much smaller than that in the case of a IL compared to water [40].

Sarkar and coworkers also studied solvent relaxation in [bmim][PF_6]/TX-100/water microemulsion using a hydrophobic probe, coumarin 153 (C153), and a hydrophilic probe, coumarin 151 (C151) [41]. For C153, with an increase in the [bmim][PF_6]/TX-100 ratio (R), the solvent relaxation time does not change much. For C151, with an increase in R, the slow component of the solvation time gradually decreases and the fast component gradually increases [41].

Correa and Levinger studied, in detail, the interactions in aqueous and non-aqueous reverse micelles using coumarin 343 (C343) as a probe [42]. They concluded that only formamide and water forms a "solvent pool". Further, in aqueous AOT microemulsions, C343 interacts neither with the sulfonate group of AOT nor with water [42].

Most recent computer simulations suggest that self-diffusion of the organic fluorophore may give rise to slow dynamics [43] and water molecules in the hydration layer of a micelle is slower by 1–2 orders [44].

Ultrafast Fluorescence Resonance Energy Transfer in Proteins and Micelles

Chattopadhyay and Mazumdar studied unfolding of cytochrome *C* using fluorescence resonance energy transfer (FRET) [45]. In the native state of cytochrome *C*, emission of the lone tryptophan unit is almost totally quenched because of ultrafast FRET to the heme unit. However, when SDS is added, the protein is partially unfolded in such a way that the distance between the tryptophan and the heme increases. This hinders FRET and gives rise to strong tryptophan emission [45]. Pal and coworkers studied homo-molecular FRET between DNA-bound acridine orange (AO) encapsulated in a reverse micelle and estimated the distance between two AO molecules [46].

Sarkar and coworkers studied energy transfer from coumarin dyes (C480, C151, C153) to rhodamine 6G (R6G) in methanol and acetonitrile solubilized AOT reverse micelles [47]. Though they observed considerable decrease in the steady state emission intensities, they did not observe substantial decrease in donor lifetimes. The apparent discrepancy between the steady state and picosecond time resolved studies may be explained as follows. In this case, in most of the micelles only the donor is present without any acceptor and those donors do not undergo FRET at all [48]. Hence, the observed picosecond decay will be dominated by the unquenched donors and the lifetime of the donor is largely unaffected [48]. However, for many donors there is an acceptor in the immediate vicinity in the same micelle. In this case, the emission of the donor is strongly quenched and the lifetime becomes shorter than the IRF of a picosecond setup. The ultrafast component of the decay of this set of donors is missed in a picosecond setup.

The components of FRET may be correctly determined if one studies the rise of the acceptor (R6G) emission using a femtosecond setup [48]. In order to study ultrafast FRET in sub-picosecond and few picosecond time scale at a very short distance, Sahu et al. studied FRET between a donor C153 and an acceptor R6G confined in two micelles – a neutral P123 triblock copolymer micelle and an anionic SDS micelle [48]. The time constants of FRET were determined from the rise time of the acceptor emission (Fig. 6) using a femtosecond setup. In SDS micelle, two ultrafast FRET components 0.7 and 13 ps correspond to the donor–acceptor distance (R_{DA}) of 12 and 19 Å, respectively. In P123 micelle, in addition to the ultrafast components (1.2 and 24 ps), a long component of 1000 ps is observed. This long component corresponds to a $R_{DA} = 44$ Å. The R_{DA} calculated from the ultrafast component is roughly equal to the donor–acceptor distance for direct contact and the R_{DA} calculated from the long component agrees well with micellar radius. For the small SDS micelle, no such long component was observed [48]. At a short D-A distance, the basic assumptions of Forster theory (weak interactions and point-dipole approximation) are questionable [49]. Still the distances calculated using Forster theory looks reasonable in this case. It seems that the surfactant chains get inserted between the donor and the acceptor and thus insulate them, and this partially justifies the Forster model [48].

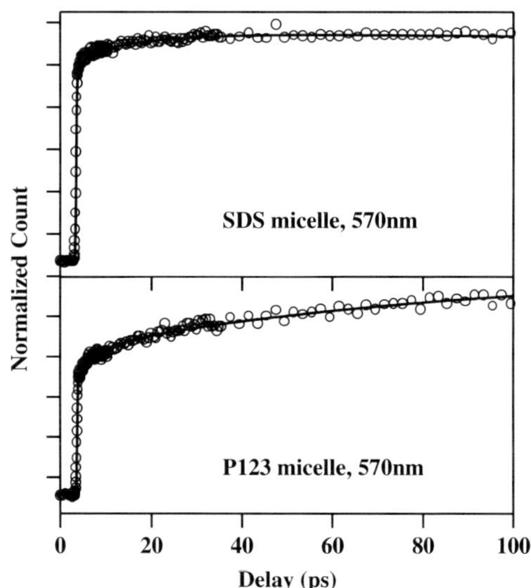

Fig. 6 FRET in a micelle. Rise of acceptor (R6G) emission in micelles. (Reused with permission from Sahu et al. [48]. Copyright 2006, American Institute of Physics)

Photoisomerization in Micelle and Reverse Micelle

Photoisomerization of olefins and polyenes plays an important role in many chemical and biological processes. Dutt and coworkers studied photoisomerization of 3,3′-diethyloxadicarbocyanine iodide (DODCI) in micellar and gel phase of aqueous triblock copolymer P123 [50]. They observed a bi-exponential decay in both the micellar and gel phase. The major (60–70%) component (1.73–1.18 ns) is same for both the micellar and the gel phase. But the minor component is slowed down by a factor of 1.4–1.1 in the gel phase compared to the micellar solution. The retardation of isomerization in the gel phase has been explained on the basis of enhancement in the friction experienced by the probe due to micelle–micelle entanglement at the interface [50].

Dutt et al. also investigated the photoisomerization of DODCI in the microemulsion of P123-butyl acetate–water in the temperature range 293–318 K [51]. Three different copolymer–oil–water compositions were chosen such that the mole ratio of water to copolymer (W) spans the range 50–150. It has been noticed that in all the three systems, the fluorescence decay of DODCI displays a long component whose contribution is 85–90%. This component has been ascribed to those solute molecules in the core region of the water pool where hydrated poly(ethylene oxide) units are present. A short-decay component is associated with the remaining fraction, and its values match with those measured in water, indicating that the water present in these reverse phases is in the form of droplets [51].

Ultrafast Photoinduced Electron Transfer (PET) in Micelles

Photoinduced electron transfer (PET) plays an important role in many chemical and biological processes, e.g., solar cell, biosynthesis, and photosynthesis [52–58]. Though PET in homogeneous solution has been studied quite extensively, such studies in an organized medium (e.g., micelle) started only recently [55–57]. Several groups have studied PET from *N*,*N*-dimethylaniline (DMA) to coumarin dyes in a micelle using a picosecond setup [55–57]. Obviously, picosecond studies are insufficient to detect the ultrafast component of ET which occurs in $\leq$10 ps time scale. Also, the earlier studies [55–56] did not consider the competition between solvation dynamics and PET in a micelle.

In a micelle, the donor (DMA) and the acceptor stay very close, and hence PET in a micelle is expected to be almost as fast as that in neat DMA. However, in a micelle, solvation dynamics is 100–1000 times slower compared to bulk water. This affects PET. Ghosh et al. detected ultrafast components ($<$10 ps) of PET in a micelle using a femtosecond up-conversion setup [57]. They observed a bell shaped dependence of ultrafast PET with the free energy (Marcus inverted region, Fig. 7) [57].

Most recently, Eisenthal and coworkers reported ultrafast dynamics of PET at DMA/water interface using femtosecond pump-probe surface SHG [58]. They detected the time constants of both forward and backward electron transfer from time resolved surface SHG resonant for the $DMA^{+\cdot}$ radical [58].

Proteins

The biological function of a protein is largely controlled by the quasi-bound water molecules at its surface which are loosely described as *biological or structured*

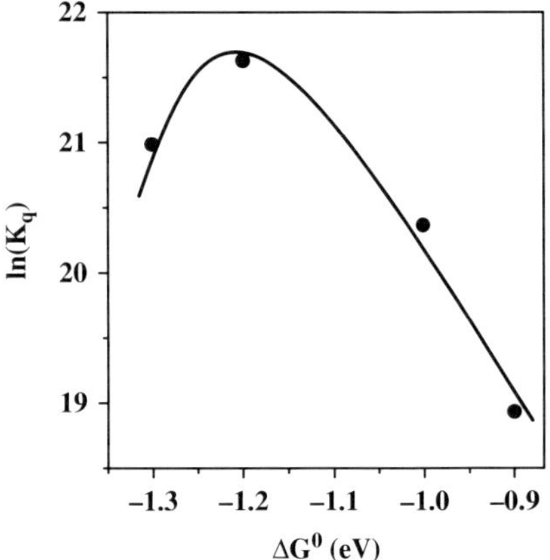

Fig. 7 Marcus inverted region in a CTAB micelle. Plot of ln K_q vs ΔG^0 for coumarin-DMA system. (Reused with permission from Ghosh et al. [57]. Copyright 2006, American Institute of Physics)

water. Compared to bulk water, the biological water molecules are substantially slower and give rise to lower polarity [59–60].

Guha et al. studied solvation dynamics at the active site of an enzyme, glutaminyl-tRNA synthetase (GlnRS), using a fluorescence probe, acrylodan, site-specifically attached at a cysteine residue C229, near the active site [61]. The picosecond time-dependent fluorescence Stokes shift indicates slow solvation dynamics at the active site of the enzyme, in the absence of any substrate. The solvation dynamics becomes still slower when the substrate (e.g., glutamine or $tRNA^{gln}$) binds to the enzyme. A mutant Y211H-GlnRS was constructed in which the glutamine binding site is disrupted. The mutant Y211H-GlnRS labeled at C229 with acrylodan exhibits significantly different solvent relaxation. This demonstrates that the slow dynamics is indeed associated with the active site. For GlnRS in the free state, the decay of $C(t)$ exhibits two components of approximately 400 and 2000 ps [54]. When glutamine binds to GlnRS, the 400 ps component becomes longer (750 ps) while the 2000 ps component is unaffected. When $tRNA^{gln}$ binds to GlnRS, the 2000 ps component slows down to 2500 ps. This suggests that at the active site of GlnRS there exist two kinds of water molecules, both much slower than bulk water (1 ps). The 400 ps component is assigned to the glutamine binding site and the 2000 ps component to the tRNA binding site [61].

The positively charged membrane protein cytochrome *C* is partially unfolded on binding to an anionic surfactant (SDS). Cytochrome *C* forms two well-defined partially folded states – one (I_S') in the presence of SDS and other (I_S'') in SDS and urea [62]. Using femtosecond time-resolved fluorescence spectroscopy, it is shown that the solvation dynamics in the two partially folded states of cytochrome *C* are very different (Fig. 8) [62]. Solvation dynamics in I_S' is relatively slower with a very minor (5%) contribution of an ultrafast component (0.5 ps) and two slow components of 90 (85%) and 400 ps (10%). Dynamics in I_S'' is much faster with a major ultrafast component of 1.3 ps (47%) and two slow components of 60 ps (12.5%) and 170 ps (10.5%) (Fig. 8). This indicates that the structure of IS″ is much more open and exposed compared to that of IS′. The difference in the dynamics of IS′ and IS″ is attributed to differences in their structures, particularly near the heme region [62].

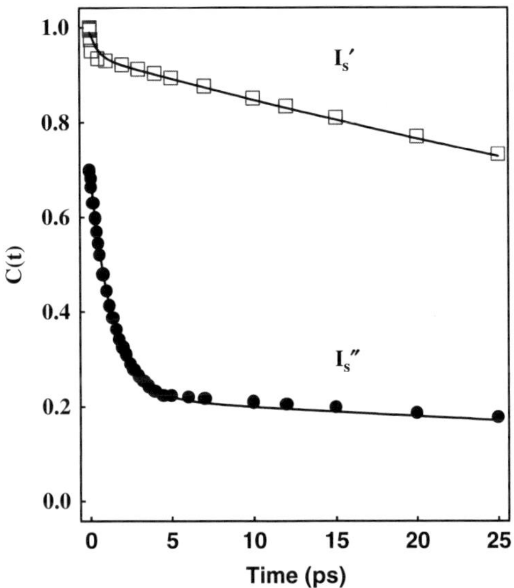

Fig. 8 Solvation dynamics of C153 in I_s' and I_s'' intermediates of cytochrome C [62]. (Copyright © 2005 American Chemical Society)

Sen et al. studied solvation dynamics in a protein, lysozyme in the presence of urea and SDS using solvation dynamics [63]. Using FRET, they showed that the probe (C153) is located near Trp 62 and Trp 108 residues of lysozyme. They found that addition of small amount of SDS (3 mM) causes partial refolding of lysozyme denatured by urea, and as a result solvation dynamics becomes native like. Addition of large amount of SDS (28 mM) causes complete unfolding of the urea denatured protein and solvation dynamics is 3.5 times slower compared to that in the native state. In this case, lysozyme resembles a polymer surrounded by the SDS micelles and the very slow solvation dynamics is attributed to dynamic exchange of the water molecules and polymer chain dynamics [63].

Sahu et al. studied temperature dependence of solvation dynamics and fluorescence anisotropy decay of 8-anilino-1-naphthalenesulfonate (ANS) bound to a protein, bovine serum albumin (BSA) [64]. Solvation dynamics of ANS bound to BSA displays a component (300 ps) which is independent of temperature in the range of 278–318 K and a long component which decreases from 5800 ps at 278 K to 3600 ps at 318 K. The temperature independent part (300 ps component) is ascribed to a dynamic exchange of bound to free water with a low barrier. The temperature variation of the long component of solvation dynamics corresponds to an activation energy of 2.1 kcal mol^{-1}. The activation energy is ascribed to local segmental motion of the protein along with the associated water molecules and polar residues. The time scale of solvation dynamics is found to be very different from the time scale of anisotropy decay. The anisotropy decays are analyzed in terms of the wobbling motion of the probe (ANS) and the overall tumbling of the protein. The entropy of activation is found to be –13 cal K^{-1} mol^{-1}. The observed negative entropy suggests an ordering of the local structure at the transition state of the protein segment during dipolar relaxation [64].

Zhong and coworkers studied solvation dynamics of melittin, in a random-coiled primary structure and detected two components – 0.62 and 14.7 ps [65]. They assigned the faster component to bulk water and the slower to surface-type hydration dynamics of proteins. In the same work, they also studied a model tripeptide (KWK). At a membrane–water interface, melittin folds into a secondary alpha-helical structure, and the interfacial water motion was found to take as long as 114 ps, indicating a well-ordered water structure along the membrane surface [65].

Recently, it is reported that a cationic surfactant CTAB binds to a cationic protein lysozyme [66–67]. This indicates that the hydrophobic interaction is stronger than the electrostatic repulsion between like charges. Sahu et al. studied lysozyme-CTAB complex by excited state proton transfer of pyranine [66] and by solvation dynamics [67]. The critical association concentration of lysozyme-CTAB system was found to be 0.4 mM [66]. The probe C153 does not bind to the native protein at a low concentration but binds to protein-CTAB complex [67]. In this study, CTAB concentration (0.8 mM) was higher than the CAC of lysozyme-CTAB but lower than the CMC of CTAB [67]. Hence, the observed dynamics does not contain any contribution from the free protein or free micelle. Solvation dynamics in lysozyme-CTAB aggregate is much slower than that in CTAB micelle [67].

The structure and dynamics of water around a protein is expected to be sensitive to the secondary structure of the protein. Several groups have applied atomistic molecular dynamics simulations to elucidate solvation dynamics in different regions of a protein. Bagchi and coworkers investigated dependence of the solvation dynamics on the secondary structure for a small 36-residue globular protein, HP-36 [68–69]. The secondary structure of this protein contains three short alpha-helices. The solvation dynamics of polar amino acid residues in helix-2 ($<\tau>=11$ ps) is found to be faster than that of the other two helices (the average time constant is smaller by a factor of 2), although the interfacial water molecules around helix-2 exhibit much slower orientational dynamics than that around the other two helices. A careful analysis shows that the origin of such a counterintuitive behavior lies in the dependence of the solvation time correlation function on the surface exposure of the probes – the more exposed is the probe, the faster the solvation dynamics [69].

In a confined environment, the folded structure of protein is much more stable than the unfolded structure. In a recent MD simulation study, Marchi and coworker showed this by confining a simple alanine zwitterionic octa-peptide (A_8) in a reverse micelle (RM) [70]. For smaller RM, they found a stable helical structure of the polypeptide which quickly assumes an extended structure as the size of RM increases [70]. Zong and coworkers studied peptide–water interaction by MD simulation [71]. They found that the relaxation dynamics occurs on several time scales ranging from femtoseconds to tens of picoseconds [71]. Heugen et al. detected marked slowing down of the water molecules near a bio-molecule (lactose) by terahertz spectroscopy [72]. They found that the hydration layer around the carbohydrate is extended to 5.13 Å from the surface and contains 123 water molecules [72].

In contrast to the large body of work demonstrating ultraslow dynamics near a proton, incoherent quasi-elastic neutron scattering and NMRD studies indicate that slowing down of the water molecules is much less significant [73–76]. The apparent discrepancy between the relaxation times of water in a protein reported by different techniques may be reconciled as follows. It should be emphasized that different techniques detect different things. First, dielectric relaxation (DR) and time dependent fluorescence stokes shift (TDFSS) represent the collective response of solvent (water) dipoles after switching on an electric field (external field in the case of DR and instantaneously created dipole for TDFSS). NMRD and QENS are governed by dynamics of a single-particle (single water molecule) in the absence of an electric field (i.e., in an equilibrium situation). Second, NMRD, QENS, and DR have no spatial resolution and all the water molecules contribute to these relaxations. TDFSS, on the other hand, has high spatial resolution with the water molecules in the first solvation shell of the fluorescence probe making a predominant contribution to the relaxation. The fluorescence probe preferentially binds to a hydrophobic pocket of a protein with a lot of slow, bound, and buried water molecules around it. Thus, TDFSS reports the dynamics of the slow water molecules. Evidently, the number of quasi-bound water molecules around a fluorescent probe is very small compared to the very large number of water molecules present around a protein. NMRD and QENS data are dominated by the huge number of water molecules, many of which are very fast (particularly those at the surface). In a protein, the relaxation dynamics varies widely from site to site. For instance, Makarov and Petit [77] detected 294 hydration sites in myoglobin using computer simulations. Among these, the buried sites (grooves, cavities or concave surfaces) are characterized by long (>80 ps) residence times [77]. In contrast, the exposed or convex sites exhibit very short residence times (<10 ps) [77]. Thus, the techniques which detect an average of the water molecules at all the sites (NMRD, QENS) often report very fast dynamics or almost bulk water like residence times with very small (sometimes negligible) contribution of the slow water molecules. In contrast, TDFSS captures predominantly the dynamics of the slow and buried water molecules in deep, hydrophobic pockets with a relaxation component slower by several orders of magnitude compared to bulk water.

DNA

Recently, many groups have studied dynamics in DNA by time dependent fluorescence Stokes shift [78–83] and MD simulation [84–85]. Zewail and coworkers studied solvation dynamics in DNA using 2-aminopurine as an intrinsic probe and a minor groove binding non-covalent probe, pentamidine [78]. They detected a bi-exponential decay with an ultrafast sub-picosecond component due to bulk water and a relatively long (~10 ps) component which they ascribed to slow "biological water"[78]. Berg and coworkers studied oligonucleotides in which a native base pair is replaced by a dye molecule (C480) [79]. They found that when the probe C480 is in the center of the helix the time scale of relaxation is broadly distributed over six decades of time scale from 40 fs to 40 ns and obeys a power-law, $(1+\tau/\tau_0)^{-\alpha}$. The very long (~40 ns) component is assigned to the reorganization dynamics of DNA. Since the interior of the double-helix is devoid of water, the observed Stokes shift seems to originate from electric field of the DNA on the probe. When the probe (C480) is attached at the end of the helix an additional very fast component of 5 ps is detected [80]. The 5 ps component and the increased mobility ("fraying") at the end of the helix are ascribed to increased exposure of the probe to bulk water and lower counter-ion concentration [80].

Sen et al. studied the effect of counter-ions on the electric-field dynamics in DNA [81]. They observed that small ions display power law decay over six decades. However, for larger counter-ions (e.g., tetraalkyl-ammonium ions) the relaxation component in the nanosecond time range exhibits an unexpected increase [81]. For DNA/protein complex, Berg and coworkers detected a difference in the absolute Stokes shift at shorter times (<40 ps) [82].

Goddard and coworkers applied molecular dynamics (MD) simulations to understand the structure and stability of various paranemic crossover (PX) DNA molecules [85]. The average dynamic structures over the last 1 ns of the 3-ns simulation preserve the Watson-Crick hydrogen bonding as well as the helical structure. The PX structures are structurally more rigid compared to the canonical B-DNA without crossover. They showed that PX65 display a helical twist and other helical structural parameters close to the values for normal B-DNA of similar length and sequence. Thus, PX65 is structurally more stable compared to other PX motifs. This is in agreement with experiments [85]. These results should aid in designing optimized DNA structures for use in nanoscale components and devices.

Finally, several recent works addressed ultrafast isomerization and high-frequency vibration of flexible fluorophores intercalated in DNA [86–88].

Conclusion and Future Outlook

In the past two years, there has been a substantial progress in the study of biological assemblies using femtosecond fluorescence spectroscopy and large scale computer simulations. It is shown that dynamics of these systems may now be interpreted in terms of motions of the individual segments of these complex systems. The ultimate goal of these studies is to correlate the dynamics with biological function. Recent rapid progress in this area suggests that this goal is not too far.

Acknowledgments Thanks are due to Department of Science and Technology (DST), Government of India, and Council of Scientific and Industrial Research (CSIR) for generous research grants. KB thanks Professor B. Bagchi for many stimulating discussions.

References

1. K. Bhattacharyya, Organized assemblies probed by fluorescence spectroscopy, *Reviews in Fluorescence* 2005, edited by C. D. Geddes, J. R. Lakowicz, (Springer, New York, 2005) p. 1–23.
2. J. R. Lakowicz, *Principles of Fluorescence Spectroscopy*, (Kluwer/Plenum, New York, 1999).
3. B. Bagchi, Water dynamics in the hydration layer around proteins and micelles, *Chem. Rev.* **105**(9), 3197–3219 (2005).
4. S. K. Pal, A. H. Zewail, Dynamics of water in molecular recognition, *Chem. Rev.* **104**(4), 2099–2124 (2004).
5. K. Bhattacharyya, Solvation dynamics and proton transfer in supramolecular assemblies, *Acc. Chem. Res.* **36**(2), 95–101 (2003).
6. N. Nandi, K. Bhattacharyya, B. Bagchi, Dielectric relaxation and solvation dynamics of water in complex chemical and biological systems, *Chem. Rev.* **100**(6), 2013–2045 (2000).
7. K. Uekama, F. Hirayama, T. Irie, Cyclodextrin drug carrier systems, *Chem. Rev.* **98**(5), 2045–2076 (1998).
8. M. Khajehpour, T. Troxler, V. Nanda, J. M. Vanderkooi, Melittin as model system for probing interactions between proteins and cyclodextrins, *Proteins* **55**(2), 275–287 (2004).
9. M. Hoshino, M. Imamura, H. Ikehara, Y. Hamai, Fluorescence enhancement of benzene derivatives by forming inclusion complexes with β-cyclodextrin in aqueous solutions, *J. Phys. Chem.* **85**(13), 1820–1823 (1981).
10. P. Sen, D. Roy, S. K. Mondal, K. Sahu, S. Ghosh, K. Bhattacharyya, Fluorescence anisotropy decay and solvation dynamics of coumarin 153 in methyl β-cyclodextrin, *J. Phys. Chem. A* **109**(43), 9716–9722 (2005).
11. M. El-Kemary, J. A. Organero, L. Santos, A. Douhal, Effect of cyclodextrin nanocavity confinement on the photorelaxation of the cardiotonic drug milrinone, *J. Phys. Chem. B* **110**(29), 14128–14134 (2006).
12. A. Douhal, M. Sanz, L. Tormo, Femtochemistry of orange II in solution and in chemical and biological nanocavities, *Proc. Nat. Acad. Sci. USA* **102**(52), 18807–18812 (2005).
13. L. Tormo, J. A. Organero, A. Douhal, Effect of nanocavity confinement on the relaxation of anesthetic analogues: Relevance to encapsulated drug photochemistry, *J. Phys. Chem. B* **109**(38), 17848–17854 (2005).
14. L. Tormo, A. Douhal, Caging anionic structure of a proton transfer dye in a hydrophobic nanocavity with a cooperative H-bonding, *J. Photochem. Photobiol. A* **173**(3), 358–364 (2005).
15. D. Roy, S. K. Mondal, K. Sahu, S. Ghosh, P. Sen, K. Bhattacharyya, Temperature dependence of fluorescence anisotropy decay and solvation dynamics of coumarin 153 in γ-cyclodextrin aggregates, *J. Phys. Chem. A* **109**(33), 7359–7364 (2005).
16. G. Pistolis, I. Balomenou, Cyclodextrin cavity size effect on the complexation and rotational dynamics of the laser dye 2,5-diphenyl-1,3,4-oxadiazole: From singly occupied complexes to their nanotubular self-assemblies, *J. Phys. Chem. B* **110**(33), 16428–16438 (2006).
17. G. Li, L. B. McGown, Molecular nanotube aggregates of β- and γ-cyclodextrins linked by diphenylhexatrienes, *Science*, **264**(5156), 249–251 (1994).
18. S. K. Mondal, D. Roy, K. Sahu, P. Sen, K. Bhattacharyya, Hydration dynamics of 4-aminophthalimide in a substituted β-cyclodextrin nanocavity, *J. Photochem. Photobiol. A* 173 (2005) 334.
19. S. Vajda, R. Jimenez, S. J. Rosenthal, V. Fidler, G. R. Fleming, E. W. Castner Jr., Femtosecond to nanosecond solvation dynamics in water and inside the γ-cyclodextrin cavity, *J. Chem. Soc. Faraday Trans.* **91**(5), 867–873 (1995).
20. N. Nandi, B. Bagchi, Ultrafast solvation dynamics of an ion in the γ-cyclodextrin cavity: Role of restricted environment, *J. Phys. Chem.* **100**(33), 13914–13919 (1996).
21. N. Nandi, S. Roy, B. Bagchi, Ultrafast solvation dynamics in water: Isotope effects and comparison with experimental results, *J. Chem. Phys.* **102**(3), 1390–1397 (1995).

22. S. K. Mondal, K. Sahu, P. Sen, D. Roy, S. Ghosh, K. Bhattacharyya, Excited state proton transfer of pyranine in a γ-cyclodextrin cavity, *Chem. Phys. Lett.* **412**(1–3), 228–234 (2005).
23. K. Mortensen, J. S. Pedersen, Structural study on the micelle formation of poly(ethylene oxide)-poly(propylene oxide)-poly(ethylene oxide) triblock copolymer in aqueous solution, *Macromolecules* **26**(4), 805–812 (1993).
24. I. Goldmints, G. Yu, C. Booth, K. A. Smith, T. A. Hatton, Structure of (deuterated PEO)-(PPO)-(deuterated PEO) block copolymer micelles as determined by small angle neutron scattering, *Langmuir* **15**(5), 1651–1656 (1999).
25. P. Alexandridis, J. F. Holzwarth, T. A. Hatton, Micellization of poly(ethylene oxide)-poly(propylene oxide)-poly(ethylene oxide) triblock copolymers in aqueous solutions: Thermodynamics of copolymer association, *Macromolecules* **27**(9), 2414–1425 (1994).
26. C. D. Grant, K. E. Steege, M. R. Bunagan, E. W. Castner Jr., Microviscosity in multiple regions of complex aqueous solutions of poly(ethylene oxide)-poly(propylene oxide)-poly(ethylene oxide), *J. Phys. Chem. B* **109**(47), 22273–22284 (2005).
27. C. D. Grant, M. R. DeRitter, K. E. Steege, T. A. Fadeeva, E. W. Castner Jr., Fluorescence probing of interior, interfacial, and exterior regions in solution aggregates of poly(ethylene oxide)poly(propylene oxide)-poly(ethylene oxide) triblock copolymers, *Langmuir* **21**(5), 1745–1752 (2005).
28. G. B. Dutt, How critical micelle temperature influences rotational diffusion of hydrophobic probes solubilized in aqueous triblock copolymer solutions, *J. Phys. Chem. B* **109**(11), 4923–4928 (2005).
29. A. P. Demchenko, The red-edge effects: 30 years of exploration, *Luminescence* **17**(1), 19–42 (2002).
30. H. Raghuraman, A. Chattopadhyay, Organization and dynamics of melittin in environments of graded hydration. A fluorescence approach, *Langmuir* **19**(24), 10332–10341 (2003).
31. P. Sen, S. Ghosh, K. Sahu, S. K. Mondal, D. Roy, K. Bhattacharyya, A femtosecond study of excitation wavelength dependence of solvation dynamics in a PEO-PPO-PEO tri-block copolymer micelle, *J. Chem. Phys.* **124**(20), 204905-1-8 (2006).
32. P. Sen, T. Satoh, K. Bhattacharyya, K. Tominaga, Excitation wavelength dependence of solvation dynamics of coumarin 480 in a lipid vesicle, *Chem. Phys. Lett.* **411**(4–6), 339–344 (2005).
33. P. Sen, S. Ghosh, S. K. Mondal, K. Sahu, D. Roy, K. Bhattacharyya, K. Tominaga, A femtosecond study of excitation wavelength dependence of solvation dynamics in a vesicle, *Chem. Asian. J.* **1**(1–2), 188–194 (2006).
34. P. Sen, S. Mukherjee, A. Halder, P. Dutta, K. Bhattacharyya, Solvation dynamics in a wormlike CTAB micelle, *Res. Chem. Intermed.* **31**(1–3), 135–144 (2005).
35. K. Sahu, D. Roy, S. K. Mondal, A. Halder, K. Bhattacharyya, Study of solvation dynamics in an ormosil. CTAB in a sol-gel matrix, *J. Phys. Chem. B* **108**(32), 11971–11975 (2004).
36. P. Sen, S. Mukherjee, A. Patra, K. Bhattacharyya, Solvation dynamics of DCM in a DPPC vesicle entrapped in a sodium silicate derived sol-gel matrix, *J. Phys. Chem. B* **109**(8), 3319–3323 (2005).
37. J. L. Anderson, V. Pino, E. C. Hagberg, V. V. Sheares, D. W. Armstrong, Solvation effects and micelle formation in ionic liquids, *Chem. Comm.* (19), 2444–2445 (2003).
38. Y. Gao, S. Han, B. Han, G. Li, D. Shen, Z. Li, J. Du, W. Hou, G. Zhang, TX-100/water/1-butyl-3-methylimidazolium hexafluorophosphate microemulsions, *Langmuir* **21**(13), 5681–5684 (2005).
39. P. Mukherjee, J. A. Crank, M. Halder, D. W. Armstrong, J. W. Petrich, Assessing the roles of the constituents of ionic liquids in dynamic solvation: Comparison of an ionic liquid in micellar and bulk form, *J. Phys. Chem. A* **110**(37), 10725–10730 (2006).
40. A. Chakraborty, D. Seth, D. Chakrabarty, P. Setua, N. Sarkar, Dynamics of solvent and rotational relaxation of coumarin 153 in room-temperature ionic liquid 1-butyl-3-methylimidazolium hexafluorophosphate confined in Brij-35 micelles: A picosecond

time-resolved fluorescence spectroscopic study, *J. Phys. Chem. A* **109**(49), 11110–11116 (2005).

41. D. Seth, A. Chakraborty, P. Setua, N. Sarkar, Interaction of ionic liquid with water in ternary microemulsions (triton X-100/water/1-butyl-3-methylimidazolium hexafluorophosphate) probed by solvent and rotational relaxation of coumarin 153 and coumarin 151, *Langmuir* **22**(18), 7768–7775 (2006).
42. N. M. Correa, N. Levinger, What can we learn from a molecular probe? New insights on the behavior of C343 in homogeneous solutions and AOT reverse micelles, *J. Phys. Chem. B* **110**(27), 13050–13061 (2005).
43. M. R. Harpham, B. M. Ladanyi, N. E. Levinger, The effect of the counterion on water mobility in reverse micelles studied by molecular dynamics simulations, *J. Phys. Chem. B* **109** (35), 16891–16900 (2005).
44. F Sterpone, G Marchetti, C Pierleoni, M. Marchi, Molecular modeling and simulation of water near model micelles: Diffusion, rotational relaxation and structure at the hydration interface, *J. Phys. Chem. B* **110**(23), 11504–11510 (2006).
45. K. Chattopadhyay, S. Mazumdar, Stabilization of partially folded states of cytochrome C in aqueous surfactant: Effects of ionic and hydrophobic interactions, *Biochemistry* **42**(49), 14606–14613 (2003).
46. A. K. Shaw, R. Sarkar, S. K. Pal, Direct observation of DNA condensation in a nano-cage by using a molecular ruler, *Chem. Phys. Lett.* **408**(4–6), 366–370 (2005).
47. D. Seth, D. Chakrabarty, A. Chakraborty, N. Sarkar, Study of energy transfer from 7-amino coumarin donors to rhodamine 6G acceptor in non-aqueous reverse micelles, *Chem. Phys. Lett.* **401**(4–6), 546–552 (2005).
48. K. Sahu, S. Ghosh, S. K. Mondal, B. C. Ghosh, P. Sen, D. Roy, K. Bhattacharyya, ltrafast fluorescence resonance energy transfer (FRET) in a micelle, *J. Chem. Phys.* **125**(4)**,** 044714-1-8 (2006).
49. H. Singh, B. Bagchi, Non-Forster distance and orientation dependence of energy transfer and applications of fluorescence resonance energy transfer to polymers and nanoparticles: How accurate is the spectroscopic ruler with $1/R^6$ rule? *Curr. Sci.* **89**(10), 1710–1719 (2005).
50. K. S. Mali, G. B. Dutt, T. Mukherjee, Polyene photoisomerization rates: Are they distinct in aqueous block copolymer micellar solutions and gels? *J. Chem. Phys.* **124**(5), 054904-1-6 (2006).
51. K. S. Mali, G. B. Dutt, T. Mukherjee, Photoisomerization of a carbocyanine derivative in the reverse phases of a block copolymer: Evidence for the existence of water droplets, *Langmuir* **22**(16), 6837–6842 (2006).
52. R. A. Marcus, Electron transfer reactions in chemistry: Theory and experiments (Nobel lecture), *Angew. Chem. Int. ed. Engl.* **32**(8), 1111–1222 (1993).
53. G. J. Kavaranos, *Fundamentals of Photoinduced Electron Transfer* (VCH, New York, 1993).
54. G. L. Closs, L. T. Calcaterra, N. J. Green, K. W. Penfield, J. R. Miller, Distance, stereoelectronic effects, and the Marcus inverted region in intramolecular electron transfer in organic radical anions, *J. Phys. Chem.* **90**(16), 3673–3683 (1986).
55. M. Kumbhakar, S. Nath, H. Pal, A. V. Sapre, T. Mukherjee, Photoinduced electron transfer from aromatic amines to coumarin dyes in sodium dodecyl sulfate micellar solutions, *J. Chem. Phys.* **119**(1), 388–399 (2003).
56. D. Chakraborty, A. Chakrabarty, D. Seth, N. Sarkar, Photoinduced electron transfer between coumarin dyes and electron donating solvents in cetyltrimethyl ammonium bromide micelles: Evidence for Marcus inverted region, *Chem. Phys. Lett.* **382**(5–6), 508–517 (2003).
57. S. Ghosh, K. Sahu, S. K. Mondal, P. Sen, K. Bhattacharyya, A femtosecond study of photoinduced electron transfer from dimethylaniline to coumarin dyes in a CTAB micelle, *J. Chem. Phys.* **125**(5), 054509-1-7 (2006).
58. E. A. McArthur, K. B. Eisenthal, Ultrafast excited-state electron transfer at an organic liquid/aqueous interface, *J. Am. Chem. Soc.* **128**(4), 1068–1069 (2006).

59. N. Nandi, B. Bagchi, Dielectric relaxation of biological water, *J. Phys. Chem. B* **101**(50), 10954–10962 (1997).
60. W. H. Qiu, L. Y. Zhang, O. Okobiah, Y. Yang, L. J. Wang, D. P. Zhong, A. H. Zewail, Ultrafast solvation dynamics of human serum albumin: Correlations with conformational transitions and site-selected recognition, *J. Phys. Chem. B* **110**(21), 10540–10549 (2006).
61. S. Guha, K. Sahu, D. Roy, S. K. Mondal, S. Roy, K. Bhattacharyya, Slow solvation dynamics at the active site of an enzyme: Implications for catalysis, *Biochemistry* **44**(25), 8940–8947 (2005).
62. K. Sahu, S. K. Mondal, S. Ghosh, D. Roy, P. Sen, K. Bhattacharyya, Femtosecond study of partially folded states of cytochrome C by solvation dynamics, *J. Phys. Chem. B* **110**(2) 1056–1062 (2006).
63. P. Sen, D. Roy, K. Sahu, S. K. Mondal, K. Bhattacharyya, Hydration dynamics of a protein in the presence of urea and sodium dodecyl sulfate, *Chem. Phys. Lett.* **395**(1–2), 58–63 (2004).
64. K. Sahu, S. K. Mondal, S. Ghosh, D. Roy, P. Sen, Temperature dependence of solvation dynamics and anisotropy decay in a protein. ANS in bovine serum albumin, *J. Chem. Phys.* **124**(12), 124909-1-7 (2006).
65. W. H. Qiu, L. Y. Zhang, Y. T. Kao, W. Y. Lu, T. P. Li, J. Kim, G. M. Sollenberger, L. J. Wang, D. P. Zhong, Ultrafast hydration dynamics in melittin folding and aggregation: Helix formation and tetramer self-assembly, *J. Phys. Chem. B* **109**(35), 16901–16910 (2005).
66. K. Sahu, D. Roy, S. K. Mondal, R. Karmakar, K. Bhattacharyya, Study of protein-surfactant interaction using excited state proton transfer, *Chem. Phys. Lett.* **404**(4–6), 341–345 (2005).
67. K. Sahu, S. K. Mondal, D. Roy, R. Karmakar, K. Bhattacharyya, Study of interaction of a cationic protein with a cationic surfactant using solvation dynamics. Lysozyme:CTAB, *Chem. Phys. Lett.***413**(4–6), 484–489 (2005).
68. S. Bandyopadhyay, S. Chakraborty, B. Bagchi, Secondary structure sensitivity of hydrogen bond lifetime dynamics in the protein hydration layer, *J. Am. Chem. Soc.* **127**(47), 16660–16667 (2005).
69. S. Bandyopadhyay, S. Chakraborty, S. Balasubramanian, B. Bagchi, Sensitivity of polar solvation dynamics to the secondary structures of aqueous proteins and the role of surface exposure of the probe, *J. Am. Chem. Soc.* **127**(11), 4071–4075 (2005).
70. S. Abel, M. Waks, W. Urbach, M. Marchi, Structure, stability, and hydration of a polypeptide in AOT reverse micelles, *J. Am. Chem. Soc.* **128**(2), 382–383 (2006).
71. A. A. Hassanali, T. P. Li, D. P. Zhong, S. J. Singer, A molecular dynamics study of Lys-Trp-Lys: Structure and dynamics in solution following photoexcitation, *J. Phys. Chem. B* **110**(21), 10497–10508 (2006).
72. U. Heugen, G. Schwaab, E. Brundermann, M. Heyden, X. Yu, D. M. Leitner, M. Havenith, Solute-induced retardation water dynamics probed directly by terahertz spectroscopy, *Proc. Natl. Acad. Sci. USA* **103**(33), 12301–12306 (2006).
73. C. Caronna, F. Natali, A. Cupane, Incoherent elastic and quasi-elastic neutron scattering investigation of hemoglobin dynamics, *Biophys. Chem.* **116**(3), 219–225 (2005).
74. D. Russo, R. K. Murarka, J. R. D. Copley, T. Head-Gordon, Molecular view of water dynamics near model peptides, *J. Phys. Chem. B* **109**(26), 12966–12975 (2005).
75. D. Russo, G. Hura, T. Head-Gordon, Hydration dynamics near a model protein surface, *Biophys. J.* **86**(3), 1852–1862 (2004).
76. K. Modig, E. Liepinsh, G. Otting, B. Halle, Dynamics of protein and peptide hydration, *J. Am. Chem. Soc.* **126**(1), 102–114 (2004).
77. V. Makrov, B. M. Petit, Solvation and hydration of protein and nucleic acids: A theoretical view of simulation and experiment, *Acc. Chem. Res.* **35**(6), 376–384 (2002).
78. S. K. Pal, L. Zhao, T. Xia, A. H. Zewail, Site- and sequence-selective ultrafast hydration of DNA, *Proc. Natl. Acad. Sci. USA* **100**(24), 13746–13751 (2003).

79. D. Andreatta, J. L. P. Lustres, S. A. Kovalenko, N. P. Ernsting, C. J. Murphy, R. S. Coleman, M. A. Berg, Power-law solvation dynamics in DNA over six decades in time, *J. Am. Chem. Soc.* **127**(20), 7270–7271 (2005).
80. D. Andreatta, S. Sen, J. L. P. Lustres, S. A. Kovalenko, N. P. Ernsting, C. J. Murphy, R. S. Coleman, M. A. Berg, Ultrafast dynamics in DNA: "fraying" at the end of the helix, *J. Am. Chem. Soc.* **128**(21), 6885–6892 (2006).
81. S. Sen, L. A. Gearheart, E. Rivers, H. Liu, R. S. Coleman, C. J. Murphy, M. A Berg, Role of monovalent counterions in the ultrafast dynamics of DNA, *J. Phys. Chem. B* **110**(26), 13248–13255 (2006).
82. S. Sen, N. A. Paraggio, L. A. Gearheart, E. E. Connor, A. Issa, R. S. Coleman, D. M. Wilson, M. D. Wyatt, M. A. Berg, Effect of protein binding on ultrafast DNA dynamics: Characterization of a DNA: APE1 complex, *Biophys. J.* **89**(6), 4129–4138 (2005).
83. M. M. Somoza, D. Andreatta, C. J. Murphy, R. S. Coleman, M. A. Berg, Effect of lesions on the dynamics of DNA on the picosecond and nanosecond timescales using a polarity sensitive probe, *Nucl. Acids Res.* **32**(8), 2494–2507 (2004).
84. S. Pal, P. K. Maiti, B. Bagchi, Anisotropic and sub-diffusive water motion at the surface of DNA and of an anionic micelle CsPFO, *J. Phys. Condens. Matter* **17**(49), S4317–S4331 (2005).
85. P. K. Maiti, T. A. Pascal, N. Vaidehi, J. Heo, W. A. Goddard, Atomic-level simulations of Seeman DNA nanostructures: The paranemic crossover in salt solution, *Biophys. J.* **90**(5) 1463–1479 (2006).
86. D. Mandal, S. K. Pal, A. Datta, K. Bhattacharyya, Photoisomerization of diethyloxadicarbocyanine iodide in DNA and proteins, *Res. Chem. Intermed.* **25**(7), 685–693 (1999).
87. V. Karunakaran, J. L. F. Lustres, L. J. Zhao, N. P. Ernsting, O. Seitz, Large dynamic stokes shift of DNA intercalation dye thiazole orange has contribution from a high-frequency mode, *J. Am. Chem. Soc.* **128**(9), 2954–2962 (2006).
88. A. Furstenberg, M. D. Julliard, T. G. Deligeorgiev, N. I. Gadjev, A. A. Vasilev, E. Vauthey, Ultrafast excited-state dynamics of DNA fluorescent intercalators: New insight into the fluorescence enhancement mechanism, *J. Am. Chem. Soc.* **128**(23), 7661–7669 (2006).

Fluorescence Signal Amplification for Ultrasensitive DNA Detection

Kim Doré, Mario Leclerc, and Denis Boudreau

Abstract Ultrasensitive and reliable DNA detection tools are currently needed for the diagnostic of infectious and genetic diseases. To achieve the required sensitivity, one can amplify either the target DNA or the signal generated by each target molecule detected. Since enzymatic amplification of DNA is prone to contamination or inhibition, sensors that can benefit from signal amplification are usually more robust. Several approaches have been developed to obtain an amplification of the fluorescence signal generated by some biosensors upon DNA detection. Some among them feature the use of conjugated polymers as biosensors; due to their collective system response, they offer an amplification of the signal as compared to the response of individual monomers. Other approaches feature the use of efficient Förster energy transfer schemes. The use of fluorophore encapsulation, to increase the number of fluorophores reporting a detection event and to protect them from quenching species, can also be used to obtain an amplified fluorescence signal. The purpose of this review is to point out and discuss how these different methods achieve the observed fluorescence signal amplification, in order to relate them to what is known about fluorescence emission and energy transfer.

Keywords Fluorescent conjugated polymers · Resonent energy transfer · Orientation and confinement in aggregates · Ultrafast energy transfer

Introduction

The rapid, sensitive, and specific detection of nucleic acids is of central importance for the diagnostic of infections, the identification of genetic mutations, and forensic analyses. In particular, the efficient and affordable detection of infectious disease agents is envisioned by the World Health Organization as the most critical biotechnological development for improving health in developing countries [1]. Moreover, as more and more links are discovered between hereditary diseases and

---

D. Boudreau (✉)
Département de chimie, Université Laval, Québec, QC, Canada G1K 7P4

C.D. Geddes (ed.), *Reviews in Fluorescence 2007*, Reviews in Fluorescence 2007,
DOI 10.1007/978-0-387-88722-7_9,

DNA mutations, their diagnostic based on DNA mutation detection is increasingly important. The recognition capabilities of DNA through hybridization reactions are well established, but adequate transducers are needed to generate a physically measurable signal from the recognition event. In this regard, many DNA hybridization transducers have been proposed (see [2–5] for selected reviews). However, such biosensors need to possess some important qualities in order to be efficient and useful in real life applications [6]. Notably, since the majority of infectious diseases need to be diagnosed promptly in order to be curable and only a few pathogens are usually present in the blood at the onset of an infection, DNA detection techniques need to be ultrasensitive to the point of being able to detect as little as a few copies of the target DNA sequence. To reach this level of sensitivity, researchers have investigated several amplification strategies, either enzymatic amplification of the target DNA material or amplification of the signal generated by each detection event. In addition, since most genetic diseases are caused by single mutations in the patient genome, the assay's selectivity must be sufficiently high to accurately distinguish the mutated DNA. Furthermore, as the targeted genetic material can be of different forms, i.e., most often double-stranded DNA (bacteria, genomic DNA) but sometimes single- or double-stranded RNA (viruses), biosensors must ideally be versatile enough to be able to detect all these different targets. The latter is particularly important in light of the fact that most biosensors developed up to now rely on the availability of the target as single-stranded DNA [7–13], which hinders their usefulness in real-life applications. Finally, as the cost, the turnaround time, and the susceptibility to contamination are major concerns for the successful implementation of any diagnostic platform, the use of simple preparation steps and robust instrumentation should be preferred. This aspect implies that the usual strategies to obtain good detection sensitivity, such as chemical tagging of DNA target and/or polymerase chain reaction (PCR) amplification, should ideally be avoided.

Out of all the existing detection methods used for the recognition of molecular interactions, fluorescence is one of the most sensitive and is relatively inexpensive and easy to implement [5]. Several strategies were used by researchers to amplify the fluorescence response of DNA biosensors. For example, the fluorescence signal amplification can be achieved by the intermediate of a biological molecule such as an allosteric enzyme which generates a fluorescent product in large quantities [14]. However, this type of biological transduction pathway shares some of the culprits of polymerase chain reaction, i.e., it can be hindered by irregularities inherent to the enzyme activity and the overall detection can be affected by contamination. For this reason, methods based on catalytic amplification will not be discussed in this review.

Conjugated Polymers

Because of their collective system response which makes them sensitive to very minor perturbations, conjugated polymers (CP) are one of the most promising classes of transduction materials [3]. The amplification effect achieved by these materials is derived from their ability to serve as a highly efficient transport conduit for electronic states. Highly fluorescent CPs have been created in this way [15]. Whereas the fluorescent properties of CPs are usually due to their conjugated backbone, their recognition properties are mostly defined by their side-chain groups. The first successful attempt at CP-based DNA detection, based on a novel, non-water-soluble hybrid biosensor made of a singly stranded DNA probe attached to an electroactive polypyrrole, was reported in 1997 by the group of Garnier [16]. CPs with highly polar or ionic side-chain groups were later developed [15], which improved their water solubility and their applicability to the biological field.

Interestingly, DNA being itself a negatively charged biopolymer, the presence of cationic side-chain groups on the repeat units of ionic CPs made possible their use for sensitive DNA detection, as was demonstrated by the Leclerc group with a cationic water-soluble polythiophene [17] (Fig. 1) which can form stoichiometric polyelectrolyte complexes (coacervates) with anionic oligonucleotide probes

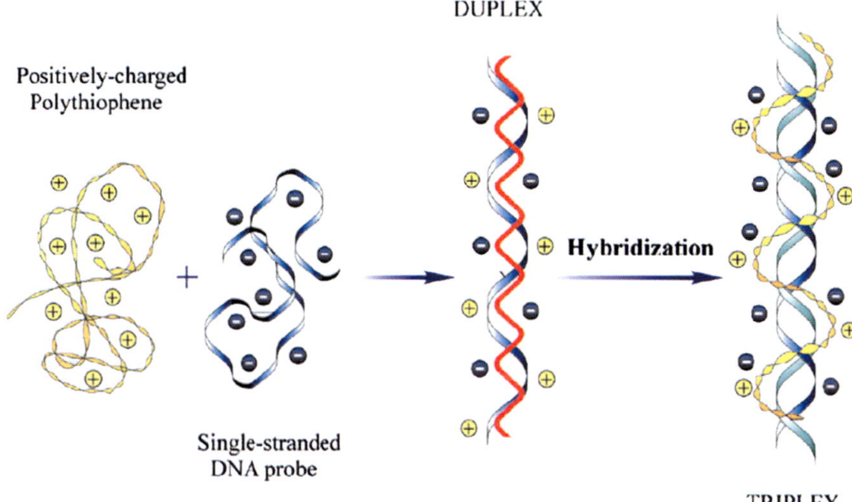

Fig. 1 Schematic description of the formation of polythiophene/single-stranded nucleic acid duplex and polythiophene/hybridized nucleic acid triplex (Reprinted with permission from [18], Copyright 2008 American Chemical Society)

(duplex). These complexes tend to form aggregates in aqueous solutions [19] and, as a result, the intrinsic fluorescence of the polymer backbone (quantum yield of 3% in its random coil conformation) is quenched, due to the close packing and stiffening of the polythiophene chains [17,20]. However, the emission intensity recovers when the probe hybridizes with the perfectly matched complementary nucleic acid single strand. Spectroscopic data suggests that this increase in fluorescence is due to the polythiophene backbone adopting a helical structure that wraps with greater affinity around the anionic phosphate backbone of double-strand nucleic acids [17] (as compared to its affinity for single-stranded nucleic acids), combined with a better solubilization of the resulting non-stoichiometric polyelectrolyte complex (triplex) which limits interchain quenching [17,21].

The detection efficiency of this polymeric biosensor also benefits from the relatively large Stokes shift between the optimal excitation wavelength (~425 nm) and the wavelength for maximum fluorescence emission (~525 nm) of the triplex, which allows the efficient rejection of dominant sources of background signals at low analyte concentration (i.e., Rayleigh and Raman scattering) using simple interference filters. A simple detection device was developed specifically for this polythiophene sensor, based on a powerful and stable LED-based excitation source, a compact InGaAs photon-counting detector and a high light throughput, non-dispersive interference filter-based optical design. Using this robust yet simple DNA sensing platform, a limit of detection of $\sim 3 \times 10^2$ molecules (0.5×10^{-21} mole) was demonstrated for 20-mer target oligonucleotides and a probed volume of approximately 150 μL, corresponding to a concentration of 4×10^{-18} M [19]. The presence as concomitant species in the sample of sequences having only one mismatch with the probe induced only a slight increase of the luminescence intensity, as did the addition of a large excess (100 equivalents) of an oligonucleotide with two mismatches. This polymeric biosensor was validated using a 32-mer oligonucleotide capture probe designed for the specific recognition of all *Influenza* A strains [22] and crude lysates from viral particles, and nucleic acids corresponding to as few as 750 genome copies of the virus were successfully detected in less than 30 min. The versatility of this fluorescent conjugated polymer in transducing the hybridization of oligonucleotides of varying lengths and of RNA material is a definite advantage for the design of molecular diagnostic tools for genetic material of diverse origin.

In 2003, the group of Inganäs also reported the use for DNA detection of a polythiophene biosensor with a chemical structure similar to that of Leclerc et al. (Fig. 2), but in which hydrogen bonding plays a greater role [20]. When the conjugated polymer is in the presence of the ssDNA probe, electrostatic interaction between the two polyelectrolytes disrupts the internal H-bonding within the polymer chain, which leads to a more planar polymer backbone and the appearance of an emission band near 595 nm. The DNA bases are also believed to form hydrogen bonds with the amino and carbonyl groups of nearby CP chains, resulting in their aggregation. When the complementary oligonucleotide target is added to the

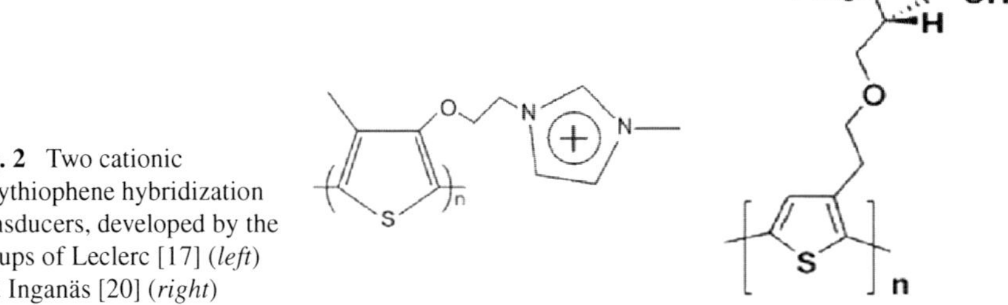

Fig. 2 Two cationic polythiophene hybridization transducers, developed by the groups of Leclerc [17] (*left*) and Inganäs [20] (*right*)

ssDNA/polymer complex, hybridization leads to the separation of the CP chains, which causes an important increase of CP emission and a shift of the fluorescence maximum to 585 nm. The authors reported a detection limit of 10^{-11} mol of DNA in a 1.5 mL liquid volume. Interestingly, this detection limit was achieved at room temperature, a sizable advantage for an eventual implementation into a dedicated portable/handheld device. The authors also stated that their method is highly sequence specific, and that a single-nucleotide mismatch can be detected within 5 min. Unfortunately, further studies on this system revealed that the DNA capture probe can be displaced by other DNA strands in solution, thus potentially limiting sensitivity and selectivity [23].

One distinct feature of the two approaches described above is that they involve an increase of emission intensity upon DNA detection, i.e., they are "turn-on" sensing methods, whereas CPs are most often used in "turn-off" sensing, where their collective response is used to obtain superquenching effects [24–28]. For example, the group of Whitten demonstrated the use of superquenching in a CP for biosensing purposes in 1999 [24]. By adding a biotinylated methyl viologen molecule to a water-soluble polyanionic CP, they demonstrated that one avidin-labeled protein could quench as many as 1000 repeat units, which is approximately equivalent to one whole polymer chain. The same group later extended this approach to DNA detection [26]. Unfortunately, a severe limitation of superquenching assays arises for samples having complex matrices, where many compounds can quench fluorescence and lead to false positive signals. Such matrix effects are less likely to happen with turn-on sensing approaches.

Conjugated Polymers and Förster Resonance Energy Transfer

The intermolecular sensing capacity of Resonant Energy Transfer (RET) is extensively used for DNA detection, for example, by using a donor-labeled DNA probe and a DNA target labeled with its acceptor, resulting in a drastic augmentation of RET between the two fluorophores upon DNA hybridization. Among the numerous improvements made to this technique in the past 20 years [29], assays based on the labeling of the DNA probe with both the donor and acceptor dyes offer the significant advantage of not needing any modification to the target DNA, a step that otherwise greatly limits sample throughput. Such doubly labeled DNA probe biosensors are also being used in real-time assays. Molecular beacons, introduced in 1996, are based on a single donor- and quencher-labeled DNA probe designed to form a stem-loop structure [30]. The loop region contains the DNA sequence complementary to the target strand and the stem region keeps the fluorophore and the quencher in close proximity. When introduced in a PCR amplification assay, such structures allow the monitoring of the number of amplified target DNA copies (amplicons) by becoming fluorescent upon hybridization with the latter. However, whereas this technique is very useful in this specific context, its use for ultrasensitive DNA detection (without enzymatic amplification of the target) is limited since it does not benefit from any significant fluorescence signal amplification.

The requirements for efficient RET include close proximity of the donor and acceptor molecules, strong spectral overlap between the absorption band of the donor and the emission band of the acceptor, high molar absorptivity of the donor, high quantum yield of the acceptor, and a favorable orientation of the dipole moments of the fluorophores. In addition, the energy transfer rate of the RET donor–acceptor pair should ideally be faster than the natural decay rate of the donor, i.e., a good RET donor molecule should be able to transfer the excitation energy to an acceptor molecule many times during its lifetime in the excited state [30]. Therefore, in principle, an amplification of fluorescence emission using RET is possible if the donor fluorophore is able to excite several acceptors within its excited state lifetime. Conjugated polymers, accurately described as "molecular wires" [15], were first used in a RET-based DNA detection scheme by Gaylord et al. in 2002 [10]. They used a cationic poly(fluorene-*co*-phenylene) as the donor and a PNA (peptide nucleic acid) probe labeled with fluorescein as the acceptor (Fig. 3). In the presence of non-complementary target DNA, the neutral PNA probe is unlikely to bind to the electrostatic complex formed by the polymer and the mismatched target, and the polymer donor will not be within range of the PNA-bound acceptor to excite it by RET (Fig. 3, at right). On the other hand, in the presence of complementary target DNA and using optimized conditions (1:1 ratio of polymer chain to DNA, 2.5×10^{-8} M), the authors observed a 25-fold amplification of the fluorescein emission with the RET scheme (via excitation of the polymer) when compared to the direct excitation of fluorescein (Fig. 3, at left), and demonstrated the successful detection of DNA at a minimum concentration of 10 pM. As the authors observed a small amount of RET to the PNA-bound acceptor in the absence of target DNA, they suggested that hydrophobic interactions may also play a role in this RET mechanism, i.e., neutral PNA probes and polymer chains (having a hydrophobic polymeric backbone) might

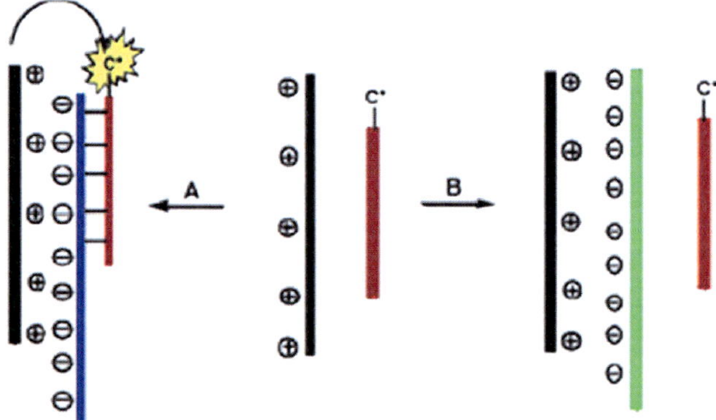

Fig. 3 Schematic representation of the PNA-ssDNA RET detection scheme reported by Gaylord et al. [10]. (**A**) Electrostatic attraction between the complementary PNA–ssDNA complex brings the polymer donor close to the acceptor dye, resulting in efficient RET; and (**B**) non-complementary ssDNA does not bring donor and acceptor within range of RET. (Reprinted with permission from [10], Copyright 2004 PNAS)

gather together in solution in order to minimize interaction with water and thus give rise to RET. This hypothesis was later confirmed when the addition of an organic solvent (10% *N*-methylpyrrolidone) resulted in the reduction of non-specific RET [31] .

The quantum yield of DNA intercalating dyes such as ethidium bromide (EB) is enhanced upon intercalating between the stacked bases of dsDNA. Because several EB can intercalate within a single dsDNA helix, the excitation of all EB acceptors with a single polymer RET donor might result in fluorescence signal amplification. Bazan et al. investigated this approach in 2004, using a polymer similar to that used in the work described above (Fig. 4) [32]. RET from the conjugated polymer to EB should only occur when the DNA probe hybridizes with the target, forming the double helix and bringing EB in close contact with the cationic polymer which is electrostatically bound to the dsDNA (Fig. 4, left). Interestingly, the authors did not observe any RET between the polymer and EB, despite the fact that the distance between the fluorophores and the spectral overlap were satisfactory for efficient RET to occur. The authors suggested that the orientation of the dipole moments was nearly orthogonal (a hypothesis confirmed by subsequent time-dependant anisotropy measurements [33]), thus preventing RET. To overcome this problem, the authors used instead a two-step RET scheme by means of a fluorescein-labeled DNA probe (Fig. 4, right). With this fluorescein "RET gate", the authors obtained a better overall energy transfer from the polymer to EB and a 8-fold amplification compared to the direct excitation of EB [32]. The use of this supplementary step in the RET mechanism also greatly improved the assay selectivity. The authors used time-resolved fluorescence measurements to confirm the two-step nature of the RET mechanism; accordingly, fluorescein lifetime decreased from 0.92 to 0.32 ns upon EB addition [33]. This study suggested an unusual and interesting means to obtain fluorescence signal amplification via a single-donor-to-multiple-acceptor RET scheme, while highlighting some theoretical aspects of RET in the process.

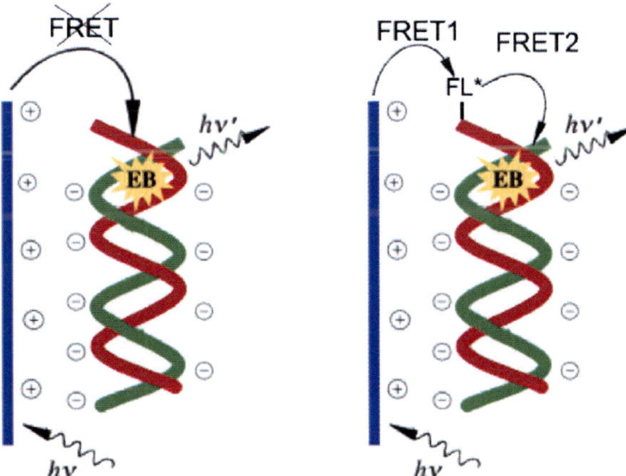

Fig. 4 Schematic representation of the RET-based DNA detection suggested by Bazan et al. [32,33]. *Left*: inefficient energy transfer from the cationic polymer to intercalated EB; and *right*: two-step energy transfer with fluorescein acting as a "gate" between the polymer donor and EB acceptor

The Leclerc group also investigated the use of their cationic polythiophene [17] as the donor in a RET pair for the purpose of ultrasensitive DNA detection. Starting from their initial detection scheme involving the polymer/DNA probe complex, they labeled the probe with a high quantum yield fluorophore (i.e., Alexa Fluor 546, AF546) whose absorption band ($\lambda_{max} = 516$ and 556 nm) suitably overlaps with the emission band of the polymeric transducer ($\lambda_{max} = 530$ nm) (Fig. 5) [21]. Obeying the chromic behavior of the polythiophene transducer, the polymer/tagged probe complex remains in its quenched form in the absence of DNA or upon addition of non-complementary DNA, which prevents the RET mechanism from occurring. However, when hybridization with a complementary oligonucleotide occurs, and upon excitation of the polymer at 420 nm, RET occurs between the polymer donor and the AF546 acceptor and fluorescence of AF546 is detected at 572 nm. Figure 6 shows calibration curves obtained by testing mutated genomic DNA (presenting the defective gene responsible for the human genetic disease *hereditary tyrosinemia type I*) using AF546-labeled DNA probes complementary with the mutated and wild type sequences [34]. This approach was shown to be able to rapidly distinguish wild-type DNA from mutated DNA – a single nucleotide polymorphism (SNP) in the entire genome could be detected in less than 5 min – at ultralow concentrations (about 5 copies in the entire 3 mL sample volume), without prior amplification or enrichment of the target. Interestingly, it was shown that the cationic polythiophene transducer also serves as a localized counter-ion for the negative charges of the DNA phosphate moieties, therefore promoting hybridization in experimental conditions (i.e., in pure water at 65°C) that would otherwise prevent denaturated (i.e., single-stand) DNA to re-hybridize. Consequently, this characteristic grants

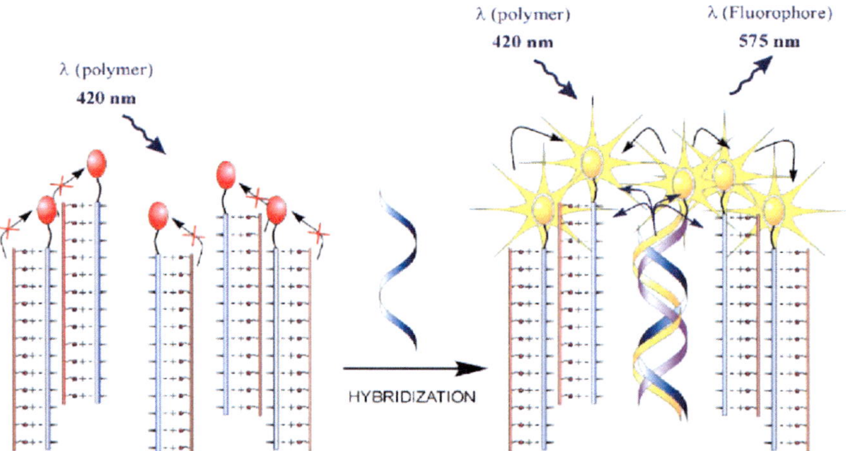

Fig. 5 Schematic description of the proposed signal amplification detection mechanism based on the conformational change of cationic polythiophene and energy transfer for ultrasensitive, selective, and rapid DNA detection. (Reprinted with permission from [18], Copyright 2008 American Chemical Society)

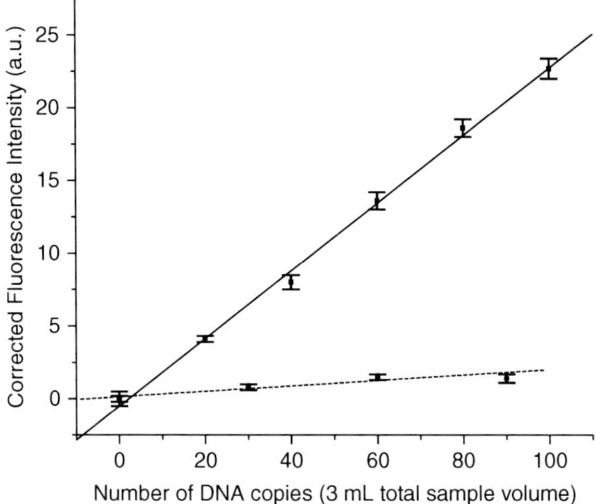

Fig. 6 Background-subtracted fluorescence intensity measured at 572 nm with excitation at 420 nm in pure water at 65°C, as a function of the number of mutated (i.e., tyrosinemia) genomic DNA copies detected with (*solid line*) tyrosinemia-specific probe (perfectly matched) and (*dashed line*) wild-type (i.e., normal) probe (one mismatch). (Reprinted with permission from [21], Copyright 2005 American Chemical Society)

exquisite detection specificity to this polymer-based biosensor by decreasing the risk of false positives that can occur in the presence of excess *ds*DNA material. As a case in point, although both 15-mer oligonucleotides used to probe tyrosinemia in the experiment described above were designed to be unique in the human genome, there are numerous loci presenting only one mismatch with these probes, notably at the extremities. Nevertheless, no significant hybridization signal was detected from these non-specific target sequences.

The degree of fluorescence signal amplification achieved with this approach was evaluated by comparing the sensitivity obtained with labeled and non-labeled 20-mer oligonucleotide probes and using the best experimental conditions for each detection scheme. It was determined that the sensitivity is ~4000 times higher and the signal RSD ~7 times lower for the labeled probe system [35] . Notably, the increase in sensitivity is much higher than the 30:1 ratio between the quantum yields of the AF546 dye and the polythiophene. The model called "fluorescence chain reaction" (FCR) that was proposed to explain this massive signal amplification involves the formation, prior to the introduction of the DNA targets, of aggregates of duplexes in solution allowing Förster resonance energy transfer (RET) among the donor (i.e., the unquenched target/probe/polymer "triplex") and neighboring AF546 acceptors. Fluorescence signal amplification then hinges on two conditions: the aggregates must be stable and compact enough to ensure sufficient proximity between the donor and the acceptors, and the energy transfer rate between the "triplex" and the AF546 must be high enough to allow the excitation of several acceptor molecules by a single donor. Fluorescence lifetime measurements performed using phase-modulation instrumentation [35] have shown that the excited lifetime of polythiophene/*ds*DNA (i.e., the donor in the RET system), at 0.86 ns, is much longer than that of the complex bearing the AF546-labeled probe (τ = 0.05 ns). The RET rate calculated from these values, at 1.88×10^{10} s^{-1} or 53 ps per transfer, is ~16 times faster than the excited state lifetime of 860 ps measured for the non-labeled triplex, and the authors suggested that faster energy transfers could even be anticipated with a lower degree of hybridization than those studied. Thus, the massive increase in detection sensitivity observed with the so-called FCR scheme, with the use of labeled probes and the subsequent formation of labeled aggregates, could result from this fast energy transfer between the polymer donor and neighboring AF546 acceptors, whereas homotransfer between AF546 chromophores, given their small Stokes shift of 20 nm, could also play a role in the fluorescence signal amplification observed.

Interestingly, a different strategy to detect DNA with a RET-based fluorescence signal amplification system was recently reported by Lee et al. [36]. Using a water soluble and highly emissive (quantum yield of 45%) polymer, poly(*p*-phenyleneethynylene) (PPE-R_1-COOH), the authors were able to selectively detect DNA hybridization using two distinct approaches. The first scheme (top half of Fig. 7), consist of a polymer bioconjugated with ssDNA probes that acts as a RET donor when the acceptor-labeled target DNA hybridizes with the probes. The

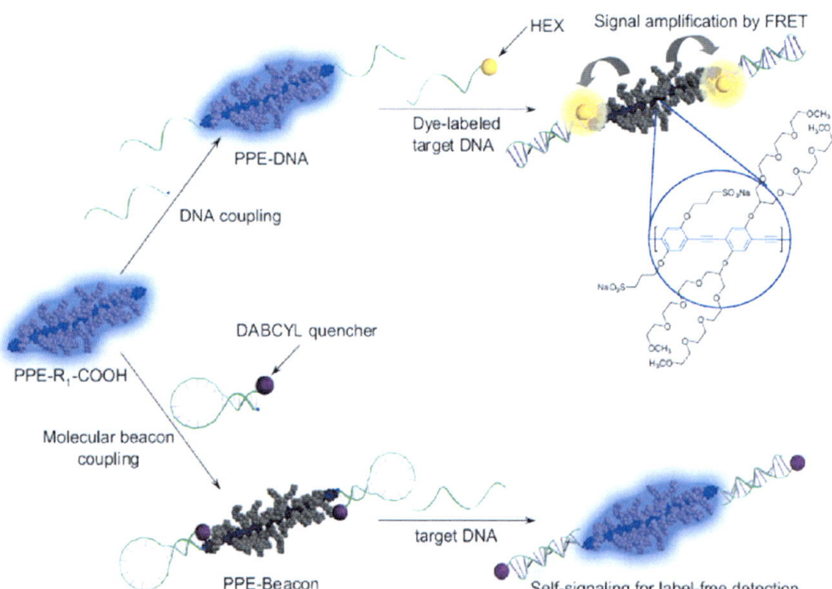

Fig. 7 DNA sensing schemes based on the transduction of hybridization by poly(*p*-phenyleneethynylene), using (*top*) RET with acceptor-labeled targets and (*bottom*) molecular beacon-type DNA probes. (Reprinted with permission from [36], Copyright 2007 Wiley-VCH)

authors found that the fluorescence intensity of the HEX acceptor dye in the PPE-DNA/DNA-HEX complex was increased by a factor of ~13 when excited via RET by the polymer donor, compared to its direct excitation. The authors attributed this amplification to the efficient energy harvesting and transport properties of the conjugated polymer, in agreement with the model proposed for the FCR scheme described previously. To overcome the drawback of using labeled target DNA, the authors demonstrated the use of their polymer as a self-signaling sensor. In that second scheme, the polymer is bioconjugated to DNA probes designed to form the stem-loop structure of a molecular beacon (bottom half of Fig. 7). Prior to the addition of target DNA, the fluorescence intensity of the PPE-beacon complex is low due to the proximity of the DABCYL quencher. Upon hybridization of target DNA and the subsequent removal of the DABCYL moieties away from PPE, the authors measured an increase in fluorescence intensity by almost two orders of magnitude. Despite the lack of quantitative information on detection sensitivity, this polymer-based detection scheme is an interesting and promising variation on the well-established molecular beacon method.

Fluorophore Encapsulation

Fluorescence emission is known to be very sensitive to environmental quenching (oxygen or collisional quenching by solvent molecules) and photobleaching [29]. Therefore, researchers have sought to protect fluorophores within solid matrices (micro- or nanoparticles) [37–40] or by incorporating them in supramolecular assemblies like micelles or aggregates [21,41–45]. A study published in 2005 showed that fluorescent guest molecules (pyrene butyric acid, PBA) confined within host capsules showed a significant increase in their fluorescence quantum yield, an effect attributed by the authors to the structural constraint imposed by the host, which minimizes radiationless decay pathways such as collisional quenching by solvent molecules [42] . Likewise, others have reported that surfactants have a beneficial influence on the luminescence of water-soluble conjugated polymers, an effect attributed to a lessening of quenching by interfacial water molecules [43,46,47]. The influence of non-ionic surfactants on the behavior of Leclerc's polythiophene-based biosensor was also investigated. In highly dilute regimes, it was found that the presence of 0.3 mM of Triton X-100 improved the limit of detection by a factor of 10 or more [19]. This result may be attributed to increases in both molar extinction coefficient and luminescence quantum yield following the release of interfacial water molecules and the formation of more extended polymer chains, as well as to the encapsulation of the fluorescent polymer by the surfactant molecules [43,47].

Particles or molecular assemblies can also be used to bring more signaling species together to enhance the fluorescence signal. In 2003, Kwakye and Baeumner used phospholipid vesicles (liposomes) to entrap thousands of marker molecules and generate significant amplification of the emission signal upon RNA detection, reporting an increase in detection sensitivity of three orders of magnitude over single fluorophore detection [48]. The authors used a sandwich assay with two DNA probes, one coupled to the signaling liposomes and the other coupled to superparamagnetic beads for target immobilization. By integrating this sensor within a microfluidic device, they were able to detect as few as 10 amol/μL of the Dengue virus RNA.

In 2004, Zhou and Zhou reported the use of dye-doped nanoparticles to improve the detection sensitivity and photostability of DNA hybridization on microarrays [37]. The authors developed novel cyanine dye-doped Au/silica core-shell nanoparticles in which the Cy3 or Cy5 fluorophores were conjugated to an Au colloidal core first and then embedded in the silica shell, this shell acting as a shield from environmental effects for the fluorophores and also as support for the bioconjugation of DNA probes. The authors claimed that this design avoids the problems of dye leakage and variations of the amount of fluorophore load from nanoparticle to nanoparticle, which are observed with fluorophore-doped silica nanoparticles commonly used as labels. Furthermore, being able to control the number of fluorophores (via conjugation with the Au colloids) allows one to tune this number in order to avoid aggregation and/or self-quenching. These core-shell fluorescent labels were compared to conventional dye-labeled DNA probes in a sandwich hybridization assay as illustrated in Fig. 8. The authors were able to detect as low as 1 pM of the 50 bp heme DNA target with the nanoparticle conjugates, or about 10 times better than using the dye-labeled DNA probes. They noted that the improvement in detection sensitivity brought by their approach might be limited by the larger size of the nanoparticles and the resulting steric crowding effect on hybridization. Interestingly, these nanoparticles also showed improved photostability: after nine consecutive scans of the microarrays, the signal from Cy5-doped nanoparticles was still at 95% of initial intensity whereas signal intensity from Cy5-labeled probes had dropped to 76%.

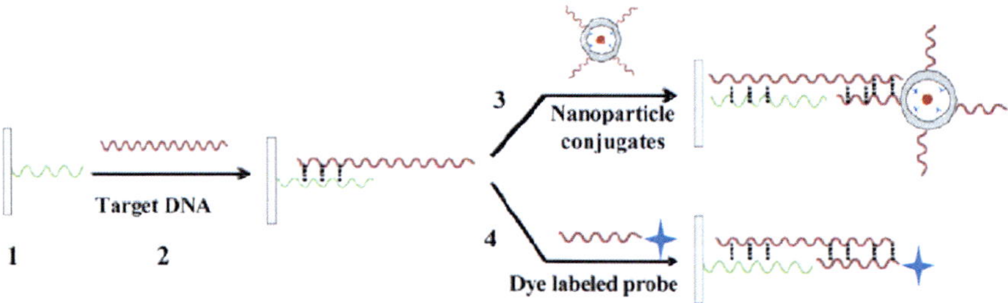

Fig. 8 Sandwich hybridization conducted on DNA microarray (1) oligonucleotide probes spotted on aldehyde slide; (2) oligonucleotide microarray hybridized with target sample; (3) hybridization with dye-doped nanoparticle/DNA conjugates; (4) hybridization with conventional dye-labeled probes. (Reprinted with permission from [37], Copyright 2004 American Chemical Society)

In a similar study, the group of Ratna attached Cy5 molecules at fixed locations on the viral capsid of a mutant version of the cowpea mosaic virus (CPMV) to use as a label for DNA microarrays [41]. The 30-nm viral particles display no cysteine residues on the capsid surface, but the engineered version of the virus contained 60 rather equally spaced cysteines inserted within the protein scaffold and available for dye grafting via the thiol groups. The authors then conjugated Cy5 fluorophores to the particles (along with neutravidin proteins in order to detect biotinylated DNA targets). They were able to introduce ~40 Cy5 molecules into the virus scaffold without observing any fluorescence quenching. Furthermore, they measured a significant increase in quantum yield for the dye molecules enclosed in the virus scaffold as compared to dye solutions of commensurable concentration (with and without virus particles present). They observed the appearance of an energy transfer process between Cy5 and the virus which they attributed to the coupling between Cy5 molecules and the virus scaffold but noted that this process alone could not explain the observed increase in quantum yield of the fluorophore, suggesting instead that this increase might be ascribed to environmental effects on the dye (i.e., encapsulation leading to protection from collisional quenchers).

Several studies such as those described above have documented the usefulness of encapsulation to quench fluorescent biosensors from environmental effects. The FCR aggregates described previously (Fig. 5) were shown to display a markedly improved photostability vs. the non-labeled polymer–ssDNA transducer (100% conserved fluorescence signal after 60 min exposition to the excitation source vs. 40% for the non-labeled complex), possibly a consequence of the confinement of RET acceptor molecules within the aggregates and away from quenching species, in agreement with the studies described above [37,42]. Furthermore, encapsulation has also been used to bring conjugated polymer donors and acceptor dye molecules within Förster range to promote RET mechanisms and fluorescence signal amplification. Detailed studies of the FCR mechanism using dynamic light scattering have revealed that aggregates were formed spontaneously when a large excess of labeled polymer+ssDNA complexes was placed in solution, and that the change in size of these aggregates upon changes in temperature betrayed a micellar structure [49]. Critical micellar concentration (cmc) values of 6.3×10^{-11} M and 2.8×10^{-9} M were calculated for the labeled and unlabeled system, respectively, suggesting that the self-assembly of labeled complexes into micelle-like structures is promoted by the AF546 label, which is more hydrophobic than the highly charged polymer-DNA complex [50]. Using MALLS (Multi-Angle Laser Light Scattering) measurements, they calculated a weight-averaged aggregate mass, A_M, of $4.7 \pm 0.3 \times 10^7$ g/mol and a radius of gyration, R_G, of 116 ± 6 nm. Assuming a low polydispersity index, dividing A_M by the number-averaged molecular weight of a single polymer-labeled probe complex (10,244 g/mol) yielded an aggregation number, N_{agg}, of ~4000–5000 polymer-labeled probe complexes. Such a high number is in line with the large number of aggregated RET acceptors needed to explain the large amplification of the fluorescence signal observed with FCR. Moreover, the ratio R_G/R_H (R_H being the hydrodynamic radius) is known to be a sensitive indicator of aggregate geometry, and the same MALLS measurements yielded a value $R_G/R_H = 2.1$, which is indicative of rod-shaped micelles, a hydrodynamic radius R_H of 55 nm and a rod length L equal to 402 nm (via the relation $R_G{}^2 = L^2/12$) [51–54]. Figure 9 presents a schematic representation of a possible rod micelle arrangement

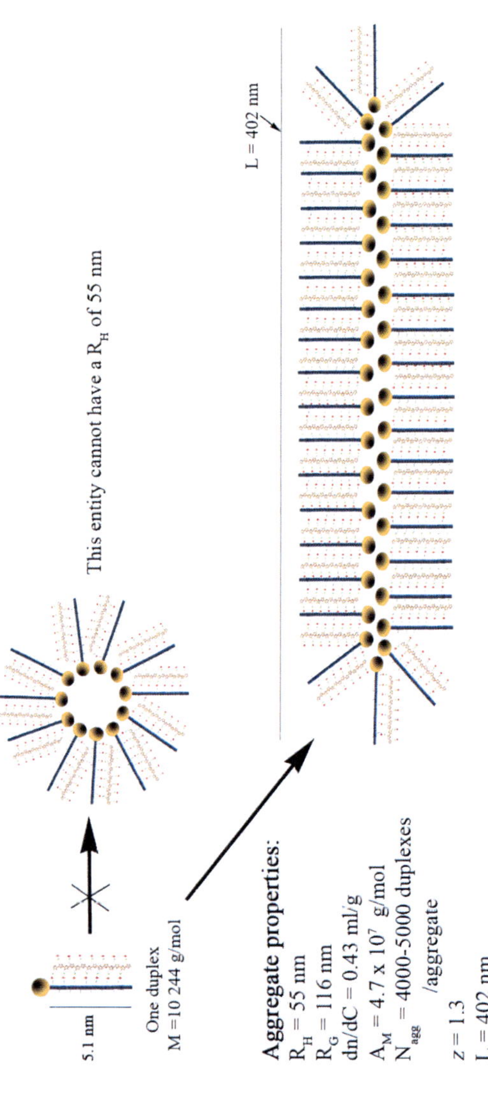

Fig. 9 Proposed physical rod-like model of the labeled aggregates (Reprinted with permission from [49], Copyright 2007 American Chemical Society)

having the dimensions given by the light scattering measurements of labeled duplex aggregates (the polymer–probe complex has approximately the same dimensions as *ds*DNA, which has a radius of 9.5 Å and a length of 3.4 Å per base for a total length of ~5 nm). Furthermore, rod-shaped structures of similar dimensions to those calculated using light scattering measurements were observed by TEM [49].

Interestingly, experimental data suggests that encapsulation within FCR micelle-like aggregates seems to have a beneficial impact also on the speed of hybridization between probe and target and, ultimately, on sample throughput. Comparison between labeled [21] and non-labeled [19] systems show that the fluorescence signal plateaus almost instantly (<5 min) for FCR aggregates, whereas close to 60 min are needed for the non-labeled polythiophene–ssDNA scheme. This enhanced transduction speed seemingly results from the high local ssDNA probe concentration within the aggregates, in line with a previous report by Park et al. [55] where inverse micelles were used to confine DNA complementary strands and speed up speed of hybridization by an order of magnitude as compared with hybridization of freely diffusing oligonucleotides in aqueous solution. In the case of FCR, the large excess of ssDNA probes vs. targets (10^9:1) in solution and the formation of polymer-probe complexes into micelles could be responsible for the extremely fast detection times observed.

Conclusion and Outlook

The development of ultrasensitive DNA detection schemes that do not rely on chemical target amplification is of critical importance for research and public health in general. Consequently, in order to address the need for biosensors able to detect a few copies of DNA targets, several methods were developed to amplify the fluorescence signal generated by a signal detection event. Among these new approaches, those based on some sort of fluorescence signal amplification are very promising because the intrinsic amplification mechanisms on which they are based tend to make them be more robust towards contamination issues that are frequently encountered in real diagnostic situations. Fluorescent conjugated polymers are very advantageous in this regard, since their collective system response is in itself a form of signal amplification, i.e., each repeating unit being a fluorescent entity but with the entire polymer chain acting as one signaling probe. In addition, because DNA is an anionic polymer, cationic fluorescent conjugated polymers can be used as direct DNA hybridization transducers if the polymer emission properties are changed upon formation of the double helix. Interestingly, conjugated polymers are generally also very efficient charge carrier transporters and are as such used in transistors and light-emitting devices [15,56,57], a property which makes them also very efficient RET donors. It seems that when in close proximity with well-oriented acceptors, these polymers can efficiently harvest the excitation energy, which then travels very quickly along the delocalized structure of the polymer chain to be channeled towards the acceptors in its proximity. This very quick energy transportation has been confirmed by time-resolved studies by two research groups who observed that the polymer lifetime was decreased by at least an order of magnitude in the presence of the acceptors [35,31]. It should be mentioned that it is still unclear as to what extent this condition of multiple available acceptors is needed to establish fluorescence amplification, compared to the role played the transport mechanism itself, and more studies of acceptor lifetime dynamics during the hybridization transduction event are needed. Moreover, encapsulation of RET fluorophores within aggregates or similar structures also enhances the energy transfer mechanism, by both bringing together donors and acceptors and protecting them from environmental quenching. Scholes [58] even proposed that donors and acceptors enclosed in supramolecular aggregates may be close enough to perturb their respective transition densities and the resulting energy transfer properties. Finally, other approaches might be used in conjunction with conjugated polymers acting as RET donors, such as the use of metal nanostructures to enhance radiative relaxation even further by plasmonic coupling [59,60].Clearly, the emerging field of conjugated polymers as RET-based optical hybridization transducers will very likely play a central role in the advent of true PCR-free fluorescence-based sensing techniques.

References

1. Daar, A. S., Thorsteinsdottir, H., Martin, D. K., Smith, A. C., Nast, S., Singer, P. A., Top ten biotechnologies for improving health in developing countries. *Nature Genetics*, **32**(2), 229–232, (2002).
2. Piunno, P. A. E., Krull, U. J., Trends in the development of nucleic acid biosensors for medical diagnostics. *Analytical and Bioanalytical Chemistry*, **381**(5), 1004–1011, (2005).
3. Thomas, S. W., Joly, G. D., Swager, T. M., Chemical sensors based on amplifying fluorescent conjugated polymers. *Chemical Reviews*, **107**(4), 1339–1386, (2007).
4. Ihara, T., Mukae, M., Homogeneous DNA-detection based on the non-enzymatic reactions promoted by target DNA. *Analytical Sciences*, **23**(6), 625–629, (2007).
5. Altschuh, D., Oncul, S., Demchenko, A. P., Fluorescence sensing of intermolecular interactions and development of direct molecular biosensors. *Journal of Molecular Recognition*, **19**(6), 459–477, (2006).
6. Boissinot, M., Bergeron Michel, G., Toward rapid real-time molecular diagnostic to guide smart use of antimicrobials. *Current Opinion in Microbiology*, **5**(5), 478–82, (2002).
7. Cao YunWei, C., Jin, R., Mirkin Chad, A., Nanoparticles with Raman spectroscopic fingerprints for DNA and RNA detection. *Science*, **297**(5586), 1536–40, (2002).
8. Dubertret, B., Calame, M., Libchaber, A. J., Single-mismatch detection using gold-quenched fluorescent oligonucleotides. *Nature Biotechnology*, **19**(4), 365–370, (2001).
9. Elghanian, R., Storhoff, J. J., Mucic, R. C., Letsinger, R. L., Mirkin, C. A., Selective colorimetric detection of polynucleotides based on the distance-dependent optical properties of gold nanoparticles. *Science*, **277**(5329), 1078–1081, (1997).
10. Gaylord, B. S., Heeger, A. J., Bazan, G. C., DNA detection using water-soluble conjugated polymers and peptide nucleic acid probes. *Proceedings of the National Academy of Sciences of the United States of America*, **99**(17), 10954–10957, (2002).
11. Gaylord, B. S., Heeger, A. J., Bazan, G. C., DNA Hybridization detection with water-soluble conjugated polymers and chromophore-labeled single-stranded DNA. *Journal of the American Chemical Society*, **125**(4), 896–900, (2003).
12. Taton, T. A., Mirkin, C. A., Letsinger, R. L., Scanometric DNA array detection with nanoparticle probes. *Science*, **289**(5485), 1757–1760, (2000).
13. Wang, J., Polsky, R., Merkoci, A., Turner, K. L., "Electroactive beads" for ultrasensitive DNA detection. *Langmuir*, **19**(4), 989–991, (2003).

14. Saghatelian, A., Guckian, K. M., Thayer, D. A., Ghadiri, M. R., DNA Detection and signal amplification via an engineered allosteric enzyme. *Journal of the American Chemical Society*, **125**(2), 344–345, (2003).
15. Swager, T. M., The molecular wire approach to sensory signal amplification. *Accounts of Chemical Research*, **31**(5), 201–207, (1998).
16. KorriYoussoufi, H., Garnier, F., Srivastava, P., Godillot, P., Yassar, A., Toward bioelectronics: Specific DNA recognition based on an oligonucleotide-functionalized polypyrrole. *Journal of the American Chemical Society*, **119**(31), 7388–7389, (1997).
17. Ho, H. A., Boissinot, M., Bergeron, M. G., Corbeil, G., Dore, K., Boudreau, D., Leclerc, M., Colorimetric and fluorometric detection of nucleic acids using cationic polythiophene derivatives. *Angewandte Chemie-International Edition*, **41**(9), 1548–1551, (2002).
18. Ho, H. A., Najari, A., Leclerc, M., Optical detection of DNA and proteins with cationic polythiophenes. *Accounts of Chemical Research*, **41**(2), 168–178, (2008).
19. Dore, K., Dubus, S., Ho, H. A., Levesque, I., Brunette, M., Corbeil, G., Boissinot, M., Boivin, G., Bergeron, M. G., Boudreau, D., Leclerc, M., Fluorescent polymeric transducer for the rapid, simple, and specific detection of nucleic acids at the zeptomole level. *Journal of the American Chemical Society*, **126**(13), 4240–4244, (2004).
20. Nilsson, K. P. R., Inganas, O., Chip and solution detection of DNA hybridization using a luminescent zwitterionic polythiophene derivative. *Nature Materials*, **2**(6), 419–U10, (2003).
21. Ho, H. A., Dore, K., Boissinot, M., Bergeron, M. G., Tanguay, R. M., Boudreau, D., Leclerc, M., Direct molecular detection of nucleic acids by fluorescence signal amplification. *Journal of the American Chemical Society*, **127**(36), 12673–12676, (2005).
22. Klimov, A. I., Rocha, E., Hayden, F. G., Shult, P. A., Roumillat, L. F., Cox, N. J., Prolonged shedding of amantadine-resistant influenza—a viruses by immunodeficient patients – detection by polymerase chain-reaction – restriction analysis. *Journal of Infectious Diseases*, **172**(5), 1352–1355, (1995).
23. Karlsson, K. F., Asberg, P., Nilsson, K. P. R., Inganas, O., Interactions between a zwitterionic polythiophene derivative and oligonucleotides as resolved by fluorescence resonance energy transfer. *Chemistry of Materials*, **17**(16), 4204–4211, (2005).
24. Chen, L. H., McBranch, D. W., Wang, H. L., Helgeson, R., Wudl, F., Whitten, D. G., Highly sensitive biological and chemical sensors based on reversible fluorescence quenching in a conjugated polymer. *Proceedings of the National Academy of Sciences of the United States of America*, **96**(22), 12287–12292, (1999).
25. Herland, A., Inganas, O., Conjugated polymers as optical probes for protein interactions and protein conformations. *Macromolecular Rapid Communications*, **28**(17), 1703–1713, (2007).
26. Kushon, S. A., Ley, K. D., Bradford, K., Jones, R. M., McBranch, D., Whitten, D., Detection of DNA hybridization via fluorescent polymer superquenching. *Langmuir*, **18**(20), 7245–7249, (2002).
27. Jones, R. M., Bergstedt, T. S., McBranch, D. W., Whitten, D. G., Tuning of superquenching in layered and mixed fluorescent polyelectrolytes. *Journal of the American Chemical Society*, **123**(27), 6726–6727, (2001).
28. Kumaraswamy, S., Bergstedt, T., Shi, X. B., Rininsland, F., Kushon, S., Xia, W. S., Ley, K., Achyuthan, K., McBranch, D., Whitten, D., Fluorescent-conjugated polymer superquenching facilitates highly sensitive detection of proteases. *Proceedings of the National Academy of Sciences of the United States of America*, **101**(20), 7511–7515, (2004).
29. Lakowicz, J. R., Principles of Fluorescence Spectroscopy. 2nd ed., Kluwer Academic/Plenum: New York, p xxiii, 698, (1999).
30. Tyagi, S., Kramer, F. R., Molecular beacons: probes that fluoresce upon hybridization. *Nature biotechnology*, **14**(3), 303–308, (1996).
31. Xu, Q. H., Gaylord, B. S., Wang, S., Bazan, G. C., Moses, D., Heeger, A. J., Time-resolved energy transfer in DNA sequence detection using water-soluble conjugated polymers: The role of electrostatic and hydrophobic interactions. *Proceedings of the National Academy of Sciences of the United States of America*, **101**(32), 11634–11639, (2004).

32. Wang, S., Gaylord, B. S., Bazan, G. C., Fluorescein provides a resonance gate for FRET from conjugated polymers to DNA intercalated dyes. *Journal of the American Chemical Society*, **126**(17), 5446–5451, (2004).
33. Xu, Q. H., Wang, S., Korystov, D., Mikhailovsky, A., Bazan, G. C., Moses, D., Heeger, A. J., The fluorescence resonance energy transfer (FRET) gate: A time-resolved study. *Proceedings of the National Academy of Sciences of the United States of America*, **102**(3), 530–535, (2005).
34. St-Louis, M., Tanguay, R. M., Mutations in the fumarylacetoacetate hydrolase gene causing hereditary tyrosinemia type I: Overview. *Human Mutation*, **9**(4), 291–299, (1997).
35. Dore, K., Leclerc, M., Boudreau, D., Investigation of a fluorescence signal amplification mechanism used for the direct molecular detection of nucleic acids. *Journal of Fluorescence*, **16**(2), 259–265, (2006).
36. Lee, K., Povlich, L. K., Kim, J., Label-free and self-signal amplifying molecular DNA sensors based on bioconjugated polyelectrolytes. *Advanced Functional Materials*, **17**(14), 2580–2587, (2007).
37. Zhou, X. C., Zhou, J. Z., Improving the signal sensitivity and photostability of DNA hybridizations on microarrays by using dye-doped core-shell silica nanoparticles. *Analytical Chemistry*, **76**(18), 5302–5312, (2004).
38. Yan, J. L., Estevez, M. C., Smith, J. E., Wang, K. M., He, X. X., Wang, L., Tan, W. H., Dye-doped nanoparticles for bioanalysis. *Nano Today*, **2**(3), 44–50, (2007).
39. Lian, W., Litherland, S. A., Badrane, H., Tan, W. H., Wu, D. H., Baker, H. V., Gulig, P. A., Lim, D. V., Jin, S. G., Ultrasensitive detection of biomolecules with fluorescent dye-doped nanoparticles. *Analytical Biochemistry*, **334**(1), 135–144, (2004).
40. Bagwe, R. P., Zhao, X. J., Tan, W. H., Bioconjugated luminescent nanoparticles for biological applications. *Journal of Dispersion Science and Technology*, **24**(3–4), 453–464, (2003).
41. Soto, C. M., Blum, A. S., Vora, G. J., Lebedev, N., Meador, C. E., Won, A. P., Chatterji, A., Johnson, J. E., Ratna, B. R., Fluorescent signal amplification of carbocyanine dyes using engineered viral nanoparticles. *Journal of the American Chemical Society*, **128**(15), 5184–5189, (2006).
42. Dalgarno, S. J., Tucker, S. A., Bassil, D. B., Atwood, J. L., Fluorescent guest molecules report ordered inner phase of host capsules in solution. *Science*, **309**(5743), 2037–2039, (2005).
43. Chen, L. H., Xu, S., McBranch, D., Whitten, D., Tuning the properties of conjugated polyelectrolytes through surfactant complexation. *Journal of the American Chemical Society*, **122**(38), 9302–9303, (2000).
44. Sautter, A., Kaletas, B. K., Schmid, D. G., Dobrawa, R., Zimine, M., Jung, G., van Stokkum, I. H. M., De Cola, L., Williams, R. M., Wurthner, F., Ultrafast energy-electron transfer cascade in a multichromophoric light-harvesting molecular square. *Journal of the American Chemical Society*, **127**(18), 6719–6729, (2005).
45. Vallotton, P., Tairi, A. P., Wohland, T., Friedrich-Benet, K., Pick, H., Hovius, R., Vogel, H., Mapping the antagonist binding site of the serotonin type 3 receptor by fluorescence resonance energy transfer. *Biochemistry*, **40**(41), 12237–12242, (2001).
46. Gaylord, B. S., Wang, S. J., Heeger, A. J., Bazan, G. C., Water-soluble conjugated oligomers: Effect of chain length and aggregation on photoluminescence-quenching efficiencies. *Journal of the American Chemical Society*, **123**(26), 6417–6418, (2001).
47. Lavigne, J. J., Broughton, D. L., Wilson, J. N., Erdogan, B., Bunz, U. H. F., "Surfactochromic" conjugated polymers: Surfactant effects on sugar-substituted PPEs. *Macromolecules*, **36**(20), 7409–7412, (2003).
48. Kwakye, S., Baeumner, A., A microfluidic biosensor based on nucleic acid sequence recognition. *Analytical and Bioanalytical Chemistry*, **376**(7), 1062–1068, (2003).
49. Dore, K., Neagu-Plesu, R., Leclerc, M., Boudreau, D., Ritcey, A. M., Characterization of superlighting polymer-DNA aggregates: A fluorescence and light scattering study. *Langmuir*, **23**(1), 258–264, (2007).
50. Myers, D., Surfaces, Interfaces, and Colloids: Principles and Applications. Wiley-VCH, New York, 433 pp (1991).

51. Ishizu, K., Toyoda, K., Furukawa, T., Sogabe, A., Electrostatic interaction of anionic/nonionic polyelectrolyte prototype copolymer brushes with cationic linear polyelectrolyte. *Macromolecules*, **37**(10), 3954–3957, (2004).
52. Appell, J., Porte, G., An investigation on the micellar shape using angular dissymmetry of light scattered by solutions of cetylpyridinium bromide. *Journal of Colloid and Interface Science*, **81**(1), 85–90, (1981).
53. Fuetterer, T., Nordskog, A., Hellweg, T., Findenegg, G. H., Foerster, S., Dewhurst, C. D., Characterization of polybutadiene-poly(ethyleneoxide) aggregates in aqueous solution: A light-scattering and small-angle neutron-scattering study. *Physical Review E*, **70**(4), (2004).
54. Hiemenz, P. C., Rajagopalan, R. (Editors), Principles of Colloid and Surface Chemistry, 3rd ed., Revised and Expanded. Taylor & Francis, New York, p. 688 , (1997).
55. Park, L. C., Maruyama, T., Goto, M., DNA hybridization in reverse micelles and its application to mutation detection. *Analyst*, **128**(2), 161–165, (2003).
56. Peng, Q., Xie, M. Q., Huang, Y., Lu, Z. Y., Cao, Y., Novel supramolecular polymers based on zinc-salen chromophores for efficient light-emitting diodes. *Macromolecular Chemistry and Physics*, **206**(23), 2373–2380, (2005).
57. Drolet, N., Morin, J. F., Leclerc, N., Wakim, S., Tao, Y., Leclerc, M., 2,7-carbazolenevinylene-based oligomer thin-film transistors: High mobility through structural ordering. *Advanced Functional Materials*, **15**(10), 1671–1682 (2005).
58. Scholes, G. D., Long-range resonance energy transfer in molecular systems. *Annual Review of Physical Chemistry*, **54**, 57–87, (2003).
59. Zhang, J., Lakowicz, J. R., A model for DNA detection by metal-enhanced fluorescence from immobilized silver nanoparticles on solid substrate. *Journal of Physical Chemistry B*, **110**(5), 2387–2392, (2006).
60. Lessard-Viger, M., Saiveng Live, L., Dupont-Therrien, O., Boudreau, D., Reduction of self-quenching in fluorescent silica-coated nanoparticles. *Plasmonics*, **3**, 33–40, (2008).

Exploring the Electrostatic Landscape of Proteins with Tryptophan Fluorescence

Patrik R. Callis

Abstract The objective of this review is to present information gained during the last decade that has transformed tryptophan (Trp) into a well-understood fluorescent probe—a probe with no more complications than most non-intrinsic probes, which are usually large dyes with incompletely explored complexity. An overview highlights and summarizes key advancements during the last decade that have increased our understanding of Trp and its fluorescence in proteins. This is followed by sections devoted to (1) the understanding of quenching (mostly caused by electron transfer for which there has been considerable progress), (2) the prediction of steady state fluorescence wavelengths in proteins, (3) nonexponential decay (which remains challenging), and (4) a section that catalogs a number of fundamental experiments and computations on the isolated Trp chromophore that have provided much of the infrastructure for the preceding sections. There is continual emphasis in this chapter on the simple—yet accurate—concept that the extremes of electrostatic fields in proteins are mirrored in the variability and sensitivity of Trp fluorescence quantum yields, lifetimes, and wavelengths.

Keywords Tryptophan · Fluorescence · Proteins · Electrostatics · Electron transfer · Quenching · Lifetimes · Wavelengths

Introduction and Overview

Intrinsic fluorescence from tryptophan (Trp) continues to be widely used to follow many transformations of proteins [1–5]. The more certain structural knowledge associated with the use of Trp fluorescence is a powerful asset, one that usually outweighs what are perceived as its complicating disadvantages. This is apparent, for example, in its extensive use in following protein folding [6–11]. One objective of this chapter is to present information gained during the last decade that presents Trp as a well-understood probe—a probe with no more complications than

P.R. Callis (✉)
Department of Chemistry and Biochemistry, Montana State University, Bozeman, MT, USA

C.D. Geddes (ed.), *Reviews in Fluorescence 2007*, Reviews in Fluorescence 2007,
DOI 10.1007/978-0-387-88722-7_10,

most non-intrinsic probes, which are usually large dyes with incompletely explored complexity.

The usefulness of Trp fluorescence stems from the sensitivity to the precise local environment of the intensity, i.e., quantum yield (ϕ_f), excited state lifetime (τ_f), and wavelength (λ_{max}). In particular, it is the local electric field strength and *direction* that determine whether the fluorescence will be red or blue shifted and whether an electron acceptor will or will not quench the fluorescence. There will be continual emphasis in this chapter on the simple—yet accurate—concept that the ruggedness of the electrostatic landscape in proteins is mirrored in the ruggedness of the Trp fluorescence quantum yield and lifetime landscape. This is quite pertinent because the diversity and strength of electric fields in proteins is key to their function as enzymes [12].

In a previous review of underlying principles of tryptophan fluorescence [13], basic properties of the nearly degenerate ^1L_a and ^1L_b [14–17] excited states of Trp pertinent to Trp fluorescence were presented, with emphasis on what were then the latest quantum chemical studies and spectroscopy of indole and other molecules closely related to Trp. Several questions were identified but left largely unanswered. A third objective of this chapter is to revisit works appearing in the last decade that have provided definitive microscopic insight into the following questions:

(1) What are the microscopic determinants of the fluorescence wavelength of Trp in proteins?
(2) Do any proteins emit from the Trp ^1L_b state?
(3) What determines the fluorescence quantum yield and average lifetimes of Trp in proteins?
(4) What causes the non-exponential fluorescence decay?

The answer to these questions depends always on the local electric field—or what is the same thing—*the change in electric potential (volts) between where the electron is initially and where it ends up during an electronic transition (including electron transfer)*. See appendix for a brief tutorial on electric fields and potentials as pertains to Trp fluorescence. Figure 1 provides the essence of what is important. For

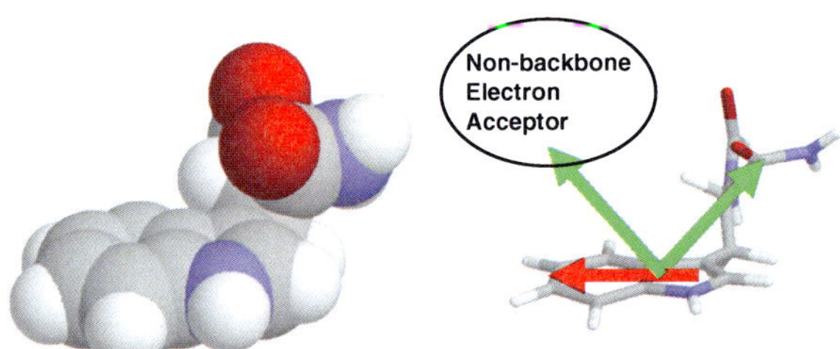

Fig. 1 N-formyltryptophanamide used for QM calculations trimmed from protein structure, also showing the two kinds of electron density shifts that control Trp fluorescence wavelength (*red*) and intensity/lifetime (*green*). The CPK form on the left emphasizes the nearness (3.15 Å) of the CG ring atom to the C of the C-terminal amide, the primary electron transfer destination

λ_{max}, it is the stabilization of the electron density shift from the pyrrole to the benzene part of the indole ring during the ground to excited state (^1L_a) that determines the red shift. For intensity, it is the stabilization of the electron density shift from the indole ring as a whole to the electron acceptor during photoinduced electron transfer (either to a local amide, another protein side chain, or external quencher).

This chapter is structured as follows: an overview highlights and summarizes key advancements during the last decade that build on the previous chapter [15] and have increased understanding of the Trp chromophore. This is followed by sections devoted to the understanding of quenching (for which there has been considerable progress), the prediction of steady state λ_{max}, and non-exponential decay (which remains challenging). The final major section catalogs a number of fundamental experiments and computations on the Trp chromophore that have provided much of the infrastructure for the preceding sections.

Overview

Resolved Spectra of Models. Milestones covered in this chapter (chronologically) begin with the work on indole derivatives in cold vacuum and in solid argon that revealed important details regarding the relative energies of the ^1L_a and ^1L_b electronic states [18–22]. This information provided confidence that in proteins it is always ^1L_a that emits, with the one lingering doubt (at the time) being Trp48 of azurin. Recently, however, Broos et al. have conclusively shown that even for Trp48 of azurin at room temperature ($\lambda_{max} = 306.5$ nm), the fluorescence is purely ^1L_a [23]. The same authors have also shown that the dI component of transhydrogenase from *Rhodospirillumrubrum*, the only other known protein showing $\lambda_{max} \approx 308$ nm, also emits purely from ^1L_a. Substitution of Met97 in this protein with Val, however, results in a ~4 nm blue shift of the λ_{max} of Trp72 fluorescence to 303.5 nm, which *does* result in substantial ^1L_b fluorescence. In addition, very recently [24] these authors reported that replacement of Met97 with Ala or Leu also result in a similar blue shift and ^1L_b fluorescence while mutations of other residues do not have this effect. It is suggested that this results from the loss of the polarizable sulfur atom in the mutants raising the energy of ^1L_a above that of ^1L_b.

Quantitative Steady State Wavelength Predictions. Vivian and Callis, using a hybrid quantum mechanics/molecular mechanics (QM-MM) method demonstrated that simple electrostatics can be used to predict Trp fluorescence wavelengths in proteins [25]. The primary message of that work is that positive charge near the benzene or negative charge near the pyrrole ring causes a red shift, and the opposite causes a blue shift (See Fig. 1). The large red shift of Trp fluorescence in water is the most dramatic manifestation of this. The shift of electron density to the benzene ring upon excitation to the ^1L_a state is followed by fast relaxation of nearby waters to create H-bond-like interactions with the π-electrons of the benzene ring. This work also pointed to extensive coupling of protein and water contributions to the steady state Stokes shift by at least two mechanisms. More recent simulations [26] find that protein–water coupling is an essential ingredient to the *dynamics* of the Stokes shift.

Quantitative Quantum Yield/Lifetime Predictions from Electron Transfer Quenching. The QM-MM method was extended and applied [27, 28] to the long-standing question as to the origin of the intensity sensitivity—a puzzle because the quantum yield is barely affected by solvent when the indole ring is not in a protein. That work provided the first quantitative evidence that the full 30-fold range of Trp fluorescence quantum yields (and lifetimes) observed in proteins is due primarily to different rates of electron transfer (ET) from the excited indole ring to the empty antibonding MO of one of two nearest backbone amides. The dependence on protein environment arises mainly from the average local electric potential difference between the Trp ring and acceptor amide and from the amplitude of potential difference fluctuation caused by protein and solvent motions. This can be strongly influenced by the *location of nearby charges* relative to the transfer direction. Figure 1 illustrates the basic idea: negative charge near the indole ring and/or positive charge near the electron acceptor will increase the electron transfer rate (decrease the ϕ_f), because these arrangements stabilize the charge transfer (CT) state. In the opposite case, there will be minimal electron transfer, and the Trp will be highly fluorescent.The method has been applied to a number of interesting proteins [29–31].

Electron Transfer Matrix Elements. Recently we have reported a reasonable way to estimate the electronic coupling matrix element between the fluorescing and CT states of Trp in proteins [32]. This provides a basis for computing electron transfer rates from our QM-MM method without resorting to empirical assumptions about the coupling strength. The ab initio matrix elements are 10–100 times stronger than the 10 cm^{-1} value we used previously and are much more consistent with expected Franck–Condon factors. A new picture emerges, in which photoinduced electron transfer from Trp is possible only during short windows of time. One of the important findings is that—at the short distances pertinent to Trp fluorescence quenching—the coupling element does not decrease exponentially with distance between the electron acceptor amide and the indole ring. In fact, we find that through-bond coupling is extremely strong, and quite dependent on rotamer conformation. The rotamer showing the strongest average coupling has $\chi_1 = -60^o$, for which the amide is *most distant* from the indole ring. There is, however, virtually no correlation between coupling strength and experimental electron transfer rate (as deduced from fluorescence quantum yields) because of the more critical dependence on the electrostatic environment.

Non-exponential Fluorescence Decay: Femtoseconds to Nanoseconds. Almost all Trp fluorescence decay curves exhibit non-exponential behavior. This is likely an unavoidable consequence of the combined sensitivity of λ_{max} and ϕ_f to the electrostatic environment in proteins. While bothersome for purposes of studying energy transfer and rotational diffusion, for example, in the long run there may be considerable information to be gained. As was the case a decade ago, there is still no consensus regarding the primary cause(s) of the non-exponential decay. Opinions—and therefore interpretations—remain divided between the view that discrete subpopulations (often assigned to different Trp χ_1 and χ_2 rotamers [2,33–36]) exhibit different *population* decay times, and at the other extreme, the view that the excited population is homogeneous, but has a time-dependent fluorescence spectrum that shifts to longer wavelengths. Nevertheless, significant new knowledge has emerged, providing hope that a better understanding of the underlying causes lies in the near future.

The development and increased use of ultrafast methods now consistently reveals the fast solvent shifts from solvent relaxation that were always known to exist for proteins at ambient temperature, but which were not accessible to the widely used time correlated single photon counting (TCSPC) method. Controversy has expanded to include the intriguing arena of 10–100 ps, wherein time-resolved Stokes shifts (TRSS) are more readily embraced than in the nanosecond region. At the same time, a better understanding of what controls electron transfer-based quenching (including experimental observation) has raised expectations of the existence of fast-decaying blue-shifted subpopulations, which also create red shifting of the fluorescence spectrum in time. Two recent observations have greatly strengthened the evidence for such subpopulations:

(1) The discovery that 5-fluorotryptophan (5FTrp), when incorporated into proteins, practically always shows exponential decay of excited state population. The finding that 5FTrp has a 0.2 eV higher ionization potential (making it a much poorer electron donor) strongly implicates heterogeneity in the electrostatic environment as a cause for non-exponential decay of Trp fluorescence [37].
(2) The non-natural amino acid, Aladan, has a large change in dipole upon excitation and, like Trp, can be used as a probe for water exposure in proteins. The free probe in solution exhibits non-exponential fluorescence decay, and an average fluorescence lifetime that increases with increasing solvent polarity. In a recent study of the protein GB1[38], the time-resolved fluorescence spectrum of Aladan, when in all but the most buried sites, exhibited the remarkable behavior of *blue shifting* at times longer than 0.5 ns. This is almost certainly a proof of heterogeneity in environment, with the bluer fluorescence of those molecules in less solvent-exposed environments persisting longer than those in the more polar environments. The certain knowledge of this type of heterogeneity for Aladan gives credibility to its existence for Trp, where it is much harder to detect.

Understanding Quenching from First Principles: Photoinduced Electron Transfer

Background

Virtually all reported quantum yields of single Trps in proteins fall in the range 0.35 down to or 0.01 or less [2,5]. This striking variability is not due to differences in the local polarity because the quantum yield is nearly independent of solvent polarity for indole chromophores. The unique and important study by Meech et al. [39]

revealed that *both* the radiative and non-radiative rates for the model of the isolated chromophore, 3-methylindole (3MI) [40], increase considerably as solvent polarity decreases. As a result, the quantum yield remains relatively independent of polarity while the lifetime decreases from ~9 ns in water to about 4 ns in cyclohexane, with intermediate values in ether and alcohol.

The idea that the large variation in quantum yields is due to variation in electron transfer rate from the ^1L_a excited state emerged very early [34, 41, 42]. The indole molecule has an exceptionally low ionization potential, so that photoinduced electron transfer to a nearby electron acceptor is facile. Understanding, however, of Trp lifetimes and quantum yields in proteins is complex for two main reasons:

(1) There are several groups in proteins that can, in principle, act as electron acceptors.
(2) Whether an acceptor will, in fact, quench depends on the distance (and perhaps orientation) from the Trp, and—equally important—the relative electrostatic potential (voltage) difference between the Trp and the acceptor (this includes the state of protonation of the acceptor).

Electron Transfer Quenchers in Water. There is compelling experimental evidence that the following groups can act as electron acceptors (quenchers) of the pertinent model chromophore, 3MI, in *water* [43–45].

Disulfide bonds
Thiolate (deprotonated cysteine pH >8)
Neutral cysteine
Histidine *cation* (pH <6)
Neutral glutamic and aspartic acids pH <5
Amides of peptide bonds
Di-amides, including N-acylated glutamine and asparagine
Neutral histidine (pH >6)

Proton Transfer Quenching in Water. In addition, the following groups are found to quench indole and 3MI in water by a not so well understood mechanism involving proton transfer to the excited indole ring in water [45]:

Lysine side chain
Neutral tyrosine side chain
N-terminal ammonium groups

This type of quenching is distinguished from ET quenching by its pronounced deuterium isotope effect. Proton transfer from the ammonium group of free Trp zwitterion at neutral pH reduces ϕ_f to 0.14 relative to the value near 0.35 found for 3MI and Trp at pH 10 [46].

What Are the Known Quenchers in Proteins? This question is of greater pertinence to this chapter. It is harder to answer because—with the likely exception of the disulfide bond—the above mentioned electron transfer quenchers require some electrostatic assistance from the protein to quench at the rapid rate required to compete with the normal ^1L_a excited state decay. There are several reports in which some of the strong quenchers in water have no apparent quenching effect in proteins, even when in close contact. The study of cyclic hexapeptides, designed to contain no quencher candidates except the backbone [47,48], clearly emphasizes that the backbone amides are effective quenchers, but some Trps in proteins do not show evidence of quenching. At least two studies of proteins with weakly fluorescing Trps in which possible quenching side chains were mutated to nonquenchers failed to increase ϕ_f, clearly implicating the amide [29,49]. In an extensive study of the weakly fluorescing Trps 68 and 156 of human gamma D crystallin mutants, Chen et al. [29] replaced Cys32 whose S is only 4 Å distant from Trp 68 ring atoms; there was no increase in quantum yield, apparently because of negative groups near the cysteine, which suppress its electron accepting power. In other cases, there is evidence of moderate quenching by Cys [50].

Histidine cation has been documented as a quencher in proteins, typically by showing a decrease in ϕ_f upon lowering pH [51,52]. However, for the Q105H mutant of T4 lysozyme, Harris and Hudson [53] found that lowering the pH *increased* ϕ_f of the very close Trp138. We have argued that this is a case where the electrostatic effect of the His cation positive charge reduces the already substantial quenching by the backbone amide [27].

Nonquenchers in Proteins. For emphasis and clarity, below is a list of those groups that are sometimes stated to be quenchers, but for which there is little or no evidence. These include cases for which there is clear evidence for quenching of 3MI in solution, yet virtually no evidence within a protein.

Glutamate and Aspartate: It is clear from the Ricci–Nesta [45] and Chen–Barkley [43] papers that this group does not quench at all in water, nor would it be expected to because of the negative charge. A primary theme of this chapter is that the electrostatic environment can enable or disable quenching by a nearby quencher, including the backbone amides, as illustrated in Figs. 1 and 2. This is why charged side chains are sometimes mistakenly identified as quenchers.

Phenylalanine: It is understandable that considerable confusion surrounds this non-quencher [2]. Nominally, the highest occupied molecular orbital (HOMO) of benzene is too low in energy and the lowest unoccupied MO (LUMO) is too high to facilitate quenching of Trp. Nevertheless, benzene and toluene do quench 3MI fluorescence effectively, but not 1-Methyl and 1,2-dimethylindole [54,55]. The latter two cannot donate an H-bond. Nanda and Brand [56,57] reported that Trp 48, a conserved, buried residue commonly found in the hydrophobic core of homeodomains, has an unusually low fluorescence quantum yield. For example, unfolding of ultrabithorax causes a 20-fold enhancement of fluorescence. Structure and sequence analysis of several homeodomains, combined with the above information on indole quenching by benzene, led to the reasonable postulate that quenching was due to a transient, excited-state NH π hydrogen bond involving Trp 48 and

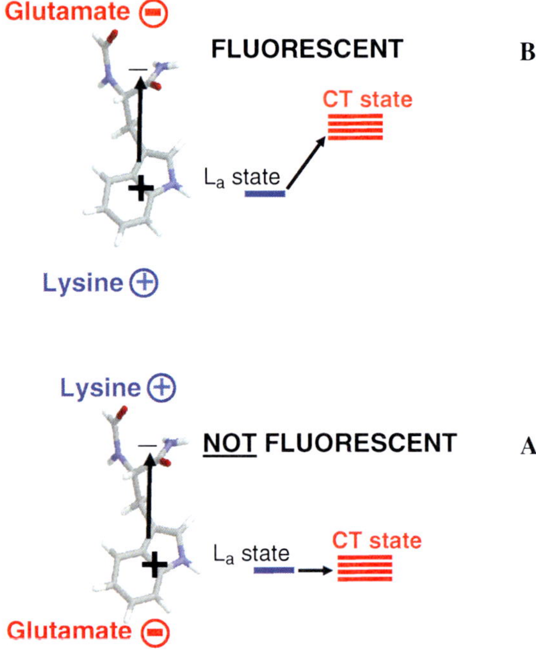

Fig. 2 How the *location* of charges determines the quenching by a backbone amide. (**A**) Charges stabilize the dipole of the CT state with electron transferred towards the positive Lys and away from the negative Glu. The CT state is in resonance with the fluorescing (^1L_a) state, electron transfer is fast and fluorescence is weak. (**B**) Charges are arranged oppositely from case A, greatly destabilizing the CT state. There can be little opportunity for resonance, and quenching is slight

a conserved aromatic residue at position 8. This phenomenon was also noticed by Subramaniam et al. [58] who, in addition, examined the F8A mutant of Bicoid-homeo domain. They found that, while important for stability, the Phe8 was not responsible for the strong quenching. The lack of quenching is also consistent with the finding that N-acetylphenyalanineamide does not quench 3MI in water [43].

We have presented evidence [59] indicating that the ubiquitous strong quenching of the conserved Trp 48 in homeo domains appears more likely due to the backbone amide as a result of several stabilizing electrostatic interactions largely due to the orientation of the Trp ring to the helix dipole, whose field strongly directs electrons from the ring to the amide. This finding is solidified by a similar finding for another case of conserved low quantum yield: Trps 68 and 156 of γD-crystallins from many species all have fluorescence quantum yields near 0.01, yet extensive mutation of nearby plausible quenchers fails to increase the quantum yield [29]. There, as in the case of the homeo domains, the conformation and surroundings result in efficient quenching by the nearby backbone amide. The *neutral* carboxylic acids, glutamic, and aspartic acids quench by electron transfer [43] in water. In proteins, there are few opportunities to observe these as quenchers because above pH 4, they are mostly in the unprotonated form (glutamate and aspartate).

Proton Transfer Quenching in Proteins? There is little evidence for proton transfer quenching in proteins. There is often association of Trp with Lys or Arg in what are called cation-π complexes. Trps in such complexes are almost always highly fluorescent. Apparently, in such complexes, the ammonium group of the Lys is either conformationally hindered from donating a proton, or a more aqueous environment is required. The high ϕ_f values in these complexes are easily understood in terms of electrostatics, as indicated in Fig. 2B. The close presence of tyrosines to Trps is common in proteins, but does not seem to correlate with quenching.

Backbone Amides. Cowgill's early experiments [41] showing the ϕ_f of N-actetyltryptophanamide (NATA) [40] is 50% lower than 3MI decisively implicated the local backbone amide as a quencher. Feitelson [42] was among the first to articulate that the quenching was probably caused by electron transfer to the amide. This was for a long time difficult to rationalize because amides are not good electron acceptors, and several studies confirmed that N-acetylformamide does not detectably quench the fluorescence of 3MI in aqueous solution [43–45]. Furthermore, it was difficult to understand why many Trps in proteins do not seem to be quenched at all, even though they are as close to amides as those which are strongly quenched.

The answer to this question lies in the enormously strong local electric fields that exist in proteins, and the diversity of strength and direction of these fields experienced by Trps in different proteins. Electron transfer—as for all processes in nature—requires that energy be conserved; there can be no electron transfer unless the CT state and the fluorescing states have the same energy at the instant of transfer. In the protein, the CT state energy can be made equal to the fluorescing state energy if positive charges are close to the acceptor, or if negative charges are near the donor. An example of this arrangement is shown in Fig. 2A for the specific case of a Lys and Glu stabilizing electron transfer to a backbone amide. Such an arrangement will lead to a very weakly fluorescing Trp, in the absence of other nearby charges. The opposite arrangement will lead to a highly fluorescent Trp because the CT state will almost never be in resonance with the fluorescent state. One could say that what usually governs Trp fluorescence intensity in proteins might be called an *internal* resonance Stark effect, and may be compared with the external resonance Stark effect described by Boxer and coworkers [60].

To put this in quantitative terms (see appendix), a realistic approximation can be made if the CT state dipole is represented by +1 and –1 elementary charges at the indole ring and amide separated by 5Å, and the Lys and Glu are treated as point charges another 5Å beyond the –1 and +1 charges, respectively. The energy of interaction of the CT dipole with each of the environment charges in panel A is $14.4\ (-1/5 + 1/10) = -1.44\ \text{eV} = -33.1\ \text{kcal/mol} = -137.4\ \text{kJ/mol} = -11600\ \text{cm}^{-1}$. In the reversed arrangement (panel B), the energy is +1.44 eV. The difference is enormous, and nicely captures the essence of what controls quenching. Recalling that a difference of 1.4 kcal in activation energy will make a change in rate by a factor of 10 shows that a ratio of quenching rates in the two cases could be on the order of 10^5 assuming a protein dielectric constant of 9. Smaller charges that are closer to the amide would have a similar effect. Most importantly, we will emphasize that the presence or absence of a water hydrogen bonded to the carbonyl oxygen of the amide can have a profound effect on the quantum yield (lifetime).

π–σ^*? In the gas phase, there is both experimental and theoretical evidence for a low-lying Rydberg-like excited state for indole, which has been extensively investigated by Sobolewski, Domcke, and coworkers [61–63]. This excited state can be roughly described as promotion of the electron from the HOMO to a σ^* molecular orbital localized on the N—H of the indole ring. It is primarily a 2 s atomic orbital on the N, but is antibonding for the N—H bond. Occupation of this orbital can lead to a conical intersection with the ground state and has been proposed as a general explanation for the variable quantum yield for Trp in proteins [64]. Because of the large separation of charge, its energy would be expected to be sensitive to local electric field as for CT states. While this mechanism is probably operative in the gas phase, in condensed phase Rydberg states tend to be forced to higher energy. At this time, we do not seriously consider this mechanism for variable quantum yield in proteins because of the strong circumstantial experimental evidence and theoretical evidence favoring the amide CT state. For 3MI, one would expect wide variations in quantum yield in different solvents, which would be expected to tune the energy of the $\pi\sigma^*$ state. This is not observed. The quantum yield of 3MI is rather insensitive to solvent polarity over the range of hydrocarbon to water. However, when amides, carboxylic acid, or ester groups are attached by alkyl (non-conjugative) links, the quantum yield is much lower and solvent sensitive [65].

Quantitative Predictions

Semiempirical Approach. Since 2001, we have logged *over 40 ns* of QM-MM simulations on over 35 different proteins + explicit water in which the electrostatic potential was computed for each atom of a tryptophan ring, tryptophan + local backbone, or tryptophan + nearby side chain. This translates to over 4 million INDO/S-CIS computations of the low-lying excited states of these systems, including ^1L_b, ^1L_a, and CT states. In addition, in most cases independent analysis has been performed in which the electrostatic contributions of individual amino acids and water molecules make to the energy separation between the lowest CT state and ^1L_a state, the key factor in assessing electron transfer rate. This has provided us with a unique perspective on the effect of the electrostatic landscape of proteins on Trp fluorescence quenching and wavelengths.

The long-range nature of Coulombic interactions and the large number of charges (including partial charge on virtually every atom) make even qualitative judgments regarding quantum yields difficult by merely inspecting an X-ray structure visually. We therefore adopted a QM-MM procedure to see if summation over all charges in the protein and solvent might give a useful estimate of the average potential difference experienced by an electron during electron transfer. We followed an intuitive approach based on Marcus's original notion that the energy of the fluorescing state and the CT state must be the same at the instant of electron transfer [66,67]. The electron transfer rate must therefore be proportional to the probability that the average electric potential of the electron in the two states is the same. This approach was articulated nicely by Tachiya [68], and shown to lead to the usual Marcus classical limit expression [66,67],

$$k_{\mathrm{ET}} = \left(\frac{4\pi^2}{h}\right) V^2 \left(4\pi\lambda k_{\mathrm{B}} T\right)^{-1/2} \exp\left(-\frac{(\Delta G_0 + \lambda)^2}{4\lambda k_{\mathrm{B}} T}\right) \quad (1)$$

We have based our calculations on the Fermi Golden rule expression for the transition rate from a well-defined initial state to a continuum of final states nominally of the same energy. The *instantaneous* electron transfer rate constant, k_{ET}, is given by

$$k_{\mathrm{ET}}(\Delta E_{00}) = \frac{2\pi}{\hbar} V^2 \rho_{\mathrm{FC}}(\Delta E_{00}) \quad (2)$$

where V is the electronic coupling constant, ΔE_{00} is the difference between the ^1L_a and CT zero-point energies, and ρ_{FC} is the Franck–Condon (FC)-weighted density of states function. Figure 3 illustrates the definitions of the latter two quantities, and the consequence of fluctuation of the CT state energy.

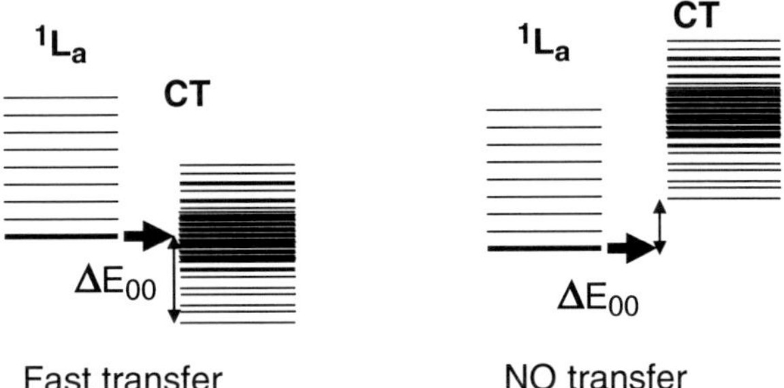

Fig. 3 *Left*: Optimal negative ΔE_{00} that has maximum ρ_{FC}, and maximum instantaneous electron transfer rate. *Right*: Positive ΔE_{00}. The instantaneous transfer from the zero point of ^1L_a is not possible

Our early simulations revealed the importance of the Coulombic attraction between the positive indole ring and the negative amide following electron transfer. That is to say, a major contribution to the positive potential on the amide *after* ET is the positive charge of the indole ring radical cation. The unanticipated consequence is that almost always the electron acceptor will be the *closest* amide. Typically the energy of the CT state for which the next closest amide is the acceptor is ~1 eV higher, making it effectively inaccessible.

Because there is always an amide within about 4 Å of the indole ring, we initially assumed that V is essentially constant and large. The picture is that during the excited state lifetime, ET occurs only during rare fluctuations of the environment which make ΔE_{00} negative. The amplitude of the fluctuations in ΔE_{00} determine the width of the Gaussian and is related to the reorganization energy by $\sigma^2 = 2\lambda k_{\mathrm{B}} T$. The average CT– ^1L_a energy gap is closely related to $\Delta G_0 + \lambda$ in Eq. (1), where σ^2 is the variance.

Taking V as an empirical constant, and using only the average energy gap and variance, a reasonable correlation between experiment and computations was achieved with only two adjustable parameters: V_{avg}, and a constant offset applied to the semiempirical QM zero-point transition energies, ΔE_{00}, so that the average k_{ET} is given by:

$$\langle k_{ET}\rangle = \frac{2\pi}{\hbar} V_{avg}^2 \langle \rho_{FC}\rangle \tag{3}$$

with the average density of states given by:

$$\langle \rho_{FC}\rangle = (2\pi\sigma^2)^{-1/2}\int \rho_{FC}(\Delta E_{00})\mathrm{e}^{-\frac{1}{2}\left(\frac{\Delta E_{00}-\langle\Delta E_{00}\rangle}{\sigma}\right)^2} d\Delta E_{00} \tag{4}$$

An ad hoc choice for the shape of ρ_{FC} (too narrow in retrospect) led to a good fit for V = 10 cm^{-1} (too small in retrospect) and an empirical shift of ΔE_{00} by –4000 cm^{-1} (physically unreasonable). A plot of ϕ_f calculated vs. observed ϕ_f is shown in Fig. 4 ([28], Table 1). This method has been applied to the understanding of Trp fluorescence quantum yields in several additional cases [29–31].

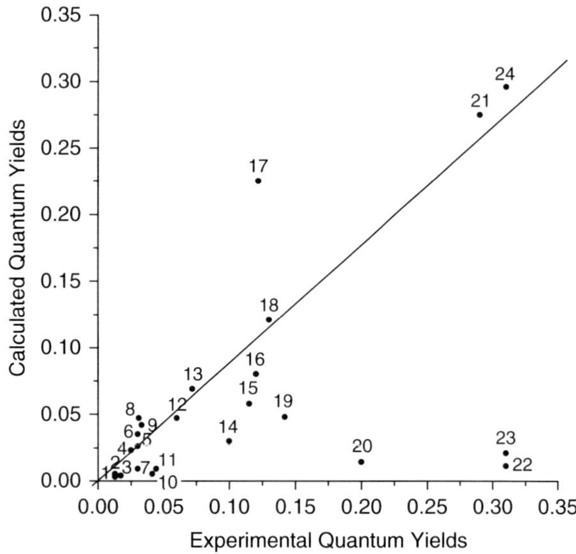

Fig. 4 Plot of Trp fluorescence quantum yields calculated using Eqs. (2–4) as a function of experimental quantum yields for the set of Trps listed in Table 1 along with the number code

Analysis of a Quenching Event. Figure 5 shows a typical QM-MM trajectory for Zindo QM energies for the ^1L_a (red) and CT (black) states relative to the ground state for Trp 126 of Dsba, a very weakly fluorescent Trp well studied by Engelborghs et al. [49]. These are vertical transitions for a fixed geometry that is that of the CT state. The protein is solvated in a 40 Å sphere of explicit water. QM is done only on the N-formyltryptophanamide NFTA (Fig. 1) portion of the protein, and the

Table 1 Experimental and calculated fluorescence quantum yields for 24 Trps in 17 proteins

Expt.	Calc.[a]	Num	Run/abbreviation	Description[b]	References[c]
0.013	0.003	1	lyd158-asn2	T4 lysozyme W158-asn2 1lyd	[53]
0.013	0.005	2	dsb126	dsBa W126 1dsb	[140]
0.017	0.004	3	barn94-his18+	barnase W94-H18 pH5 1a2p	[89]
0.025	0.023	4	cpl	human cyclophilinA 2cpl	[141]
0.03	0.026	5	trpcage	TrpCage 1l2y (nmr structure 1&2)	[142][d]
0.03	0.035	6	d6o	fkb506 binding protein 1d6o	[143][e]
0.03	0.009	7	sbc	subtilisin C 1sbc	[5][f]
0.031	0.047	8	bpp	phospholypase A2 2bpp	[5]
0.033	0.042	9	nscp	NSCP W57 W4F W170F	[144]
0.041	0.005	10	dsbQ74A/N127A	dsba W126 Q74A, N127A mutanted 1dsb	[140]
0.044	0.009	11	lyd138	T4 lysozyme W138 1lyd	[53]
0.06	0.047	12	lyd126	T4 lysozyme W126 1lyd	[53]
0.072	0.069	13	barn94	barnase W94 pH8 1a2p	[89]
0.1	0.030	14	ctx	cobra toxin 1ctx	[5]
0.115	0.058	15	mlt	melittin 2mlt	[5]
0.12	0.080	16	gcn	glucagon 1gcn	[5]
0.122	0.225	17	barn71	barnase W71 his neutral 1a2p	[89]
0.13	0.121	18	b8r	parvalbumin 1b8r	[5]
0.142	0.048	19	barn35	barnase W35 his neutral 1a2p	[89]
0.2	0.014	20	dsb76	dsba W76 1dsb,1fvk average result	[140]
0.29	0.275	21	stn	staph. nuclease 1stn	[5]
0.31	0.011	22	pfk	phosphofructokinase 6pfk	[5]
0.31	0.021	23	rnt	ribnuclease T1 9rnt	[5]
0.31	0.296	24	azb	apo-azurin W48 1azb	[5]

[a] From Eq. (5) using $|V_{el}| = 10.0\ \text{cm}^{-1}$, $D = -4000\ \text{cm}^{-1}$.
[b] Name, Trp sequence number or Trp-acceptor numbers, PDB code (all X-ray unless noted).
[c] Reference for experimental quantum yield.
[d] Estimated relative to NATA.
[e] Upper limit based on the 12-fold increase upon denaturation.
[f] Estimated from lifetimes.

electrons of this fragment are subjected to an addition electrostatic potential from *all* point charges of solvent and the remainder of the protein. Throughout the first 150 ps of the trajectory, the system is in equilibrium with the ^1L_a state. That is, the MM part of the system is provided charges on the Trp as produced by the QM calculations. The enormous fluctuations in the CT state energy (black) come from thermal fluctuations of the positions of solvent and protein. They are so large for the CT state because it is so far out of equilibrium. At the 150 ps point, the charges are no longer constrained to be those of the ^1L_a state, but are now allowed to be those of the *lowest* excited state, whatever that may be. For the next 6 ps, the ^1L_a state continues to be the lowest excited state, at which time a fluctuation in the environment caused the CT state to be lower than ^1L_a, the charges were accordingly switched to those given by the CT state, i.e., characteristic of an indole radical cation and amide

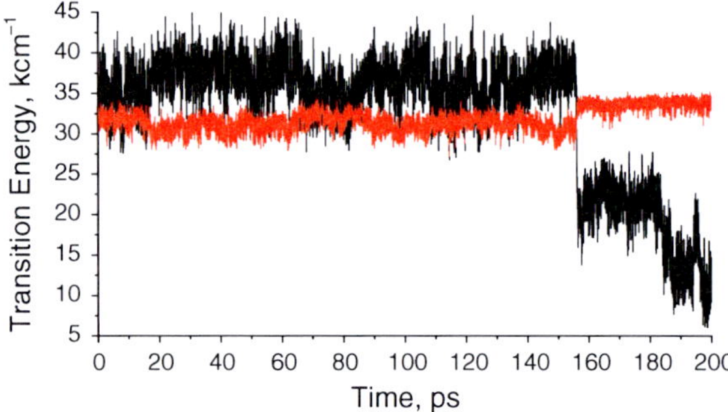

Fig. 5 QM-MM trajectories of Zindo transition energies from ground to ^1L_a (*red*) and lowest CT state (*black*) for Dsba showing a quenching event in which the system spontaneously switched to the CT state at 156.05 ps. The fast relaxation is complete after 100 fs. The second step of relaxation near 185 ps is primarily due to Lys132 becoming directly H-bonded to the Trp amide carbon

radical anion (mostly has the electron on the C atom of the Trp amide). The large sudden change of charge distribution puts the system far out of equilibrium, and the surroundings respond by quickly relaxing. The relaxation is at first dominated by the nearest charges in the surrounding that are free to move. Positive charges—such as the hydrogens on water—move toward the amide and away from the indole ring; negative charges—such as the oxygen on water—move towards the indole ring and away from the amide. The nearby charges change the energy very quickly with small motion. For example, if a + charge that is initially 2 Å from a negative point charge moves 0.1 A closer, the energy change is –0.38 eV or –8.7 kcal/mol. The huge array of point charges representing the entire protein and solvent shell also respond, but more slowly because the forces are much less and the motions dominated by the thermal agitation.

The quenching event is simply the transition from the ^1L_a state to the CT state during the time they are in resonance, followed by the sudden drop of the CT state due to relaxation. Once in the CT state, fluorescence is essentially absent because of the small transition dipole of the CT –> ground state transition. When the relaxation is large, as in this case, there is no possibility for the electron to return to the ^1L_a state, and the electron most likely returns to the empty HOMO of the indole by tunneling well before there is time to emit a photon.

It is of interest to examine the microscopic details of the relaxation process. Figure 6 shows the motions of the key groups involved in the relaxation, namely several nearby waters and Asp71 (a negative charge near the indole ring) and Lys132 (a positive charge near the amide). Panel A shows the configuration just prior to the instant of the CT–^1L_a crossing. The groups shown are stabilizing the CT energy much more than the average simply because of a random thermal fluctuation. The frequency of such stabilizing configurations is the reason the Trp 126 in Dsba is so weakly fluorescent. Panel B shows the configuration of the same groups at 157.9 ps, where the CT energy has dropped to 16,500 cm^{-1}. It is seen that further relaxation occurs typically in fast jumps. The main such event corresponds to Lys132 becoming much closer to the Trp amide C because of becoming directly H-bonded to the amide carbonyl O instead of being connected by a bridging water. The number of direct H-bonds to the O has increased from 2 to 4, responsible for lowering the energy from another 10,000 cm^{-1} to 6,000 cm^{-1} at 197.8 ps. The precise details shown in Fig. 6 are just one of many ways the quenching can play out.

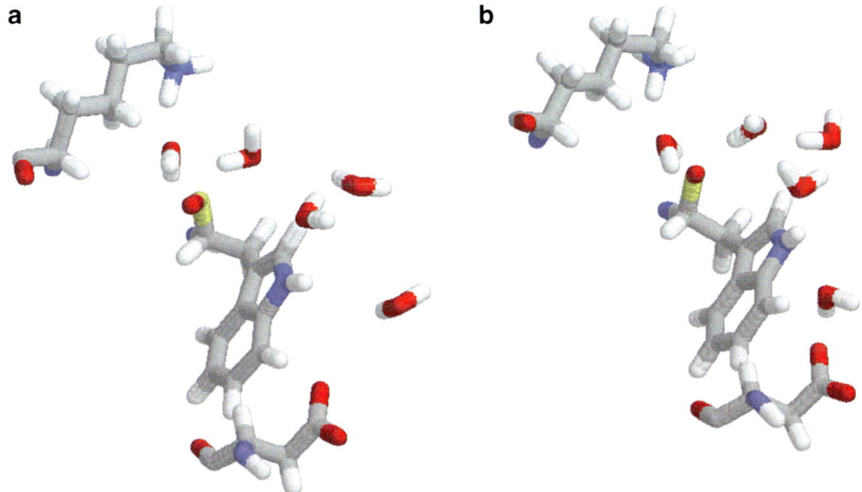

Fig. 6 Primary motions involved in relaxation immediately following electron transfer to amide. (**A**) Configuration of Lys132, Glu127, and nearby waters at the time of electron transfer (156.0 ps). (**B**) Configuration of same groups at the first deep minimum during relaxation (157.90 ps). Each of the groups contributes large incremental electrostatic stabilization through small displacements. The yellow atoms are C on Trp amide

An Ab Initio (Almost) Approach. The main purpose of the previous method described above [27] was to communicate the *concept* behind the previously unexplained variability of Trp fluorescence yields and lifetimes in proteins: that the local electrostatic field magnitude and direction dictate whether electron transfer to the local amide (or other nearby acceptor) is fast or slow. All the variables in Eq. (2) were ad hoc and empirically adjusted.

Recently we reported results of an ab initio method for computing V values [32], which are now extracted from the Hamiltonian matrix for interaction between singly excited configurations (CIS) using Gaussian03 [69]. The matrix elements so obtained are reasonably independent of basis set, allowing good estimates of V in times as short as 10 s per element for the STO-3G basis.

Now, instead of using a constant V for all proteins, the electron transfer rate for a given Trp is estimated as:

$$k_{\mathrm{ET}} = \frac{2\pi}{\hbar}\left\langle V^2\right\rangle \langle \rho \rangle \tag{5}$$

where the $<V^2>$ average comes from 500 sets of coordinates generated during a 150-ps QM-MM trajectory for a particular Trp. We continue to use Zerner's INDO/S-CIS (Zindo) [70,71] because CIS energies are too crude to provide reasonable estimates of energy gaps needed for $<\rho>$.

Figure 7 compares the average calculated $<|V|>$ for electron transfer to the C-terminal amide (Fig. 8A) using three different basis sets and the semiempirical values for 20 of the 24 Trps for which predictions were made previously [27]. It is seen that the average ab initio values fall in the range 140–1000 cm^{-1}, i.e., 14–100 times larger than the 10 cm^{-1} empirical fit used previously, with dependence on

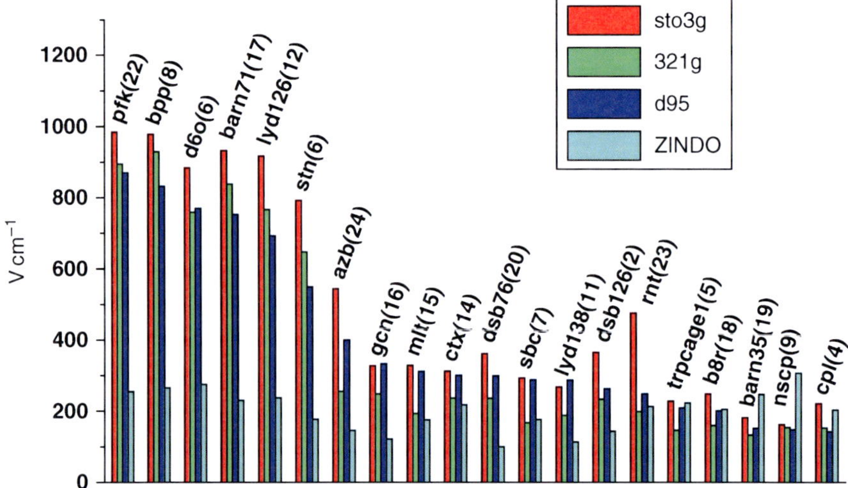

Fig. 7 Comparison of average ab initio computed CIS electronic coupling elements using different basis sets and those from INDO/S-CIS. *Numbers in parentheses* refer to Table 1, Figs. 4, and 13

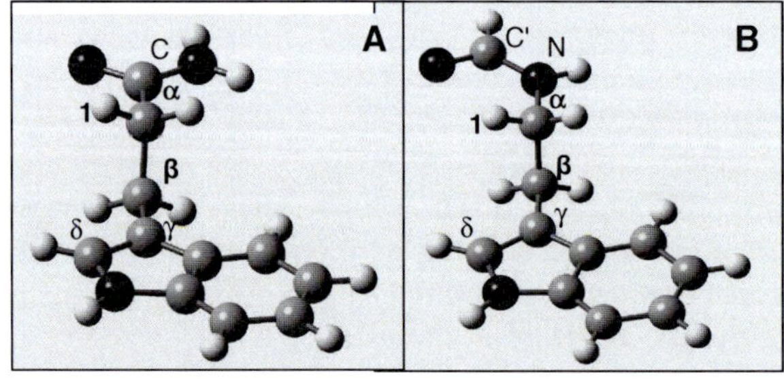

Fig. 8 Single-amide models for closest tryptophan ring-backbone amide interactions, with key atoms labeled for defining dihedral angles: (**A**) model of tryptophan and C-terminal amide; (**B**) model of tryptophan and N-terminal amide. C = tryptophan amide carbonyl carbon; C′ = amide carbonyl carbon of preceding amino acid; N = tryptophan amide N; α = alpha carbon. Dihedral angles: χ_1 = torsion about α–β, numbering from 1 to γ; χ_2 = torsion about β–γ, numbering from α to δ; ψ= torsion about α–C, numbering from 1 to *N* of the following amino acid

protein that is fairly insensitive to basis set. (<|V|> values using the STO-3G basis for the N-terminal amide are slightly smaller, falling in the range 120–500 cm^{-1}). The semiempirical Zindo values, in contrast, show a smaller range, and the averages are not correlated with the ab initio values.

Figure 9 shows values of V plotted every 10 fs over 3.5 ps segments of MD trajectories for representatives of two distinct classes of behavior observed. Charmm (version 31b1) [72] was used as in previous work [27] except that the length of the simulations was 150 ps instead of 50 ps, and the proteins were typically solvated in 40 Å radius spheres of TIP3 explicit water instead of 30 Å radius spheres. Results are for one of the two molecules shown in Fig. 8. The geometry of these systems is precisely that obtained from MD trajectories for the proteins, except that only one amide is retained and capped with H atoms, and the calculations are done for vacuum. Uniformly, <V> is much larger and the relative standard deviation (amplitude of fluctuations) is smaller for those proteins for which χ_1 is nominally near –60°, a conformation which is close to that in Fig. 8B. This large interaction is virtually all through-bond for the –60° conformation. When χ_1 is nominally +60° or 180°,

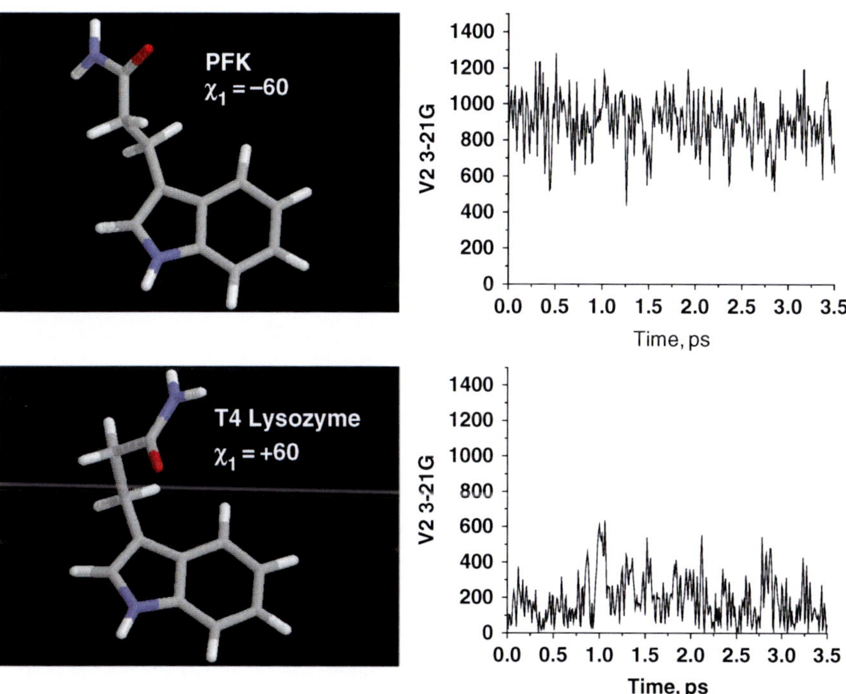

Fig. 9 Representatives of the two classes of electronic coupling behavior for electron transfer from the excited indole ring to the C-terminal amide. For $\chi_1 \sim -60^{\circ}$ the average interaction is near 900 cm^{-1} and relatively constant. For $\chi_1 \sim +60^{\circ}$ and 180° the average is close to 200 cm^{-1} and the relative standard deviation much larger, apparently because of interference between through-space and through-bond interactions

$<V>$ is always found to be ~4 times smaller, and fluctuations often reduce instantaneous values to <5 cm^{-1}. In these conformations, direct (through-space) interactions most likely exhibit quantum mechanical interference with the through-bond component.

We have established that the main source of fluctuation in V during dynamics simulations is due to large non-Condon behavior from strong dependence on torsional angles.

Significantly, there is no correlation of $<|V|>$ with the experimental quantum yields, as shown in Fig. 10. This result is consistent with our previous assumption that V is not the rate determining parameter for amide quenching of Trp fluorescence because of the large inherent energy gap (small averaged FC densities) except when the protein environment provides sufficiently stabilizing electric potential difference to create transient resonance between the fluorescing and CT states.

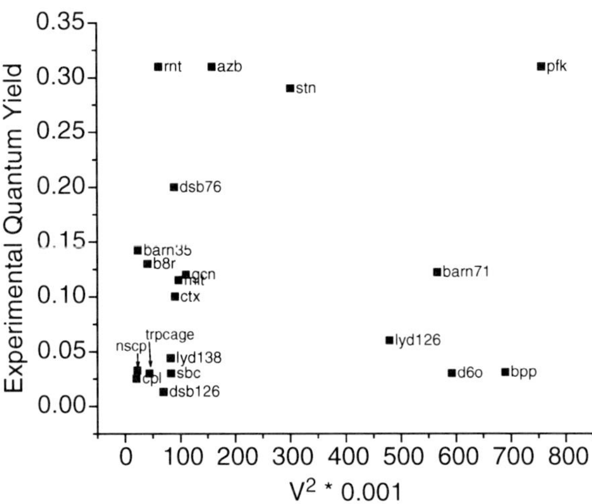

Fig. 10 Plot of observed quantum yield vs. average square of electron transfer coupling element, illustrating the lack of correlation. This is because the electrostatically modulated CT–^1L_a energy gap is the controlling factor. *Names* are keyed to Table 1, Figs. 4, 7, and 13

As an aid for interpretation and understanding the fluctuations, we performed a large set of calculations on rigid forms of the model molecules (Fig. 8) so that the effect of varying the dihedral angles only could be observed. An informative contour plot showing the variation of V with χ_1 and ψ with χ_2 fixed at +90° is shown in Fig. 11, displaying computed (RHF/3-21G) V values. The numbers on the plot refer to the dynamics-averaged rotamer conformation of the proteins studied having χ_2 ~90°. It is seen that broad maxima occur when $\chi_1 = \sim -60°$ and $\psi \sim -40°$ or +140° and that all proteins studied here have ψ values either near –40° or +140°. These positions happen to maximize the through-bond "π conjugation" between the amide and indole ring. When ψ is near –130° or + 50°, the amide carbon π AO is perpendicular to the direction of the π AOs of the ring LUMO. For $\chi_1 \sim$ +60° and 180°, V is seen in Fig. 11 to be uniformly small and in regions for which V will vary rapidly with small changes in χ_1 and ψ, as seen in MD trajectories. This is consistent with the large fluctuations of V seen in the MD trajectories for these cases.

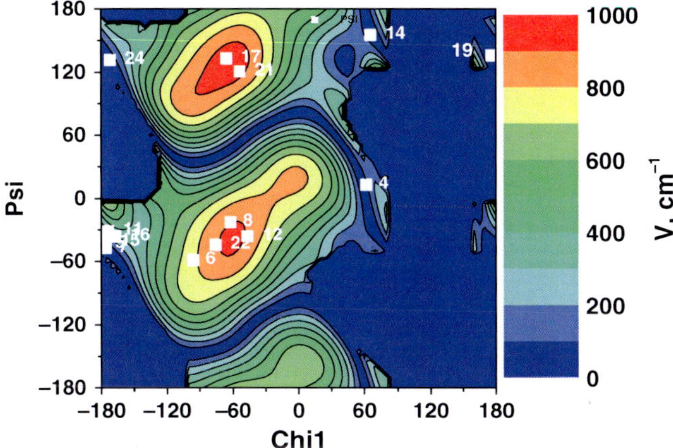

Fig. 11 Contour plot showing the computed (3-21G) electronic coupling values as a function of χ_1 and ψ with χ_2 held at +90°, for the otherwise rigid molecule in Fig. 8A. The numbers on the plot refer to the dynamics-averaged rotamer conformations of proteins in Table 1. The cluster of points near (–160, –40) contains proteins trpcage, mlt, sbc, lyd138, and gcn. The large blue areas outside the contour lines represent angles not sampled due to steric hindrance. A very similar plot is found for $\chi_2 = -90°$ for proteins b8r, dsb76, dsb126, nscp, and rnt

Density of States. The trajectories were conducted with the QM bond lengths fixed using the SHAKE routine at values representative of the CT state. CASPT2 [73,74] calculations for several formamide–indole pairs show that the ^1L_a state energy increases by ~4000 cm^{-1} relative to that of CT as the geometry is increased from that at the ^1L_a energy minimum to that of the CT state minimum [27,37]. Therefore one expects that Zindo ΔE_{00} values computed at the CT geometry should be increased by about 4000 cm^{-1} to make a realistic estimate of the true ΔE_{00}. Figure 12 speaks to the Franck–Condon weighted density of states, and the issue of the remaining empirical correction in our method: shifting the Zindo ΔE_{00} energies to optimize agreement with experiment. We model ρ_{FC} as the overlap integral of FC densities for the electron detachment from an indole ring (curves b–d) in Fig. 12 and the electron attachment to formamide (curve a in Fig. 12). Curves b–d represent $\Delta E_{00} = -4000$, 0, and +4500 cm^{-1}, respectively. These three cases lead to $\rho_{FC} = 9\times10^{-5}$, 1×10^{-5}, and 9×10^{-8} (cm^{-1})$^{-1}$ respectively with the functions displayed. With our earlier ad hoc fit of the experimental data, the use of $|V| = 10$ cm^{-1} required *subtracting* 4000 cm^{-1} from the ΔE_{00} values obtained from the QM-MM trajectories (giving overlaps similar to curve b), whereas with the present scheme ($\langle V_i \rangle$ = 140–1000 cm^{-1}) the best fit is obtained by *adding* 4500–5000 cm^{-1} (similar to curve d), i.e., values that are consistent with expectations from CASPT2 calculations [27,37].

The large average magnitude allows for a physically reasonable choice of ρ_{FC} which, when used in Eq. (5), retains semiquantitative agreement with experiment for Trp quantum yields in proteins. Part of the refinement was to use a broader, more

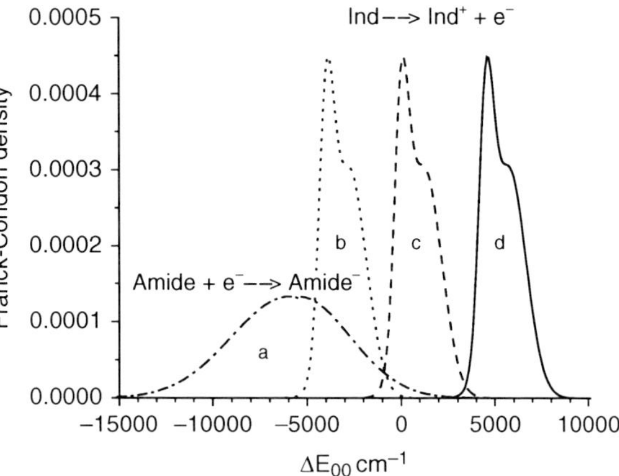

Fig. 12 Franck–Condon density curves used in Eq. (1) for electron attachment to the amide (curve a, *dash-dot*), ionization of indole with $\Delta E_{00} = -4000\ \mathrm{cm}^{-1}$ (curve b, *dots*), ionization of indole with $\Delta E_{00} = 0$ (curve c, *dashed*), and ionization of indole with $\Delta E_{00} = +4500\ \mathrm{cm}^{-1}$ (curve d, *solid*)

accurate Franck-Condon density function. The $<\rho>$ is generated from an average over 15,000 points during a 150 ps trajectory.

As before, the quantum yield is estimated with the expression

$$\phi_{\mathrm{f}} = k_{\mathrm{r}}/(k_{\mathrm{r}} + k_{\mathrm{nr}} + k_{\mathrm{ET}}),$$

where k_r and k_{nr} are the radiative and the non-electron-transfer non-radiative rate constants and k_{ET} is the electron transfer rate constant for the particular Trp/protein. We use values found in aqueous solution: $k_r = 4 \times 10^7\ \mathrm{s}^{-1}$ and $k_{nr} = 9 \times 10^7\ \mathrm{s}^{-1}$ [43]. Figure 13 shows an example of fitting predicted and experimental quantum yields using the "ab initio" scheme presented above. The larger V values tend to exaggerate the lack of precision for those proteins with quantum yields in the intermediate range. At least two general reasons can be identified as possible causes for the imprecision. Almost certainly, 150-ps trajectories do not capture the relevant configuration space. Single molecule experiments on the quenching of flavin fluorescence in proteins suggest this possibility [75]. Another possibility is that the large V values lead to breakdown of thc golden rule. Indeed, preliminary estimation of the Laudau–Zener curve crossing probabilities suggest that the weak adiabatic regime [76] is more relevant for some of the proteins.

The use of CIS diabatic interaction elements is one of several techniques that have been accepted as providing reasonable values in estimating ET rates [77–79]. Given the modest ET rates considered in typical Trp quenching ($<1\times10^{10}\ \mathrm{s}^{-1}$), it is clear that the system spends considerable time ($>$100 ps) prior to an ET event following excitation. This means that initial state preparation becomes an issue. Given the large values and rapid fluctuations of V found in the present work, one anticipates a spectrum of pathways that lead to quenching through the CT state. It is commonly accepted that the excited state is created in a time short relative to nuclear motion time scales. Immediately following excitation, the wavefunction will

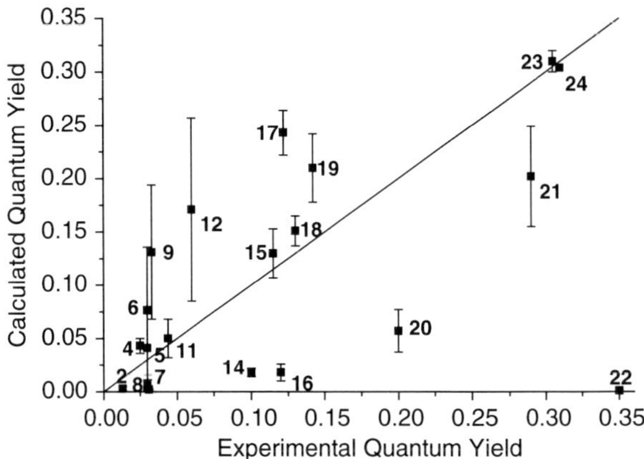

Fig. 13 Representative plot of calculated vs experimental fluorescence quantum yields for 20 Trp's in 17 proteins using the D95 ab initio $\langle V_i^2 \rangle$ values and the new Franck–Condon density of states scheme described in the text. The error bars are the standard deviations for three 50 ps MD trajectories. Here, 4700 cm^{-1} is added to the Zindo CT energy computed at CT geometry, and a Gaussian with standard deviation = 3000 cm^{-1} is used for the amide electron attachment FC density spectrum (see Fig. 12). For the buried Trp 59 of rnt only, the quantum yield was computed from the small equilibrium constant (positive ΔG_0) for the electron transfer process stemming from its small reorganization energy (the relaxed CT state lies above the 1La state). For dsb76 and pfk, the present accuracy of our ΔG_0 estimates leaves open the possibility that equilibrium will favor the 1La state, and the quantum yield predictions will be much higher

be localized on the indole ring, and the diabatic interaction elements presented here should be valid for use in the Fermi Golden rule for those systems in resonance at the instant of excitation. From the small FC factors consistent with the large $<V_i>$ values, the vast majority of cases will, however, not find the system in resonance at the moment of excitation, and the system will have relaxed electronically and vibrationally to the thermalized zero-point of the ^1L_a electronic eigenstate. This will contain a CT component determined by its energy and interaction relative to the CT state. In principle, if resonance is approached slowly enough and V is changing slowly, the adiabatic limit will prevail during the avoided crossing with the CT state. Quenching depends on whether the solvent near the adiabatically forming CT state is poised for inertial relaxation (~50 fs) or not. If the solvent relaxation does not trap the CT state, recrossing to the ^1L_a state is likely. Our MD simulations suggest, however, that both the energy gap and the interaction often change drastically in 10 fs, but that they are not correlated. This time scale borders on the dephasing time associated with electronic excitation. Conceivably, rapid changes in either the gap or interaction will create an electronic state sufficiently non-stationary that the non-adiabatic regime is reached. The picture that emerges points to a continuum of possibilities for an individual excited Trp, ranging from the limits of purely non-adiabatic (fast change in energy gap and small V) to purely adiabatic (slow energy gap change and large V), with many cases falling into an intermediate behavior. Ratner and coworkers have recently presented an illuminating discussion of these issues [80,81]. We are presently investigating schemes in which charge densities are updated on the 1 fs time scale in our QM MM simulations to identify adiabatic behavior and to estimate transmission coefficients.

Discussion of Selected Papers

Ultrafast Trp Fluorescence Decays. Qui et al. [82] have very recently reported an extensive systematic probe of ultrafast quenching of tryptophan fluorescence in a survey of $\sim$40 proteins, looking particularly at decay times $<$100 ps. With site-directed mutagenesis, they placed Trp in desired positions or altered its neighboring residues, especially in apomyoglobin. By observation at wavelengths $>$360 nm, they have identified four cases wherein a Trp within 3.5–5 Å of a cysteine or disulfide residue has substantial fluorescence decay components in the range 10–30 ps. That the thiol and disulfide groups are found to produce short lifetimes is consistent with their potent diffusional quenching abilities in water [43].

Four cases were identified with fast components in the range 3–85 ps for which the quencher is identified as "the carbonyl group" [82]. This assignment was based on MD simulations revealing that a glutamine or glutamate side chain is within van der Waals distance of the indole ring. Classifying these efficient quenchers under the single umbrella of carbonyl group, however, ignores the several "flavors" of carbonyl group with vastly different electron quenching ability. It ignores the considerable effect on the electron accepting power by what is bonded to the C of the C=O bond. This was well articulated in the early studies by Ricci and Nesta [45], and refined more recently by Chen and Barkley [43].

In the case of K34W, glutamate was assigned as the quencher. As noted above, the complete absence of quenching by the carboxylate group in water and its negative charge, which would profoundly lower the electric potential at the indole ring, strongly suggest that the role of the nearby Glu52 in the fast quenching is the strong promotion of electron transfer to the backbone amide, as illustrated in Figs. 1 and 2.

The carbonyls of backbone peptide bonds and the side chains of glutamine and asparagine are all part of *amides*. The π electrons of the O=C—N— group are profoundly affected by electron donation from the N, greatly reducing the intrinsic electron accepting power. Indeed both Ricci–Nesta and Chen–Barkley show that a single amide does not quench the indole ring in water. (However, molecules with two amide groups do weakly quench in water. Apparently one amide electrostatically assists the electron transfer to the other.) [44]. We have consistently argued that only by electrostatic stabilization can an electron be easily transferred to an amide, an argument also recently made for electron attachment to peptides in vacuum [83].

Because of the very close chemical similarity of the amides of the backbone and glutamine, when a glutamine side chain is in close contact with the indole ring, whether the excited electron will transfer to the backbone amide or the glutamine amide will be determined by which group has the more positive electric potential relative to the ring. In both cases, the electronic coupling should be large enough to support fast transfer. Without specific information on the electrostatic landscape, either result is reasonable. Either way, a strong electrostatically bound "exciplex" could be formed immediately following electron transfer. If the electron goes to the glutamine, the π clouds of the indole ring and amide can be expected to be stabilized by both covalent and electrostatic means. If the electron goes to the backbone, the positive indole ring can be strongly stabilized by the partially negative carbonyl O of the glutamine. Qiu et al. [82] consider the glutamine side chain and the peptide bond (backbone amide) as separate classes of quencher. This may be true, but there is no clear chemical basis for such a distinction at this time.

Xu et al. [30] have connected the ultrafast scale (0–100 ps) with the conventional TCSPC scale (100 ps–10 ns) for Trp 3 of monellin and Trp 21 of the E21W mutant of IIA^{Glc}, obtaining decay associated spectra (DAS) spanning time constants from 1.3 ps to 5 ns. For monellin, they report a 16 ps decay component, which they attribute to a subpopulation of Trps undergoing fast ET. An earlier report by Peon et al. [84] attributed a 16 ps decay component to hydration dynamics, and will be discussed later in this chapter. Lifetime heterogeneity on a short time scale is not unreasonable for Trp3 of monellin. The environment near Trp3 is unusually rich in possibilities for lifetime heterogeneity, with at least *four* plausible candidates for electron acceptors in an electron transfer fluorescence quenching process. In addition to the usual nearby backbone amides, two other amides are transiently quite close to Trp3: the backbone amide of Gly1 and the side chain amide of Gln59. The latter two are particularly potent because of the positioning of nearby positive charges that are close to the acceptor yet fairly distant from the Trp ring. Furthermore, these quenchers on relatively disordered groups provide large fluctuations in ET rate. QM-MD trajectories to predict the ET quenching propensity for the four amides using our recently published method [27] show that the energy gap varies considerably during the trajectory. About 65% of the time, the lowest CT state is on Gly1. The remainder of the time it resides mainly on the Trp and occasionally on Glu2. Most pertinent, however, is that 100% of large downward fluctuations in the CT energy, which signify resonance with ^1L_a, are for electron transfer to the amide on Gly1. Comparison with a trajectory in which the energy of charge transfer to the side chain amide of Gln59 is surveyed shows similar low-energy CT states for transfer to the side chain amide of Gln59, although the occurrence is much less than for transfer to Gly1 [30].

For the E21W mutant of IIA^{Glc}, Xu et al. [30] found no transients in the range 5–100 ps. The longest decay time was fit well by 210 ps and is similar to that attributed to relaxation in the TCSPC study by Toptygin et al. [85]. Neither side of the spectrum exhibited terms slower than 2 ps that could be attributed to the solvent shift correlation function in the range 1–100 ps.

Compared to monellin, IIA^{Glc} exhibits a remarkably long lifetime, nearly the longest expected, based on that of 3-methylindole. A simulation similar to that for monellin shows that the average CT–^1L_a gap is considerably larger than for Trp 3 of monellin. This is reasonable considering the unusual electrostatic environment found for Trp21, which is in what could be considered a cation-π complex with Arg165 on one side of the indole ring. There are no nearby charged groups in the direction of electron transfer from the Trp ring toward the Trp backbone amide. Another large difference between the two is the amplitude of fluctuation in the CT energy. During the trajectory, the CT energy for monellin varies by about 20000 cm^{-1} (2.5 eV) in contrast to that for IIAGlc, where the variation is only 14000 cm^{-1}, with the extremes being considerably more transient than for monellin. The large fluctuations for monellin arise because Trp3 resides so close to the N-terminus. The terminal positively charged ammonium group greatly stabilizes electron transfer to the Gly1 amide group (typically 1.5–2 eV). Significantly, during the monellin trajectory, the CT energy is much lower during the first 25 ps than during the last 25 ps, with the transition occurring over about 10 ps. This suggests that a significant subpopulation of the excited ensemble may undergo electron transfer quenching at much higher rate than the average. This provides an alternative source of decay microheterogeneity to the strict "rotamer" mechanism.

Distance Dependence of Interaction at Short Range. The scope of our earlier paper [27] was to address the broad question of how a wide range of fluorescence quantum yields and lifetimes is observed for single Trp *proteins*, where the ubiquitous presence of multiple rotamers is more controversial. For that goal, we used as a working hypothesis the assumption that the detailed local Trp rotamer conformation in *proteins* is given correctly by the published X-ray or NMR structures and that this rotamer is almost always the one with the major impact on the quantum yield in aqueous solution. At the other extreme is the approach exemplified by that of Hellings et al. in their interpretation of the non-exponential fluorescence decay from a number of proteins [36]. The X-ray structure assignment is given little weight compared to molecular modeling computational predictions of relative rotamer stability. By linking the effect of acrylamide quenching on lifetime decay components, electron-transfer rates were found to fit an exponential dependence on the distance from the Trp amide carbonyl C atom to the C4 (CE3) atom of the indole ring. This is a remarkable result, given the delocalized nature of the indole electrons and the other nearby amides are ignored. Although exponential distance dependence is expected and observed for *long range* electron transfer when normalized to zero activation energy, there is no particular reason to expect this at the distances necessary for quenching the Trp excited state, i.e., distances of 4–6 Å. We find no correlation (Fig. 10) between coupling strength and experimental electron transfer rate (as deduced from fluorescence quantum yields) because of the more critical dependence on the electrostatic environment [32].

Use of Trp Fluorescence to Follow Protein Folding. Changes in Trp quantum yield are used extensively for following protein folding, primarily because they are often a natural part of the protein, thereby avoiding structural uncertainties introduced by extrinsic probes. Here it is particularly of interest to understand the mechanism underlying the change of quantum yield (ET rate) accompanying folding.

The fast folding, 35 residue villin headpiece, HP35, has been at the center of numerous protein folding rate simulations [9,86], The Eaton group [87,88] has experimentally followed the folding with the N27H mutant, plausibly because the protonated His27 quenches the fluorescence from Trp23 in the folded form, but not when unfolded. Because of this, at least in some simulations of the folding a major criterion for the folded form is close proximity of His27 to Trp 23 [9]. Protonated His is indeed a potent quencher of Trp fluorescence in solution [43] and in proteins [51,89,90]. The quenching by His^{+} in proteins, however, can be tricky. For Trp94 of barnase, which is in close contact with His18, the ϕ_f decreases substantially when the pH is lowered below 5. Unexpectedly, for the T4 lysozyme (T4Lyso) mutant Q105H, lowering the pH has the *reverse* effect, i.e., the ϕ_f increases [91]. In our QM-MM simulations [27], for the Trp94–His18 pair in barnase, the lowest CT state is found to be on the ring of His18 most of the time and the energy of this CT state is quite low. For the W138–H105 pair in the T4Lyso mutant, negative charges from Asp and Glu residues near His105 destabilize the His CT state and simultaneously stabilize the Trp amide CT state to the point that the lowest CT state has the transferred electron on the Trp amide. The Trp amide CT state is considerably destabilized by the presence of the protonated His18 next to the Trp ring, thereby increasing the Trp amide CT–^1L_a state gap, which increases the quantum yield.

For HP35 (N27H), preliminary studies indicate [92] that the low quantum yield is caused mainly by electron transfer to the Trp amide, because the CT state for transfer to the His27 cation is destabilized by its salt bridge with Glu31. ET to the amide appears to be facilitated by the helix dipole and electrostatic stabilization by the His27 cation.

In the actual folded state of HP35, His27 and Trp23 are in close proximity because they are part of the same alpha helix. In this case, there is no practical difference whether the His27 proximity makes it the electron acceptor or whether its *stabilizing* influence on the amide CT state enhances quenching by the backbone. One could imagine other cases where quenching by the backbone amide instead of a side chain from another residue could have divergent structural implications, and incorrect conclusions might result.

Quantitative Steady State Wavelength Predictions

The numerous quantum chemical computations dating back many years and summarized previously [13] consistently showed that the increased dipole in the ^1L_a state of indole and 3MI have—without exception—predicted that electron density is shifted from the pyrrole ring to the benzene ring upon excitation to the ^1L_a state. This means that positively charged residues near the benzene end or negative charges near the pyrrole end of the Trp ring will shift λ_{max} to longer wavelengths (produce a red shift), with the opposite configuration producing a blue shift. Because these shifts are due to the electric field imposed by the protein and solvent, they may be termed an internal Stark effect, by analogy to the familiar shifting of energy levels via an applied (external) field. The internal Stark effect (ISE) has emerged as a useful concept to understand spectral shifts for a wide range of chromophores embedded in a "host" medium, including polyenes [93,94], porphyrins [95], and other probes [96–98], but has seen little attention for the explanation of Trp fluorescence. This has at least in part been because of uncertainties surrounding the nature of the fluorescing state, which we feel is no longer an issue. Although not generally recognized as such, the large red shift observed in fluid solvents of high dielectric constant is also a Stark effect, with the strong electric field (reaction field) at the solute being a manifestation of the partial orientation of solvent dipoles around the large solute dipole. A primary contention in the extension of the ISE hypothesis to proteins is that the source of the field is irrelevant; only the magnitude and sign of its projection on the long axis of the indole ring matters in determining the spectral shift.

Following an exploratory study [99], Vivian and Callis [25] have predicted the fluorescence wavelengths of 19 tryptophans in 16 proteins, starting with crystal structures and using a hybrid quantum mechanical-classical molecular dynamics method, with the assumption that only electrostatic interactions of the tryptophan ring electron density with the surrounding protein and solvent affect the transition energy. With only one adjustable parameter, the scaling of the quantum mechanical atomic charges as seen by the protein/solvent environment, the mean absolute deviation between predicted and observed fluorescence maximum wavelength is 6 nm, as shown in Fig. 14. The modeling of electrostatic interactions, including hydration,

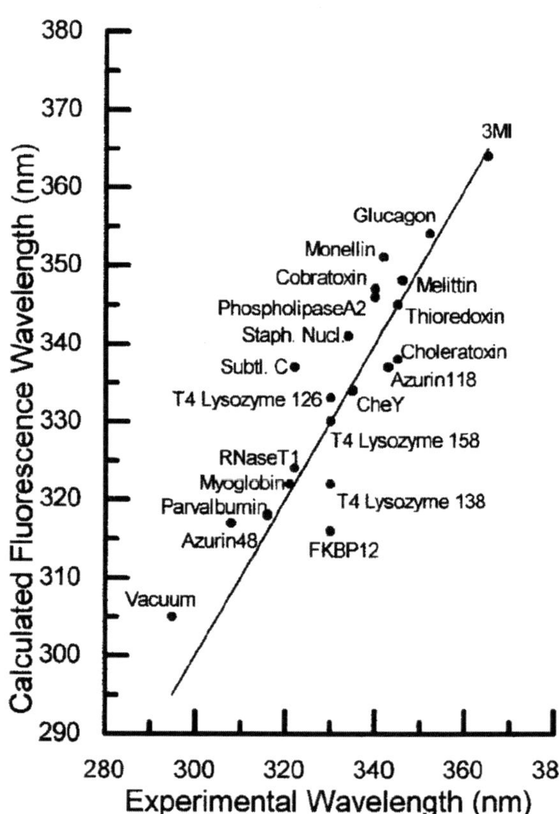

Fig. 14 Plots of calculated vs. experimental fluorescence maximum wavelengths for 19 Trps in 16 proteins and for 3-methylindole in water. Charges on the Trp ring are multiplied by 0.80 and the calculated values are averages over the 2400 values calculated during the last 24 ps of 30-ps QM-MM trajectories

in proteins is vital to understanding function and structure, and this study helps to assess the effectiveness of current electrostatic models.

This study has solidified the notion that the wavelength is determined primarily by the electric potential difference across the long axis of the indole ring. This means that the relative direction of a charge from the Trp ring is crucial: positive charges create a red shift when on the benzene ring end and a blue shift when on the pyrrole ring end, with the size of the shift being inversely proportional to the distance from the center of the Trp ring. The reverse is true for negative charges.

The combination of the CHARMM22 force field, explicit water and INDO/S Löwdin charges on Trp (scaled by 0.80), and ab initio geometry difference (CIS-HF/3–21G) gives a good quantitative prediction for the fluorescence wavelength maximum, with the only fitting parameter being the charge scaling factor for the Trp ring. No protein dielectric constant was imposed.

Both water and protein contribute to the shifts in widely varying ratios for different Trps, and the ratio is in most cases difficult to anticipate. When charged groups lie close to Trp, they usually dominate the mechanism of the shift, and waters may even create a blue shift in such environments. Water exposure, per se, is not sufficient for a red shift. If the exposure is only along one edge of the Trp, only modest red shifts from water are possible. However, if one or both faces of the benzene ring are water-exposed, the wavelength of the fluorescence peak is usually near 350 nm. In such cases, a few waters make particularly large red shifts (~20 nm) due to exciplex-like H-bonds with negatively charged C atoms of the benzene ring in the ^1L_a excited state.

A surprising finding is that for 16 of the 19 Trps, protein contributes a red shift to the steady state Stokes shift, a result that is statistically very improbable. The extreme bias toward red shifts for the protein contributions suggests that protein electric fields relative to the modest ground state dipole of the Trp residue may be important in the evolution of the protein folds.

Orientational Dielectric Compensation. Our simulations indicate that water may cause significant (10–20 nm) red shifts for Trps that are essentially buried. This appears to be due to the collective action of regions of water up to 25 nm distant that are probably oriented by the charges and/or shape of the protein. This was particularly evident in the case of S. nuclease, for which Trp140 is sandwiched between two lysines. Protein was found to make a ~ 90 nm red shift, primarily due to the two close lysines, K133 and K110, but this was largely compensated by a large blue shift contribution from water, which combine to give a reasonable prediction of the steady state λ_{max}. This intriguing result inspired a revealing study by Qiu et al. [50], in which ultrafast TRSS was determined for the WT and the four mutants K133A, K110A, E129A, and K110C. They found, remarkably, that all five proteins gave a steady state λ_{max} of ~332.5 ± 0.5 nm, and concluded that the observed Stokes shift cannot be due to solvation by the neighboring charged residues, but is dominantly due to hydration. This statement, however, overlooks the large polarization of the solvent by the charged groups, which were shown to cause the water to contribute an overall *blue shift* to the steady state Stokes shift [25]. The mutations that delete a nearby charged residue have two large, compensating effects (dielectric compensation), assuming the crystal structure positions are correct. There is a large direct change in Coulombic stabilization/destabilization on the electron density shift in the indole ring, and there is the collective effect due to change in orientational polarization of dozens of water molecules. The conclusions of Qiu et al. are neither contrary to those of Vivian and Callis [25], nor are they necessarily incorrect. It is possible that the *dynamics* of the Stokes shift is virtually all due to water relaxing from contributing a large blue shift to a smaller blue shift while the protein contribution remains a large constant red shift.

Indole Ring Electronic Polarization. In addition to the dielectric compensation just described, protein and solvent are also coupled through the considerable polarization of the ^1L_a excited state caused by the large relaxation of the solvent/protein reaction field. This is not included in most simulations. In our QM-MM simulations, the computed dipole changes from ~5 D to ~8 D while in the excited state, following excitation as the reaction field created by the excited dipole forms and causes the Stokes shift dynamics. The increasing dipole creates a shift due to the field from protein charges as well as from water—even in the complete absence of protein motion.

The method described here can be applied to gain insight about environmental effects on the absorption and fluorescence of other chromophores and environments.

Non-exponential Decay: Relaxation and Heterogeneity

Introduction. The sensitivity of Trp fluorescence intensity to local environment apparently comes with a price. Almost all Trp fluorescence decay curves exhibit non-exponential behavior, which is bothersome for purposes of studies relying on FRET and rotational relaxation, for example. No consensus exists as to the cause of the non-exponential decay. Opinions are divided between the view that discrete subpopulations (most often assigned to different Trp χ_1 and χ_2 rotamers [2,33–36]) exhibit different decay times, and at the other extreme, the view that the excited population is homogeneous, but has a time-dependent fluorescence spectrum that shifts to longer wavelengths on a nanosecond time scale [2,100]. Other more general views have also been expressed [101–103]. An important paper by Ababou and Bombarda [104] articulated the idea that heterogeneity could arise from electron transfer or lack thereof to neighboring amide carbonyls, following the formalism of Bajzer and Prendergast [101]. As noted above, other than a few exceptions due to histidine cation [51] and disulfide quenching [105,106], the intrinsic quenching of Trp fluorescence in proteins has long been thought to be caused by photo-induced electron transfer in which an electron is transferred from the indole ring to a nearby amide of the peptide backbone. As discussed above, our QM-MM simulations show that electron transfer to the somewhat unlikely electron acceptor amide can be sensitively controlled by the electric field strength and direction caused by local charged and polar groups [27]. Because of this, other reasonable candidates for lifetime heterogeneity are

1) varying number of waters H-bonded to the local amide carbonyl oxygen;
2) distribution of protonation state of nearby by acidic/basic groups, especially when the pH ~ p*K*a; and
3) varying position of salt bridges.

5-Fluorotryptophan Suppresses Electron Transfer Quenching. Broos et al. have reported that 5-fluorotryptophan (5FW), when incorporated into proteins, almost always shows purely exponential fluorescence decay [107]. 5FW also has little intensity sensitivity, but does retain wavelength sensitivity. The single exponential and electron transfer diagnostic aspects have been usefully applied [31,108,109].

We followed the hunch that the fluoro group was withdrawing electrons from the indole ring, which would lower the energy of the HOMO, thereby increasing the ionization potential. This would make electron transfer quenching more difficult because the 5-FTrp would be a poor electron donor. We first computed the unknown ionization potentials for 3MI and 5F-3MI and several related systems [37]. As shown in Fig. 15, subsequent experimental measurements showed excellent agreement [37]. This work substantially clarifies the origin of the ubiquitous non-exponential fluorescence decay of tryptophan in proteins. The results strongly suggest that the extent of non-exponential fluorescence decay is governed primarily by the efficiency of electron transfer quenching by a nearby amide group in the peptide bond. Fluoro substitution increases the ionization potential (IP) of indole, thereby suppressing the ET rate, leading to a longer average lifetime and therefore a more homogeneous decay. The results predict the IP of 5-fluorotryptophan to be 0.19 eV higher than that of tryptophan. 5-Fluoro substitution does not measurably alter the excitation-induced change in permanent dipole moment nor does it change the fluorescent state from ^1L_a to ^1L_b. In combination with electronic structure information this argues that the increased IP and the decreased excitation energy of the ^1L_a state, together 0.3 eV, are solely responsible for the strong reduction of

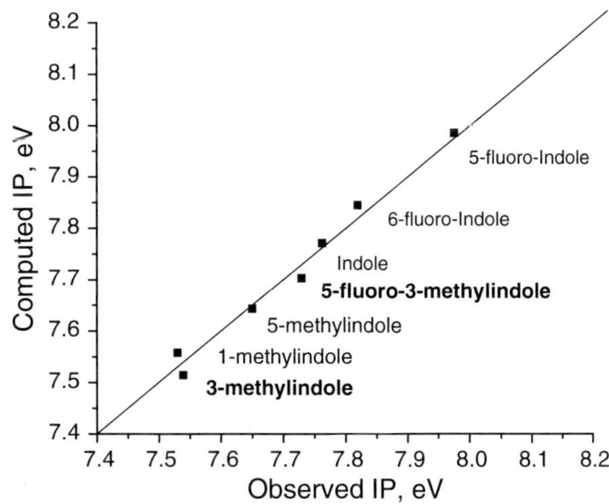

Fig. 15 Plot of computed vs. observed gas phase ionization potentials (IP) showing the effect of 5- and 6-fluoro substitution for selected indoles in vacuum. Note that 5-fluoro substitution increases the IP of indole and 3-methylindole by 0.2 eV, greatly reducing the electron donating strength of the indole ring

electron transfer quenching. 6-Fluoro substitution is predicted to increase the IP by only 0.09 eV. In agreement with our conclusions, the fluorescence decay curves of 6-fluorotryptophan-containing proteins are well fit using only 2 decay times compared to 3 required for Trp [37].

Regardless of the cause of lifetime heterogeneity and average lifetime variations between proteins, this work forces the realization that the $^{1}L_{a}$ – CT energy gap for Trp in proteins is optimally positioned so that small changes in gap will make a significant impact on the ET rate, and therefore on lifetime and quantum yield. The usefulness of the sensitivity of Trp fluorescence intensity for monitoring protein structural changes depends upon this optimal gap. If heterogeneity in protein structure exists for a particular protein, this will carry a high liability for lifetime heterogeneity. Conversely, probes such as 5FTrp and 7-azaTrp [110], whose energy gap is too large to enable ET in most proteins, will not show much variation in quantum yield and lifetime in different protein environments [107]. They will not be so useful for monitoring protein structural changes, e.g., folding/unfolding, but they will be much more useful for monitoring fluorophore-quencher proximity by Förster quenching (FRET) or by extremely potent quenchers, e.g., acrylamide, because there will not be the interfering background of lifetime heterogeneity from sources that are difficult to assign. For example, one impact of eliminating ET in 5FTrp is that lifetime and wavelength of currently available 5FTrp-containing mutants are well correlated [37], and in a manner very close to that observed by Meech et al. [39] for 3MI in different solvents, where ET is not an issue.

The effect of the fluorine in 5FTrp is apparently to remove total population decay heterogeneity that is specifically exposed by controlling the average lifetime by controlling the ET quenching rate, as illustrated in Fig. 16. There are many possible types of structural variations in proteins that can modulate the ET rate; these include solvent and protein hydrogen bond arrangement, relative positions of charged groups, and relative donor acceptor geometry. Potentially all of these variables are affected by local (rotameric) and global protein conformation changes, including fluctuations in solvation.

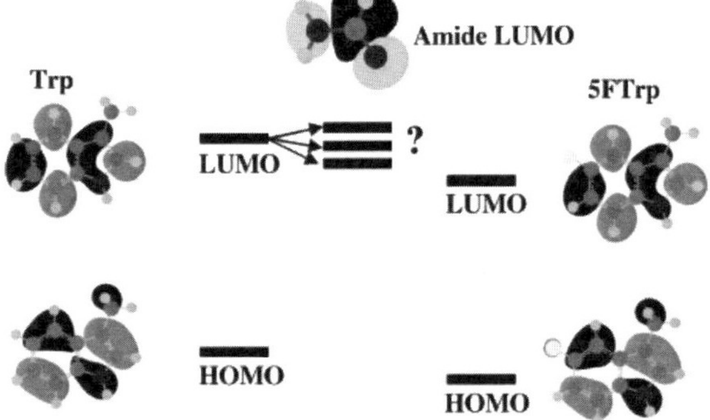

Fig. 16 Symbolic representation of the effect of 5-fluoro substitution on the ionization potential, excitation energy, and molecular orbitals of Trp. The empty amide acceptor MO is shown with charge-transfer energy levels for different presumed discrete structural subpopulations

In addition, most of these same mechanisms contribute to an *ET-independent* source of non-exponential decay caused by solvent relaxation about the excited state dipole. Given sufficient time resolution (~1 ps), a wavelength-dependent, non-exponential decay will be evident for almost all proteins because of rapid shifting of the fluorescence spectrum due to protein and solvent response to the greatly changed electronic charge distribution in the ^1L_a state [13,84,111,112]. On the nanosecond time scale, the latter type of non-exponential decay is largely obscured when a wide band of fluorescence wavelengths is observed and the time resolution is >>1 ps, but at least two well-documented cases have been reported [85,109]. One advantage of the 5FTrp probe is that the solvent relaxation time heterogeneity is expected to be virtually the same as for Trp, thereby permitting a clear view of the ET-based heterogeneity. Toptygin et al. [109] have beautifully demonstrated this kind of non-exponential decay for the single Trp GB1, which apparently has no ET quenching, as deduced from its τ_f value of 7 ns. In the same study, the authors found that 5FTrp, which is expected to show even less ET, displayed the same behavior.

The great similarity of the fluorescing state electronic structure for 5FTrp and Trp rules out the possibility that 5-fluoro substitution diminishes the electron interaction matrix element responsible for electron transfer, and also rules out the possibility that non-exponential decay has its sole source in the relaxation of the fluorescence spectrum on a nanosecond time scale.

The behavior of 5FTrp, therefore, greatly strengthens the case that Trp fluorescence lifetime heterogeneity (non-exponential decay) and the large variation of average lifetimes between Trps in different protein environments is almost entirely due to the sensitivity to local structure of fluorescence quenching by electron transfer.

Single Molecule Studies. A fresh look at the question of non-exponential decay is provided by the elegant single molecule studies of FAD fluorescence in Escherichia coli flavin reductase (Fre) by Xie and coworkers [75]. The chromophore in FAD is the isoalloxazine (ISO) ring, which by itself in the form of FMN or FAD in aqueous buffer or in the Tyr35Ser mutant of Fre has a high quantum yield and single exponential decay with $\tau_f \sim 4$ ns. In the wt protein, the fluorescence is considerably quenched by the nearby Tyr35, and shows non-exponential decay requiring four terms, with decay times ranging from 10 ps to 4 ns. Fluorescence lifetimes of single molecules of the protein-FAD complex were measured every 0.2 s using maximum likelihood estimation over time spans of up to 5 min per molecule. A continuous distribution of lifetimes was observed, covering a range of 10 ps to 3 ns, with a mean of ~1 ns and standard deviation of ~0.5 ns. The distribution is reasonably fit by a sum of three exponential terms with decay time constants quite similar to those for the ensemble. The stochastic variation of τ_f in time was modeled by dynamic interaction heterogeneity, wherein V varies in time due to distance fluctuations between Tyr35 and the ISO ring. QM-MM simulations [113] suggest that electrostatic environment fluctuations may also contribute to the variations in quenching rate. The connection to this chapter on Trp fluorescence is that neither rotamers nor large time-dependent Stokes shifts are suspected as complications to the flavin fluorescence dynamics, yet there is tremendous lifetime heterogeneity.

Ultrafast Measurements. During the past several years, ultrafast techniques involving upconversion, which provide sub-picosecond time resolution, have been developed and applied to the fluorescence of Trp in solution [114,115,111,116,117] and Trp in proteins [30,84,112,82,118]. A number of these investigations of single-Trp proteins have invariably revealed roughly double exponential time profiles for the λ_{max} correlation function in the essentially unexplored 10–100 ps range.

One protein, monellin, has been the subject of two ultrafast studies [30,84]. Xu et al. [30] have connected the ultrafast scale (0–100 ps) with the conventional TCSPC scale (100 ps to 10 ns) for Trp 3 of monellin and Trp 21 of the E21W mutant of IIA^{Glc}, obtaining DAS spanning time constants from 1.3 ps to 5 ns. Given the difficulty of these experiments, the two results agree fairly well, yet the interpretations differ considerably. Whereas Peon et al. [84] find a slightly rising signal at 380 nm, Xu et al. find a slightly decaying signal at 390 nm and a definitive decay at all times longer than 5 ps, with a ~16 ps decay time. The small difference leads to completely different interpretation. Because of the slightly rising signal at 380, Peon et al. concluded that the 16 ps time constant is the consequence of solvation dynamics on a time scale similar to that observed for subtilisin Carlsberg (sC) [112]. The slow time scale was thought to reflect the slow exchange time between bulk and surface bound water [119].

This interpretation was challenged by Nilsson and Halle [120], whose MD simulations suggested that slow protein motions were at the root. In the absence of any rise time on this time scale at 400 nm, however, Xu et al. [30] concluded that the 16 ps decay time is because of excited state population decay rate heterogeneity. Their DAS contain a large component with λ_{max} ~332 nm (considerably blue of that for the 2 ns component) that is associated with a 16 ps decay time. The rapid loss of the 332 nm component creates an overall spectrum that shifts to the red with a 16 ps time constant.

The non-normalized spectra from Xu et al. [30] are decidedly consistent with this interpretation, with the integrated intensity at 30 ps clearly diminished in accord with a 16 ps population decay. The corresponding curves from Peon et al. [84] are less clear in this regard, and appear consistent with the absence of a 16 ps component in the population decay.

In contrast to monellin, the TRSS of the E21W mutant of IIA^{Glc} was found by Xu et al. to show no transients in the range 5–100 ps. Single wavelength transients of the E21W protein did yield the characteristic blue positive/red negative 1.2 ps exponential previously seen for Trp solvation in bulk water [111,112,121]. Neither side of the spectrum exhibited terms slower than 2 ps that could be attributed to the exchange of bound water, except for decay terms well over 50 ps in the range previously studied by TCSPC[85]). They questioned whether the partially exposed nature of E21W could have rendered it less sensitive to such signals. Although the femtosecond response of this protein mimics Trp in water, steady-state iodide quenching of IIAGlc-E21W yielded a Stern–Volmer constant only approximately twice that for monellin.

More recently, Li et al. [26] have reported experimental and computational studies on the ultrafast fluorescence decay from Trp 7 of the W14Y mutant of *Sperm whale* myoglobin. As with earlier studies the experimental observation is a biphasic distribution in a few ps and tens of ps. Their extensive MD simulations revealed quite interesting results. A transition from one isomeric protein configuration to another was observed after 10 ns during a 30 ns ground-state simulation. Both linear response and non-equilibrium (average of 360 140-ns simulations) analyses were carried out, and gave similar results. For one isomer, the surface hydration energy dominates the slow component of the total relaxation energy. For the other isomer, the slow component is dominated by protein interactions with the chromophore. Quite importantly, however, companion simulations revealed that the slow component is *absent* when either protein or water was held fixed. Thus, they conclude that *coupled* water–protein motion is shown to be necessary for observation of the slow dynamics. Other interesting details from this work are that the total Stokes shifts from the non-equilibrium calculations are 2–3 times larger than the experimental value, ~50% of which comes from the ubiquitous sub-picosecond (inertial) relaxation, but which was *not* seen in the experiment. The authors suggest that this may be a short coming of the commonly used MD force fields.

The authors distance themselves somewhat from the Bagchi-Zewail [119] and Nilsson-Halle [120] mechanisms for the slow water relaxation.

The former revealed the possible relationship between the observed solvation time with water residence time and emphasized the role of water fluctuations. The latter did emphasize the role of protein fluctuations in producing a slow time component in the protein contribution to the total Stokes shift, but did not anticipate that even when a slow component in the protein contribution to the Stokes shift is not evident, *both* protein and water flexibility is required to observe the slow component in the total Stokes shift for Trp7 of Mb.

The simulations of Li et al. [26] directly illustrate the possibility of multiple protein substates with profoundly different Trp fluorescence properties because of quite different electrostatic microenvironments. Although no assessment was made on the difference in ET quenching rates for the two isomers, by inference from studies that show a wide diversity of quenching rate with environment [29,31,82], it is clear that population decay heterogeneity is also implied from this study. Subsequent studies by Zhong and coworkers [26,82,117,118] recognize the possibility of heterogeneity and ultrafast decaying components, and address them separately.

Obviously, a study of monellin containing 5-fluoroTrp in place of Trp3 would help to resolve the question of heterogeneity vs. relaxation. Presumably, if the 16 ps component is due to fast population decay, this component would be diminished considerably in the 5FTrp protein, but not if it is due to relaxation.

Heterogeneity Exposed. Boxer and coworkers [38] have incorporated the non-natural amino acid Aladan (the chromophore is essentially PRODAN) into a variety of locations in the small compact protein GB1. The results are profoundly interesting—not so much for what they reveal about solvent relaxation in buried vs. exposed sites (which are qualitatively as expected)—but for how they expose the spectre that often haunts interpretation of fluorescence decay studies: population decay heterogeneity.

The motivation for using Aladan or other extrinsic probes instead of Trp is the concern of complications due to the near degeneracy of the ^1L_a and ^1L_b states. But, given that there is virtually no question, *now*, about the nature of the emitting state (always ^1L_a unless λ_{max} <306 nm), one could argue that Aladan is at least as complicated as Trp. Not only does it have biexponential population decay in buffer (0.1–10 ns time scale), it may have a twisted excited state in the most polar environments [122]. The λ_{max} of the free Aladan in buffer actually red-shifts by 500 cm^{-1} from 0.1 to 10 ns, which would cloud any attempt to interpret TRSS in terms of protein behavior in that time range. The good news is that Aladan can reveal heterogeneity. In all but the most non-polar sites of GB1, λ_{max} red-shifts by ~2000 cm^{-1} from 0 to about 0.5 ns, then starts *blue shifting* for the next 10 ns. The authors used TRANES (Time-Resolved Area Normalized Emission Spectra) [123,124] decay curves to show that the average lifetime of the free Aladan decreases with increasing solvent polarity. This is opposite the behavior of Trp, and can unequivocally expose the heterogeneity because molecules in the more polar environments relax faster and red-shift the most, but also disappear first. As they fade the spectrum becomes dominated by longer lived, blue-shifted species [38]. It is just the reverse of what is seen for Trp, for which it is apparently the blue-shifted species that usually disappear first, creating an ambiguous interpretation. While there is no ambiguity about the existence of heterogeneity in population decay when the wavelength blue shifts in time, a deeper understanding of this probe would seem to be necessary to separate the effect of the heterogeneity and true relaxation effects.

Tryptophan Fluorescence of Denatured Proteins. There is increasing interest in the denatured (unfolded) state of proteins, but little systematic study of Trp fluorescence of intrinsically unordered proteins. This would seem important especially for using Trp to follow folding/unfolding. Recently, Pace and coworkers [125,126], acting on the belief that characterizing the denatured state ensemble is crucial to understanding protein stability and the mechanism of protein folding, have contributed a revealing study on the denatured state ensemble of five engineered mutants of ribonuclease Sa (Rnase Sa), each containing a single Trp in a location commonly found in other members of the microbial ribonuclease family. To better understand the protein denatured state, they also studied the fluorescence properties of the following peptides: N-acetyl-Trp-amide (NATA), N-acetyl-Ala-Trp-Ala-amide (AWA), N-acetyl-Ala-Ala-Trp-Ala-Ala-amide (AAWAA), and the five pentapeptides with the same sequence as the Trp substitution sites in Rnase Sa. They examined the relative Trp fluorescence intensity (I_F), and quenching by acrylamide and iodide, in both urea and guanidine hydrochloride denaturants.

Overall, they find that nonlocal effects in the denatured states of proteins influence Trp fluorescence and accessibility significantly. More specifically, they found the following: (1) there is good correlation between the I_F values of the proteins and the corresponding peptides. However, the I_F values for the proteins are always substantially (~30%) higher than those of the peptides. (2) Adjacent side chains represented in their study that are known to quench Trp fluorescence (neutral His, Tyr, and Gln) do not seem to make a large contribution to the differences among the I_F values for the peptides. They suggest that differences in the I_F values of the individual peptides depended more on the effect of sequence on local conformations on the quenching by backbone amides than on side chain quenching. (3) The wavelength of maximum fluorescence intensity, λ_{max}, does not differ significantly for the peptides and the denatured proteins. (4) fluorescence quenching of Trp by acrylamide and iodide is more than 50% greater in the peptides than in the denatured proteins, showing that long-range effects limit the accessibility of the quenchers to the Trp side chains in the proteins even in denatured proteins.

In a much earlier study, Swaminathan et al. [127] reported on the similarity of Trp fluorescence lifetime distributions in 6 M guanidine hydrochloride denatured proteins compared to when folded. In the denatured state, the average lifetimes formed a narrow distribution ranging from 1.5 to 2.9 ns, whereas for the native proteins the range was from 1.3 to 4.0 ns.

As a side note, both Swaminathan et al. [127] and Pal et al. [112] studied sC obtained from a commercial source without subsequent purification. Neither work seemed to be aware of the studies on sC by Willis and Szabo [128,129]. Unpurified sC routinely gives λ_{max} ~340 nm when excited at 300 nm, which has not been considered surprising given the location of Trp 113 on the surface. The long wavelength fluorescence is likely due to short, unordered peptides. However, immediately following HPLC purification, the Trp fluorescence is found to be at 322 nm. The authors conclude that autolysis by this serine protease yields Trp-containing fragments that are more solvent exposed.

Hexapeptides with No Quenchers. Barkley and coworkers have reported unique Trp fluorescence lifetime studies on a set of seven single-Trp cyclic hexapeptides, consisting solely of amino acids whose side chains are non-quenching in water [47,48]. This work was designed specifically to examine lifetime heterogeneity that arises plausibly only from different rates of quenching by backbone amides due to different rotameric states of the Trp, there being no other source of quenching nor is non-exponential decay due to solvent relaxation expected on the nanosecond time scale for such small structures. In addition, NMR information was obtained that partially defined the rotameric states. This series of small peptides vividly demonstrates the dominant role of peptide bond quenching in tryptophan fluorescence.

The peptides have fluorescence emission maxima of 350–355 nm, quantum yields of 0.04–0.24, and triple exponential fluorescence decays with lifetimes of 4.4–6.6, 1.4–3.2, and 0.2–1.0 ns at 5°C, suggesting that different tryptophan rotamers have different emission maxima even in the case of solvent-exposed tryptophans. This conclusion is supported by quantum mechanical/molecular dynamics simulations of the six canonical side chain rotamers of tryptophan of the seven solvated hexapeptides, using our published procedure [25]. The calculated range of emission maxima for the tryptophan rotamers of the seven peptides is 344–365 nm. Water molecules were found to make a major contribution to this wavelength dispersion.

A remarkable correlation of DAS lifetimes and λ_{max} was found in this study, and deserves comment. At present, this correlation is not understood in the context of ground state heterogeneity. In the absence of correlation, the heterogeneity model would predict equal probability for each of the six possible associations of three ordered lifetimes with three ordered λ_{max} values. For the seven peptides, there are therefore a total of $6^7 = 279{,}936$ possible outcomes. The probability of the singular result of progressive association (longest wavelength with longest lifetime) for each peptide would be 4×10^{-6}, yet this scenario is nearly realized. Clearly a strong physical principle is dictating this correlation. The absence of any plausible hypothesis for the nature of this principle at present is often presented as evidence against the ground state heterogeneity model.

The relaxation model, in contrast, naturally predicts this correlation. However, in its usual form, one expects that negative amplitudes will be observed for some of the components. These were not observed for the seven peptides studied here, and are seen only rarely for proteins.

The simulation results presented speak only to the existence of a rationale for a substantial dependence of λ_{max} on rotamer conformation, a requirement for the validity of both rotamer-based mechanisms. A single rotamer of the highly solvent-exposed Trps in these peptides is not expected to exhibit nanosecond-scale fluorescence shifts due solely to dielectric relaxation of solvent in response to the excitation induced dipole change, a process that experimentally is known to be complete within 5 ps [111].

Another question deserving comment is whether at least three rotamers states are substantially populated. No attempt was been made to assess the equilibrium rotamer populations with MD simulations because the parameterization is not likely to be accurate enough. However, a number of computational studies have been directed at exploring the energy surfaces of Trp and N-actetyltryptophan-methylamide (NATMA). These calculations have correctly predicted the most stable conformers found in cold expansion jets. The study by Bombasaro et al. [129A] using B3LYP/6-31 g* found 34 stable rotamers. Such computations are not directly applicable to this investigation because the hexapeptide backbone adopts conformations that do not naturally arise for NATMA, and because vacuum calculations do not model solvation realistically. However, the relative energies of rotamers are probably representative. Of interest is that the ab initio result predicts that only three rotamers would be found with population greater than 0.1 at 5°C. For the alpha-L backbone conformation, only three stable conformers were found, with the dominant one representing 96% of the population. This raises the question as to the validity of assuming that at least three rotamers will exist for all cases. The reason that three are always found may be that there is a continuous distribution that cannot be fit better than with three.

Basic Studies of Electronic Structure

Resolved Spectra of Model Molecules

Crossing of ^1L_a and ^1L_b Due to Methyl Substitution. Polarized one-photon fluorescence excitation and dispersed fluorescence spectra of 3-methylindole (3MI),

5-methylindole, and 2,3-dimethylindole in solid Ar at 20 K under site selective conditions unambiguously revealed the ^1L_a origin locations relative to the ^1L_b origin [22]. Figure 17 compares the dispersed fluorescence of these molecules along with the fluorescence and phosphorescence of indole, all in solid argon, with the origins aligned. The important information contained in Fig. 17 is that indole, 3MI, and 5MI all exhibit ^1L_b fluorescence, whereas that of 2,3-DMI bears much closer resemblance to the phosphorescence of indole, which is known to be ^3L_a. Methylation at the 3 and 2 positions of indole are well established to shift the ^1L_a state to lower energy [130]. In the inert argon matrix environment, the extra methyl in 2,3-DMI produces an inversion of the ^1L_a and ^1L_b states relative to 3MI.

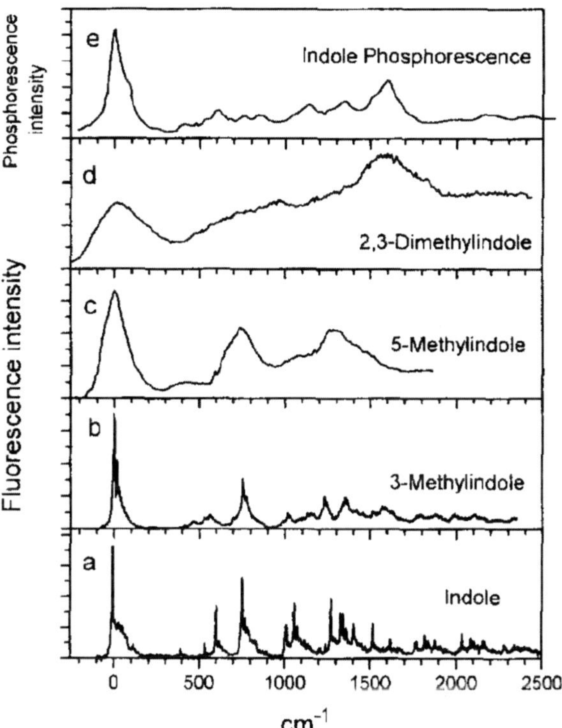

Fig. 17 (**a**) Comparison of the Ar matrix isolated (20 K) dispersed fluorescence spectra of indole; (**b**) 3-methylindole; (**c**) 5methylindole; (**d**) 2,3-dimethylindole; (**e**) dispersed *phosphorescence* of indole

The fluorescence excitation spectrum of 3MI in argon shown in Fig. 18b is considerably different from the jet spectrum [131] seen in Fig. 18a. This is quite different from indole, which shows a close correspondence between jet and argon. In the jet spectrum of 3MI, the ^1L_a appears to be split in 4 or 5 major components involving lines at 356, 409, 467, 609–617, and 739–749 cm^{-1}, as revealed by the ratio of two-photon absorptivities with circularly and linearly polarized light. In solid argon, however, the one-photon anisotropy in combination with the signature ^1L_b fluorescence spectrum clearly identifies the ^1L_a origin as the intense feature only 260 cm^{-1}

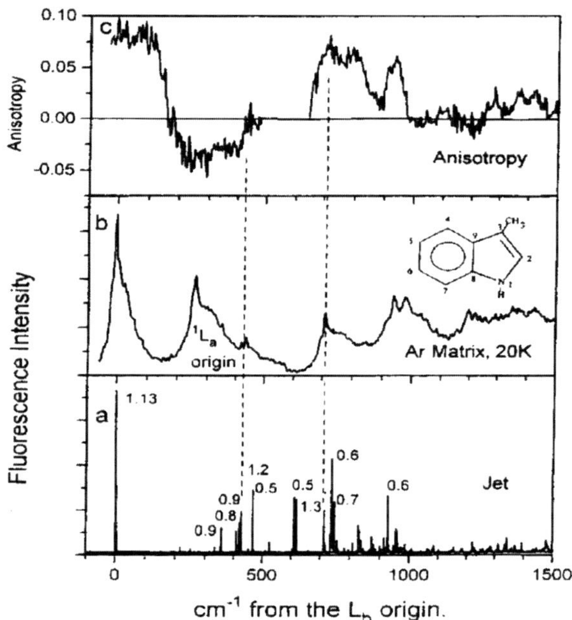

Fig. 18 Comparison of jet-cooled (**a**) and Ar matrix isolated (**b**) fluorescence excitation spectra of 3-methylindole. The fluorescence anisotropy in the Ar matrix is shown in panel (**c**). In (a), numbers >1.1 = ^1L_b; <1.0 = ^1L_a

above the origin. That it is shifted a few hundred cm^{-1} closer to the ^1L_b origin is consistent with what is found for indole.

That the two state origins can approach each other this close in energy is consistent with earlier findings deduced by Strickland and coworkers from spectra in perfluorinated hydrocarbon solvent [130], and by ab initio quantum chemical calculations [132].

These experiments afford a rare observation of the difference between ^1L_b and ^1L_a fluorescence under vibronically resolved conditions. The conclusions are well supported by ab initio calculated spectra, which are discussed below.

Ab Initio Computed ^1L_b, ^1L_a, and ^3L_a Spectra. We discovered an unlikely recipe for computing vibronic fluorescence spectra of molecules that has evolved during the last few years to be quite useful for identifying the nature of vibronically resolved emission spectra [19,20,133,134]. Since first reported, the procedure has improved considerably due to the availability of hybrid DFT methods.

Using GAUSSIAN98, surprisingly accurate geometry differences between the ground and excited state were obtained using Hartree–Fock/4-31G for the ground state optimization, and CIS/4-31G for the excited state. The geometry difference was projected onto the normal modes obtained for the ground state using two density functional methods, BLYP and B3LYP, which have been found to be particularly effective for vibrational frequencies. Both methods were used with basis sets ranging from 6-31G** to 6-311++ G** in a program [135] that computes the Franck–Condon factors [20]. These modes were used in both the ground and excited state (parallel mode approximation). When using 6-31G**, the B3LYP frequencies were scaled by a factor of 0.975 in order to closely match experimental results. The BLYP frequencies were scaled by 1.013. Both methods gave a marked improvement in accuracy over previous computed ^1L_b and ^3L_a emission spectra of indole using MP2/6-31G** modes. B3LYP and BLYP gave equally good fits with experimental spectra.

Figure 19 shows comparisons of vibronically resolved ^1L_b fluorescence and ^1L_a phosphorescence with computed spectra. The fits with B3LYP/6-311++G** modes are nearly perfect, showing the crucial importance of accounting for electron correlation at a high level. Earlier published results using MP2/6-31 g** modes (scaled by 0.962) were impressive, but failed badly for a few modes. In the ^1L_b calculated fluorescence, mode 15 was essentially absent and 14 was too active. Experimentally, both 15 and 14 have equal intensities. In the early calculated phosphorescence, mode 14 was predicted to one of the most active modes in the spectrum, yet it is completely absent in experiment. As seen in Fig. 19, both of these defects are nicely eliminated with the higher quality wavefunction.

Crossing of ^1L_a and ^1L_b in Polar Complexes. Armed with a clear picture of the vibronic spectral signatures of the ^1L_a and ^1L_b fluorescence in addition to the two-photon absorption polarization criteria, we were encouraged by the fluorescence lifetime study of van der Waals complexes in cold jets of 3MI with a series of H-bond acceptors with increasing proton acceptor strengths by the Wallace group [136]. The study by Demmer et al. gave strong support to the notion that the ^1L_a $-^1L_b$ gap decreases in proportion to the proton affinity of the complexing partner, with the crossing occurring for ammonia; interestingly, the NH_3 complex has a very short lifetime, and does not fluoresce significantly. Thus, for 3MI complexed with H_2O, methanol, ethanol, and diethyl ether, the ^1L_b origin was implicated as lowest, whereas for trimethyl and triethylamine, ^1L_a was assigned as the lowest. We therefore tuned our experiments to observe the well-resolved dispersed fluorescence of this series of complexes to see if indeed the two-photon polarization and resolved spectra coincided with these conclusions.

Figure 20 compares the experimental dispersed fluorescence of 3MI and the 3MI-triethylamine complex with the corresponding calculated spectra for ^1L_b and ^1L_a fluorescence, respectively. The spectra for the complex were augmented with an extra progression belonging to the low frequency intermolecular mode due to a geometry change involving the amine upon ^1L_a excitation, and such bands are denoted by an asterisk. The striking agreement, as evidenced by the complete lack of activity in the high-frequency C—C stretching bands (modes 10, 9 and 8 of indole) for the bare 3MI, and the strong intensity seen for the complex, is in complete accord with the assignment that true ^1L_a fluorescence is being observed. This is completely confirmed by the TP polarization we observe for excitation of the origin in the two cases [19].

The Effect of Inhomogeneous Broadening. As noted previously [13], early assignments of vibrationally structured fluorescence of Trp in certain protein environments were made as ^1L_b. This was understandable given that in most cases ^1L_a fluorescence was broad and diffuse.

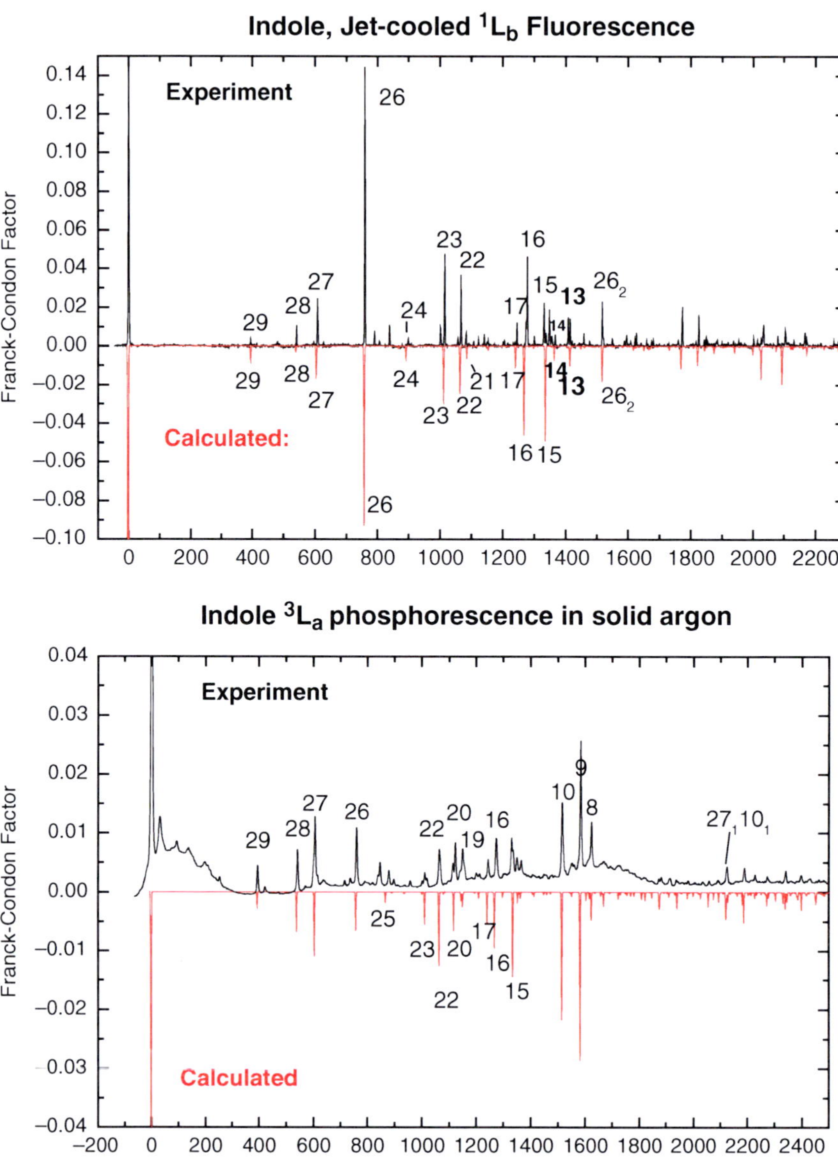

Fig. 19 Comparison of ab initio computed indole ^1L_b fluorescence (in jet) and ^3L_a phosphorescence (solid Ar) with experiment. The computed spectra used geometry differences computed with HF/3-21G and b3lyp/6-311++g** vibrational modes and frequencies for both ground and excited states

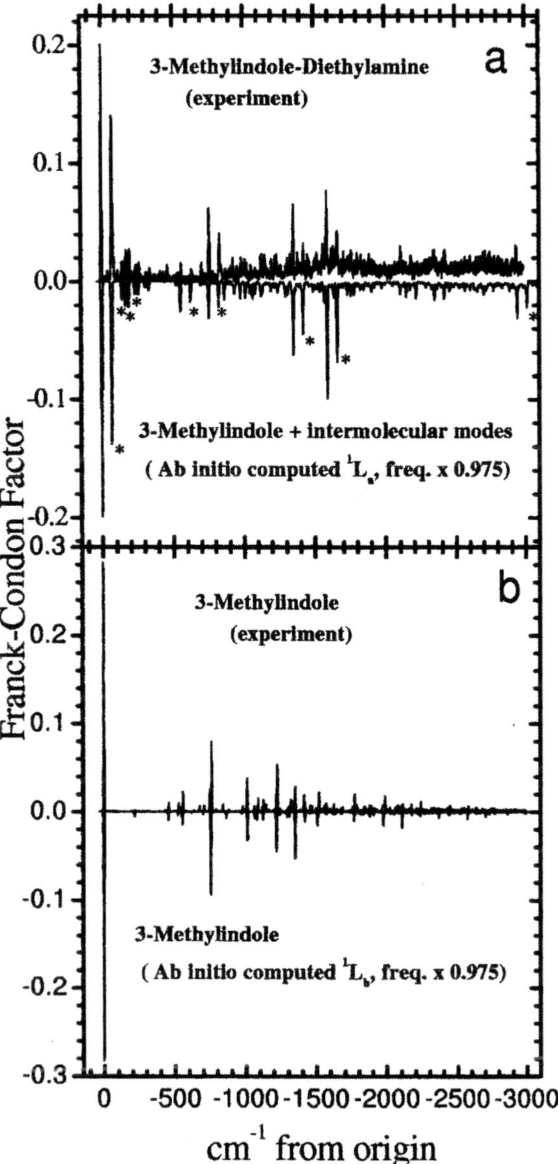

Fig. 20 (**a**) Comparison of the experimental dispersed fluorescence spectrum for the 3MI-diethylamine complex with the ^1L_a Franck–Condon factors predicted from ab initio calculations on 3MI, augmented by a synthetic intermolcular mode progression. (**b**) Comparison of the experimental dispersed fluorescence spectrum for bare 3MI with the ^1L_b Franck–Condon factors predicted from ab initio calculations on 3MI, For both figures, the frequency factors have been scaled by 0.975 and the monochromator resolution is about 14 cm^{-1}

As seen in Fig. 21, there is no reason to think that ^1L_a fluorescence is inherently unstructured. The relatively large dipole of the ^1L_a state compared to that of ^1L_b, however, makes the ^1L_a emission much more sensitive to the electrostatic landscape than ^1L_b emission. Consequently, in environments that are sufficiently polar where emission is from ^1L_a instead of ^1L_b, the spectrum tends to be diffuse and broad. As seen above, most environments that are homogeneous enough to show structure

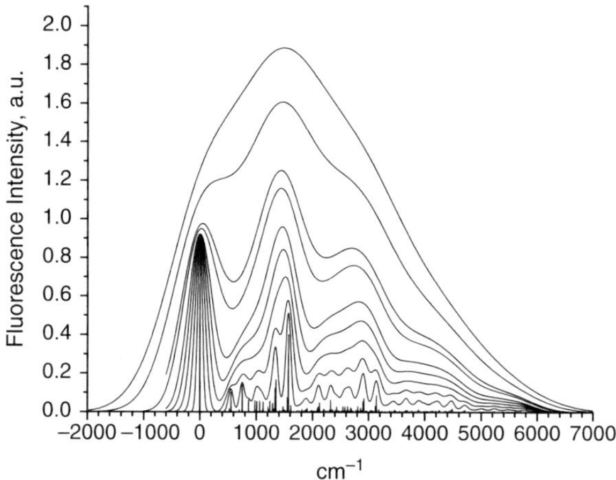

Fig. 21 Broadened 3MI ^1L_a calculated spectra. The spectral width of each line of the upwardly displaced spectra is determined by a Gaussian with FWHM of 3, 100, 200, 300, 400, 500, 700, 800, 1200, and 1500 cm^{-1}, respectively. The area under each curve is proportional to the line width

result in ^1L_b emission. Given the accurate ^1L_a computed fluorescence, it is possible to demonstrate the effect of broadening. Figure 21 shows the evolution of the calculated ^1L_a fluorescence as a function of line width. The area under each curve is proportional to the line width, for ease of viewing. The fifth spectrum from the bottom is indicative of the appearance of the fluorescence from ribonuclease T1 at 77 K [137] and of Trp in ethanol at 2 K [138], both of which were at one time suspected of being ^1L_b emission.

In contrast to ^1L_a fluorescence, phosphoresence from Trp, which is from ^3L_a, shows vibrational structure even in quite polar environments. A detailed analysis shows that small changes in the weighting of configurations making up the excited state have a surprisingly large effect on the permanent dipole of the ^3L_a state, while having little impact on the emission spectrum vibronic envelope [139].

Final Remarks

Tryptophan is perhaps the best understood of the fluorescent probes of protein dynamics and structure and has the great advantage of being intrinsic. Probes that emit in the visible region are necessarily larger, and the extended conjugation often brings complexities with respect to internal rotation.

The largest remaining challenge for interpretation of time-resolved Trp fluorescence experiments is clearly how to sort out the underlying cause of the decay constants and wavelength shifts. To solve this problem, ideally one would like to have three complementary probes; a probe (1) which has λ_{max} that is sensitive to environment but whose population decay is not; (2) which has reversed sensitivities relative to 1; (3) for which neither λ_{max} nor population decay is environment sensitive. These probes should all be in the form of non-natural amino acids that can be efficiently incorporated into proteins with minimal consequence to the protein structure and dynamics. A fourth type which would be useful has the properties of Aladan, which is similar to Trp in showing environment sensitivity to both λ_{max} and population decay, but has the opposite behavior regarding the λ_{max} dependence of population decay.

5FTrp comes close to filling the criteria for type (1), and its use is encouraged for the interpretation of ultrafast fluorescence decay experiments aiming to pull specific information from TRSS. In addition, if the decay of area normalized intensity (TRANES) [38,123,124] was routinely reported, the presence of population decay heterogeneity would, in principle, be evident.

Appendix: Notes on Electrostatics

Potential has the units of volts, i.e., joules/coulomb. An electron has lower energy in a more positive potential (closer to positive charge or farther from negative charge). From Coulombs law, one can show that at a point 1 Å distant from a proton, the potential is +14.4 V. Therefore, the energy of an electron 1 Å from a proton is –14.4 eV or –331 kcal/mol or –1385 kJ/mol. The electric fields in proteins are on the order of a few times 10^7 V/cm, i.e., a few tenths of V/Å [64,99]. The dipole change during excitation to ^1L_a is 5–10 Debye, meaning that a field of 10^7 V/cm can cause a 10 nm shift in the fluorescence.

For back-of-envelope calculations of wavelength shift, one may use the shift of electron density for excitation to ^1L_a as ~0.25 electron from the pyrrole to benzene ring. Therefore if the average potential of the benzene ring is 1 V more positive than the pyrrole ring, the energy difference of will be –0.25 eV = –2000 cm^{-1} = a ~20 nm red shift. For electron transfer, if the amide C atom is more positive than the indole ring by 1 V, the energy to create the CT state will be 1 eV = 8000 cm^{-1} = 331 kcal/mol less than if there were no potential difference.

Acknowledgements NIH Grant No. GM31824 and NSF Grants MCB- 9817372, MCB-0133064, and MCB-0446542. The author thanks Drs. Mary Barkley, Jaap Broos, and Greg Gillispie for kindly readding the manuscript.

References

1. J. B. A. Ross, W. R. Laws, and M. A. Shea, Protein Structures: Methods in Protein Structure and Structure Analysis, V. Uversky and Permyakov, E. A., Editors. Nova Science Publishers, Inc., New York, 1–18 (2007)
2. J. R. Lakowicz, Principles of Fluorescence Spectroscopy., 3rd Ed., Springer (2006)
3. A. S. Ladokhin, S. Jayasinghe, and S. H. White, How to measure and analyze tryptophan fluorescence in membranes properly, and why bother? *Anal.Biochem.* **285,** 235–245 (2000)
4. A. S. Ladokhin, Encyclopedia of Analytical Chemistry, R. A. Myers, Editor. John Wiley and Sons, Ltd., Chichester, 5762–5799 (2000)
5. M. R. Eftink, Fluorescence techniques for studying protein structure. *Methods Biochem. Anal.* **35,** 127–205 (1991)
6. C. A. Royer, Probing protein folding and conformational transitions with fluorescence. *Chemical Reviews* **106**(5), 1769–1784 (2006)
7. J. Kubelka, W. A. Eaton, and J. Hofrichter, Experimental tests of villin subdomain folding simulations. *J. Mol. Biol.* **329**(4), 625–630 (2003)

8. R. Godoy-Ruiz, E. R. Henry, J. Kubelka, J. Hofrichter, V. Munoz, J. M. Sanchez-Ruiz, and W. A. Eaton, Estimating free-energy barrier heights for an ultrafast folding protein from calorimetric and kinetic data. *J. Phys. Chem. B* **112**(19), 5938–5949 (2008)
9. D. L. Ensign, P. M. Kasson, and V. S. Pande, Heterogeneity even at the speed limit of folding: Large-scale molecular dynamics study of a fast-folding variant of the villin headpiece. *J. Mol. Biol.* **374**(3), 806–816 (2007)
10. T. Kimura, J. C. Lee, H. B. Gray, and J. R. Winkler, Site-specific collapse dynamics guide the formation of the cytochrome c ′ four-helix bundle. *Proc. Natl. Acad. Sci. U.S.A.* **104**(1), 117–122 (2007)
11. J. E. Kim, G. Arjara, J. H. Richards, H. B. Gray, and J. R. Winkler, Probing folded and unfolded states of outer membrane protein A with steady-state and time-resolved tryptophan fluorescence. *J. Phys. Chem. B* **110**(35), 17656–17662 (2006)
12. A. Warshel, Computer modeling of chemical reactions in enzymes and solutions, Ed., Wiley-Interscience, New York, NY (1991)
13. P. R. Callis, 1La and 1Lb transitions of tryptophan: Applications of theory and experimental observations to fluorescence of proteins. *Meth. Enzym.* **278,** 113–150 (1997)
14. L. S. Slater and P. R. Callis, Molecular orbital theory of the ^1L_a and ^1L_b states of indole. 2. An ab initio study. *J. Phys. Chem.* **99,** 8572–8581 (1995)
15. P. R. Callis, Molecular orbital theory of the 1La and 1Lb states of indole. *J. Chem. Phys.* **95,** 4230–4240 (1991)
16. M. R. Eftink, L. A. Selvidge, P. R. Callis, and A. A. Rehms, Photophysics of indole derivatives: Experimental resolution of La and Lb transitions and comparison with theory. *J. Phys. Chem.* **94**(9), 3469–3479 (1990)
17. P. R. Callis, Transition density topology of the La and Lb states of indoles and purines. *Int. J. Quantum. Chem.* **S18,** 579–588 (1984)
18. K. W. Short and P. R. Callis, Evidence of pure 1Lb fluorescence from redshifted indole-polar solvent complexes in a supersonic jet. *J. Chem. Phys.* **108,** 10189–10196 (1998)
19. K. W. Short and P. R. Callis, Studies of jet-cooled 3-methylindole-polar solvent complexes: Identification of ^1L_a fluorescence. *J. Chem. Phys.* **113,** 5235–5244 (2000)
20. K. W. Short and P. R. Callis, One- and two-photon spectra of jet-cooled 2,3-dimethylindole: ^1L_b and ^1L_a assignments. *Chem. Phys.* **283**(1–2), 269–278 (2002)
21. B. J. Fender, D. M. Sammeth, and P. R. Callis, Site selective photoselection study of indole in argon matrix: Location of the 1La origin. *Chem. Phys. Lett.* **239,** 31–37 (1995)
22. B. J. Fender and P. R. Callis, 1La origin Locations of methylindoles in argon matrices. *Chem. Phys. Lett.* **262,** 343–348 (1996)
23. J. Broos, K. Tveen-Jensen, E. de Waal, B. H. Hesp, J. B. Jackson, G. W. Canters, and P. R. Callis, The emitting state of tryptophan in proteins with highly blue-shifted fluorescence. *Angewandte Chemie-International Edition* **46**(27), 5137–5139 (2007)
24. T. J. Jensen, G. Strambini, G. Gonnelli, J. Broos, and J. B. Jackson, Mutations in transhydrogenase change the fluorescence emission state of trp-72 from ^1L_a to ^1L_b. *Biophys. J,* **95,** 3419–3428 (2008)
25. J. T. Vivian and P. R. Callis, Mechanisms of tryptophan fluorescence shifts in proteins. *Biophys. J.* **80** 2093–2109 (2001)
26. T. P. Li, A. A. P. Hassanali, Y. T. Kao, D. P. Zhong, and S. J. Singer, Hydration dynamics and time scales of coupled water-protein fluctuations. *J. Am. Chem. Soc.* **129**(11), 3376–3382 (2007)
27. P. R. Callis and T. Liu, Quantitative prediction of fluorescence quantum yields for tryptophan in proteins. *J. Phys. Chem. B* **108,** 4248–4259 (2004)
28. P. R. Callis and J. T. Vivian, Understanding the variable fluorescence quantum yield of tryptophan in proteins using QM-MM simulations. Quenching by charge transfer to the peptide backbone. *Chem. Phys. Lett.* **369,** 409–414 (2003)

29. J. J. Chen, S. L. Flaugh, P. R. Callis, and J. King, Mechanism of the highly efficient quenching of tryptophan fluorescence in human gamma D-crystallin. *Biochemistry* **45**(38), 11552–11563 (2006)
30. J. H. Xu, D. Toptygin, K. J. Graver, R. A. Albertini, R. S. Savtchenko, N. D. Meadow, S. Roseman, P. R. Callis, L. Brand, and J. R. Knutson, Ultrafast fluorescence dynamics of tryptophan in the proteins monellin and IIA(Glc). *J. Am. Chem. Soc.* **128**(4), 1214–1221 (2006)
31. L. C. Kurz, B. Fite, J. Jean, J. Park, T. Erpelding, and P. Callis, Photophysics of tryptophan fluorescence: Link with the catalytic strategy of the citrate synthase from Thermoplasma acidophilum. *Biochemistry* **44**(5), 1394–1413 (2005)
32. P. R. Callis, A. Petrenko, P. L. Muino, and J. R. Tusell, Ab initio prediction of tryptophan fluorescence quenching by protein electric field enabled electron transfer. *J. Phys. Chem. B* **111**(35), 10335–10339 (2007)
33. J.-P. Privat, P. Wahl, and J.-C. Auchet, Rates of deactivation processes of indole derivatives in water-organic solvent mixtures—Application to tryptophyl fluorescence of proteins. *Biophys. Chem.* **9,** 223–233 (1979)
34. J. W. Petrich, M. C. Chang, D. B. McDonald, and G. R. Fleming, On the origin of nonexponential fluorescence decay in tryptophan and its derivatives. *J. Am. Chem. Soc.* **105,** 3824–3832 (1983)
35. T. E. S. Dahms, K. J. Willis, and A. G. Szabo, Conformational heterogeneity of tryptophan in a protein crystal. *J. Am. Chem. Soc.* **117,** 2321–2326 (1995)
36. M. Hellings, M. DeMaeyer, S. Verheyden, Q. Hao, E. J. M. VanDamme, W. J. Peumans, and Y. Engelborghs, The dead-end elimination method, tryptophan rotamers, and fluorescence lifetimes. *Biophys. J.* **85,** 1894–1902 (2003)
37. T. Q. Liu, P. R. Callis, B. H. Hesp, M. de Groot, W. J. Buma, and J. Broos, Ionization potentials of fluoroindoles and the origin of nonexponential tryptophan fluorescence decay in proteins. *J. Am. Chem. Soc.* **127**(11), 4104–4113 (2005)
38. P. Abbyad, X. H. Shi, W. Childs, T. B. McAnaney, B. E. Cohen, and S. G. Boxer, Measurement of solvation responses at multiple sites in a globular protein. *J. Phys. Chem. B* **111**(28), 8269–8276 (2007)
39. S. R. Meech, A. Lee, and D. Phillips, On the nature of the fluorescent state of methylated indole derivatives. *Chem. Phys.* **80** 317–328 (1983)
40. 3MI and NATA are useful model compounds. NATA has two amides connected to the indole ring just as in the protein, and does not have the ammonium and carboxylate characteristic of the free amino acid Trp at pH 7. The amides are separated from the system of the indole ring by two saturated carbons. Therefore, 3MI has very nearly the same electronic properties, without the possibility of electron transfer to the amides.
41. R. W. Cowgill, Fluorescence and the structure of proteins. I. Effects of substituents on the fluorescence of indole and phenol compounds. *Arch. Biochm. Biophys.* **100,** 36–44 (1963)
42. J. Feitelson, Environmental effects on the fluorescence of tryptophan and other indole derivatives. *Israel J. Chem.* **8,** 241–252 (1970)
43. Y. Chen and M. D. Barkley, Toward understanding tryptophan fluorescence in proteins. *Biochemistry* **37,** 9976–9982 (1998)
44. Y. Chen, B. Liu, H.-T. Yu, and M. D. Barkley, The peptide bond quenches indole fluorescence. *J. Am. Chem. Soc.* **118,** 9271–9278 (1996)
45. R. W. Ricci and J. M. Nesta, Inter- and intramolecular quenching of indole fluorescence by carbonyl compounds. *J. Phys. Chem.* **80,** 974–980 (1976)
45A. H. Shizuka, M. Serizawa, T. Shimo, I. Saito, and T. Matsuura, Fluorescence quenching mechanism of tryptophan. Remarkably efficient internal proton-induced quenching and charge-transfer quenching. *J. Am. chem. soc.* **110**, 1930–1934 (1988)
46. M. R. Eftink, Y. Jia, D. Hu, and C. A. Ghiron, Fluorescence studies with tryptophan analogues: Excited state interactions involving the side chain amino group. *J. Phys. Chem.* **99,** 5713–5723 (1995)

47. C. P. Pan, P. R. Callis, and M. D. Barkley, Dependence of tryptophan emission wavelength on conformation in cyclic hexapeptides. *J. Phys. Chem. B* **110**(13), 7009–7016 (2006)
48. P. D. Adams, Y. Chen, K. Ma, M. G. Zagorski, F. D. Sonnichsen, M. L. McLaughlin, and M. D. Barkley, Intramolecular quenching of tryptophan fluorescence by the peptide bond in cyclic hexapeptides. *J. Am. Chem. Soc.* **124,** 9278–9286 (2002)
49. A. Sillen, J. Hennecke, D. Roethlisberger, R. Glockshuber, and Y. Engelborghs, Fluorescence quenching in the DsbA protein from Escherichia coli: The complete picture of the excited-state energy pathway and evidence for the reshuffling dynamics of the microstates of tryptophan. *Proteins Struct Funct Genet* **37,** 253–263 (1999)
50. W. H. Qiu, Y. T. Kao, L. Y. Zhang, Y. Yang, L. J. Wang, W. E. Stites, D. P. Zhong, and A. H. Zewail, Protein surface hydration mapped by site-specific mutations. *Proc. Natl. Acad. Sci. U.S.A.* **103**(38), 13979–13984 (2006)
51. R. Vos and Y. Engelborghs, A fluorescence study of tryptophan-histidine interactions in the peptide anantin and in solution. *Photochem. Photobiol.* **60,** 24–32 (1994)
52. K. Willaert, R. Loewenthal, J. Sancho, M. Froeyen, A. Fersht, and Y. Engelborghs, Determination of the excited-state lifetimes of the tryptophan residues in barnase, via multifrequency phase fluorometry of tryptophan mutants. *Biochemistry* **31,** 711–716 (1992)
53. D. L. Harris and B. S. Hudson, Fluorescence and molecular dynamics study of the internal motion of the buried tryptophan in bacteriophage T4 lysozyme: effects of temperature and alteration of nonbonded networks. *Chem. Phys.* **158,** 353–382 (1991)
54. C. D. Borsarelli, S. G. Bertolotti, and C. M. Previtali, Exciplex-type behavior and partition of 3-substituted indole derivatives in reverse micelles made with benzylhexadecyldimethylammonium chloride, water and benzene. *Photochem. Photobiol.* **73**(2), 97–104 (2001)
55. C. Rivarola, S. G. Bertolotti, C. D. Borsarelli, J. J. Cosa, C. M. Previtali, and M. G. Neumann, Hydrogen bonding and charge transfer interactions in exciplexes formed by excited indole and monosubstituted benzenes in cyclohexane. *Chem. Phys. Lett.* **262**(1–2), 131–136 (1996)
56. V. Nanda, S. M. Liang, and L. Brand, Hydrophobic clustering in acid-denatured IL-2 and fluorescence of a Trp NH–pi H-bond. *Biochem. Biophys. Res. Commun.* **279,** 770–778 (2000)
57. V. Nanda and L. Brand, Aromatic interactions in homeodomains contribute to the low quantum yield of a conserved, buried tryptophan. *Proteins* **40,** 112–125 (2000)
58. V. Subramaniam, T. M. Jovin, and R. V. Rivera-Pomar, Aromatic amino acids are critical for stability of the bicoid homeodomain. *J. Biol. Chem.* **276,** 21506–21511 (2001)
59. T. Liu and P. R. Callis, Quantitative prediction of tryptophan fluorescence quenching in proteins, III: Quenching by disulfide, aromatics, and acrylamide. *Biophys. J.* **88**(1), 161A–162A (2005)
60. T. P. Treynor and S. G. Boxer, Probing excited-state electron transfer by resonance stark spectroscopy: 3. Theoretical foundations and practical applications. *J. Phys. Chem. B* **108**(35), 13513–13522 (2004)
61. A. L. Sobolewski, W. Domcke, C. Dedonder-Lardeux, and C. Jouvet, Excited-state hydrogen detachment and hydrogen transfer driven by repulsive $1\pi\sigma*$ states: A new paradigm for nonradiative decay in aromatic biomolecules. *Phys. Chem. Chem. Phys.* **2002,** 1093–1100 (2002)
62. A. Sobolewski and W. Domcke, Photoinduced charge separation in indole-water clusters. *Chem. Phys. Lett.* **329,** 130–137 (2000)
63. A. L. Sobolewski and W. Domcke, Ab initio investigations on the photophysics of indole. *Chem. Phys. Lett.* **315**(3–4), 293–298 (1999)
64. C. donder-Lardeux, C. Jouvet, S. Perun, and A. L. Sobolewski, External electric field effect on the lowest excited states of indole: Ab initio and molecular dynamics study. *Phys. Chem. Chem. Phys.* **5**(22), 5118–5126 (2003)
65. P. L. Muiño and P. R. Callis, Solvent effects on the fluorescence quenching of tryptophan by amides via electron transfer. experimental and computational studies. *J. Phys. Chem. B* **113,** 2572–2577 (2008)
66. R. A. Marcus and N. Sutin, Electron transfers in chemistry and biology. *Biochim. Biophys. Acta* **811,** 265–322 (1985)

67. R. A. Marcus, On the theory of oxidation-reduction reactions involving electron transfer. I. *J. Chem. Phys.* **24,** 966–978 (1955)
68. M. Tachiya, Generalization of the Marcus equation for the electron-transfer rate. *J. Phys. Chem.* **97,** 5911–5916 (1993)
69. M. J. Frisch, G. W. Trucks, H. B. Schlegel, G. E. Scuseria, M. A. Robb, J. R. Cheeseman, V. G. Zakrzewski, J. A. Montgomery, R. E. Stratmann, J. C. Burant, S. Dapprich, J. M. Millam, A. D. Daniels, K. N. Kudin, M. C. Strain, O. Farkas, J. Tomasi, V. Barone, M. Cossi, R. Cammi, B. Mennucci, C. Pomelli, C. Adamo, S. Clifford, J. Ochterski, G. A. Petersson, P. Y. Avala, Q. Cui, K. Morokuma, N. Rega, P. Salvador, J. J. Dannenberg, D. K. Malick, A. D. Rabuck, K. Raghavachari, J. B. Foresman, J. Cioslowski, J. V. Ortiz, A. G. Baboul, B. B. Stefanov, G. Liu, A. Liashenko, P. Piskorz, I. Komaromi, R. Gomperts, R. L. Martin, D. J. Fox, T. Keith, M. A. Al-Laham, C. Y. Peng, A. Nanayakkara, M. Challacombe, P. M. W. Gill, B. Johnson, W. Chen, M. W. Wong, J. L. Andres, C. Gonzalez, M. Head-Gordon, E. S. Replogle, and J. A. Pople, Gaussian 98 (Revision A.11.3) (2002)
70. J. Ridley and M. Zerner, Intermediate Neglect of Differential Overlap (INDO) technique for spectroscopy: Pyrrole and the azines. *Theor. Chim. Acta(Berl)* **32,** 111–134 (1973)
71. M. A. Thompson and M. C. Zerner, A theoretical examination of the electronic structure and spectroscopy of the photosynthetic reaction center from *Rhodopseudomonas viridis*. *J. Am. Chem. Soc.* **113,** 8210–8215 (1991)
72. A. D. MacKerell, Jr., D. Bashford, M. Bellott, R. L. Dunbrack, J. D. Evanseck, M. J. Field, S. Fischer, J. Gao, S. Ha, D. Joseph-McCarthy, L. Kuchnir, K. Kuczera, F. T. K. Lau, C. Mattos, S. Michnick, T. Ngo, D. T. Nguyen, B. Prodhom, W. E. Reiher III, B. Roux, M. Schlenkrich, J. C. Smith, R. Stote, J. Straub, M. Watanabe, J. Wiorkiewicz-Kuczera, D. Yin, and M. Karplus, All atom empirical potential for molecular modeling and dynamics studies of proteins. *J. Phys. Chem. B* **102,** 3586–3616 (1998)
73. K. Andersson, P.-Å. Malmqvist, and B. O. Roos, 2nd-order perturbation-theory with a complete active space self-consistent field reference function. *J. Chem. Phys.* **96**(2), 1218–1226 (1992)
74. K. Andersson, M. Barysz, A. Bernhardsson, M. R. A. Blomberg, D. L. Cooper, M. P. Fülscher, C. de Graaf, B. A. Hess, G. Karlström, R. Lindh, P.-Å. Malmqvist, T. Nakajima, P. Neogrády, J. Olsen, B. O. Roos, B. Schimmelpfennig, M. Schütz, L. Seijo, L. Serrano-Andrés, P. E. M. Siegbahn, J. Stålring, T. Thorsteinsson, V. Veryazov, and P.-O. Widmark, MOLCAS5.4, Lund University, Sweden (2002)
75. H. Yang, G. Luo, P. Karnchanaphanurach, T.-M. Louie, I. Rech, S. Cova, L. Xun, and X. S. Xie, Protein conformational dynamics probed by single-molecule electron transfer. *Science* **302,** 262–266 (2003)
76. J. T. Hynes, Outer-sphere electron-transfer reactions and frequency-dependent friction. *J. Phys. Chem.* **90**(16), 3701–3706 (1986)
77. M. M. Toutounji and M. A. Ratner, Testing the condon approximation for electron transfer via the Mulliken-Hush model. *J. Phys. Chem. A* **104**(37), 8566–8569 (2000)
78. L. Y. Zhang, R. A. Friesner, and R. B. Murphy, Ab initio quantum chemical calculation of electron transfer matrix elements for large molecules. *J. Chem. Phys.* **107**(2), 450–459 (1997)
79. M. D. Newton, Quantum chemical probes of electron-transfer kinetics – the nature of donor-acceptor interactions. *Chemical Reviews* **91**(5), 767–792 (1991)
80. G. Ashkenazi, R. Kosloff, and M. A. Ratner, Photoexcited electron transfer: Short-time dynamics and turnover control by dephasing, relaxation, and mixing. *J. Am. Chem. Soc.* **121**(14), 3386–3395 (1999)
81. R. Kosloff and M. A. Ratner, Rate constant turnovers: Energy spacings and mixings. *J. Phys. Chem. B* **106**(33), 8479–8483 (2002)
82. Qiu, W, T. Li, L. Zhang, Y. Yang, Y.-T. Kao, L. Wang, and D. Zhong, Ultrafast quenching of tryptophan fluorescence in proteins:Interresidue and intrahelical electron transfer. *Chem. Phys.* **350,** 154–164 (2008)

83. M. Sobczyk, W. Anusiewicz, J. Berdys-Kochanska, A. Sawicka, P. Skurski, and J. Simons, Coulomb-assisted dissociative electron attachment: Application to a model peptide. *J. Phys. Chem. A* **109**(1), 250–258 (2005)
84. J. Peon, S. K. Pal, and A. H. Zewail, Hydration at the surface of the protein Monellin: Dynamics with femtosecond resolution. *Proc. Natl. Acad. Sci. U.S.A.* **99**(17), 10964–10969 (2002)
85. D. Toptygin, R. S. Savtchenko, N. D. Meadow, and L. Brand, Homogeneous spectrally- and time-resolved fluorescence emission from single-tryptophan mutants of IIA^{Glc}. *J. Chem. Phys.* **105,** 2043–2055. (2001)
86. B. Zagrovic, C. D. Snow, S. Khaliq, M. R. Shirts, and V. S. Pande, Native-like mean structure in the unfolded ensemble of small proteins. *J. Mol. Biol.* **323**(1), 153–164 (2002)
87. J. Kubelka, J. Hofrichter, and W. A. Eaton, The protein folding 'speed limit'. *Curr. Opin. Struct. Biol.* **14**(1), 76–88 (2004)
88. T. K. Chiu, J. Kubelka, R. Herbst-Irmer, W. A. Eaton, J. Hofrichter, and D. R. Davies, High-resolution x-ray crystal structures of the villin headpiece subdomain, an ultrafast folding protein. *Proc. Natl. Acad. Sci. U.S.A.* **102**(21), 7517–7522 (2005)
89. K. De-Beuckeleer, G. Volckaert, and Y. Engelborghs, Time resolved fluorescence and phosphorescence properties of the individual tryptophan residues of barnase: Evidence for protein-protein interactions. *Proteins-.July* **36,** 42–53 (1999)
90. K. Willaert and Y. Engelborghs, The quenching of tryptophan fluorescence by protonated and unprotonated imidazole. *Eur. Biophys. J.* **20,** 177–182 (1991)
91. M. Van Gilst and B. S. Hudson, Histidine-tryptophan interactions in T4 lysozyme: 'Anomalous' pH dependence of fluorescence. *Biophys. Chem.* **63,** 17–25 (1996)
92. R. Tusell and P. R. Callis, Ab initio prediction and mechanism of the tryptophan fluorescence quantum yield for the Villin Headpiece Mutant N27H. *Biophys. J.* 329A (2008)
93. B. Honig, U. Dinur, K. Nakanishi, V. Balogh-Nair, M. A. Gawinowicz, M. Arnaboldi, and M. G. Motto, An external point-charge model for wavelength regulation in visual pigments. *J. Am. Chem. Soc.* **101** 7084–7086 (1979)
94. B. E. Kohler and J. C. Woehl, Measuring internal fields with atomic resolution. *J. Chem. Phys.* **102,** 7773–7781 (1995)
95. R. Varadarajan, D. G. Lambright, and S. G. Boxer, Electrostatic interactions in wild-type mutant recominant human myoglobins. *Biochemistry* **28** 3771–3781 (1989)
96. D. Sitkoff, D. J. Lockhart, K. A. Sharp, and B. Honig, Calculation of electrostatic effects at the amino terminus of an alpha helix. *Biophys. J.* **67** 2251–2260 (1994)
97. D. J. Lockhart and P. S. Kim, Electrostatic screening of charge and dipole interactions with the helix backbone. *Science* **260,** 198–202 (1993)
98. D. J. Lockhart and P. S. Kim, Internal stark effect measurement of the electric field at the amino terminus of an alpha helix. *Science* **257,** 947–951 (1992)
99. P. R. Callis and B. K. Burgess, Tryptophan fluorescence shifts in proteins from hybrid simulations: An electrostatic approach. *J. Phys. Chem. B* **101**(46), 9429–9432 (1997)
100. J. Lakowicz, On spectral relaxation in proteins. *Photochem. Photobiol.* **72,** 421–437 (2000)
101. Z. Bajzer and F. G. Prendergast, A model for multiexponential tryptophan fluorescence intensity decay in proteins. *Biophys. J.* **65,** 2313–2323 (1993)
102. J. Wlodarczyk and B. Kierdaszuk, Interpretation of fluorescence decays using a power-like model. *Biophys. J.* **85**(1), 589–598 (2003)
103. B. S. Hudson, J. M. Huston, and G. Soto-Campos, A reversible "dark state" mechanism for complexity of the fluorescence of tryptophan in proteins. *J. Phys. Chem. A* **103,** 2227–2234 (1999)
104. A. Ababou and E. Bombarda, On the involvement of electron transfer reactions in the fluorescence decay kinetics heterogeneity of proteins. *Protein Sci.* **10,** 2102–2113 (2001)
105. J. J. Prompers, C. W. Hilbers, and H. A. M. Pepermans, Tryptophan mediated photoreduction of disulfide bond causes unusual fluorescence behaviour of Fusarium solani pisi cutinase. *FEBS Letters* **456** 409–416 (1999)

106. P. C. M. Weisenborn, H. Meder, M. R. Egmond, T. J. W. G. Visser, and A. van Hoek, Photophysics of the single tryptophan residue in *Fusarium solani cutinase*: Evidence for the occurrence of conformational substates with unusual fluorescence behaviour. *Biophys. Chem.* **58,** 281–288 (1996)
107. J. Broos, F. Maddalena, and B. H. Hesp, In vivo synthesized proteins with monoexponential fluorescence decay kinetics. *J. Am. Chem. Soc.* **126,** 22–23 (2004)
108. G. R. Winkler, S. B. Harkins, J. C. Lee, and H. B. Gray, alpha-synuclein structures probed by 5-fluorotryptophan fluorescence and F-19 NMR spectroscopy. *J. Phys. Chem. B* **110**(13), 7058–7061 (2006)
109. D. Toptygin, A. M. Gronenborn, and L. Brand, Nanosecond relaxation dynamics of protein GB1 identified by the time-dependent red shift in the fluorescence of tryptophan and 5-fluorotryptophan. *J. Phys. Chem. B* **110**(51), 26292–26302 (2006)
110. A. V. Smirnov, D. S. English, R. L. Rich, J. Lane, L. Teyton, A. W. Schwabacher, S. Luo, R. W. Thornburg, and J. W. Petrich, Photophysics and biological applications of 7-azaindole and its analogs. *J. Phys. Chem. B* **101**(15), 2758–2769 (1997)
111. X. H. Shen and J. R. Knutson, Subpicosecond fluorescence spectra of tryptophan in water. *J. Phys. Chem. B* **105**(26), 6260–6265 (2001)
112. S. K. Pal, J. Peon, and A. H. Zewail, Biological water at the protein surface: Dynamical solvation probed directly with femtosecond resolution. *Proc. Natl. Acad. Sci. U.S.A.* **99**(4), 1763–1768 (2002)
113. P. R. Callis and T. Q. Liu, Short range photoinduced electron transfer in proteins: QM-MM simulations of tryptophan and flavin fluorescence quenching in proteins. *Chem. Phys.* **326**(1), 230–239 (2006)
114. A. J. Ruggiero, D. C. Todd, and G. R. Fleming, Subpicosecond fluorescence anisotropy studies of tryptophan in water. *J. Am. Chem. Soc.* **112**(3), 1003–1014 (1990)
115. J. E. Hansen, S. J. Rosenthal, and G. R. Fleming, Subpicosecond fluorescence depolarization studies of tryptophan and tryptophanyl residues of proteins. *J. Phys. Chem.* **96**(7), 3034–3040 (1992)
116. O. F. A. Larsen, I. H. M. van Stokkum, A. Pandit, R. van Grondelle, and H. van Amerongen, Ultrafast polarized fluorescence measurements on tryptophan and a tryptophan-containing peptide. *J. Phys. Chem. B* **107**(13), 3080–3085 (2003)
117. W. Y. Lu, J. Kim, W. H. Qiu, and D. P. Zhong, Femtosecond studies of tryptophan solvation: Correlation function and water dynamics at lipid surfaces. *Chem. Phys. Lett.* **388**(1–3), 120–126 (2004)
118. L. Y. Zhang, L. J. Wang, Y. T. Kao, W. H. Qiu, Y. Yang, O. Okobiah, and D. P. Zhong, Mapping hydration dynamics around a protein surface. *Proc. Natl. Acad. Sci. U.S.A.* **104**(47), 18461–18466 (2007)
119. S. K. Pal, J. Peon, B. Bagchi, and A. H. Zewail, Biological water: Femtosecond dynamics of macromolecular hydration. *J. Phys. Chem. B* **106**(48), 12376–12395 (2002)
120. L. Nilsson and B. Halle, Molecular origin of time-dependent fluorescence shifts in proteins. *Proc. Natl. Acad. Sci. U.S.A.* **102**(39), 13867–13872 (2005)
121. X. H. Shen and J. R. Knutson, Femtosecond internal conversion and reorientation of 5-methoxyindole in hexadecane. *Chem. Phys. Lett.* **339**(3–4), 191–196 (2001)
122. M. Vincent, B. de Foresta, and J. Gallay, Nanosecond dynamics of a mimicked membrane-water interface observed by time-resolved stokes shift of LAURDAN. *Biophys. J.* **88**(6), 4337–4350 (2005)
123. A. S. R. Koti and N. Periasamy, Application of time resolved area normalized emission spectroscopy to multicomponent systems. *J. Chem. Phys.* **115**(15), 7094–7099 (2001)
124. A. S. R. Koti, M. M. G. Krishna, and N. Periasamy, Time-resolved area-normalized emission spectroscopy (TRANES): A novel method for confirming emission from two excited states. *J. Phys. Chem. A* **105**(10), 1767–1771 (2001)
125. R. W. Alston, M. Lasagna, G. R. Grimsley, J. M. Scholtz, G. D. Reinhart, and C. N. Pace, Peptide sequence and conformation strongly influence tryptophan fluorescence. *Biophys. J.* **94**(6), 2280–2287 (2008)

126. R. W. Alston, M. Lasagna, G. R. Grimsley, J. M. Scholtz, G. D. Reinhart, and C. N. Pace, Tryptophan fluorescence reveals the presence of long-range interactions in the denatured state of ribonuclease Sa. *Biophys. J.* **94**(6), 2288–2296 (2008)
127. R. Swaminathan, G. Krishnamoorthy, and N. Periasamy, Similarity of fluorescence lifetime distributions for single tryptophan proteins in the random coil state. *Biophys. J.* **67**(5), 2013–2023 (1994)
128. K. J. Willis, A. G. Szabo, J. Drew, M. Zuker, and J. M. Ridgeway, Resolution of heterogeneous fluorescence into component decay-associated excitation-spectra – application to subtilisins. *Biophys. J.* **57**(2), 183–189 (1990)
129. K. J. Willis and A. G. Szabo, Resolution of tyrosyl and tryptophyl fluorescence emission from subtilisins. *Biochemistry* **28**(11), 4902–4908 (1989)
129A. J.A. Bombasaro, A.M. Rodriguez, and R.D. Enviz, Comprehensive conformational analysis of N-acetyl-L-tryptophone-N-methylamide. An abinitio and DFT study. *Theochem* 724, 173–184 (2005)
130. E. H. Strickland, J. Horwitz, and C. Billups, Near-ultraviolet absorption bands of tryptophan. Studies using indole and 3-methylindole as models. *Biochemistry* **9,** 4914–4920 (1970)
131. D. M. Sammeth, S. S. Siewert, L. H. Spangler, and P. R. Callis, 1La transitions of jet-cooled 3-methylindole. *Chem. Phys. Lett.* **193**, 532–538 (1992)
132. L. S. Slater and P. R. Callis, Molecular-orbital theory of the ^{1}La and ^{1}Lb states of indole .2. An ab-initio study. *J. Phys. Chem.* **99**, 8572–8581 (1995)
133. P. R. Callis, J. T. Vivian, and L. S. Slater, Ab initio calculations of vibronic spectra for indole. *Chem. Phys. Lett.* **244,** 53–58 (1995)
134. B. J. Fender, K. W. Short, D. K. Hahn, and P. R. Callis, Vibrational assignments for indole with the aid of ultrasharp phosphorescence spectra. *Int. J. Quantum Chem.* **72**, 347–356 (1999)
135. J. T. Vivian and P. R. Callis, Vibronic band shapes for indole from scaled bond order changes. *Chem. Phys. Lett.* **229**(1–2), 153–160 (1994)
136. D. R. Demmer, G. W. Leach, and S. C. Wallace, 1La-1Lb coupling in the excited state of 3-methylindole and its polar clusters. *J. Phys. Chem.* **98,** 12834–12843 (1994)
137. J. W. Longworth, Excited state interactions in macromolecules. *Photochem. Photobiol.* **7,** 587–596 (1968)
138. T. W. Scott, B. F. Campbell, R. L. Cone, and J. M. Friedman, Line narrowing and site selectivity in tryptophan fluorescence from proteins and glasses: Cryogenic studies of conformational disorder and dynamics. *Chem. Phys.* **131,** 63–79 (1989)
139. D. K. Hahn and P. R. Callis, The lowest triplet state of indole: An ab initio study. *J. Phys. Chem.* **101,** 2686–2691 (1997)
140. J. Hennecke, A. Sillen, W. M. Huber, and Y. Engelborghs, Quenching of tryptophan fluorescence by the active-site disulfide bridge in the DsbA protein from Escherichia coli. *Biochemistry* **36,** 6391–6400 (1997)
141. M. Gastmans, G. Volckaert, and Y. Engelborghs, Tryptophan microstate reshuffling upon the binding of cyclosporin A to human cyclophilin A. *Proteins .June* **35,** 464–474 (1999)
142. L. Qiu, S. A. Pabit, A. E. Roitberg, and S. J. Hagen, Smaller and faster: The 20-residue Trp-cage protein folds in 4μs. *J. Am. Chem. Soc.* **124,** 12952–12953 (2002)
143. D. A. Egan, T. M. Logan, H. Liang, E. Matayoshi, S. W. Fesik, and T. F. Holzman, Equilibrium denaturation of recombinant human FK binding protein in urea. *Biochemistry* **32,** 1920–1927 (1993)
144. A. Sillen, J. F. Diaz, and Y. Engelborghs, A step toward the prediction of the fluorescence lifetimes of tryptophan residues in proteins based on structural and spectral data. *Protein-Sci* **9,** 158–169 (2000)

Fluorescent Probes for Two-Photon Excitation Microscopy

Christoph J. Fahrni

Abstract Two-photon excitation microscopy (TPEM) is a noninvasive imaging technique that is increasingly utilized in biological research to visualize fluorescently labeled biomolecules and cellular structures.

Compared to traditional fluorescence microscopy, TPEM offers intrinsic three-dimensional resolution combined with reduced phototoxicity, increased depth penetration, and lower background fluorescence. At present, the majority of fluorophores used in TPEM have been adopted from linear microscopy, and their photophysical properties have not been optimized for nonlinear two-photon excitation. This review offers an overview of the two-photon absorption properties of fluorophores commonly applied in biological research, and discusses design principles and recent advances in optimizing the nonlinear optical properties of fluorescent labels and probes for TPEM.

Keywords Two photon excitation microscopy · Nonlinear absorption · Fluorescent probes · TPA cross section · Metal cation sensors · Cellular imaging

Introduction

Two-photon excitation microscopy (TPEM) is a three-dimensional imaging technique that has found widespread use in biological and biomedical research [1,2]. As in conventional fluorescence microscopy, fluorescent labels and probes are used to visualize cellular structures, proteins, lipids, nucleic acids, metabolites, and other biomolecules. By using an ultrafast pulsed laser as excitation source, two photons are absorbed simultaneously by the fluorophore, which undergoes a transition to an electronically excited state followed by emission of a single photon leading back to the ground state. Because the two-photon absorption process scales nonlinearly with the squared intensity of the laser beam, excitation occurs only within

C.J. Fahrni (✉)
School of Chemistry and Biochemistry and Petit Institute for Bioengineering and Bioscience, Georgia Institute of Technology, 901 Atlantic Drive, Atlanta, Georgia 30332, USA

C.D. Geddes (ed.), *Reviews in Fluorescence 2007*, Reviews in Fluorescence 2007,
DOI 10.1007/978-0-387-88722-7_11,

a spatially confined volume at the focal point of the microscope objective. By acquiring images at equally spaced positions across the vertical axis of the specimen, a three-dimensional volume rendering can be mathematically generated. Compared with one-photon excitation, the two-photon process requires only half the energy or twice the wavelength to excite the fluorophore, thus offering improved depth penetration in scattering media and reduced phototoxicity. These properties are particularly beneficial for imaging live cells and tissues in biomedical research [3–7].

In principle, any fluorescent dye can be used in TPEM; however, most fluorophores employed in conventional fluorescence microscopy offer only a modest brightness. The latter is defined by the product of quantum yield (η) and absorption cross section (δ), also referred to as action cross section ($\eta\delta$). While the quantum yield is typically assumed to be independent of the excitation mode, the cross section follows different quantum mechanical rules for the nonlinear two-photon compared with a one-photon absorption process. Optimizing the two-photon absorption cross section is therefore critical to improving the optical sensitivity of organic fluorophores for TPEM applications. With an increased brightness, the fluorophore can be detected at a lower concentration and with decreased laser intensity, which will reduce not only photobleaching and phototoxicity, but also improve the imaging contrast by reducing the cellular background fluorescence. The development of new fluorophores with improved brightness is particularly vital for biological sensing of metal cations. A fluorescent probe with low optical sensitivity must be administered at a high concentration, and therefore, requires also a larger amount of analyte to yield the same differential fluorescence response between bound and unbound probe. Increased binding of endogenous metal ions, however, might significantly alter the cellular physiology and interfere with the actual process that is being studied. In addition, the fluorescent probe itself might exert some toxicity or biological activity that is further enhanced at higher concentrations. Ideally, the intracellular concentration of a fluorescent probe should remain below the dissociation constant (K_d) of the metal ion it is designed to detect. Under these conditions, the sensor response is truly reflecting dynamic changes in metal concentrations without acting as a sequestration agent by mobilizing intracellular pools of more tightly bound metal. While calcium and magnesium are present at micromolar concentrations in the cytosol, the concentration of chelatable or "free" transition metal ions has been estimated to be in the subnanomolar range [8]. The detection of metal cations in this concentration regime within the complex environment of a live cell poses significant challenges with respect to the brightness of a fluorescent probe, both in TPEM and conventional fluorescence microscopy.

Given the advantages of TPEM over conventional fluorescence microscopy, the development of new fluorophores with optimized TPA cross sections is of great importance and is pursued by many research groups. This review is intended to provide an overview of the nonlinear optical properties of fluorophores commonly used in biological imaging as well as more recent advances in designing labels and probes with improved TPA cross sections. Special emphasis is given to fluorescent probes for in vivo imaging of metal cations in live cells.

Two-Photon Absorption Cross Sections of Biomolecular Probes and Fluorophores

Literature values of two-photon cross sections for a given fluorophore may vary significantly depending on the chosen experimental conditions. A direct comparison of cross sections reported by different research groups may therefore be not meaningful, unless the data were acquired by the same technique and under identical conditions. Differences may be particularly large when comparing cross sections acquired with different excitation pulse widths. Due to excited state absorption processes, cross sections measured in the nano-second regime appear often two- to three orders of magnitude larger compared to data acquired with femtosecond excitation pulses [9].

Routine measurements of two-photon absorption cross sections can be performed by comparing the two-photon induced fluorescence intensity at a given wavelength with a reference compound, such as fluorescein [10], whose cross section has been determined accurately. The cross section δ_s of the unknown sample is then calculated according to equation (1)

$$\delta_s = \frac{S_s \eta_r \Phi_r C_r}{S_r \eta_s \Phi_s C_s} \delta_r, \tag{1}$$

where the subscripts s and r refer to the sample and reference compound, respectively, η is the emission quantum yield, C is the concentration of the molecules in solution, S is the detected signal intensity, and Φ is the overall fluorescence collection efficiency of the experimental setup (which depends on the geometry of the setup, filters, the detector response, and the refractive index of the solution) [11]. Giving tribute to Maria Göppert-Mayer, who based on theory predicted molecular two-photon absorption already in 1931 [12], cross sections δ are typically reported in units of Göppert-Mayer (1 GM $= 10^{-50}$ cm^4 s/photon). Cross sections calculated by this method are effectively fluorescence excitation cross sections and thus depend on the accuracy of the fluorescence quantum yield determined for the unknown sample. Photon fluxes must be chosen such that photodegradation and optical saturation are absent, which would reduce otherwise the measured emission intensity and apparent cross section. Furthermore, the presence of even slight one-photon absorption can lead to grossly exaggerated cross sections. This is particularly a concern for two-photon transitions into higher lying excited states.

Table 1 gives an overview of two-photon cross sections of fluorescent probes and fluorophores commonly used in biological research. Although design strategies for molecules with large δ are well established [11,13], the majority of these fluorophores have been developed for one-photon excitation and exhibit modest action cross sections ranging between 1 and 50 GM. With a cross section of 180 GM, rhodamine B ranks among the brightest of the traditional fluorophores used in TPEM. The widely used calcium indicator fura-2 was originally designed for dual wavelength excitation ratiometric imaging (see also "Metal Cation Responsive Fluorescent Probes") [14], and does not undergo significant changes in the

Table 1 Two-photon cross sections of selected fluorophores commonly used in biological research

Fluorophore	Two-photon cross section (δ) [GM]	Action cross section (ηδ) [GM]	Excitation wavelength [nm]	Literature references
Fluorescein, pH 13	37		780	[32]
BODIPY		17	920	
Nile Blue A	0.6		800	[33]
Lucifer Yellow	2.6		850	[34]
Rhodamine 6G	55		750	[34]
Rhodamine B	180		850	[34]
Coumarin 485	35		750	[34]
Metal Ion Probes:				
Fura-2		11	700	[35]
with Ca(II)		12	700	[35]
Indo-1		3.5	700	[35]
with Ca(II)	2.1	1.5	700	[35]
Indo-1	12	4.5	700	[10]
Calcium Green		2	820	[35]
with Ca(II)		30	820	[35]
Mag-fura2 with Mg(II)	56	17	780	[36]
Endogenous Fluorophores:				
NADH	0.02		690–730	[1]
Flavins	0.1–0.8		700–730	[1]
EGFP	41		920	[37]
DsRed	11		960	[37]

emission spectrum. Nevertheless, after careful calibration the dye has been successfully applied to two-photon excitation imaging of intracellular calcium within the mammalian cerebral cortex [15], an epithelial cell line [16], and ventricular cardiomyocytes [17]. Unlike fura-2, indo-1 exhibits strong calcium-dependent changes of the fluorescence emission spectrum [18], rendering the dye well suited for ratiometric TPEM with a single excitation source. Using this technique, voltage induced intracellular calcium waves were visualized in fish keratocytes [19] and calcium transients were followed in the course of the mouse sperm–egg fusion [20].

Mostly based on established fluorophore platforms, numerous cation sensors for the detection of other biologically important metal ions have been developed [21]. While their photophysical properties were not specifically optimized for two-photon excitation, some of the probes have been also successfully employed in TPEM. For example, intracellular zinc stores were visualized in cell culture and in mossy fiber synapses of live hippocampal slices by TPEM with fluorescein-based probes [22,23]. Zinquin, one of the early zinc-selective fluorescent probes, was utilized for imaging zinc-rich intracellular compartments in primary cultured hippocampal neurons [24]. Emission ratiometric imaging of intracellular zinc with TPEM was also accomplished using a benzoxazole derivative [25].

A number of biomolecules such as NAD(P)H or flavoproteins are intrinsically fluorescent and can be directly visualized by TPEM [26,27]. While their cross sections are significantly smaller compared to exogenous fluorescent labels (Table 1), TPEM of biomolecules offers the opportunity to noninvasively study biological processes under physiological conditions. Taking advantage of its fluorescent properties, the in vivo distribution of NAD(P)H has been visualized in human skin cells as well as in mitochondria of skeletal muscles [28,29]. Last but not least, green fluorescent protein (GFP) and its variants exhibit substantial two-photon cross sections and provide particularly exciting possibilities to label and track non-fluorescent proteins for live cell imaging studies [30,31].

Design Principles for Fluorophores with Large Two-Photon Absorption Cross Sections

A number of molecular design principles have been identified that yield fluorophores with large two-photon absorption cross sections [11,13,38–40]. From a photophysical point of view, the TPA cross section is directly proportional to the imaginary part of the second hyperpolarizability [41]:

$$\delta(\omega) \propto \mathrm{Im}[\gamma(-\omega; \omega, \omega, -\omega)]. \tag{2}$$

Two of the key parameters that lead to enhanced cross sections are therefore the change of the molecular dipole moment between ground and excited state as well as the magnitude of the transition dipole moment connecting the two states. On a molecular level, these parameters can be maximized by combining three distinct structural elements that are composed of electron donating groups (D), electron accepting groups (A), and conjugated π-bridges connecting the first two components. As illustrated with Fig. 1, depending on their spatial arrangement, simple unsymmetrical dipolar (a) or more complex quadrupolar (b) or octupolar structures can be built. Moreover, the architectures are not limited to two-dimensional arrangements, but can be further expanded to even more complex 3D designs. Despite the structural versatility, all of these architectures are intended to maximize the degree of charge relocalization upon photoexcitation while maintaining a strong transition dipole moment.

Centrosymmetric Quadrupolar Fluorophores

Linear quadrupolar conjugated molecules substituted with electron donating and accepting groups typically exhibit large TPA cross sections [9]. The two-photon properties of these types of molecules can be rationalized with a simple three state model that includes the electronic ground state (S_0), the lowest excited singlet state (S_1), and an energetically higher lying two-photon state (S_2) [11,13]. Denoting the transition energies from the ground to the lowest excited and two-photon state as E_1 and E_2, respectively, the peak cross section δ_{max} at the two-photon resonance energy

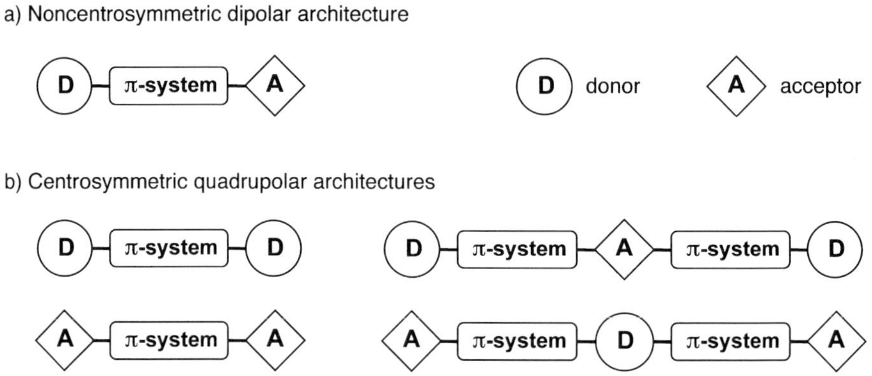

a) Noncentrosymmetric dipolar architecture

b) Centrosymmetric quadrupolar architectures

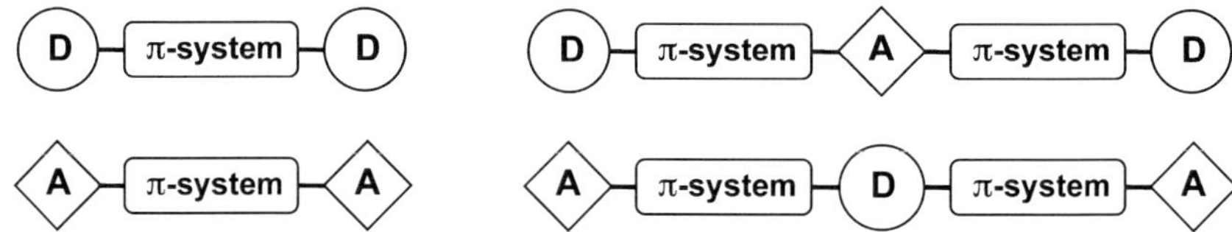

Fig. 1 Molecular architectures for designing fluorophores with large two-photon absorption cross sections

($2\hbar\omega = E_2$) can be expressed as a function of the transition dipole moments M_{01} and M_{12} according to equation (3)

$$\delta_{\max} \propto \frac{M_{01}^2 M_{12}^2}{\left(E_1 - \frac{1}{2}E_2\right)^2 \Gamma}, \tag{3}$$

where Γ corresponds to a damping factor that represents the overall bandwidth (full width at half-maximum) of the $S_0 \rightarrow S_2$ transition. For many quadrupolar organic chromophores, Γ can be approximated with ≈ 0.1 eV [13].

Considering the importance of the transition dipole moments, it can be readily understood why electron donating and withdrawing substituents favorably influence the TPA cross section. As illustrated with the stilbene derivatives **1** and **2** (Chart 1), the cross section increases by nearly a factor of 20 upon substitution with the electron donating dibutylamino-groups. Quantum chemical calculations indicate that photoexcitation of **2** into S_1 and S_2 is accompanied by a substantial charge redistribution from the terminal amino donors towards the central portion of the conjugated π-system, thus resulting in a significantly increased quadrupole moment [13].

Chart 1

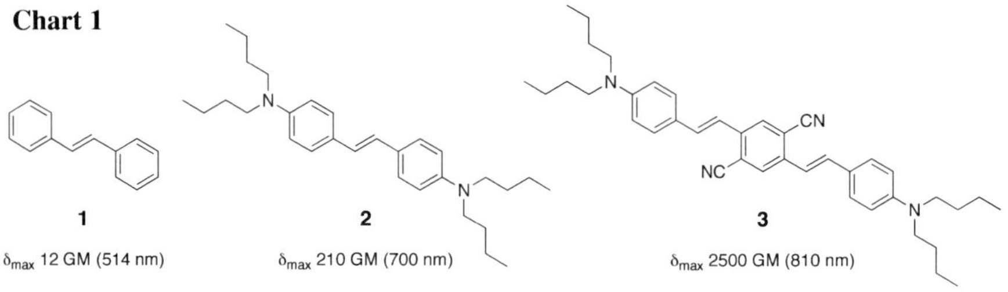

1 δ_{max} 12 GM (514 nm)

2 δ_{max} 210 GM (700 nm)

3 δ_{max} 2500 GM (810 nm)

A comparison of the calculated transition dipole moments reveals that, upon substitution with the amino groups, M_{01} and M_{12} indeed increase from 7.1 to 8.8 D and 3.1 to 7.2 D, respectively [13]. In addition, the donor substituents exert also a favorable influence on the one-photon detuning term $E_1 - \frac{1}{2}E_2$, which decreases from 1.8 to 1.5 eV. Hence, structural features that maximize the transition dipole moments, e.g., through an increased conjugation length or the presence of additional donor or acceptor groups, are expected to yield further improved TPA cross sections. For example, fluorophore **3** combines an extended π-system with additional electron withdrawing cyano-substituents located at the central benzene ring. The resulting TPA cross section is larger by two orders of magnitude compared to the unsubstituted parent compound **1** (Table 2) [42]. Since its initial conception [13,40], this design strategy has been applied to a steadily increasing number of quadrupolar multiphoton absorbing materials [9].

Table 2 Photophysical properties of metal ion sensors based on quadrupolar fluorophores

Fluorophore/probe	Two-photon cross section (δ) [GM]	Action cross section ($\eta\delta$) [GM]	Excitation wavelength [nm]	Solvent	References
3	2500	300	810	acetonitrile	[42]
9	2150	430	810	acetonitrile	[42]
9+Mg(II) (55 mM)	45	8.8	810	acetonitrile	
10	320	150	740	toluene	[52]
	120		640	acetonitrile	
10+Ca(II)	~10		740	acetonitrile	
11	360±15	112	750	DMSO	[53]
11+Zn(II)	68±10	22	750	DMSO	
12		14	780	acetonitrile	[54]
12+Zn(II)		4	780	acetonitrile	
13+K(I)+Pb(II)	998	100	780	acetonitrile	[55]

Noncentrosymmetric Dipolar Fluorophores

Simple donor–acceptor substituted linear π-systems may also exhibit significant two-photon absorption cross sections. Applying a simplified two state model that includes the electronic ground state (S_0) and the lowest excited state (S_1), the peak cross section δ_{max} both is related to both the transition dipole moment M_{01} and the change in permanent dipole moment $\Delta\mu = \mu_1 - \mu_0$ between ground and excited state according to equation (4)

$$\delta_{max} \propto \frac{M_{01}^2(\Delta\mu)^2}{\Gamma}, \tag{4}$$

where Γ corresponds again to a damping factor that represents the overall bandwidth of the corresponding $S_0 \rightarrow S_1$ transition [38,39]. Similar to centrosymmetric fluorophores, the TPA cross section of dipolar architectures can be also enhanced by lengthening the bridging conjugate π-system. Because the TPA cross section depends also on the change in dipole moment upon photoexcitation, the initial ground state polarization plays an important role. With increasing difference between donor and acceptor strength the ground state increasingly resembles the charge-separated zwitter-ionic resonance structure (Fig. 2). If both resonance structures contribute equally to the ground state, the so-called cyanine limit is reached for which the degree of π-bond alteration (BOA) is vanishingly small. Theoretical studies have shown that the TPA cross section peaks at an intermediate value of BOA; however, with increasing ground state polarization, the associated excitation energy is substantially reduced and lies beyond the spectral window of the Ti-sapphire laser [39]. Hence, finding the optimal combination of donor and acceptor strength for a given π-system is therefore pivotal for maximizing the TPA cross section of dipolar fluorophores.

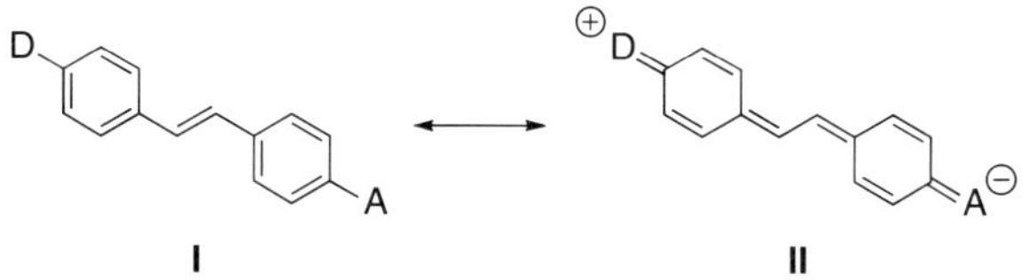

Fig. 2 Resonance structures for donor (D)–acceptor (A) substituted stilbene

Fluorescent Labels with Enhanced TPA Cross Sections for Cellular Imaging

For biological imaging applications, fluorescent probes and labels are immersed in the aqueous environment of a living cell. While very large TPA cross sections have been reported for centrosymmetric fluorophore architectures in non-protic solvents, the values are often drastically reduced in aqueous media [43]. Molecules with large TPA cross sections inherently exhibit highly polarized excited states, which inevitably interact with water molecules [44]. Hydrogen bonding to donor and acceptor substituents as well as undesired aggregation of the fluorophore are presumably responsible for the significantly reduced TPA cross sections in aqueous solution [43]. Solvent interactions with the excited state are also responsible for promoting additional nonradiative deactivation pathways, which in turn result in a decreased quantum yield and thus an overall reduced brightness. Hence, the development of fluorescent labels and probes optimized for TPEM applications is a challenging task. The following section is intended to highlight recent advances in developing such materials with a few selected examples.

Building upon a centrosymmetric molecular architecture, fluorophore **4** was specifically designed for biological imaging applications [45,46]. In aqueous solution, the dye exhibits a TPA cross section of 160 GM when excited at 720 nm.

With a quantum yield of 0.72, the dye exceeds the brightness ($\eta\delta$) of fluorescein by more than three fold. Despite the larger molecular size compared with traditional fluorophores, the dye is readily taken up into live cells within 10 min revealing a perinuclear staining pattern (Fig. 3).

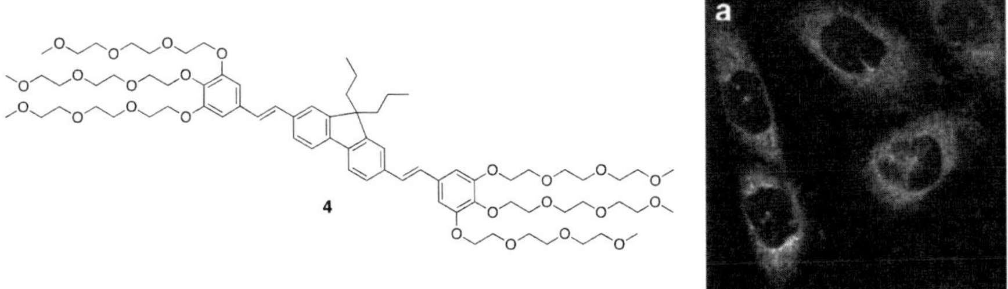

Fig. 3 TPEM micrograph showing the intracellular distribution of fluorophore **4** in live CHO cells after 10 min incubation in complete culture medium containing 5 μM of the dye. Reprinted with permission from reference [46]

The dipolar fluorene-based dye **5** was chemically modified for bioconjugation with proteins (Chart 2) [47]. In DMSO solution, a TPA action cross section of ~19 GM was measured. The dye also produced bright two-photon excited intracellular staining pattern in rat cardiomyoblasts [47]. The Y-shaped fluorophores of type **6** were acquired by Hep-G2 cells and stained efficiently the cytoplasmic region [48]. While no data in aqueous solution were reported, fluorophore **6** showed a large TPA cross section of 523 GM (820 nm) in chloroform. Rylene dyes, including perylene and terrylene derivatives, exhibit remarkable photophysical properties which render them also suitable for single molecule applications [49]. The water-soluble diimide derivative **7** exhibited a cross section of ~50 GM at 840 nm. Upon incubation of CHO cells with 5 μM of **7** for 2 h, a perinuclear intracellular staining pattern was observed. Photobleaching experiments demonstrated that **7** is more stable compared to rhodamine B. Using a [2.2] paracyclophane core to combine two distyrylbenzene fragments (**8**), exceptionally high TPA cross sections of around 700 GM were achieved in aqueous solution [50]. The study also demonstrated that stronger donors do not necessarily translate into increased TPA cross sections and brightness. In fact, weaker electron donors appear, in aqueous solution, to be the better choice, because the reduced charge transfer character of the excited fluorophore results in weaker solvent interactions and thus leads to a higher quantum yield and enhanced overall brightness.

Metal Cation Responsive Fluorescent Probes

Metal-cation selective fluorescent sensors can be divided into two classes based on their response mode. The first group undergoes a change in fluorescence intensity, either an increase or decrease upon binding of the metal cation, whereas the

Chart 2

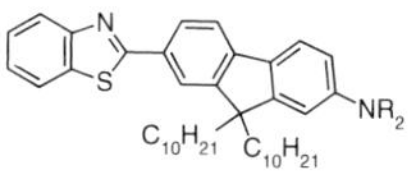

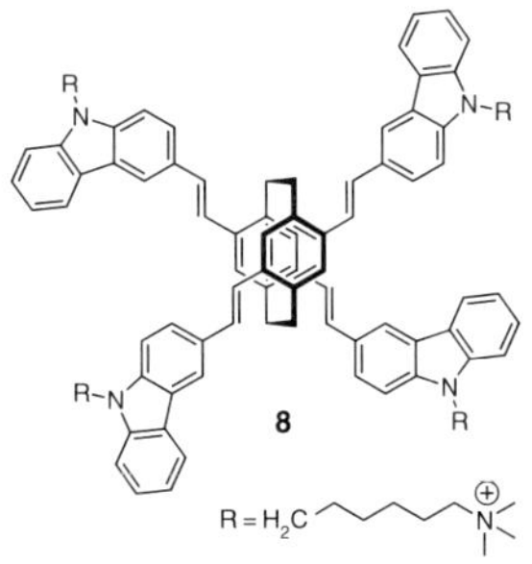

second group yields a fluorescence response in form of a spectral shift [51]. For a 1:1 complex stoichiometry, the change in fluorescence intensity F at a fixed fluorophore concentration can be related to the free metal concentration according to equation (5):

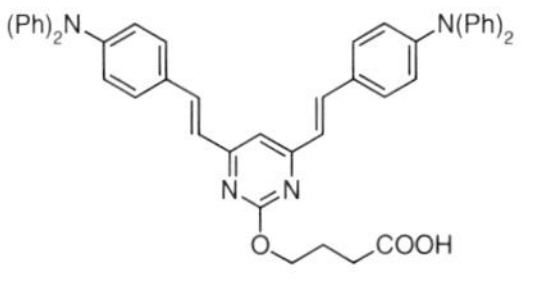

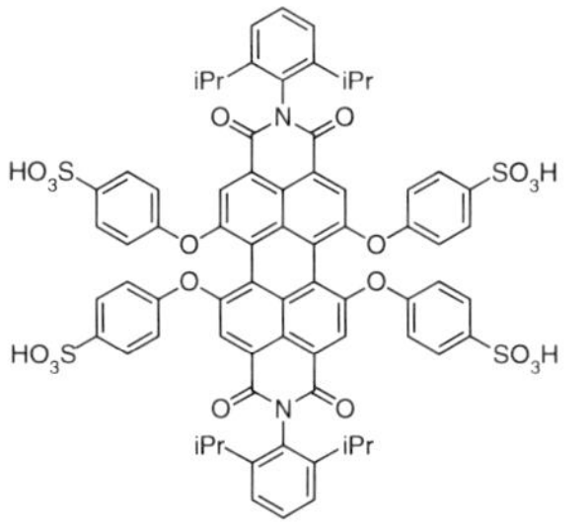

$$[M^{n+}] = K_{\mathrm{d}} \left(\frac{F - F_{\mathrm{min}}}{F_{\mathrm{max}} - F} \right), \tag{5}$$

where K_{d} is the dissociation constant of the metal-probe complex, and F_{min} and F_{max} are the minimum and maximum intensity for the free and saturated probe. In cellular applications, the fluorophore is exposed to a heterogeneous environment, and its concentration will not only vary from cell to cell, but also within different regions of the cytoplasm. Reliable measurements of intracellular metal ion concentrations based on intensity data according to equation (5) are therefore difficult to achieve. Furthermore, if the analyte-free probe exhibits significant background fluorescence, particular care is advised to avoid artifacts and data misinterpretations due to accumulation of the probe in subcellular compartments. In contrast, fluorescent probes that undergo a spectral shift upon analyte binding offer the advantage that the free probe can be chromatically distinguished from the analyte bound form. By measuring the ratio R of fluorescence intensities at two distinct excitation or emission wavelengths, the probe response can be quantitatively related to the metal ion concentration according to equation (6) [14]:

$$[M^{n+}] = K_{\mathrm{d}} \left(\frac{R - R_{\mathrm{min}}}{R_{\mathrm{max}} - R} \right) \left(\frac{S_{\mathrm{f}}}{S_{\mathrm{b}}} \right), \tag{6}$$

where K_{d} is the dissociation constant of the metal-probe complex (with 1:1 stoichiometry), R_{min} and R_{max} are the minimum and maximum intensity ratios for the free and saturated probe, and S_{f} and S_{b} are instrument-dependent calibration constants. Ratiometric concentration measurements have the advantage that intensity differences due to uneven intracellular distribution, photodegradation, or instrument-dependent fluctuations are cancelled out. The instrument parameters S_{f} and S_{b} as well as the limiting intensity ratios R_{min} and R_{max} can be obtained from calibrations with metal-buffered solutions. The dissociation constant of the probe-metal complex is best determined based on intensity data alone according to equation (5). While the ratiometric approach is principally suitable to visualize temporal changes in analyte concentrations, the technique is still not devoid of potential artifacts. The acquired ratio–intensity data are a reflection of the fractional saturation of the fluorescent probe, and therefore depend also on the overall analyte availability. Within this context, the measurement of low analyte concentrations remains a particularly challenging task. For example, although for the average eukaryotic cells the total concentration of transition metal cations such as zinc or iron lies in the high micromolar range, their cytosolic availability is believed to be tightly buffered at subnanomolar concentrations or lower [8]. To reliably detect the cytosolic metal pool, a suitable fluorescent probe should have a dissociation constant in the subnanomolar range otherwise no analyte binding would take place. At the same time, for a reliable signal/noise ratio of the ratiometric emission readout, an intracellular probe concentration of 1 μM or higher should be used. Consequently, a fractional saturation of the dye of 10% or higher would require that most of the analyte be mobilized from intracellular buffer sites. Such an equilibration process might not only impact the cellular physiology, but may also introduce significant uncertainties. Meaningful data may be obtained only if the probe is entirely equilibrated with the endogenous buffer sites. Hence, low analyte concentrations are still best visualized using "turn-on" fluorescent probes with a high contrast ratio, while ratiometric probes would be the preferred choice for the detection of analytes in the micro- to millimolar concentration range.

An additional caveat to consider when applying ratiometric fluorescent sensors in TPEM is that the majority of the probes commonly used in biological research undergo a large shift of the excitation but not emission peak. Given the significant costs of femtosecond-pulsed lasers, only a single excitation source is typically available for TPEM, and therefore such probes are not well suited for dynamic imaging with temporal resolution. Hence, a fluorescent probe can be utilized for ratiometric TPEM only if metal ion binding induces a substantial shift of the peak emission energy.

Fluorescent Cation Sensors with Centrosymmetric Architecture

For the design of metal-ion selective two-photon absorbing fluorophores, the terminal electron donating groups of quadrupolar fluorophores, initially introduced to maximize the TPA cross section, may conveniently serve as Lewis bases for metal-ion coordination. This strategy has been explored for the construction of magnesium [42], calcium [52], and zinc [53,54] responsive sensors. To improve the metal ion selectivity and affinity, the amine donor has been further modified with specific chelating moieties as illustrated with fluorophores **9-12** (Chart 3).

While the uncoordinated fluorophore **9** exhibits a TPA cross section that is comparable with the dibutylamino-substituted analog **3**, titration with Mg(II) ions in acetonitrile solution leads to a substantial decrease (Table 2). As shown in Fig. 4, the two-photon action spectrum of **9** is strongly weakened across the entire

Chart 3

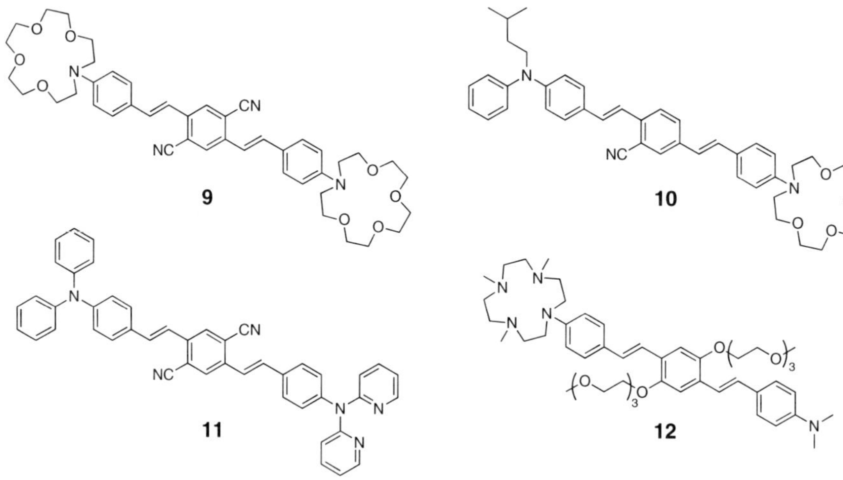

spectral window in the presence of Mg(II) compared to the free fluorophore [42]. Although metal coordination to the terminal amino groups was expected to reduce the degree of charge relocalization in the excited state, thus adversely affecting the TPA cross section, the observed reduction is surprisingly pronounced. Quantum chemical calculations indicated that the reduced TPA cross section is in fact

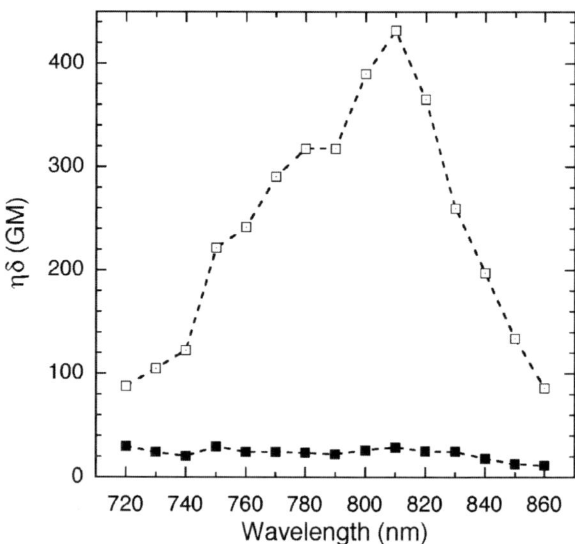

Fig. 4 Effect of metal-ion coordination on the two-photon absorption properties of a quadrupolar D-π-A- π -D fluorophore. Two-photon action spectrum of **9** in acetonitrile measured using coumarin 307 as a reference. The open squares are data for the fluorophore in the absence of magnesium and the filled squares are data obtained with 10 μM fluorophore and 2 mM Mg^{2+}. Reprinted with permission from reference [42]

primarily a consequence of a hypsochromic shift of the TPA maximum from 1.9 to 2.2 eV upon metal coordination, rather than a substantial decrease of the TPA cross section [42]. Contrary to the spectral shift in the two-photon as well as linear absorption spectra, metal coordination had no significant effect on the fluorescence emission maximum. Similar observations have been made for the structurally related fluorescent sensors **10-12** [52–54], all of which exhibit a substantial decrease of the TPA cross section upon metal coordination within the experimental spectral window (Table 2).

While fluorophores with centrosymmetric architectures offer substantially improved TPA cross sections over traditional fluorophore platforms, the design of cation sensors based on this strategy remains challenging. Metal coordination to the terminal donors results in a hypsochromic shift of the TPA maximum, which lies then typically outside the spectrally available window of the Ti-sapphire laser. As a consequence, the brightness of the sensors is strongly reduced upon metal coordination. The strong background fluorescence of the free sensor compared to the weak signal of its cation-bound form renders this type of sensor a less ideal choice for biological imaging applications.

By utilizing a different sensing mechanism, fluorescence enhancement with centrosymmetric fluorophores can be still achieved as recently demonstrated with the calix[4]arene-based probe **13** (Fig. 5) [55]. In absence of metal cations, the fluorescence is quenched presumably due to resonance energy transfer. Upon binding of K^+ and Pb^{2+}, the probe emits with a quantum yield of 0.1 and exhibits an excellent two-photon cross section of 998 GM. Coordination of K^+ to the crown-ether moiety is necessary to achieve the fluorescence enhancement due to an allosteric effect.

Cation Sensors with Dipolar Fluorophore Architecture

The construction of cation sensors based on simple dipolar structures offers some advantages. While quadrupolar fluorophores require two cation binding sites to restore the overall symmetry, a single coordination site is sufficient for the inherently noncentrosymmetric dipolar fluorophores. Furthermore, coordination of the metal cation to the acceptor rather than donor site may increase the charge transfer character, which in turn would yield an increased rather than decreased TPA cross section along with a bathochromic shift in the absorption and emission spectrum. At first glance, coordination of an electropositive metal cation to an electron deficient organic group might appear as a contradiction; however, several Lewis bases commonly used for metal coordination have in fact electron withdrawing properties. For example, pyridine forms a wide range of coordination complexes with transition metal cations while simultaneously exhibiting electron withdrawing properties that are comparable to carboxylic acid esters as judged from their similar Hammett constants σ_p of 0.44 and 0.45, respectively [56].

In the design of fluorescent probe **14,** both of the above features were explicitly incorporated: a dipolar donor–acceptor structure for an enhanced TPA cross section combined with a single cation binding site comprised of an electron withdrawing

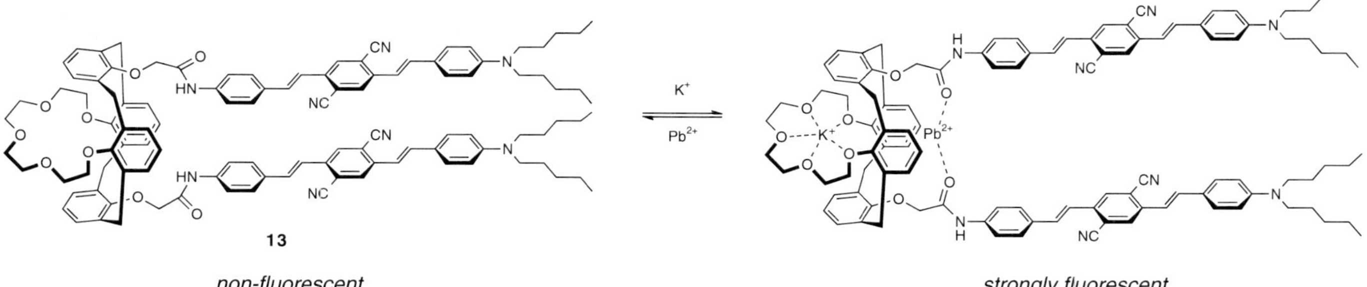

Fig. 5 Fluorescence switching with quadrupolar fluorophores. The free probe **13** is quenched through resonance energy transfer. Coordination of K^+ allosterically renders the conformation of the probe more favorable for Pb^{2+} binding, which inhibits fluorescence quenching through an additional conformational change

Chart 4

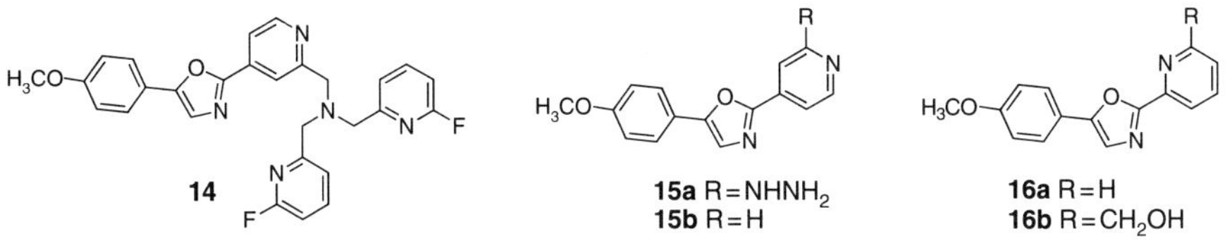

14

15a R = $NHNH_2$
15b R = H

16a R = H
16b R = CH_2OH

pyridine acting as acceptor (Chart 4) [57]. The probe binds Zn(II) with a 1:1 complex stoichiometry and has a dissociation constant of $K_d = 2.4\ \mu M$. Upon saturation with Zn(II), the TPA cross section increased from 31 to 77 GM (Fig. 6) and a substantial shift of the emission maximum from 441 to 497 nm was observed. Contrary to fluorophores with D-π-A-π-D architecture, the increased cross section upon Zn(II) coordination results in a fluorescence enhancement rather than attenuation. While the spectral shift of the fluorescence emission can be utilized for ratiometric measurements of Zn(II) concentrations, the probe can be also employed in a "turn on" mode. At wavelengths around 800 nm, the TPA cross section of **14** is vanishingly small in the absence but strongly enhanced in the presence of Zn(II), thereby resulting in a fluorescence enhancement by a factor of more than 10,000.

The oxazole derivatives **15–16** were designed by Jullien and coworkers as probes for ratiometric pH measurements in aqueous solution (Chart 4) [58,59]. For example, the hydrazine-substituted fluorophore **15a** in its neutral form exhibits a maximum TPA cross section of 15 GM, which then increases to 60 GM upon quantitative protonation of the pyridine nitrogen at pH 2 (Table 3) [58]. Consistent with an increased charge transfer character, the fluorescence emission maximum is also substantially shifted from 465 to 530 nm upon protonation. By modifying the substituent in the 2-position of the pyridine ring, the pK_a can be adjusted over a wide

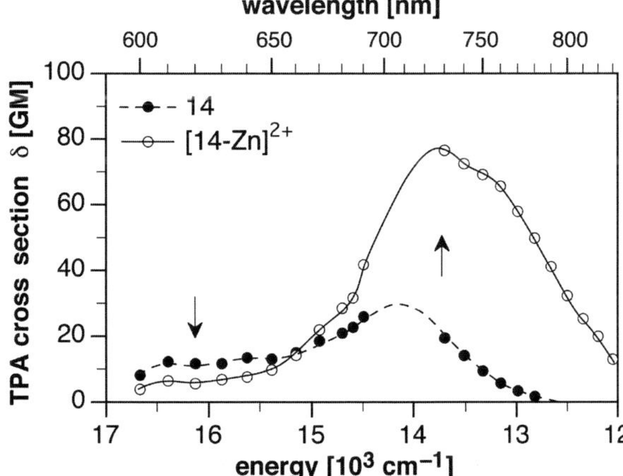

Fig. 6 Two-photon excitation spectra of fluorophore **14** (100 μM solution in methanol) in the absence and presence of $Zn(II)(OTf)_2$. Reprinted with permission from reference [57]

Table 3 Photophysical properties of metal ion probes based on fluorophores with dipolar architecture

Fluorophore/probe	Two-photon cross section (δ) [GM]	Action cross section ($\eta\delta$) [GM]	Excitation wavelength [nm]	Solvent	References
14	31	11	690	MeOH	[57]
14-Zn(II)	77	55	730	MeOH	[57]
15a (pH 2)	60		710	aq. buffer	[58]
15a (pH 9)	15		710	aq. buffer	[58]
15b (pH 2)	65		800	aq. buffer	[59]
15b (pH 12)	5		710	aq. buffer	[59]
16a (pH 2)	40		780	aq. buffer	[59]
16a (pH 12)	20		710	aq. buffer	[59]
16b (pH 2)	35		780	aq. buffer	[59]
16b (pH 12)	20		710	aq. buffer	[59]
17a (pH 3.2)	143	86	780	aq. buffer	[60]
17b (pH 3.2)	128	92	780	aq. buffer	[60]
18	290	84	820	Tris pH 7.05	[61]
18-Mg(II)	382	107	880	Tris pH 7.05	[61]
19-Mg(II)	215	125	780	Tris pH 7.05	[36]
20-Ca(II)		110	780	MOPS pH 7.2	[63]

range while maintaining the increased two-photon sensitivity [59]. For example, derivative **15b** lacking the hydrazine moiety has a pK_a of 4.3, which is 1.1 units lower compared to **15a**. The basicity of the pyridine nitrogen is even further lowered in the ortho-substituted derivatives **16a** and **16b**, which exhibit pK_as of 2.6 and 1.8, respectively.

Building around a donor–acceptor substituted naphthalene core, Cho and coworkers developed a series of cation-responsive two-photon sensors (Chart 5) [60–63]. Fluorophores **17a** and **17b** were designed as pH responsive probes and

Chart 5

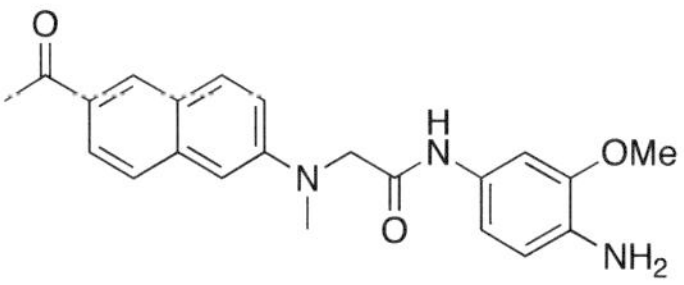

17a 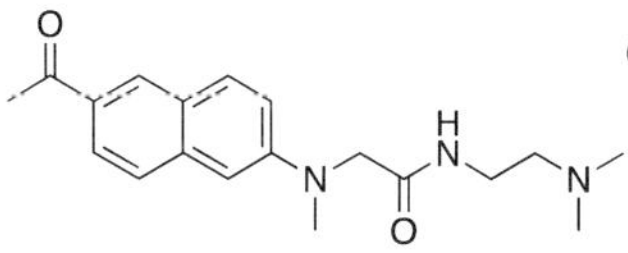 **17b** 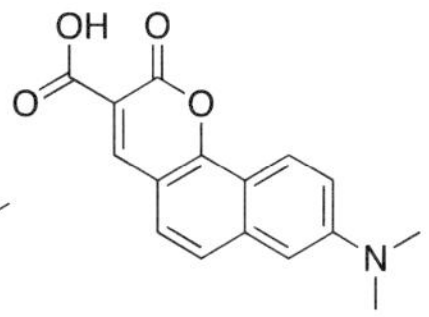 **18**

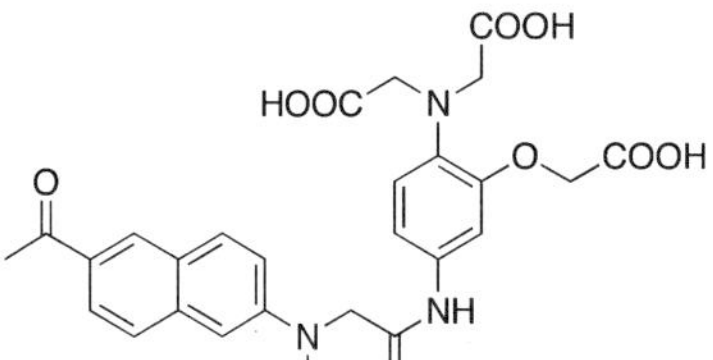

19 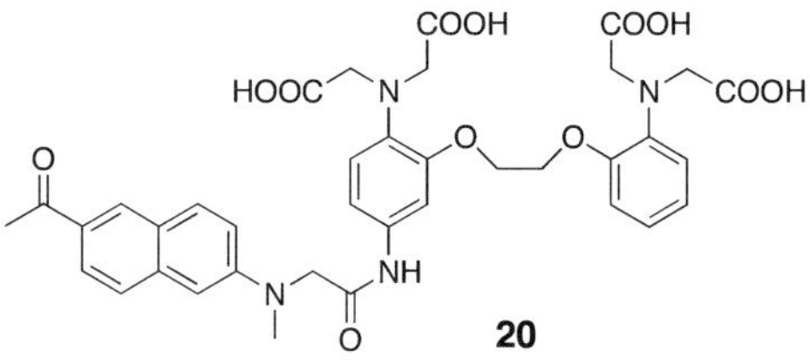 **20**

exhibit substantial two-photon action cross sections of 88 and 92 GM, respectively (Table 3) [60]. The aniline moiety in **17a** quenches the fluorescence emission through a photoinduced electron transfer mechanism, which is inhibited upon protonation at acidic pH, resulting in a 64-fold fluorescence emission increase. The probe was successfully applied to visualize acidic compartments in macrophages using TPEM. Although fluorophore **17b** does not exhibit pH dependent emission changes, the probe is effectively trapped upon protonation in acidic compartments and was successfully used in macrophages and acute rat hippocampal slices [60]. Fluorophore **18** selectively responds to Mg(II) over Ca(II) resulting in a 2.5-fold enhancement of the two-photon action cross section at 880 nm [61]. The probe is readily loaded into live cells and tissues, and with a dissociation constant of $K_d = 2.5$ mM, it is suitable for dynamic imaging of endogenous magnesium stores. Presumably also due to a photoinduced electron transfer switching mechanism, the fluorescence emission of fluorophore **19** is essentially quenched in neutral buffer and dramatically increases upon saturation with Mg(II) [36]. With an action cross section of 125 GM, the probe exhibits a high two-photon sensitivity rendering it useful for the visualization of magnesium stores in live cells and tissues by TPEM. Similar to fluorophore **19**, derivative **20** acts as a "turn-on" sensor with an excellent two-photon action cross section of 100 GM at 780 nm [62,63]. The probe selectively responds to Ca(II) over other divalent metal ions and has been used to visualize calcium waves in astrocytes and acute rat hypothalamic slices.

The often considerable hydrophobicity of fluorescent probes can lead to their accumulation within intracellular membranes. The lower polarity environment results typically in an increase in fluorescence quantum yield, which may produce artifacts, thus limiting their application in cellular studies. This problem has been elegantly addressed by taking advantage of the polarity-dependent emission spectrum of donor–acceptor substituted fluorophores [61,62,64]. The membrane-bound probes **18–20** exhibit a blue-shifted emission maximum compared to their metal-ion complexed form localized in the cytosol. The selective detection of intracellular Ca(II) and Mg(II) is then readily accomplished by collecting the red-shifted fluorescence emission of the cytosolic probe with an appropriate filter set.

Conclusions

The broad utility of two-photon excitation microscopy in biological research underscores the importance for developing new fluorescent labels and probes with enhanced brightness. For the design of cation-responsive fluorescent sensors, centrosymmetric quadrupolar as well as simple dipolar molecular architectures have been explored. These studies have demonstrated that optimizing the fluorophore brightness amounts to a challenging balancing act between several opposing effects. On the one hand, the TPA cross section increases with increasing charge transfer character upon photoexcitation. This could be principally realized with molecular designs that are comprised of larger conjugation paths as well as stronger electron donating and withdrawing substituents. On the other hand, in polar protic solvents the increasingly polarized nature of the excited state of such molecules will trigger additional nonradiative deactivation pathways that substantially reduce the emission quantum yield. Furthermore, with increasing conjugation length and charge transfer character, the orbital overlap between ground and excited state is gradually diminished, thus leading to smaller transition dipole moments and therefore reduced TPA cross sections. Finally, these structural modifications are inherently associated with a bathochromic shift of the TPA maximum, which ultimately will fall beyond the spectrally available window of the Ti-sapphire laser. At present, these factors appear to set physical limits for achieving truly large two-photon action cross sections in polar solvents. Nevertheless, many of the described probes and labels offer already sufficient two-photon sensitivity for imaging endogenous analytes in live cells with excellent signal-to-noise ratio and are expected to find broad utility in biological research.

Acknowledgment Financial support by NIH (DK68096) is gratefully acknowledged.

References

1. Diaspro, A.; Chirico, G.; Collini, M., Two-Photon Fluorescence Excitation and Related Techniques in Biological Microscopy, *Q. Rev. Biophys.*, 38, 97–166, 2005.
2. So, P. T. C.; Dong, C. Y.; Masters, B. R.; Berland, K. M., Two-Photon Excitation Fluorescence Microscopy, *Annu. Rev. Biomed. Engin.*, 2, 399–429, 2000.
3. Zipfel, W. R.; Williams, R. M.; Webb, W. W., Nonlinear Magic: Multiphoton Microscopy in the Biosciences, *Nat. Biotechnol.*, 21, 1369–1377, 2003.
4. Niggli, E.; Egger, M., Applications of Multi-Photon Microscopy in Cell Physiology, *Frontiers Biosci.*, 9, 1598–1610, 2004.
5. Helmchen, F.; Denk, W., Deep Tissue Two-Photon Microscopy, *Nat. Met.*, 2, 932–940, 2005.
6. Svoboda, K.; Yasuda, R., Principles of Two-Photon Excitation Microscopy and Its Applications to Neuroscience, *Neuron*, 50, 823–839, 2006.
7. Kerr, J. N. D.; Denk, W., Imaging in Vivo: Watching the Brain in Action, *Nat. Rev. Neurosci.*, 9, 195–205, 2008.
8. Finney, L. A.; O'Halloran, T. V., Transition Metal Speciation in the Cell: Insights from the Chemistry of Metal Ion Receptors, *Science*, 300, 931–936, 2003.
9. He, G. S.; Tan, L. S.; Zheng, Q.; Prasad, P. N., Multiphoton Absorbing Materials: Molecular Designs, Characterizations, and Applications, *Chem. Rev.*, 108, 1245–1330, 2008.
10. Xu, C.; Webb, W. W., Measurement of Two-Photon Excitation Cross Sections of Molecular Fluorophores with Data from 690 to 1050 nm, *J. Opt. Soc. Am. B*, 13, 481–491, 1996.
11. Rumi, M.; Ehrlich, J. E.; Heikal, A. A.; Perry, J. W.; Barlow, S.; Hu, Z. Y.; McCord-Maughon, D.; Parker, T. C.; Röckel, H.; Thayumanavan, S.; Marder, S. R.; Beljonne, D.; Brédas, J. L., Structure-Property Relationships for Two-Photon Absorbing Chromophores: Bis-Donor Diphenylpolyene and Bis(Styryl)Benzene Derivatives, *J. Am. Chem. Soc.*, 122, 9500–9510, 2000.
12. Göppert-Mayer, M., Über Elementarakte Mit Zwei Quantensprüngen, *Ann. Phys. (Leipzig)*, 5, 273–294, 1931.
13. Albota, M.; Beljonne, D.; Brédas, J. L.; Ehrlich, J. E.; Fu, J. Y.; Heikal, A. A.; Hess, S. E.; Kogej, T.; Levin, M. D.; Marder, S. R.; McCord-Maughon, D.; Perry, J. W.; Röckel, H.; Rumi, M.; Subramaniam, C.; Webb, W. W.; Wu, X. L.; Xu, C., Design of Organic Molecules with Large Two-Photon Absorption Cross Sections, *Science*, 281, 1653–1656, 1998.
14. Grynkiewicz, G.; Poenie, M.; Tsien, R. Y., A New Generation of Ca^{2+}-Indicators with Greatly Improved Fluorescence Properties, *J. Biol. Chem.*, 260, 3440–3450, 1985.

15. Stutzmann, G. E.; LaFerla, F. M.; Parker, I., Ca^{2+}-Signaling in Mouse Cortical Neurons Studied by Two-Photon Imaging and Photoreleased Inositol Triphosphate, *J. Neurosci.*, 23, 758–765, 2003.
16. Ricken, S.; Leipziger, J.; Greger, R.; Nitschke, R., Simultaneous Measurements of Cytosolic and Mitochondrial Ca^{2+} Transients in HT_{29} Cells, *J. Biol. Chem.*, 273, 34961–34969, 1998.
17. Wokosin, D. L.; Loughrey, C. M.; Smith, G. L., Characterization of a Range of Fura Dyes with Two-Photon Excitation, *Biophys. J.*, 86, 1726–1738, 2004.
18. Szmacinski, H.; Gryczynski, I.; Lakowicz, J. R., Calcium-Dependent Fluorescence Lifetimes of Indo-1 for One-Photon and 2-Photon Excitation of Fluorescence, *Photochem. Photobiol.*, 58, 341–345, 1993.
19. Brust-Mascher, I.; Webb, W. W., Calcium Waves Induced by Large Voltage Pulses in Fish Keratocytes, *Biophys. J.*, 75, 1669–1678, 1998.
20. Jones, K. T.; Soeller, C.; Cannell, M. B., The Passage of Ca^{2+} and Fluorescent Markers Between the Sperm and Egg After Fusion in the Mouse, *Development*, 125, 4627–4635, 1998.
21. Domaille, D. W.; Que, E. L.; Chang, C. J., Synthetic Fluorescent Sensors for Studying the Cell Biology of Metals, *Nat. Chem. Biol.*, 4, 168–175, 2008.
22. Nolan, E. M.; Jaworski, J.; Okamoto, K. I.; Hayashi, Y.; Sheng, M.; Lippard, S. J., QZ1 and QZ2: Rapid, Reversible Quinoline-Derivatized Fluoresceins for Sensing Biological Zn(II), *J. Am. Chem. Soc.*, 127, 16812–16823, 2005.
23. Chang, C. J.; Nolan, E. M.; Jaworski, J.; Okamoto, K. I.; Hayashi, Y.; Sheng, M.; Lippard, S. J., ZP8, a Neuronal Zinc Sensor with Improved Dynamic Range; Imaging Zinc in Hippocampal Slices with Two-Photon Microscopy, *Inorg. Chem.*, 43, 6774–6779, 2004.
24. Love, R.; Salazar, G.; Faundez, V., Neuronal Zinc Stores are Modulated by Non-Steroidal Anti-Inflammatory Drugs: An Optical Analysis in Cultured Hippocampal Neurons, *Brain Res.*, 1061, 1–12, 2005.
25. Taki, M.; Wolford, J. L.; O'Halloran, T. V., Emission Ratiometric Imaging of Intracellular Zinc: Design of a Benzoxazole Fluorescent Sensor and Its Application in Two-Photon Microscopy, *J. Am. Chem. Soc.*, 126, 712–713, 2004.
26. Kierdaszuk, B.; Malak, H.; Gryczynski, I.; Callis, P.; Lakowicz, J. R., Fluorescence of Reduced Nicotinamides Using One- and Two-Photon Excitation, *Biophys. Chem.*, 62, 1–13, 1996.
27. Huang, S. H.; Heikal, A. A.; Webb, W. W., Two-Photon Fluorescence Spectroscopy and Microscopy of NAD(P)H and Flavoprotein, *Biophys. J.*, 82, 2811–2825, 2002.
28. Masters, B. R.; So, P. T. C.; Gratton, E., Multiphoton Excitation Fluorescence Microscopy and Spectroscopy of In Vivo Human Skin, *Biophys. J.*, 72, 2405–2412, 1997.
29. Rothstein, E. C.; Carroll, S.; Combs, C. A.; Jobsis, P. D.; Balaban, R. S., Skeletal Muscle NAD(P)H Two-Photon Fluorescence Microscopy In Vivo: Topology and Optical Inner Filters, *Biophys. J.*, 88, 2165–2176, 2005.
30. Giepmans, B. N. G.; Adams, S. R.; Ellisman, M. H.; Tsien, R. Y., Review – the Fluorescent Toolbox for Assessing Protein Location and Function, *Science*, 312, 217–224, 2006.
31. Shaner, N. C.; Steinbach, P. A.; Tsien, R. Y., A Guide to Choosing Fluorescent Proteins, *Nat. Met.*, 2, 905–909, 2005.
32. Albota, M. A.; Xu, C.; Webb, W. W., Two-Photon Fluorescence Excitation Cross Sections of Biomolecular Probes from 690 to 960 nm, *Appl. Opt.*, 37, 7352–7356, 1998.
33. Fisher, J. A. N.; Salzberg, B. M.; Yodh, A. G., Near Infrared Two-Photon Excitation Cross-Sections of Voltage-Sensitive Dyes, *J. Neurosci. Met.*, 148, 94–102, 2005.
34. Makarov, N. S.; Drobizhev, M.; Rebane, A., Two-Photon Absorption Standards in the 550-1600 nm Excitation Wavelength Range, *Opt. Expr.*, 16, 4029–4047, 2008.
35. Xu, C.; Zipfel, W.; Shear, J. B.; Williams, R. M.; Webb, W. W., Multiphoton Fluorescence Excitation: New Spectral Windows for Biological Nonlinear Microscopy, *Proc. Natl. Acad. Sci. USA*, 93, 10763–10768, 1996.
36. Kim, H. M.; Jung, C.; Kim, B. R.; Jung, S. Y.; Hong, J. H.; Ko, Y. G.; Lee, K. J.; Cho, B. R., Environment-Sensitive Two-Photon Probe for Intracellular Free Magnesium Ions in Live Tissue, *Angew. Chem. Int. Ed.*, 46, 3460–3463, 2007.

37. Blab, G. A.; Lommerse, P. H. M.; Cognet, L.; Harms, G. S.; Schmidt, T., Two-Photon Excitation Action Cross-Sections of the Autofluorescent Proteins, *Chem. Phys. Lett.*, 350, 71–77, 2001.
38. Barzoukas, M.; Blanchard-Desce, M., Molecular Engineering of Push-Pull Dipolar and Quadrupolar Molecules for Two-Photon Absorption: A Multivalence-Bond States Approach, *J. Chem. Phys.*, 113, 3951–3959, 2000.
39. Kogej, T.; Beljonne, D.; Meyers, F.; Perry, J. W.; Marder, S. R.; Brédas, J. L., Mechanisms for Enhancement of Two-Photon Absorption in Donor-Acceptor Conjugated Chromophores, *Chem. Phys. Lett.*, 298, 1–6, 1998.
40. Reinhardt, B. A.; Brott, L. L.; Clarson, S. J.; Dillard, A. G.; Bhatt, J. C.; Kannan, R.; Yuan, L. X.; He, G. S.; Prasad, P. N., Highly Active Two-Photon Dyes: Design, Synthesis, and Characterization toward Application, *Chem. Mater.*, 10, 1863–1874, 1998.
41. Dick, B.; Hochstrasser, R. M.; Trommsdorff, H. P., Resonant Molecular Optics, In *Nonlinear Optical Properties of Organic Molecules and Crystals*; Chemia, D. S., Zyss J., Eds.; Academic Press: Orlando, 1987; Vol. 2, p 159–212.
42. Pond, S. J. K.; Tsutsumi, O.; Rumi, M.; Kwon, O.; Zojer, E.; Brédas, J. L.; Marder, S. R.; Perry, J. W., Metal-Ion Sensing Fluorophores with Large Two-Photon Absorption Cross Sections: Aza-Crown Ether Substituted Donor-Acceptor-Donor Distyryl Benzenes, *J. Am. Chem. Soc.*, 126, 9291–9306, 2004.
43. Woo, H. Y.; Liu, B.; Kohler, B.; Korystov, D.; Mikhailovsky, A.; Bazan, G. C., Solvent Effects on the Two-Photon Absorption of Distyrylbenzene Chromophores, *J. Am. Chem. Soc.*, 127, 14721–14729, 2005.
44. Halpern, A. M.; Ruggles, C. J.; Zhang, X. K., Complex-Formation Between Excited-State Saturated Amines and Water in Normal-Hexane Solution, *J. Am. Chem. Soc.*, 109, 3748–3754, 1987.
45. Hayek, A.; Bolze, F.; Nicoud, J. F.; Baldeck, P. L.; Mely, Y., Synthesis and Characterization of Water-Soluble Two-Photon Excited Blue Fluorescent Chromophores for Bioimaging, *Photochem. Photobiol. Sci.*, 5, 102–106, 2006.
46. Hayek, A.; Grichine, A.; Huault, T.; Ricard, C.; Bolze, F.; Van Der Sanden, B.; Vial, J. C.; Mély, Y.; Duperray, A.; Baldeck, P. L.; Nicoud, J. F., Cell-Permeant Cytoplasmic Blue Fluorophores Optimized for In Vivo Two-Photon Microscopy with Low-Power Excitation, *Microsc. Res. Techn.*, 70, 880–885, 2007.
47. Schafer-Hales, K. J.; Belfield, K. D.; Yao, S.; Frederiksen, P. K.; Hales, J. M.; Kolattukudy, P. E., Fluorene-Based Fluorescent Probes with High Two-Photon Action Cross-Sections for Biological Multiphoton Imaging Applications, *J. Biomed. Opt.*, 10, 2005.
48. Liu, B.; Zhang, H. L.; Liu, J.; Zhao, Y. D.; Luo, Q. M.; Huang, Z. L., Novel Pyrimidine-Based Amphiphilic Molecules: Synthesis, Spectroscopic Properties and Applications in Two-Photon Fluorescence Microscopic Imaging, *J. Mater. Chem.*, 17, 2921–2929, 2007.
49. Margineanu, A.; Hofkens, J.; Cotlet, M.; Habuchi, S.; Stefan, A.; Qu, J. Q.; Kohl, C.; Mullen, K.; Vercammen, J.; Engelborghs, Y.; Gensch, T.; De Schryver, F. C., Photophysics of a Water-Soluble Rylene Dye: Comparison with Other Fluorescent Molecules for Biological Applications, *J. Phys. Chem. B*, 108, 12242–12251, 2004.
50. Woo, H. Y.; Hong, J. W.; Liu, B.; Mikhailovsky, A.; Korystov, D.; Bazan, G. C., Water-Soluble [2.2]Paracyclophane Chromophores with Large Two-Photon Action Cross Sections, *J. Am. Chem. Soc.*, 127, 820–821, 2005.
51. de Silva, A. P.; Gunaratne, H. Q. N.; Gunnlaugsson, T.; Huxley, A. J. M.; McCoy, C. P.; Rademacher, J. T.; Rice, T. E., Signaling Recognition Events with Fluorescent Sensors and Switches, *Chem. Rev.*, 97, 1515–1566, 1997.
52. Kim, H. M.; Jeong, M. Y.; Ahn, H. C.; Jeon, S. J.; Cho, B. R., Two-Photon Sensor for Metal Ions Derived from Azacrown Ether, *J. Org. Chem.*, 69, 5749–5751, 2004.
53. Bozio, R.; Cecchetto, E.; Fabbrini, G.; Ferrante, C.; Maggini, M.; Menna, E.; Pedron, D.; Ricco, R.; Signorini, R.; Zerbetto, M., One- and Two-Photon Absorption and Emission Properties of a Zn(II) Chemosensor, *J. Phys. Chem. A*, 110, 6459–6464, 2006.

54. Ahn, H. C.; Yang, S. K.; Kim, H. M.; Li, S. J.; Jeon, S. J.; Cho, B. R., Molecular Two-Photon Sensor for Metal Ions Derived from Bis(2-Pyridyl)Amine, *Chem. Phys. Lett.*, 410, 312–315, 2005.
55. Kim, J. S.; Kim, H. J.; Kim, H. M.; Kim, S. H.; Lee, J. W.; Kim, S. K.; Cho, B. R., Metal Ion Sensing Novel Calix[4]Crown Fluoroionophore with a Two-Photon Absorption Property, *J. Org. Chem.*, 71, 8016–8022, 2006.
56. Hansch, C.; Leo, A.; Taft, R. W., A Survey of Hammett Substituent Constants and Resonance and Field Parameters, *Chem. Rev.*, 91, 165–195, 1991.
57. Sumalekshmy, S.; Henary, M. M.; Siegel, N.; Lawson, P. V.; Wu, Y.; Schmidt, K.; Brédas, J. L.; Perry, J. W.; Fahrni, C. J., Design of Emission Ratiometric Metal-Ion Sensors with Enhanced Two-Photon Cross Section and Brightness, *J. Am. Chem. Soc.*, 129, 11888–11889, 2007.
58. Charier, S.; Ruel, O.; Baudin, J. B.; Alcor, D.; Allemand, J. F.; Meglio, A.; Jullien, L., An Efficient Fluorescent Probe for Ratiometric pH Measurements in Aqueous Solutions, *Angew. Chem. Int. Ed.*, 43, 4785–4788, 2004.
59. Charier, S.; Ruel, O.; Baudin, J. B.; Alcor, D.; Allemand, J. F.; Meglio, A.; Jullien, L.; Valeur, B., Photophysics of a Series of Efficient Fluorescent pH Probes for Dual-Emission-Wavelength Measurements in Aqueous Solutions, *Chem. Eur. J.*, 12, 1097–1113, 2006.
60. Kim, H. M.; An, M. J.; Hong, J. H.; Jeong, B. H.; Kwon, O.; Hyon, J. Y.; Hong, S. C.; Lee, K. J.; Cho, B. R., Two-Photon Fluorescent Probes for Acidic Vesicles in Live Cells and Tissue, *Angew. Chem. Int. Ed.*, 47, 2231–2234, 2008.
61. Kim, H. M.; Yang, P. R.; Seo, M. S.; Yi, J. S.; Hong, J. H.; Jeon, S. J.; Ko, Y. G.; Lee, K. J.; Cho, B. R., Magnesium Ion Selective Two-Photon Fluorescent Probe Based on a Benzo[*H*]Chromene Derivative for In Vivo Imaging, *J. Org. Chem.*, 72, 2088–2096, 2007.
62. Kim, H. M.; Kim, B. R.; An, M. J.; Hong, J. H.; Lee, K. J.; Cho, B. R., Two-Photon Fluorescent Probes for Long-Term Imaging of Calcium Waves in Live Tissue, *Chem. Eur. J.*, 14, 2075–2083, 2008.
63. Kim, H. M.; Kim, B. R.; Hong, J. H.; Park, J. S.; Lee, K. J.; Cho, B. R., A Two-Photon Fluorescent Probe for Calcium Waves in Living Tissue, *Angew. Chem. Int. Ed.*, 46, 7445–7448, 2007.
64. Kim, H. M.; Fang, X. Z.; Yang, P. R.; Yi, J. S.; Ko, Y. G.; Piao, M. J.; Chung, Y. D.; Park, Y. W.; Jeon, S. J.; Cho, B. R., Design of Molecular Two-Photon Probes for In Vivo Imaging. 2*H*-Benzo[*H*]Chromene-2-one Derivatives, *Tetrahedron Lett.*, 48, 2791–2795, 2007.

High-Resolution Fluorescence Studies on Excited-State Intra- and Intermolecular Proton Transfer

Joost S. de Klerk, Arjen N. Bader, Freek Ariese, and Cees Gooijer

Abstract High-resolution fluorescence spectroscopies are useful for studying excited-state intramolecular proton transfer (ESIPT) processes, as well as excited-state double proton transfer (ESDPT) in the case of dimeric species. Recent results are shown for four classes of compounds, i.e., hydroxychromones, azaindoles, pyrazoloquinolines, and amino-substituted pyridine-*N*-oxides in liquid solutions, polycrystalline matrices, and/or a supersonic jet. These species can exist as monomers, dimers, complexes with polar solvent molecules, or mixtures of these. Proton/deuteron exchange, which strongly slows down those transfer reactions that take place through tunneling, is an important tool for elucidating the reaction mechanisms. In this chapter, emphasis is on spectral information in the wavelength domain, but time-resolved femtosecond studies (for example, on the ESDPT in dimers of 7-azaindole) are also included. Cryogenic high-resolution (Shpol'skii) spectroscopy – not only in the fluorescence but also in the absorbance mode – receives special attention. It can be used to characterize the various chemical species involved in the photochemistry/physics of the above compounds, but in favorable cases (as demonstrated for hydroxychromones) it can also detailed kinetic information on excited-state proton transfer processes that can be extracted from the homogeneous spectral linewidths.

Keywords shpol'skii spectroscopy · Supersonic jet spectroscopy · Tunneling · Cryogenic techniques · Proton/deuteron exchange

Introduction

Both intramolecular and intermolecular proton transfer reactions in the electronically excited-state have received extensive interest in the chemical literature. At the time of writing, the number of articles dealing with 'excited-state proton transfer'

F. Ariese (✉)
Analytical Chemistry and Applied Spectroscopy, Laser Centre, Vrije Universiteit Amsterdam, The Netherlands
e-mail: ariese@few.vu.nl

C.D. Geddes (ed.), *Reviews in Fluorescence 2007*, Reviews in Fluorescence 2007, 271
DOI 10.1007/978-0-387-88722-7_12, © Springer Science+Business Media, LLC 2009

published in refereed journals since 2000 is larger than 1500. In this chapter, we will limit ourselves to a selection of fluorophoric compounds in the condensed phase, focusing on highly resolved spectral information in the wavelength domain that can be obtained at cryogenic temperatures. In addition, time-resolved femtosecond or picosecond fluorescence studies on proton transfer will be included since they provide complementary information.

Intramolecular proton transfer can occur in molecules in which a hydrogen bond donating group and a H-bond acceptor are geometrically sufficiently close. During electronic excitation, the position of the nuclei will not change (Franck Condon state), but during the lifetime of the excited states significant changes in molecular structure such as the transfer of a proton may occur. In certain classes of molecules, the acid/base properties of the groups involved are significantly different in the excited state from those in the ground state, and excited state intramolecular proton transfer (ESIPT) may lead to a tautomeric product T^* that is more stable than the normal excited state N^*. Fluorescence spectroscopy can be used to study these phenomena if N^* or T^* or both are fluorescent. After relaxation to the ground state, back proton transfer (BPT) will normally occur, which is the reconversion to the normal ground state ($T \rightarrow N$).

Although in most cases proton transfer reactions occur without direct involvement of the surrounding molecules, the solvent may influence the relative energy levels and therefore the most favorable conformations of the tautomers. In flexible molecules, the polar groups involved in the proton transfer will adopt a conformation that allows intramolecular hydrogen bonding; the resulting energy stabilization will be more important in a non-polar solvent than in a polar environment. Proton transfer may also occur via a nearby solvent molecule that acts both as a H-bond acceptor and donor, especially in cases where the distances are too large for intramolecular bonds to be formed. Therefore, certain probe molecules based on proton transfer reactions can give information on the local solvent characteristics.

Particularly interesting are molecules with H-bond donating and accepting groups that can form dimers with dual intermolecular hydrogen bonds. In such systems, excited state double proton transfer (ESDPT) can take place, in which two protons are exchanged after excitation, resulting in the formation of an excited tautomeric dimer ($NN^* \rightarrow TT^*$). The extent of dimerization will of course depend on the concentration, the temperature, and the solvent. Under certain conditions mixtures of monomers and dimers may be present, which makes spectroscopic studies of such systems more complex.

A complicating factor in proton transfer studies is that fluorescence spectra of organic molecules in the condensed phase tend to be rather broad, and subtle spectral changes often remain undetected. In such cases, high-resolution fluorescence spectroscopic techniques can prove extremely valuable. Several of the systems discussed in this chapter were studied in low-temperature polycrystalline matrices. In "High-Resolution Low-Temperature Fluorescence Spectroscopy", the most characteristic features of this so-called Shpol'skii method are discussed. This technique is most often applied in the fluorescence mode, but it can also be used to record high-resolution phosphorescence and absorption spectra. An alternative approach to reduce spectral broadening is to remove the influence of the solvent matrix and to study individual molecules or small clusters in a supersonic jet.

The following sections deal with specific examples of classes of compounds that undergo excited-state proton transfer processes. The chemical interest in these compounds originates from different areas. For example, we refer to the work of Alexander Demchenko and coworkers dealing with the development of dual ratiometric fluorescence probes [1]. They focused on derivatives of 3-hydroxychromone, a class of compounds characterized by a single ground-state monomeric species (and thus by a single excitation spectrum) and by two clearly distinguishable emission bands. These bands belong to the initially excited normal form N^* and the reaction product T^*. Interestingly, the probe is excited at the optimum absorption wavelength of N and the intensity ratio of the emission bands is highly sensitive to the interaction with the environment. Obviously, the suitability of a particular 3-hydroxychromone derivative as a local polarity probe depends on how the substituents affect the energy barriers and thus the kinetics of ESIPT. In "Hydroxychromones", we will show how high-resolution Shpol'skii spectra can be used to obtain information on lifetimes in the femtosecond range. From the homogeneous spectral linewidths in excitation and emission, detailed kinetic information about proton transfer could be inferred both for the excited state ($N^* \rightarrow T^*$) and for the BPT reaction in the electronic ground state ($T \rightarrow N$).

Intermolecular proton transfer processes received also wide attention, partly because in DNA proton translocating tautomerization of base pairs has been suggested as a cause of mutations [2]. In this regard the dimers of 7-azaindole (7-AI) are prototypical because they are structurally similar to the adenine–thymine (A-T) pair with two H-bonds. In the dimer of 7-AI, excited state double proton transfer can take place and a major question here is whether the two protons are exchanged in a concerted or a stepwise mechanism. Significant progress has been achieved in recent years, as will be illustrated in "Aza-Indoles". By deuteration experiments (deuteron transfer being much slower than proton transfer in a tunneling mechanism), one can influence the emission intensities of the normal and tautomeric dimer. Of major importance is the femtoseconds kinetic study of Douhal, Kim and Zewail that provided a comprehensive potential energy profile for the proton transfer processes being operative [3]. In our opinion, for this system the total picture is now fairly complete.

As will be shown in "Pyrazoloquinolines", the Shpol'skii fluorescence spectroscopy mode has been quite successful in elucidating the photochemical and photophysical behavior of pyrazolo[3,4-b]quinoline (PQ). Similar to 7-AI, this compound forms dimers in non-polar solvents, but unfortunately $(PQ)_2$ does not emit fluorescence and is therefore less readily accessible [4]. Careful Shpol'skii spectroscopic studies on non-deuterated and deuterated PQ revealed that also for this compound excited-state double proton transfer processes play a main role. Furthermore, an unusual photochemical reaction was observed: in low-temperature *n*-octane matrices the "dark" dimer can be reversibly converted to brightly fluorescent monomers.

"Alkylamino Pyridine-*N*-Oxides" focuses on amino-substituted pyridine-*N*-oxides, in particular the 2-butylamino-6-methyl substituted derivative denoted as 2B6M. This compound was studied in *n*-alkane under fluid and frozen conditions. In the liquid state, the monomeric species is strongly preferred so that only a single species is dealt with, but very unexpectedly this system shows dual emission of which the relative intensity depends on the excitation wavelength, in violation of the Kasha–Vavilov rule. For this molecule, intramolecular excited-state proton transfer plays a role, but apparently this process does not only start at the lowest excited S_1-state. ESIPT seems to occur also starting at higher excited states, being fast enough to compete with non-radiative decay [5]. Low-temperature spectra indicated that in a frozen *n*-octane matrix a mixture of 2B6M monomer and two different dimers is formed.

In "Concluding Remarks", some general conclusions are drawn regarding intra- and intermolecular proton transfer reactions and the role of high-resolution fluorescence methods to study these fascinating phenomena.

High-Resolution Low-Temperature Fluorescence Spectroscopy

General

The low-resolution spectra typically obtained for solutions at room temperature are composed of broadened vibronic transitions. The origin of this broadening can be homogeneous or inhomogeneous. Homogeneous broadening affects all molecules to the same extent; it comes from vibronic coupling to the rapidly fluctuating surrounding matrix and from the limited lifetime of the states involved. A very short lifetime implies an uncertainty in energy and therefore an increased linewidth. Broadening is called inhomogeneous when it affects all molecules in an ensemble differently, for instance when we are dealing with a distribution of solute molecules with different vibronic transition energies. In the condensed phase, these energy variations are often caused by different interactions with the solvent environment. As noticed above, in a liquid or amorphous solid matrix any fluorophore molecule has its own particular interaction energy with its direct environment at the moment of excitation. Generally, inhomogeneous broadening is the main contributor to the total spectral linewidth.

Thus, in order to obtain high-resolution molecular fluorescence spectra, the inhomogeneous broadening should be reduced as much as possible. For this purpose, several methods have been developed and amply described in the literature: molecular beam spectroscopy (also denoted as supersonic jet spectroscopy, SSJ) [6], matrix isolation spectroscopy, fluorescence line-narrowing spectroscopy (FLNS), and Shpol'skii spectroscopy [7,8].

In supersonic jet spectroscopy (SSJ), experiments are performed in a gas flow. In a gas at ambient temperature, the molecules are virtually isolated. Inhomogeneous broadening due to interactions with the solvent and other molecules can be neglected. Also electron–phonon interactions are absent in the gas phase. Further, line-narrowing effects in spectra of gaseous samples require low temperatures. Very low temperatures are achieved by expanding the sample (diluted with an inert gas) through a nozzle into a vacuum. The molecules are cooled down to a few kelvin;

physical separation precludes condensation of the molecules. Typical linewidths that can be achieved are of the order of 0.01 cm^{-1} [9]. An application of SSJ is described in "Aza-Indoles".

In Shpol'skii spectroscopy (see Fig. 1), solid solutions are studied at cryogenic temperatures ($T = 25$ K or lower). It should be emphasized that freezing alone does not imply that high-resolution will be achieved. The main prerequisite in Shpol'skii spectroscopy is that the solvent should form a regular (poly)crystalline matrix upon cooling. In the ideal situation, the analyte molecule displaces a distinct number of solvent molecules in the lattice and fits in a well-defined manner in the available space thus provided. As a result, individual fluorophore molecules will experience practically the same interaction with the regular matrix, and therefore their S_0–S_1 energy differences will become (practically) identical. Almost exclusively *n*-alkane solvents are used as Shpol'skii matrices. In practice, finding the most appropriate matrix for the fluorophore to be studied is not always evident and requires some trial and error. The properties of the fluorophore play an important role here, in particular the three-dimensional shape of the molecule, its rigidity, and the number (and positions) of polar substituents. In general, the difficulty in achieving Shpol'skii spectra increases with the number of polar groups present in the fluorophore since they reduce the solubility in *n*-alkanes and tends to promote the formation of aggregates and clusters during cooling. It should be realized that in Shpol'skii spectroscopy the

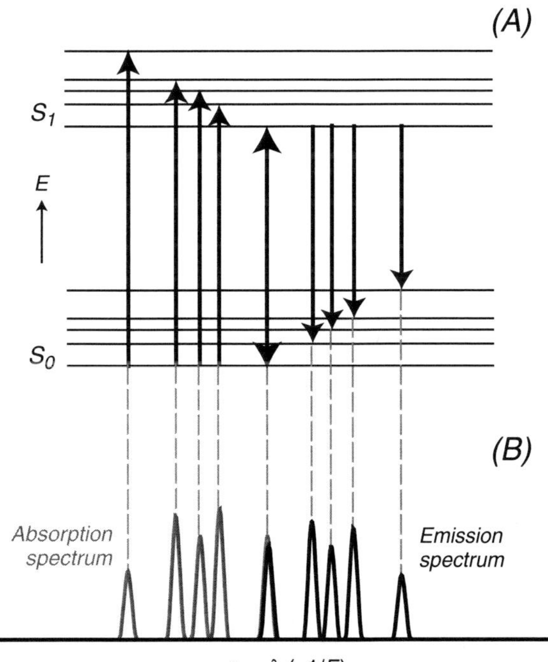

Fig. 1 Jablonski diagram illustrating the concept of Shpol'skii spectroscopy (**A**) and the resulting spectrum (**B**). Only five vibrations are shown; in the absence of reorientation during the fluorescence lifetime, there is no Stokes' shift between the longest-wavelength absorption band and the shortest-wavelength emission band

solidification of the solution should happen as rapidly as possible, so that in practice under cryogenic conditions one deals with a non-equilibrated distribution of "trapped" solute molecules in a matrix.

Often, there is more than one way in which the fluorophore molecules fit in the matrix; in that case the analyte will be present in a (limited) number of distinct sites. A multiplet will appear in the spectrum for every vibronic transition, both in excitation and emission. Within these multiplet clusters, the energy (wavenumber) differences and the intensity ratios are constant; the latter reflect the relative populations of the different sites. As will be obvious from Fig. 1, to obtain high-resolution in Shpol'skii spectroscopy a laser is not strictly needed. Nevertheless, the use of (tunable) laser excitation in fluorescence is quite advantageous, not only in terms of sensitivity, but also in terms of selectivity. By using a narrow-bandwidth light source such as a (tunable) laser, a particular site can be selectively excited so that only this site will appear in the emission spectrum.

Shpol'skii spectroscopy has been almost exclusively applied in the luminescence (fluorescence and phosphorescence) mode. Shpol'skii absorption spectra could be recorded with the system described by Nakhimovsky et al., which was based on two coupled scanning high-resolution monochromators [10]. Interestingly, quite recently more user friendly high-resolution absorption setups have been reported, based on a scanning dye laser [11], or on multiplex CCD detection [12]. High-resolution absorption spectra are of main importance if one deals with non-fluorescent species in the sample that need to be structurally characterized.

Spectral Linewidths Under Shpol'skii Conditions

In favorable situations (excellent host–guest compatibility), all solute molecules experience practically the same interactions with the environment and the broadening due to matrix inhomogeneity is reduced to about 1 cm^{-1}. In the absence of instrumental contributions, that is often also the total spectral bandwidth observed. However, significant extra broadening can be observed if one of the states involved in excitation or emission (usually the final state) has a lifetime shorter than approximately 1 ps. The origin of this so-called lifetime broadening is the time–energy uncertainty relationship: when the time can be determined accurately, its energy is poorly defined and vice versa.

From the broadened spectral lines, the lifetime can be calculated according to:

$$\tau = \frac{1}{2 \cdot \pi \cdot \Delta\nu \cdot c} \tag{1}$$

where τ is the lifetime of the short-living state, $\Delta\nu$ the homogeneous broadening (in cm^{-1}) and c the speed of light (in cm/s). Often the lifetime of the other state involved is much longer and does not contribute to the spectral width.

Usually, in Shpol'skii spectroscopy these effects are only observed in excitation when higher electronic states (S_2, S_3...) are involved, since subsequent vibrational relaxation and/or internal conversion is fast. As a result, the bands observed in the excitation and/or absorption spectrum tend to be broader at shorter wavelengths [10,11]. That is why for selective excitation the S_1-band is utilized.

In particular cases wherein the fluorophore undergoes extremely fast photochemical reactions, line-broadening effects are also observed upon excitation in the lowest excited singlet state. In that case Eq. (1) can be used to obtain detailed kinetic information. Consider a molecule that upon excitation undergoes an intramolecular proton transfer reaction on a femtosecond timescale. This means that the lifetime of the excited state is strongly reduced and consequently homogeneous broadening will be observed in the Shpol'skii absorption/excitation spectrum. Similarly, the ground state reverse reaction indicated as BPT can result in homogeneous broadening in the Shpol'skii emission spectrum of the tautomeric product.

Since in Shpol'skii spectroscopy inhomogeneous broadening is minimal and we can readily observe additional broadening effects, it can be used to determine the rates of fast intramolecular reactions. There are, however, some conditions to be fulfilled. First of all, a suitable Shpol'skii matrix must be found for the compound of interest. Secondly, the reaction rate of interest – under the cryogenic sample conditions dealt with in Shpol'skii spectroscopy – should be faster than 10^{12} s^{-1} in order to observe significant homogeneous broadening. This means that only barrierless reactions or fast tunneling have to be considered; activated processes would be too slow under cryogenic conditions. Finally, it should be possible to distinguish the broadening due to lifetime effects from inhomogeneous broadening and instrumental contributions to the observed bandwidth. As will be shown below, all these conditions are fulfilled in case of 3-hydroxyflavone ("3-Hydroxychromones"). Other applications of Shpol'skii spectroscopy will be highlighted in "Pyrazoloquinolines" and "Alkylamino Pyridine-*N*-Oxides".

3-Hydroxychromones

High-resolution Shpol'skii spectroscopy has been successfully invoked to investigate the excited-state intramolecular reactions of 3-hydroxyflavone (3HF), the phenyl-substituted 3-hydroxychromone depicted in Fig. 2 [13–15]. Upon excitation, a proton can be transferred from the 3-hydroxy to the 4-carbonyl group, so that tautomer T* is formed (see Fig. 2) [16]. For 3HF, this process is favorable since these functional groups are part of a five-membered ring so that in principle only a minor molecular rearrangement is required. Both the normal form N* and the tautomeric form T* of 3HF emit fluorescence; the emission in the blue originates from the N* state, whereas the T* state emits in the green. Of particular interest within the context of fluorescence probing is the fact that the rate of ESIPT in parent 3HF and in other 3-hydroxychromone derivatives is solvent dependent. In non-polar solvents, only green tautomeric emission is observed, whereas in more polar H-bonding solvents also blue emission originating from the N* form is

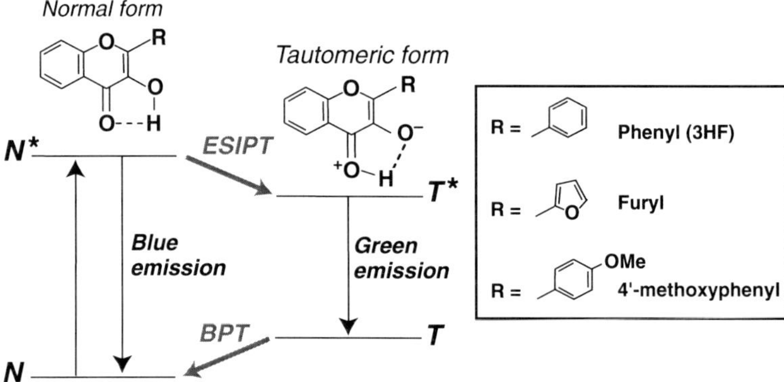

Fig. 2 Scheme explaining the excited state intramolecular proton transfer (ESIPT) and back proton transfer (BPT) reactions studied for 3-hydroxyflavone (3HF) and other 3-hydroxychromone derivatives

found [1]. As noted above, this is why these compounds are attractive as polarity probes.

This dual emission has made 3HF one of the most intensively studied molecules undergoing ESIPT reactions. Most of these studies deal with the associated reaction dynamics [17,18]. From Arrhenius plots, the reaction barrier of this reaction was determined to be 2.9 kcal/mol [18]. The time constant of the ESIPT reaction is more difficult to determine experimentally. In the 1980s, the tautomer fluorescence rise times were reported to be on the picosecond timescale [19–24]. In non-polar solvents, two rise times were observed and the slowest was attributed to torsion of the phenyl ring prior to the ESIPT reaction [21–24]. Improvement in the time-resolution of the available instrumentation resulted in a more accurate estimation of the ESIPT rate in 3HF. In 1992, Schwartz et al. performed femtosecond transient absorption experiments on 3HF; an ESIPT time constant of 240 fs was found in non-polar solvents [25]. From transient absorption spectra [26,27], other probe wavelengths were selected, resulting in an even shorter time constant for ESIPT: for 3HF in methylcyclohexane an instrument-limited value of 35 fs was estimated [28].

Alternatively, attempts have been made to determine the time constants from the extent of homogeneous broadening in the spectra. Brucker and Kelley recorded the spectra of 3HF and its deuterated analogue 3-deuteroxyflavone (3DF) in a 10 K Ar matrix [22]. From their poorly resolved emission spectra, they were able to determine the time constants of BPT (Fig. 2) at 60 and 260 fs for 3HF and 3DF, respectively [22]. However, in such spectra the inhomogeneous broadening is still substantial, so that the contribution from homogeneous broadening may be overestimated (and therefore the actual lifetimes are probably longer). Mühlpfordt et al presented the fluorescence excitation spectrum of 3HF and 3DF in a supersonic neon jet [26,29]. The obtained bandwidths were much narrower than predicted on the basis of the time resolved data. This difference was attributed to 2-phenyl and

3-hydroxy torsion that should take place before ESIPT can occur in the solvent-free molecule [29].

Shpol'skii spectroscopy provides a very elegant way of determining the ESIPT rates of 3HF and its derivatives [13,15]. The major advantage of the Shpol'skii technique is that the inhomogeneous broadening can be reduced to a few wavenumbers. Furthermore, phenyl torsions are not expected in a Shpol'skii matrix, which allows studying the proton transfer as such. Because deuteron transfer is much slower than proton transfer, the spectra of 3DF were used to determine the contributions from inhomogeneous (and instrumental) broadening. The additional homogeneous broadening in the spectra of 3HF can then be used to determine the rates of ESIPT and BPT. Figure 3 and Table 1 show that for 3-hydroxyflavone the time constant of ESIPT thus obtained is 93 ± 10 fs, whereas BPT takes place in 210 ± 20 fs [15].

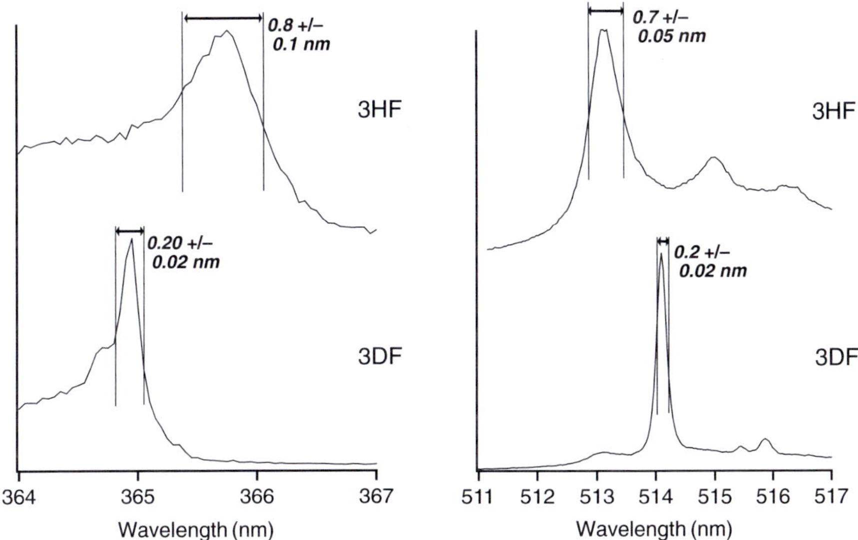

Fig. 3 4 K Shpol'skii excitation (*left*) and green emission (*right*) spectra (0-0 regions) of 3HF (*top*) and its deuterated analogue 3DF (*bottom*); solvent: *n*-octane. From the indicated values of the bandwidth, the rates of ESIPT and BPT are calculated (after Bader et al. [15])

This concept was applied to study the effects of 3-hydroxychromone substituents that change the distribution of electronic density on the chromophore and their influence on these reaction rates [15]. The focus was on three compounds, i.e., 3HF, 3-hydroxy-4′-methoxyflavone (3HF-4′OMe), and 2-furyl-3-hydroxychromone (3HC-F). Compared to 3HF, the rate for the ESIPT reaction was reduced to $(210\ \text{fs})^{-1}$ for 3HF-4′OMe and to less than $(\sim 600\ \text{fs})^{-1}$ for 3HC-F. The BPT rates were reduced to $(470\ \text{fs})^{-1}$ and $(>2000\ \text{fs})^{-1}$, respectively. It was concluded that the electron donating substituents have a strong influence on the proton tunneling

Table 1 Lifetimes of excited state intramolecular proton transfer (ESIPT) and back proton transfer (BPT) determined for 3HF. The rate of ESIPT is obtained from the total bandwidth of the site-selected band in the excitation spectrum (see Fig. 3, left). The rate of BPT was obtained from the total bandwidth of the site-selected band in the emission spectrum (see Fig. 3, right). The homogeneous broadening (Γ_{hom}) was calculated from the total broadening (Γ_{tot}), the instrumental broadening (Γ_{instr}), and the inhomogeneous broadening (Γ_{inh}) via $\Gamma_{tot}{}^2 = \Gamma_{instr}{}^2 + \Gamma_{inh}{}^2 + \Gamma_{hom}{}^2$ (after Bader et al. [15])

	Γ_{tot} (cm^{-1})	Γ_{instr} (cm^{-1})	Γ_{inh} (cm^{-1})		Γ_{hom}(cm^{-1})		τ_{pt} (ps)
			Min.	Max.	Min.	Max.	
3HF							
ESIPT	52–67	4	0	16	50	67	0.093 ± 0.01
ESIDT	14–17	4	0	16	0	16	> 0.3
BPT	24–28	4	0	7.3	23	28	0.21 ± 0.02
BDT	6.7–8.2	4	0	7.3	0	7.3	> 0.7

rates of 3HF and its derivatives. They significantly reduce the rate of both ESIPT and BPT.

Besides the dynamics of the ESIPT reaction, also its solvent dependency has been the subject of a large number of studies. In an anhydrous methylcyclohexane (MCH) glass, the fluorescence emission of 3HF is predominantly green (maximum at 523 nm) [30–32]. However, in the presence of minor amounts of H-bonding solvents, two additional emission bands are observed: a blue band with a maximum at 400 nm, and an additional green emission with a maximum at 497 nm [30–34]. These three bands were attributed to tautomeric emission of 3HF (523 nm band), tautomeric emission of 3HF/monohydrate (497 nm band), and normal emission of 3HF/polyhydrate (400 nm band). In other words, 3HF can form complexes with solvent molecules in which the ESIPT reaction is either affected or completely prohibited [30–34]. Attempts to elucidate the structure of the 3HF/monosolvate complex have been reported [22,33,34]. Besides H-bonding, electrostatic effects also seem to influence the ESIPT reaction [35]. In solvents such as acetonitrile (polar but without H-bonding properties), the rate of ESIPT can be limited by either the solvent-induced barrier or the rate of solvent reorganization [35].

The matrices used in Shpol'skii spectroscopy are always non-polar. However, by adding minor traces of H-bonding impurities, the mechanism of the proton transfer reaction can be changed [13,14]. Surprisingly, the complex of 2-octanol and 3HF appears to fit in an *n*-octane matrix (Fig. 4) [13]. A high-resolution spectrum is obtained, in which the tautomeric emission is blue shifted over 10 nm. Thus, an isolated 3HF molecule in a 1:1 complex with a hydroxyl group could be studied, and the influence of H-bonding on the rate and mechanism of the ESIPT reaction could be determined. For one furyl derivative of 3HF, the mono-complexation with octanol even blocked the ESIPT and consequently only high-resolution blue emission was observed [14].

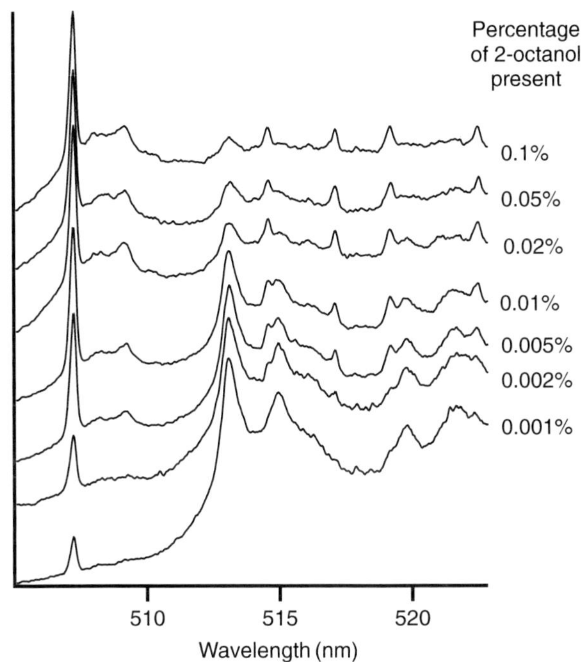

Fig. 4 10 K Shpol'skii emission spectra of 3HF in *n*-octane with increasing concentration of 2-octanol added. The sample was excited at 365 nm. A 2-octanol concentration of 0.001% corresponds to 70 μM. The concentration of 3HF was 10 μM (after Bader et al. [13])

Aza-Indoles

Within the context of studying proton transfer reactions in electronic excited states, 7-azaindole (7-AI) received wide interest in the chemical literature over a period of some four decades. 7-AI exhibits a particularly well-documented example of cooperative double proton transfer in its excited state. In H_2O solution, the tautomerization of the monomer form has been reported: upon UV excitation, the visible emission from the tautomer is observed [36]. However, main interest has been devoted to the dimer of 7-AI. It is considered a model system for the adenine-thymine (A-T) pair in DNA, because of their structural similarity. Proton transfer processes in DNA are a major issue since tautomerization of base pairs has been suggested as a cause of mutations [2].

In order to study the kinetics and mechanisms of very fast excited state proton transfer processes quantitatively in detail, one of the main requirements is that the chemical species involved are known and structurally fully characterized. This is a main challenge. High-resolution fluorescence studies are quite appropriate in this context.

Unfortunately, in the condensed phase such spectra are difficult to obtain. As far as we know, Shpol'skii spectra of 7-AI have not been reported. Nevertheless, some information can be extracted from low temperature spectra, as they show somewhat more detail than room temperature spectra, as will be discussed below. To gain high-resolution spectra, molecular beam experiments have been performed by Nakajima et al. [37].

Molecular Beam Experiments

Nakajima et al. studied the structures of 7-AI-$(H_2O)_n$ complexes with $n = 1,2,3$ by means of high-resolution laser-induced fluorescence (LIF) in a supersonic jet [37]. Surprisingly, in a jet-cooled system no tautomerization (visible emission) can be observed, in contrast to the behavior of 7-AI in liquid H_2O solutions. In the SSJ system, high-resolution was achieved (about 0.01 cm^{-1}) and for the three H_2O complexes the spectra showed partially resolved rotational structure, enabling a detailed structural analysis that was supported by theoretical calculations. As an illustration, the rotationally resolved LIF spectrum of 7-AI-$(H_2O)_1$ is depicted in Fig. 5b, together with simulated spectra assuming a planar (Fig. 5a) and a

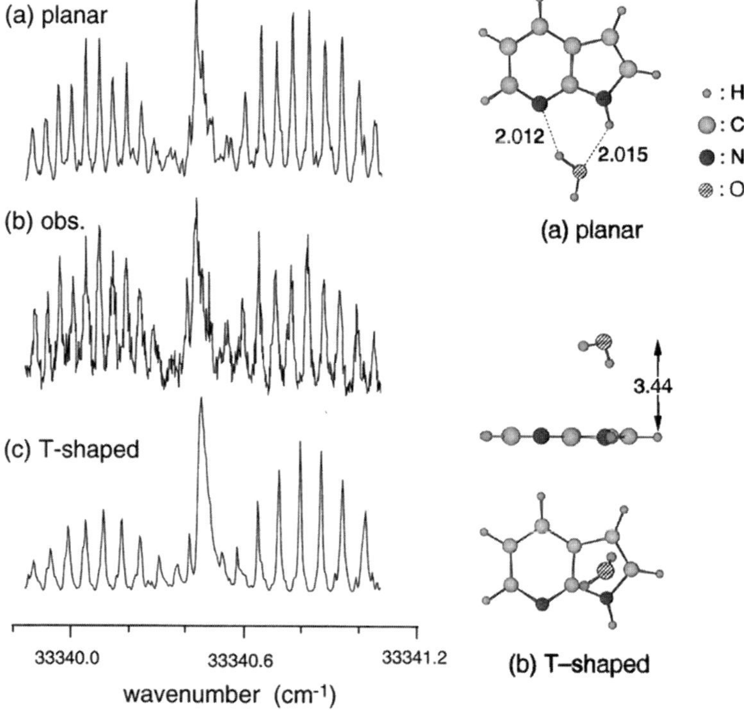

Fig. 5 Rotationally resolved SSJ LIF spectrum of the 7-AI-$(H_2O)_1$ complex: (**a**) simulation assuming a planar structure, (**b**) observed spectrum, and (**c**) simulation assuming a T-shaped structure. At the *right*, the optimized geometries of the 7-AI-$(H_2O)_1$ complex in the ground state from an ab initio calculation (basis set 6-31G): (**a**) a planar structure and (**b**) a T-shaped structure with side and top views. The planar structure of (**a**) is 40 kJ/mol more stable than the T-shaped one. The lengths of the hydrogen bonds are shown in angstroms (after Nakajima et al. [37])

T-shaped structure (Fig. 5c), respectively. The planar structure is formed via two hydrogen bonds, one between the N-H of 7-AI and the O-atom of H_2O, and the other between the N-atom of 7-AI and the H-atom of the molecule water. Conversely, in the T-shaped structure the interaction between the H-atom in water and the π-electrons of the aromatic ring dominates. Theoretical calculations revealed that the planar structure is 40 kJ/mol more stable than the T-shaped structure. Also the simulated spectrum of the last structure is not compatible with the experimental spectrum as is obvious from Fig. 5: the calculated spacing of the peaks is much too large.

Thus, the spectra point to a planar structure of 7-AI-$(H_2O)_1$. By comparing the rotational constants in the S_0-state and the S_1-state, it was concluded that in the excited state H_2O is some 0.1 Å closer to the 7-AI molecule (some 5% change): apparently, the hydrogen bonds become stronger.

The wavenumbers of the 0-0 bands indicate a red shift upon complexation: the 0-0 transition is located at 33 340.4 cm^{-1} for the free 7-AI, implying a shift of 1288 cm^{-1} for 7-AI-$(H_2O)_1$, a further shift of 708 cm^{-1} for 7-AI-$(H_2O)_2$, and an additional shift of only 78 cm^{-1} for the third water molecule in the 7-AI-$(H_2O)_3$ complex. Presumably, both the first and the second H_2O are attached to a sensitive site of 7-AI, whereas the third one is not. Detailed analysis of the rotationally resolved spectra revealed that the 7-AI-$(H_2O)_2$ and 7-AI-$(H_2O)_3$ complexes have eight- and ten-membered planar ring structures, respectively.

According to the authors, the fact that ESIPT is observed in the liquid state (H_2O) but not in SSJ can be addressed as follows: the complex 7-AI-$(H_2O)_1$ has a planar structure which is incompatible with proton transfer without considerable nuclear rearrangements. Such rearrangements cannot be achieved at the extremely low temperatures dealt with under SSJ conditions.

The same paper deals even with the problem of a reactive and a non-reactive 7-AI dimer. In contrast to the former, the latter does not exhibit proton transfer; in other words does not undergo tautomerization after being excited. The LIF spectra obtained could be simulated with a T-shaped geometry. However, there is still some uncertainty since the presence of H_2O could not be fully excluded as 7-AI is strongly hygroscopic.

Condensed Phase Experiments

Unfortunately, condensed phase spectra of 7-AI do not show sufficient spectral details for an unambiguous characterization of the various ground-state and excited-state species that (might) play a role in the photophysics/chemistry of 7-AI. In fact, as can for instance be read in the 2000-paper of Catalan and Kasha [38], there has been considerable variation and interpretation of experimental results. Points of discussion are for instance the 77 K luminescence spectra in various solvents (the freezing technique applied presumably affects the results) and the stability of the 7-AI tautomer dimer in the electronic ground state.

As regards the structural characterization of species, quite some progress has been achieved by Catalan and Kasha [38]. They focused on 7-AI solutions using the non-polar aprotic solvent 2-methylbutane (2 MB), a favorable solvent forming a clear glass upon solidification (but consequently not a Shpol'skii matrix), still providing relatively sharp vibronic transitions in the spectra, especially at low temperatures.

In Fig. 6, the various species possibly involved in fluorescence and in proton transfer are depicted. The first question to be answered is: "does the normal dimer emit fluorescence or not". To answer this question, first of all the existence of a monomer–dimer equilibrium in 2 MB was unambiguously demonstrated by concentration and temperature-dependent absorption spectroscopy measurements: at 10^{-4}

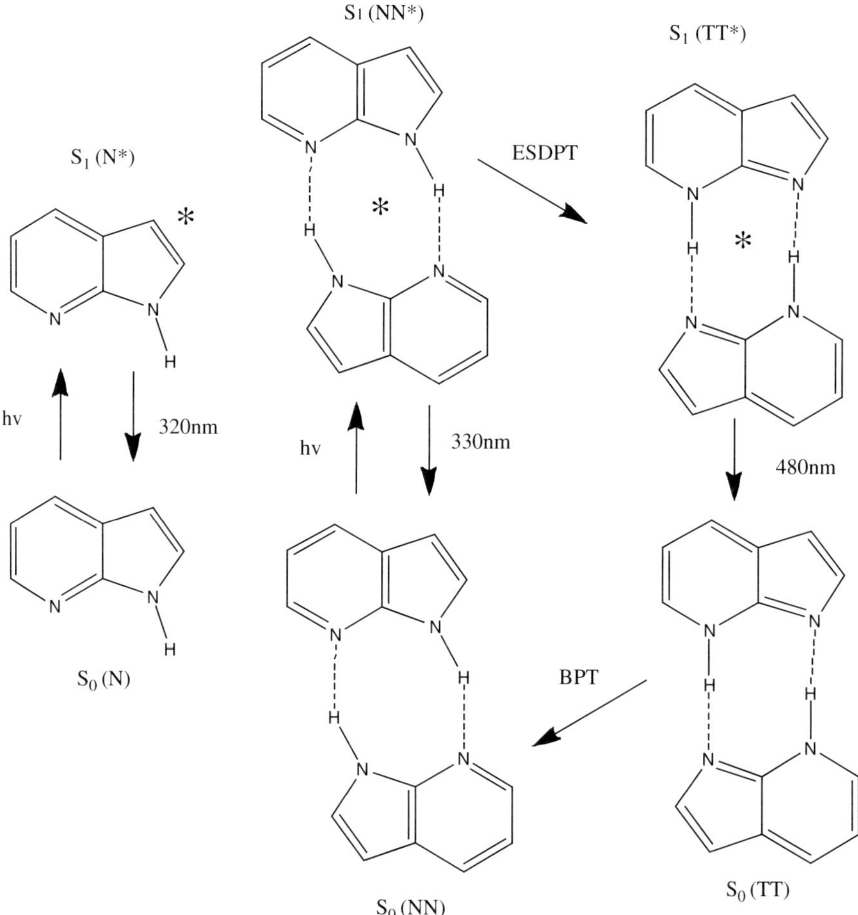

Fig. 6 The various species of 7-AI possibly involved in fluorescence and in proton transfer (after Catalan et al. [38])

M the solution of 7-AI in 2 MB is composed of pure dimer over a wide temperature range, so that it can be readily studied. Furthermore it can be selectively excited at 315 nm (its first vibronic peak), the onset of the monomer absorption being at about 305 nm.

Consequently the dimer fluorescence could be readily distinguished from monomer fluorescence: exclusively tautomer fluorescence from the dimer TT* (see Fig. 6) was observed (with onset at 413 nm) while no fluorescence from the normal dimer NN* was detectable.

Deuteration of 7-AI gave rise to major spectroscopic changes. In that case for the dimer only normal tautomer (NN)* fluorescence was observed, while emission from dimer TT* was not detected (dual fluorescence was an artifact stemming from incomplete deuteration). Apparently, the double proton transfer is essentially blocked by the deuteration. According to the authors, the lack of temperature influence on these spectra point to "tunneling" as the excited-state double proton transfer mechanism.

The question whether the dimers in the normal or tautomeric forms also emit phosphorescence could not be readily solved. For many years, a structured luminescence spectrum starting at 430 nm with a long lifetime (observed using a delay of 220 μs) could not be unambiguously identified. Recently, Catalan [39] addressed this issue by studying the bromine-substituted compound, i.e., 3-bromo-7-azaindole (3Br7-AI) in 2 MB. In a 10^{-5} M solution at 77 K almost exclusively dimers appear to be present. Also for the brominated dimer the proton transfer process dominated in the S_1-state: its speed is so high (estimated at 5×10^{11} s^{-1}) that fluorescence and heavy-atom enhanced intersystem crossing (ISC) induced by the bromine atom are still negligible.

On the contrary, the presence of the bromine substituent affects the decay of the tautomeric dimer in the S_1-state dramatically. Its fluorescence is fully quenched as a result of enhanced ISC. Secondly, the heavy atom favors the yield of triplet formation and enables the phosphorescence emission of the tautomeric dimer.

To summarize, the phosphorescence transitions observed for 3Br7-AI (with the onset at 430 nm and a 0-0 component at 444 nm) can be assigned to the tautomeric form of the dimer that preserves its C_{2h} symmetry. This assignment is supported by theoretical calculations.

Femtosecond Kinetic Studies

So far, in the above discussion on 7-AI no special attention was devoted to the fast dynamics of the excited double proton transfer that happens not only in dimers but also in monomers interacting with H-bonded solvent molecules. In fact, elucidation of the dynamics has been the subject of various studies (see [40] and references therein). One of the fundamental issues is whether the transfer processes of the two hydrogens proceed in a concerted manner (with a single transition state) or stepwise by forming intermediate species.

Douhal et al. were the first studying the kinetics of isolated dimers in a molecular beam [3]. They made use of femtosecond pulse excitation, followed by another pulse at different delay times to observe the mass spectra. The parent ion showed a decay with two time constants $(650\ \text{fs})^{-1}$ and $(3300\ \text{fs})^{-1}$ pointing to a non-concerted reaction pathway. However, in another study the opposite was concluded [41], maybe due to a limited time-resolution. Also theoretical studies did not result in uniform conclusions, but recent advanced calculations showed the presence of stable intermediate states having zwitterionic and covalent characters [42]. It was also shown that the transition state is 3.5 kcal/mol higher in energy than the stepwise pathway, a difference which should be even larger in polar media.

Also condensed-phase kinetic studies have been published. Fiebig et al. [43] made use of transient absorption spectroscopy with a time-resolution of 130 fs and fluorescence upconversion with a time-resolution of 300 fs. Again two lifetimes were obtained: the first proton transfers within 130 fs and the second one within 1200 fs (for the deuterated species the respective lifetimes being 280 and 5000 fs). Other researchers obtained similar lifetimes using fluorescence upconversion (200 and 1100 fs for the undeuterated species) but gave another interpretation: the shortest lifetime was assigned to internal conversion from the ^1L_a to the ^1L_b electronic state and the longest one to a concerted proton transfer process [44].

Very recently Kwon and Zewail published a femtosecond fluorescence study on 7-AI dimers in the condensed phase examining various parameters: solvent polarity and viscosity, solute concentration, and isotopic fractionation [45]. The results obtained provide conclusive evidence for the presence of an ionic intermediate species that is formed on a femtosecond time scale and decays to the final tautomeric form on a picosecond scale. This is fully in line with a stepwise proton transfer mechanism.

A detailed discussion of the results is beyond the scope of this paper, but some points are well worth to be highlighted. It was found that the viscosity of the medium in the range of $0.3 \leq \eta \leq 3$ Pa s has no significant influence on the proton transfer rates, whereas the medium polarity does. The rate of the slow component changed from $(1.1\ \text{ps})^{-1}$ in *n*-heptane (relative dielectric constant $\epsilon = 1.92$) to $(2.1\ \text{ps})^{-1}$ in acetonitrile ($\epsilon = 36.64$). From these results it was concluded, by using the electrostatic model of Kirkwood et al. [46], that the intermediate has an electrical dipole moment of 12 D.

Furthermore, quite interestingly, a systematic kinetic study of the four dimeric isotopomers HH, HD, DH, and DD was performed, where the successive two letters denote the protic hydrogen atom on the pyrrolic nitrogen (N_1) and the pyridinic nitrogen (N_7) atoms, respectively. In the excited state, HD and DH are not identical if the excitation is localized on a monomeric part. Strong isotope effects were observed, as illustrated in Fig. 7. This figure shows the fast rise and the picosecond decay of the 380 nm intermediate and the picosecond formation of the tautomeric product, fluorescing at 530 nm. Careful analysis revealed the rate constants for the four isotopomers to be $(1.1\ \text{ps})^{-1}$, $(1.9\pm0.2\ \text{ps})^{-1}$, $(4.7\pm0.4\ \text{ps})^{-1}$ and $(9.9\pm2.8\ \text{ps})^{-1}$, respectively.

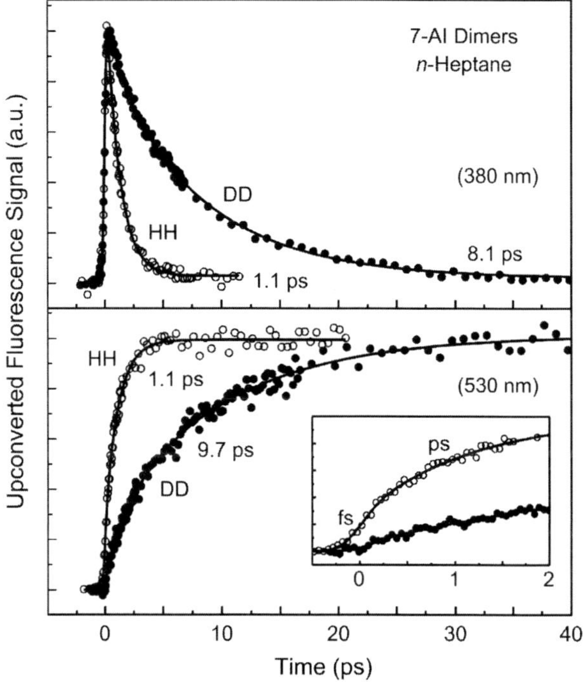

Fig. 7 Isotope effects. Fluorescence transients of protiated (*open circles*) and deuterated 7-AI (*filled circles*) in *n*-heptane. The fluorescence was collected at 380 nm (*upper*) and 530 nm (*lower*). (*Inset*) An enlargement of the early-time response. *Solid curves* are best-fitted exponential decays (after Kwon et al. [45])

For the elucidation of the mechanism, it is important to note that k_{HD} differs significantly from k_{DH}. In a concerted transfer mechanism, these rate constants should be identical, equal to the geometric mean of k_{HH} and k_{DD} which is $(3.3 \pm 0.8\ \text{ps})^{-1}$. Thus the results support the stepwise pathway mechanism, where the first step takes place on a femtosecond time scale, and the second one on the picosecond timescale determines the overall rate. The observed kinetic isotopic effect of a factor of 9 for DD is composed of a primary effect of 4.7 – the tunneling process – and a secondary effect of 1.9 originating from cooperativity, since the N—N distance of the deuterated bridge is longer than that of the protonated one. Accordingly, upon deuteration the barrier for N—N motion increases and the rate decreases by a factor of 1.9.

Combination of the experimental and theoretical findings led to the potential energy profile for the double proton transfer dynamics of 7-AI dimers depicted in Fig. 8. It gives a complete picture clearly showing the reaction coordinates and structures involved and highlights the energetics and the influence of solvent polarity.

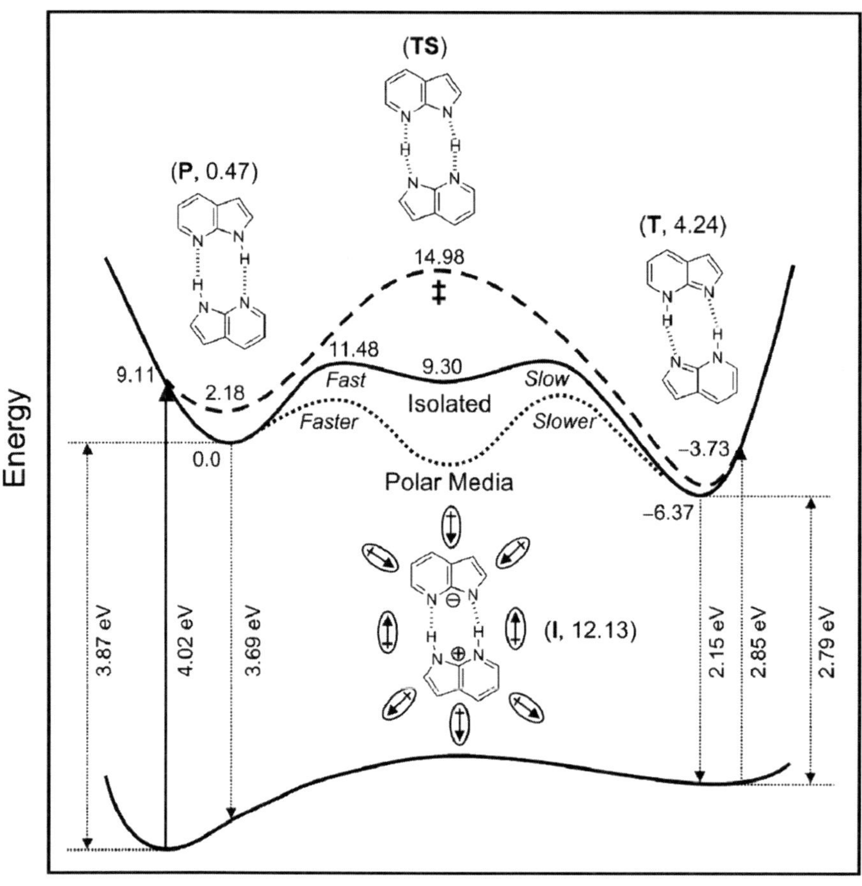

Fig. 8 A potential energy profile. For the proton transfer processes of 7-AI dimers, we adapt the potential energy curves calculated [42], and include the influence of polar solvents. Energy differences of involved species in kcal/mol are also given near reaction paths; the values are taken relative to that of the lowest excited state of the parent dimer. The *numbers in parentheses* give the dipole moment in Debye. The effect of polarity is schematically superimposed on the calculated curves to illustrate the observed changes with dielectric constant (after Kwon et al. [45])

Pyrazoloquinolines

Pyrazoloquinolines – compounds composed of fused quinoline and pyrazolo rings – have been of interest in various chemical disciplines, as summarized in a recent paper by Zapotoczny et al. [47]. After the synthesis of several new pyrazoloquinoline derivatives, Rechthaler and coworkers reported on the photochemical and photophysical properties of these molecules in 1997 [48]. Pyrazoloquinolines are highly fluorescent in liquid solvents as well as in the solid state. This property makes them good candidates for several optical devices. Examples are brightly fluorescent molecular sensors [49] and organic light emitting diodes (OLEDs) [50].

In 2007, Zapotoczny and coworkers [47] reported on the tautomerism phenomenon of pyrazolo[3,4-b]quinoline (PQ). Obviously, PQ shows some structural similarity with 7-AI (see Fig. 6 in "3-Hydroxychromones"), but its structure is more complicated due to the presence of a second nitrogen atom in the five-membered ring. Theoretical calculations showed that for PQ three tautomeric forms are possible, as depicted in Fig. 9. PQ might show tautomerism, not only in the electronic ground state, but also in excited-state intramolecular proton transfer (ESIPT). Furthermore, in view of its similarity with 7-AI, also intermolecular (double) proton transfer reactions might play a role, provided that dimers are formed [51].

Fig. 9 Structure of the three tautomeric forms of pyrazolo[3,4-b]quinoline

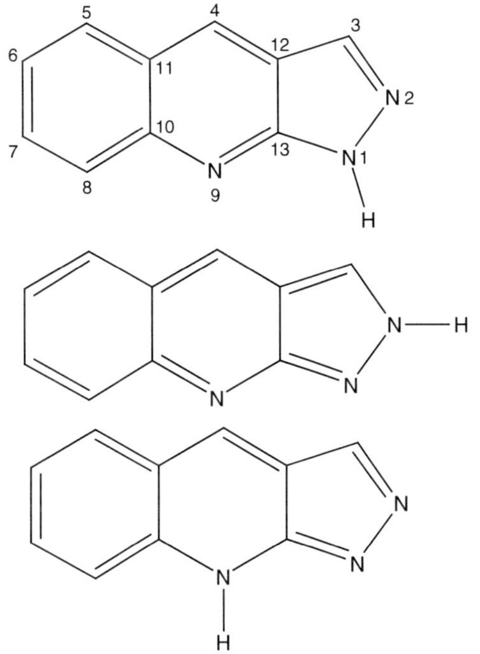

1H-PQ

2H-PQ

9H-PQ

In collaboration with Zapotoczny's group, we conducted an extensive study on the photochemistry and photophysics of PQ in *n*-alkane solvents, by utilizing absorption and fluorescence spectroscopy, both under "normal" and high-resolution conditions [4]. Temperature- and concentration-dependent measurements pointed to the presence of PQ dimers, indeed. However, in contrast with 7-AI dimers, PQ dimers do not exhibit any fluorescence emission, not from the normal and not from the possible tautomeric form. This even applies for spectra recorded under Shpol'skii conditions at temperatures as low as 5 K (see Fig. 10A). Remarkably, in contrast to regular PQ, for the deuterated PQ dimer Shpol'skii emission spectra can be recorded (see Fig. 10C). This points at the following mechanism: the

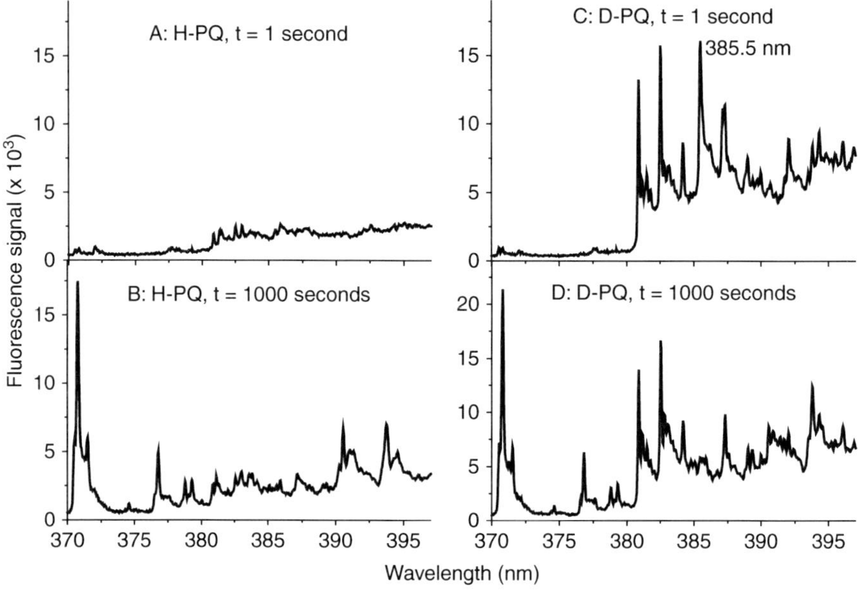

Fig. 10 Shpol'skii spectra of non-deuterated and deuterated PQ (concentrations are equal) in *n*-octane at 5 K excited with an XeCl excimer laser (308 nm) after 1 and 1000 s of illumination (after de Klerk et al. [4])

undeuterated dimer (in its normal form) is not able to emit fluorescence since the radiative transition cannot compete successfully with the very fast excited-state double proton transfer process; this creates the tautomeric dimer, which does not exhibit native fluorescence. In contrast, deuterated PQ dimers do show fluorescence under cryogenic conditions, since excited-state deuteron transfer is much slower.

The detailed high-resolution fluorescence spectra in Fig. 10A and C were recorded immediately after starting the experiment (using laser excitation at 308 nm). The sharp features at 381.0 and 382.5 nm – weak in Fig. 10A and stronger in Fig. 10C – are not relevant in the present discussion: they were assigned to a PQ monomer – alcohol or water complex. The crucial difference between Fig. 10A and Fig. 10C is the sharp emission at 385.5 nm in the latter spectrum, pertaining to the deuterated dimer. An analogous dimer emission is absent in Fig. 10A.

The immediate recording of spectra in Fig. 10A and C is important since PQ solutions under cryogenic conditions undergo interesting photochemical effects. It was found that a narrow-banded spectrum appears gradually in time with 0-0 transitions at 370.50, 370.75 and 371.48 nm as shown in Fig. 10B after 1000 s illumination with a laser. All these transitions have the same vibrational structure and therefore belong to the same species in three different sites, as explained in "High-Resolution Low-Temperature Fluorescence Spectroscopy ". The sharp spectrum disappears if the illumination is stopped and the sample is subsequently heated to approximately

100 K. The spectrum with the three sites around 370 nm only reappears after cooling back to below 40 K and subsequently starting laser illumination.

In Fig. 10C and D, two Shpol'skii spectra of deuterated PQ (D-PQ) in *n*-octane at 5 K are shown, excited at 308 nm after 1 and 1000 s of illumination, conditions exactly identical to those used for H-PQ in Fig. 10A and 10B. Also for the deuterated species, during illumination the monomeric species with 0-0 transitions at 370.55, 370.81 and 371.55 nm is formed (see Fig. 10D), very close (but not fully identical) to those observed in Fig. 10B. However, in the deuterated system, the increase of this monomer spectrum can be related with the decrease of the 385.5 nm line; after 1000 s, it has fully disappeared (whereas the 381.0 and 382.5 nm lines are not affected by illumination). The increase at 370.8 nm and the decrease at 385.5 nm show similar half life times, as expected.

It is concluded that the photochemically generated emissions around 370 nm in Fig. 10B and D can be assigned to the monomer emission of PQ. Upon irradiation of the contact dimer, it is rapidly converted to the tautomeric form and in this state it can undergo a minor molecular rearrangement, ultimately leading to two separate monomers in the electronic ground state. In this configuration, the mutual interaction between the constituents is negligible, so that monomer emission is observed. In view of the rigidity of the pseudo crystalline *n*-octane matrix at 5 K, the rearrangement should be qualified as "minor", indeed. These photochemically produced monomers, separated by a very small distance and energy barrier, are reconverted into the more stable dimeric form as soon as sample temperatures of about 100 K are applied [4].

Alkylamino Pyridine-*N*-Oxides

Pyridine-N-oxides are of interest in various chemical disciplines, especially in photochemistry and biochemistry. They may exhibit valuable properties such as efficient second harmonic generation of electromagnetic radiation [52]. The alkylamino nitropyridine-*N*-oxides are particularly interesting because they can undergo not only intramolecular proton transfer, but also intermolecular proton transfer in case dimers are present [53–55]. In the spectroscopic literature the focus is mainly on 2-butylamino-6-methyl-4-nitropyridine-*N*-oxide (2B6M) in aprotic polar solvents [56]. In the crystal structure of 2B6M also dimers are observed that are formed via a double hydrogen bond between the amino-hydrogens and the other molecules' NO-oxygen atoms. The molecular structures of the monomer and a possible dimer in their normal state are shown in Fig. 11.

In order to investigate the possible excited-state proton transfer reactions mentioned above, one needs to know whether there are more than one 2B6M species in the electronic ground state. Absorption spectra in liquid-state *n*-octane did not show any concentration dependency (over the range of 5×10^{-7} M to 1×10^{-3} M) nor any temperature dependency (from 293 to 240 K) of the liquid-state absorption spectra [5]. From these results, we concluded that in the electronic ground state

Monomer Dimer

Fig. 11 The molecular structure of 2-butylamino-6-methyl-4-nitro-pyridine-N-oxide (2B6M). Two normal state conformations (a monomer and a possible dimer) are given

only one species exists, i.e., the monomeric form of 2B6M [5]. Under those conditions, no other monomeric structures or dimeric species play a role. This monomer can undergo excited state intramolecular proton transfer (ESIPT). In the room temperature emission spectra (Fig. 12), two emission bands can be discerned indeed; after excitation, the normal form, emitting in the blue (at 452/471 nm), can undergo

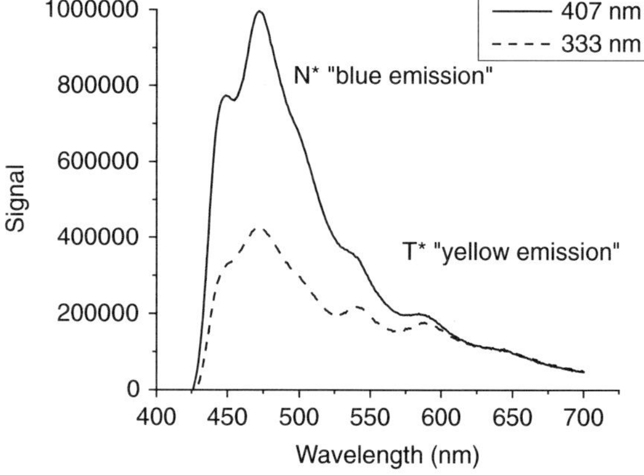

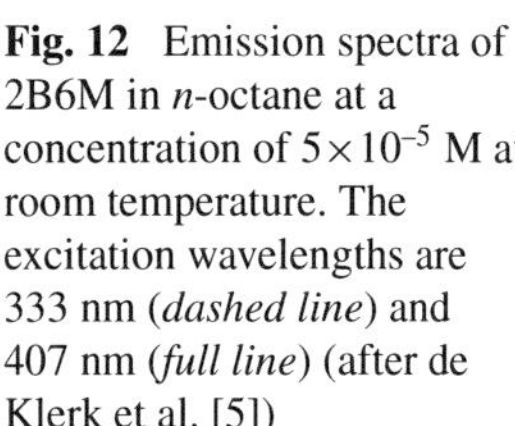

Fig. 12 Emission spectra of 2B6M in *n*-octane at a concentration of 5×10^{-5} M at room temperature. The excitation wavelengths are 333 nm (*dashed line*) and 407 nm (*full line*) (after de Klerk et al. [5])

ESIPT so that a tautomeric form is created, emitting in the yellow range at 545 nm and longer.

However, rather unexpectedly, the relative intensities of the emission in the room temperature spectra depend on the excitation wavelength, as is obvious from Fig. 12. Using 407 nm excitation, the blue emission at 452/471 nm is relatively much more intense than the band obtained using 333 nm excitation. This is also reflected by the difference between the absorption and the excitation spectrum. Their band profiles are quite similar [5], but the relative intensities of the bands in absorption differ strongly from those in excitation.

These findings were unexpected since a difference between excitation and absorption spectra normally points to the presence of more species coexisting in the ground state. Because only one ground-state species is present, another explanation has to be found. The excitation wavelength dependency was attributed to exceptional excited-state internal proton transfer, which can start not only at the lowest excited singlet state (S_1) of the normal monomeric species state, but also at higher electronic excited states S_2 and S_3 or a hot state of S_1. The crucial point is that the latter processes should be fast enough to compete efficiently with internal conversion from the S_3-state and the S_2-state to the S_1-state.

A scheme to account for the results in *n*-octane is depicted in Fig. 13A. By including fluorescence lifetime measurements and careful analysis of the data, we were able to estimate the various corresponding rate constants given in Fig. 13B.

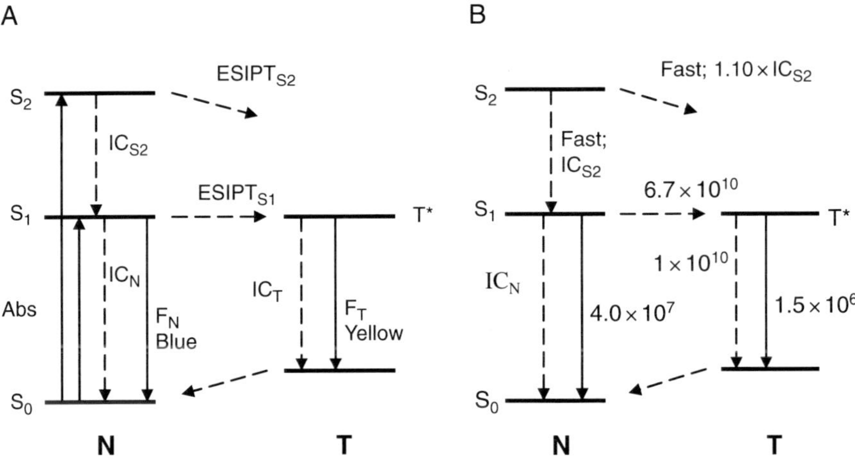

Fig. 13 **A**: Excited state processes in 2B6M; and **B**: corresponding rate constants (in s^{-1}). Solvent, *n*-octane; room temperature (after de Klerk et al. [5])

Upon freezing the sample, the absorption spectrum changes significantly as shown in Fig. 14. The most striking difference is the appearance of a novel, broad band extending to about 520 nm with its maximum around 460 nm, upon which a narrow line-spectrum is clearly observed at 14 K. It is obvious that more than one

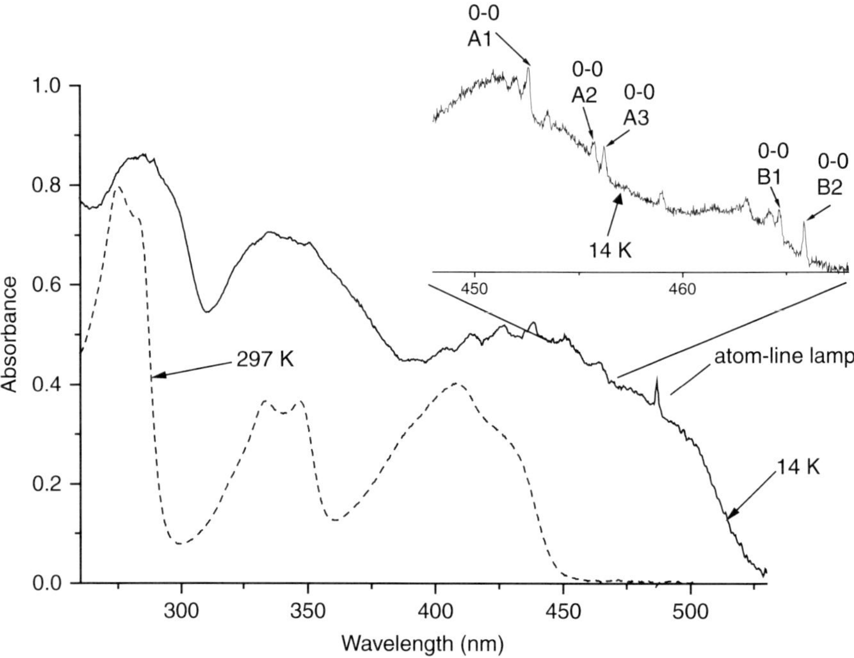

Fig. 14 Absorption spectra of 2B6M in *n*-octane at a concentration of 5×10^{-4} M at temperatures of 297 and 14 K. The 0-0 transitions of the different sites of species A and B are indicated in the high-resolution spectrum (after de Klerk et al. [12])

2B6M species plays a role. The narrow lines are attributed to the Shpol'skii effect (in absorption spectroscopy!): the strong reduction in inhomogeneous linewidth because of a well-defined fit of the guest molecules in the crystalline matrix.

With the help of low-temperature absorption and fluorescence experiments at 77 and 5 K, a detailed analysis was presented [12]. In frozen *n*-octane, three different ground-state species of 2B6M can be distinguished, denoted as A, B, and C. The Shpol'skii spectrum of 2B6M in *n*-octane, non-selectively excited at 420 nm, is shown in Fig. 15. Under these conditions, species A and B provide high-resolution emission spectra with their main 0-0 transitions at 452.62 nm (A1) and 464.60 nm (B1). Species C gives rise to a broad-banded emission with its maximum at 535 nm (not shown). For species A and B, several sites are observed; as an example selectively excited spectra of species B1 and B2 are depicted in Fig. 16.

The authors assigned the structureless band to a dimer structure of two strongly interacting molecules that is formed during the solidification of the *n*-octane solution (species C). Species A corresponds with the monomer of 2B6M. Species B was tentatively assigned to a dimeric form as in Fig. 11, held together by intermolecular H-bonds but with only minor effects on the electronic spectra.

An important question is whether under these cryogenic conditions excited-state proton transfer processes are still operative. Since the tautomeric emission was

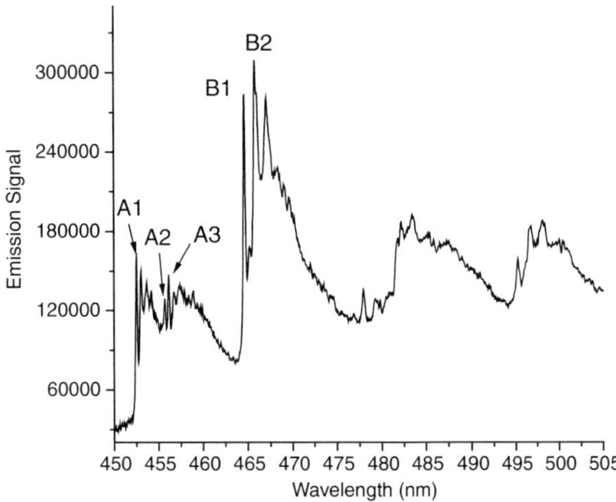

Fig. 15 Shpol'skii spectrum of 2B6M in *n*-octane at a concentration of 5×10^{-5} M, non-selectively excited at 420 nm ($T = 5$ K). The 0-0 transitions of different sites of species A and B are indicated (after de Klerk et al. [12])

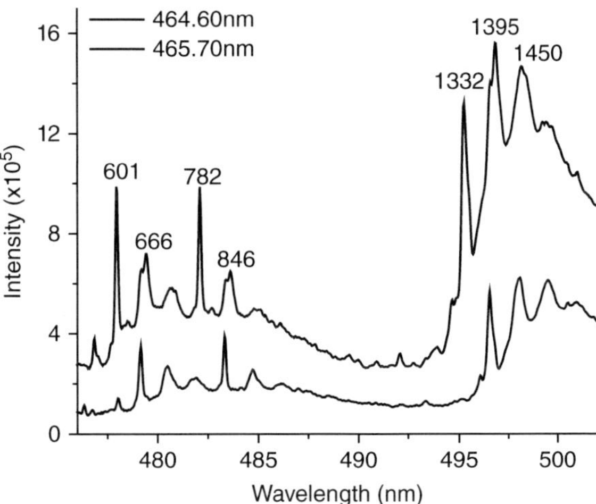

Fig. 16 Shpol'skii spectra of 2B6M (species B) in *n*-octane at a concentration of 5×10^{-5} M, selectively excited at 464.60 nm (*upper line*; species B1) and 465.70 nm (*lower line*; species B2) at a temperature of 5 K. Peak labels indicate ground-state vibrational frequencies (in wavenumbers) for B1. The spectrum of B2 is red-shifted but otherwise identical. Because of excitation overlap, there is a contribution of species B2 in the spectrum of B1 (after de Klerk et al. [12])

too weak for unambiguous detection, ESIPT starting at the S_1-state could not be detected. However, from the excitation spectra monitored at $\lambda_{em} = 485$ nm (the S_0-S_1 band being much stronger than in absorption) it can be concluded that the intramolecular mode is still operative for species A under cryogenic conditions, at least upon excitation into the S_2- or S_3-state.

Concluding Remarks

Excited-state intramolecular proton transfer and excited-state double proton transfer processes play a major role in various areas of chemistry. They constitute one of the most basic reaction steps, and in some systems they can occur at femtosecond timescales, even at cryogenic temperatures. Fluorescence spectroscopy is often employed to study these reactions, and the high-resolution methods described in this chapter are especially suitable for obtaining detailed information on the species present and/or the lifetimes of the states involved.

In the previous sections, ESIPT, ESDPT and/or BPT processes are described for four different classes of compounds. Monomer–dimer equilibria and the occurrence of fluorophore-solvent complexes strongly affect the photophysical/photochemical processes. Several interesting conclusions emerge from the above discussions. First of all it should be noted that for both 3-hydroxyflavone and 7-azaindole, representing the first two classes, at least for non-protic solvents the photochemistry is well-known and detailed information on the operative transfer reactions has recently become available. In fact, the mechanisms of the excited-state double proton transfer in 7-azaindole dimers have been completely elucidated both theoretically and experimentally. Influences of protic solvents deserve further attention and we are looking forward to the implementation of the results to the study of adenine–thymine interactions in DNA. As regards the 3-hydroxychromones, the effects of chemical substitution and furthermore the influences of solvent parameters require further investigation.

For the two other classes, the situation is very different. Here we are dealing with complex photochemistry and monomer–dimer equilibria, so that identification of the chemical species playing a role is a major challenge. Furthermore, not all species involved exhibit native fluorescence. Nonetheless, quite some progress has been achieved and fascinating results have been obtained. This holds for instance for the dimer PQ that undergoes excited-state fast double proton transfer while both the normal and the tautomeric forms are not detectable by fluorescence. Furthermore, under cryogenic conditions in a crystalline *n*-octane matrix, the "dark" dimer structure can be photochemically converted into strongly fluorescent monomers, separated by a minor energy barrier. Further research is required here and other spectroscopy modes will be needed in order to achieve full elucidation of the photochemistry and photophysics taking place in this compound.

The same holds for the alkylamino pyridine-*N*-oxides, where the liquid-state results for non-polar solvents differ strongly from the results obtained for frozen solutions. A monomeric and two different types of dimeric species were observed in a low-temperature polycrystalline *n*-octane matrix. Identification of the various species involved is a main challenge here. Furthermore, the exceptional phenomenon of excited-state internal proton transfer starting at higher excited states needs further investigations, both experimentally and theoretically.

SSJ was used to study 7-AI in clusters with 1, 2, or 3 water molecules, in the absence of solvent influences that would otherwise have caused extensive spectral broadening. In a Shpol'skii experiment, matrix inhomogeneity is strongly diminished by working in a polycrystalline *n*-alkane matrix. For all four classes of compounds, the application of high-resolution fluorescence was essential in elucidating the underlying mechanisms and the transfer rates relative to fluorescence and/or internal conversion.

References

1. A.P. Demchenko, FEBS Lett. 580 (12), 2951–2957, 2006.
2. P.D. Lowdin, Adv. Quantum Chem 2, 213–361, 1965.
3. A. Douhal, S.K. Kim, A.H. Zewail, Nature 378, 260–263, 1995.
4. J.S. de Klerk, A. Bader, S. Zapotoczny, M. Sterzal, M. Pitch, A. Danel, C. Gooijer, J. Phys. Chem. A, in press.
5. J.S. de Klerk, A. Szemik-Hojniak, F. Ariese, C. Gooijer, J. Phys. Chem. A 111, 5828, 2007.
6. G. Scoles (ed), Atomic and Molecular Beam Methods, Volume 2, Oxford University Press, Oxford, 1992.
7. C. Gooijer, F. Ariese, J.W. Hofstraat (eds), Shpol'skii Spectroscopy and Other Site-Selection Methods, Wiley Interscience, New York, Chapter 8, 2000.
8. J.W. Hofstraat, C. Gooijer, N.H. Velthorst, Molecular Luminescence Spectroscopy, Part 3, Schulman, S.G. (ed), John Wiley and Sons, New York, Chapter 9, 1993.
9. D.H. Levy, Ann. Rev. Phys. Chem. 31, 197–225, 1980.
10. I. Nakhimovsky, M. Lamotte, J. Joussot-Dubien, Handbook of Low Temperature Electronic Spectra of Polycyclic Aromatic Hydrocarbons, Elsevier, Amsterdam, 1989.
11. A.D. Campiglia, S.J.Yu, A.J. Bystol, H.Y. Wang, Anal. Chem. 79 (4), 1682–1689, 2007.
12. J.S. de Klerk, A. Szemik-Hojniak, F. Ariese, C. Gooijer, Spectrochim. Acta Part A: Molecular Biomol. Spectrosc. 72, 144–150, 2009.
13. A.N. Bader, F. Ariese, C. Gooijer, J. Phys. Chem. A 106, 2844–2849, 2002.
14. A.N. Bader, V.G. Pivovarenko, A.P. Demchenko, F. Ariese, C. Gooijer, Spectrochim. Acta A 59, 1593–1603, 2003.
15. A.N. Bader, V.G. Pivovarenko, A.P. Demchenko, F. Ariese, C. Gooijer, J. Phys. Chem. B 108 (29), 10589–10595, 2004.
16. P.K. Sengupta, M. Kasha, Chem. Phys. Lett. 68, 382, 1979.
17. G.J. Woolfe, P.J. Thistlethwaite, J. Am. Chem. Soc 103, 6916–6923, 1981.
18. M. Itoh, K. Tokumura, Y. Tanimoto, Y. Okada, H. Takeuchi, K. Obi, I.J. Tanaka, Am. Chem. Soc. 104, 4146–4150, 1982.
19. A.J.G. Strandjord, S.H. Courtney, D.M. Friedrich, P.F. Barbara, J. Phys. Chem. 87, 1125–1133, 1983.
20. D. McMorrow, T.P. Dzugan, T.J. Aartsma, Chem. Phys. Lett. 103, 492–496, 1984.
21. M. Itoh, Y. Fujiwara, M. Sumitani, K. Yoshihara, J. Phys. Chem. 90, 5672–5678, 1986.
22. G.A. Brucker, D.F. Kelley, J. Phys. Chem. 91, 2856–2861, 1987.
23. B. Dick, N.P. Ernsting, J. Phys. Chem. 91, 4261–4265, 1987.
24. G.A. Brucker, D.F. Kelley, Phys. Chem. 92, 3805–3809, 1988.
25. B.J. Schwarz, L.A. Peteanu, C.B. Harris, J. Phys. Chem. A 96, 3591–3598, 1992.
26. A. Muhlpford, T. Bultmann, N.P. Ernsting, B. Dick, Femtosecond Reaction Dynamics, Wiersma, D.A. (ed), Royal Netherlands Academy of Arts and Sciences, Amsterdam, 1993.
27. S.M. Ormson, D. LeGourrierec, R.G. Brown, P. Foggi, J. Chem. Soc., Chem. Commun., 2133, 1995.
28. S. Ameer-Beg, S.M. Ormson, R.G. Brown, P. Matousek, M. Towrie, E.T.J. Nibbering, P. Foggi, F.V.R. Neuwahl, J. Phys. Chem. A 105, 3709–3718, 2001.
29. A. Muhlpford, T. Bultmann, N.P. Ernsting, B. Dick, Chem. Phys. 181, 447–460, 1994.
30. D. McMorrow, M. Kasha, J. Am. Chem. Soc. 105, 5133–5134, 1983.
31. D. McMorrow, M. Kasha, Proc. Natl. Acad. Sci. USA 81, 3375–3378, 1984.
32. D. McMorrow, M. Kasha, J. Phys. Chem. 88, 2235–2243, 1984.
33. A.J.G. Strandjord, P.F. Barbara, J. Phys. Chem. 89, 2355–2361,1985.
34. P.F. Barbara, P.K. Walsh, L.E. Brus, J. Phys. Chem. 93, 29–34, 1989.
35. G.A. Brucker, T.C. Swinney, D.F. Kelley, J. Phys. Chem. 95, 3190–3195, 1991.
36. C.A. Taylor, M.A. El-Bayoumi, M. Kasha, Proc. Nat. Acad. Sci. USA 63, 253, 1969.
37. A. Nakajima, M. Hirano, R. Hasumi, K. Kaya, H. Watanabe, C.C. Carter, J.M. Williamson, T. A. Miller, J. Phys. Chem. 101, 392–398, 1997.

38. J. Catalan, M. Kasha, J. Phys. Chem. 104, 10812–10820, 2000.
39. J. Catalan, Chem. Phys. Letters 423, 395–400, 2006.
40. O.-H. Kwon, D.-J. Jang, J. Phys Chem. B 109, 20479–20484, 2005.
41. H. Sekiya, K. Sakota, Bull. Chem. Soc. Jpn. 79, 373–385, 2006.
42. L. Serrano-Andrés, M. Merchán, Chem. Phys. Lett. 418, 569–575, 2006.
43. T. Fiebig, M. Chachisvilis, M. Manger, A.H. Zewail, A. Douhal, I. Garcia-Ochoa, A. de La Hoz Ayuso, J. Phys. Chem. A 103, 7419–7431, 1999.
44. S. Takeuchi, T. Tahara, Chem. Phys. Lett. 347, 108–114, 2001.
45. O.-H. Kwon, A.H. Zewail, Proc. Natl. Acad. Sci. U S A. 104 (21), 8703–8708, 2007.
46. C. Reichardt, Solvents and Solvent Effects in Organic Chemistry, VCH, Weinheim, 2nd Ed, 1988.
47. S. Zapotoczny, A. Danel, M. T. Sterzel, M. Pilch, J. Phys. Chem. A 111, 5408–5414, 2007.
48. K. Rechthaler, K. Rotkiewicz, A. Danel, P. Tomasik, K. Khatcharian, G. Kohler, J. Fluoresc. 7, 301, 1997.
49. K. Rurack, A. Danel, K. Rotkiewicz, D. Grabka, M. Spieles, W. Rettig, Organic Lett. 4, 4647, 2002.
50. J.-H. Pan, Y.-M. Chou, H.-L. Chiu, B.-C. Wang, Tamkang, J. Sci. Eng. 8, 175, 2005.
51. J. Catalan, M. Kasha, J. Phys. Chem. A 104, 10812–10820, 2000.
52. J. Zyss, D.S. Chemia, J.F Nicoud, J. Chem. Phys. 74, 4800, 1981.
53. A. Szemik-Hojniak, T. Głowiak, I. Deperasińska, A. Puszko, Z. Talik, J. Non-Linear Opt., Quantum Opt. 30, 215, 2003.
54. A. Szemik-Hojniak, T. Głowiak, I. Deperasińska, A. Puszko, J. Mol. Struct. 597, 279, 2001.
55. A. Szemik-Hojniak, T. Głowiak, A. Puszko, Z. Talik, J. Mol. Struct. 449, 77, 1998.
56. B. Poór, N. Michniewicz, M. Kállay, W.J. Buma, M. Kubinyi, A. Szemik-Hojniak, I. Deperasinska, A. Puszko, H. Zhang, J. Phys. Chem. A 110, 7086, 2006.

Hydrocarbon Fluid Inclusion Fluorescence: *A Review*

Nigel J.F. Blamey and Alan G. Ryder

Abstract Geological fluid inclusions are small voids that can contain a variety of liquids which are often found in natural minerals and rocks. Typically they are less than 10 micrometres in size that host fossil fluids which existed when the minerals grew or healed after fracture. Of particular interest to the petroleum industry are inclusions that contain hydrocarbon fluids, which originated from petroleum that once migrated through the rocks before becoming trapped. These hydrocarbon-bearing fluid inclusions (HCFI) are useful for learning about the processes, fluid compositions, temperatures and pressure conditions in geologic systems such as the migration of hydrocarbon fluids in petroleum basins. The accurate characterisation of the petroleum fluid entrapped in inclusions presents the analyst with considerable challenges. HCFI samples are very valuable (usually obtained from core drilling) and thus a non-contact, non-destructive, analytical method is required. The small size of HCFI necessitates the use of microscopy based techniques while spectroscopic methods are needed to characterise the chemical composition. Fluorescence based methods offer the best combination of high sensitivity, diagnostic potential, and relatively uncomplicated instrumentation. It is the fluorescence of HCFI and the spectroscopic methods employed for their analysis which is the focus of this review. Specific sections focus on the description of HCFI, petroleum fluorescence, and microscopic techniques. The review and discussion focuses primarily on advances and studies reported in the literature from 1980's onwards, and outlines some of the issues that need to be addressed to make fluorescence methods more reproducible and quantitative for HCFI analysis.

Keywords Fluid inclusion · Hydrocarbon · Petroleum · Fluorescence · Microscopy · Spectroscopy

---

A.G. Ryder (✉)
Nanoscale Biophotonics Laboratory, School of Chemistry, National University of Ireland – Galway, Galway, Ireland
e-mail: alan.ryder@nuigalway.ie

C.D. Geddes (ed.), *Reviews in Fluorescence 2007*, Reviews in Fluorescence 2007,
DOI 10.1007/978-0-387-88722-7_13,

Introduction

In a geological context, fluid inclusions are small voids, which are found in natural minerals and rocks, that can contain a variety of liquids. They may be regarded as small sealed vials, often less than 10 μm in size, that host fossil fluids which existed when the minerals grew or healed after fracture [1]. The composition of fluid inclusions varies considerably, and may comprise liquid, solid, and/or gaseous phases, depending on the fluid source and the pressures–temperatures experienced. These phases commonly include water, dissolved gases, and salts; in extreme cases, inclusions may host daughter minerals, high-pressure vapour phases, and complex organic mixtures. Of particular interest to the petroleum industry are those inclusions that contain hydrocarbon fluids, which originated from petroleum that once migrated through the rocks before becoming trapped within inclusions [2].

Fluid inclusions, despite their small size, are highly valuable to understanding many geological processes. When these cavities within the rock were sealed, they trapped the original fluid at these fossil pressure-volume-temperature (PVT) conditions. This PVT data can be used for modelling fluid phase behaviour and give an indication of the "oil window" at which oil formation occurs [2]. Geologists can learn much about the processes, fluid compositions, temperatures, and pressure conditions in geologic systems from fluid inclusions that form at a key time, such as during ore formation in base-metal or gold deposits, or from the migration of hydrocarbon fluids in petroleum basins. Not all fluid inclusions present within a sample may represent the key event under study and therefore an experienced fluid inclusion analyst is needed to examine the paragenetic relationships between fluid inclusion assemblages, mineral growth, or fracture healing. By so doing, the geologist is able to study the migration of fluids, including petroleum fluids, and understand what processes led to the migration or trapping of petroleum.

Hydrocarbon-bearing fluid inclusions (HCFI) generally occur in diagenetic cements or grains and contain complex mixtures of mainly organic compounds depending on their source/s. Accurate analysis of the chemical composition of the entrapped hydrocarbons in HCFI can yield vital information about the history, evolution, and migration of petroleum fluids, and is thus crucial data for the petroleum exploration industry. Studying HCFI is advantageous because the trapped fluids are representative of the actual hydrocarbon fluids that existed when the inclusions were sealed in the mineral. This sealing in process preserves the petroleum fluid, thus isolating it from subsequent infiltration of petroleum fluids and events in oil reservoirs such as loss of charge, water washing, or biodegradation. It also preserves the fluid from contamination during the drilling processes used to extract samples from the ground [3].

The accurate characterisation of the fluid entrapped in inclusions presents the analyst with considerable challenges. HCFI samples are very valuable (usually obtained from core drilling) and thus a non-contact, non-destructive, analytical method is required. The small size of HCFI necessitates the use of microscopy-based techniques and fluorescence spectroscopy offers the best combination of high sensitivity, diagnostic potential, and relatively uncomplicated instrumentation. It is the fluorescence of HCFI and the spectroscopic methods employed for analysis which is the focus of this review. Specific sections focus on the description of HCFI, petroleum fluorescence, and microscopic techniques. The review and discussion focuses primarily on advances and studies reported in the literature from 1980s onwards, and outlines some of the issues that need to be addressed to make fluorescence methods more reproducible and quantitative for HCFI analysis.

Inclusions: *A Brief Description*

Fluid inclusions are found in many different types of natural minerals and rocks. They may be regarded as small sealed vials, often less than 10 μm in size, that host fossil fluids which existed when minerals grew or healed after fracture [1]. Fluid inclusion shapes vary widely from highly irregular to "negative crystals" where the host mineral crystal habit is mimicked by the inclusion. Fluid inclusion composition varies considerably, depending on the fluid source and temperatures, and may be host to aqueous or hydrocarbon liquids, solids, and gaseous phases. Aqueous fluid inclusions comprise mainly water with dissolved salts or daughter minerals along with minor dissolved inorganic, organic, and trace noble gases whereas hydrocarbon-bearing fluid inclusions (HCFI) generally comprise liquid hydrocarbons, low carbon-number gases, and occasionally dark solid phases (see Fig. 1).

HCFI may occur both in reservoirs and in migration pathways, the most commonly documented are associated with reservoirs, and are trapped in diagenetic cements, overgrowths, or secondary fractures in quartz and feldspar. HCFI may develop at any time from the onset of reservoir filling to the present day, with trapping most likely during reservoir filling rather than later stages when the water is displaced [4]. This may be attributed to quartz cementation being a slow, temperature mediated, process [5, 6] and that the micron-sized oil droplets sticking to the quartz grains inhibit quartz growth. The trapping of oil within a quartz cement may take millions of years whereas healing of secondary fractures is much more rapid, and may preserve the later stages of reservoir filling [3]. The abundance of HCFI in clastic sediments may be correlated with porosity and permeability, [7] whereas in chalk or limestone reservoirs the HCFI may be more irregular favouring cemented fractures, recrystalised fossils, or coarse-grained cement [2, 8, 9].

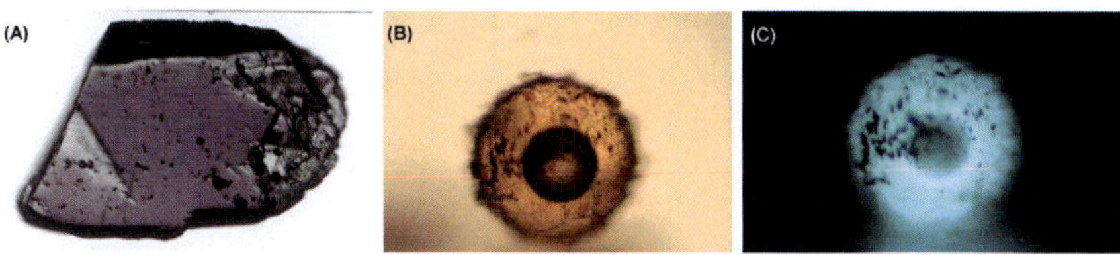

Fig. 1 Images of a rock wafer containing inclusions (**A**), white light image of a liquid-vapour HCFI (**B**), and an epifluorescence image of the same inclusion using 366 nm excitation (**C**)

HCFI are not limited to petroleum reservoirs and they may also be found in a diversity of geological environments. For example, there is increasing evidence for the presence of HCFI within metal deposits [10, 11]. Whether the presence of abundant organic compounds is an integral part of the mineralisation process may depend on specific ore deposit styles. For example, Mississippi-Valley-type (MVT) deposits commonly host HCFI, although their presence may be attributed to deep basinal brines that were primarily the source of Pb- and Zn-bearing fluids.[1] Very old, Archaean rocks are also known to host HCFI, and they have been reported from the gold-bearing Witwatersrand Basin in South Africa [12, 13] and from the Pilbara Craton in Australia [14]. The origins of hydrocarbons in these Archaean rocks is thought to be abiogenic [14]. Other unusual geologic settings include impact craters [15, 16] and hydrothermal vents [17].

The Inclusion Fluids

The hydrocarbon fluid in HCFI may have highly complex compositions but are principally liquid with variable amounts of gas (light alkanes, CO_2, N_2, H_2S), paraffins, napthalenes, aromatics, resins, and sometimes waxes [2, 3, 9, 18]. The apparent HCFI colour depends on a combination of the wavelengths absorbed by the various chromophores present in the trapped hydrocarbon liquid and the associated mineral environment. Consequently, their colour, when viewed under a microscope with plain white light illumination, may vary considerably from transparent to yellow, dark brown, or black. The mineralogical properties which can also influence the perceived colour include birefringence, polarisation, reflectivity, and variations in refractive index.

One of the key methodologies in fluid inclusion study is micro-thermometry [1, 19]. The sample containing the fluid inclusion(s) is placed in a chamber that can be either heated or cooled and then observed using a microscope while the temperature is varied. The temperatures at which phase changes occur are recorded, thus providing information about homogenisation temperature brine salinity (for aqueous inclusions), or other data. HCFI provide fewer phase changes but they do fluoresce under UV or visible light illumination which can provide another source of information about fluid composition. The petroleum composition of HCFI is typically obtained by crushing a small amount of the HCFI containing material and extracting the petroleum fluid for analysis by gas chromatography [20] and mass spectrometry [21]. The key disadvantages of the crush method are the destruction of the sample, probable mixing of fluids from different source inclusions, and potential contamination. Therefore, the oil composition data derived from these bulk analyses have to be carefully considered as the mixing of heterogeneous fluid inclusion populations may have occurred. The non-destructive quantitative analysis of individual

[1] Basinal brines are responsible for transporting hydrocarbons from deep basins where the hydrocarbons were generated from plant or animal matter.

HCFIs is therefore a desirable goal for those studying petroleum migration and is best achieved via optical methods. Optical methods have several general advantages that include: non-destructive/ non-contact analysis, reasonable sensitivity, ability to undertake micron-scale analysis of single inclusions, mature technology and methodologies, high diagnostic potential, and relatively simple instrumentation [22, 23].

Fluorescence of Crude Oils

The fluorescence of crude petroleum oils has been reviewed in detail previously [24]. Briefly, the fluorescence emission is due to the presence of a multitude of aromatic hydrocarbons in varying concentrations. The nature of the emission is governed by the complex interplay between reabsorption, energy transfer, and quenching caused by the high concentrations of fluorophores and quenchers in petroleum oils. The complexity of crude oils usually prevents the resolution of any specific chemical component in terms of fluorescence emission parameters. Factors such as the specific chemical composition (concentration of fluorophores and quenching species) and physical (viscosity and optical density) control emission properties such as emission and excitation intensity, emission and excitation wavelength, and fluorescence lifetime. Generally, light oils (high-API gravity) have relatively narrow, intense fluorescence emission bands with small Stokes' shifts. In contrast, heavy oils (low API gravity) have emissions that tend to be broader, weaker, and red shifted. The differences in emissions are attributed to a higher concentration of fluorophores and quenchers present in the heavier oils, thus leading to an increased rate of energy transfer and quenching to produce broader, weaker, red-shifted emissions [25, 26].

Since crude petroleum oils have very complex compositions, the excitation wavelength is of fundamental importance because changing wavelengths will result in the excitation of different fluorophore populations, which in turn can result in dramatic changes in the emission properties. Longer excitation wavelengths result in a narrowing of the fluorescence emission bands, along with a reduction in Stokes' shifts, fluorescence lifetime, and quantum yields [24–28]. This is attributed to excitation of different fluorophores which changes the rates of fluorescence emission, quenching, and energy transfer [26, 28, 29]. There have been many fluorescence studies of crude petroleum oils and several sources have correlated steady-state emission properties to oil composition. Using a 365 $\pm$ 30 nm excitation source, the wavelength of maximum fluorescence emission (λ_{max}) was shown to correlate with API gravity [30] and the aromatic composition [31] of crude oils. A ratio of fluorescence intensity at 650 and 500 nm (known as $Q_{R/G}$ or red–green quotient) was seen to correlate with viscosity [30], API gravity and composition from selected Canadian crude oils [31, 32]. A similar parameter (Q-535) based on the ratio between the 535–750 nm flux and 430–535 nm flux using 365 nm excitation was also used to characterise crude oils and HCFI but no quantitative correlation to petroleum composition was described [22, 33]. The Q-535 value decreases with increasing oil maturity, i.e. bluer fluorescing inclusions.[2]

One bulk optical method used for analysis of trace fluorescence on grain surfaces as well as oil inclusions is the Quantitative Grain Fluorescence (QGF) and Quantitative Grain Fluorescence on Extract (QGF-E) methods developed by CSIRO [34]. This is a bulk fluorescence method that uses a slightly modified standard fluorescence spectrometer, where a narrow band interference filter is added in the excitation path length to provide a precise excitation wavelength in the UV (254 nm for QGF and 260 nm for QGF-E). The continuous emission spectrum from 300 to 600 nm is generally measured, and these data are used to calculate a QGF index, which is defined as the average spectral intensity between 375 nm (I_{375}) and 475 nm (I_{475}) normalised to the spectral intensity at 300 nm. The authors also, rather confusingly define a QGF ratio which is the ratio between the average spectral intensity of 375–475 nm and the spectral intensity at 350 nm. The spectrum of condensate and light oils are generally skewed towards the shorter wavelengths (~400 and ~450 nm peak maxima, respectively) whereas the heavier oils have broader and more red-shifted emission maxima (~475 nm for an oil with an API gravity of 25). Isolated tetra-aromatic and polar fractions (from liquid chromatography) have spectral maxima which are even further red shifted to ~475–500 and 550 nm, respectively. Dilution of the isolated fractions in hexane results in blue shifted emission spectra particularly for the tetra-aromatics. The QGF index has been used to infer relative palaeo-oil saturation [34] and discriminate different oil types according to maturity [35].

HCFI Fluorescence

HCFI may easily be mistaken for aqueous-bearing fluid inclusions during a first-pass inspection using optical petrography in transmitted light mode because many HCFI are colourless or slightly brown with a vapour bubble and thus resemble aqueous-dominated fluid inclusions. Excitation by UV illumination is the routine preliminary method for identifying HCFI in thin section thus discriminating them from aqueous inclusions. Aqueous fluid inclusions (AFI) generally do not fluoresce although weak scattering of UV light may, in rare cases, mislead the observer into mistaking AFI for a weak violet fluorescing HCFI. The petroleum fluids within HCFI fluoresce when illuminated by light of a sufficiently energetic wavelength, in practice this can be achieved with light from the UV to the NIR region of the spectrum [24]. The earliest paper to describe HCFI fluorescence may be one by Murray in 1957 [36], but it was the increased availability of fluorescence microscopes dur-

[2]Hydrocarbon gas content generally indicates the maturity of an oil, controlled by the degree to which it has been heated to induce breakdown of large organic molecules to form oil and gas. Immature oils have low gas content (methane is dominant), whereas mature oils have moderate gas content with high ethane and propane concentrations: finally overmature oils have high methane contents.

ing the 1980s that empowered those studying HCFI to commence active research using fluorescence methods [37, 38]. Since then there has been an extensive literature on the subject, and the analysis of HCFI by fluorescence methods is now very common [2, 39].

However, despite the fact that the fluorescent properties of crude petroleum oils have been widely studied [24], the linking of fluid inclusion composition to crude oil composition is not always straightforward [2, 20, 40]. The most important caveat is that most bulk oil studies involve the use of "dead" oils where there has been a loss (intentional or otherwise) in the light alkane and gaseous fractions from the oils prior to analysis. A loss in light fraction alkanes tends to result in additional fluorescence quenching, which manifests itself as weaker emission and shorter lifetimes. When one considers HCFI, many of the inclusions, because they are sealed, should retain a significant light gaseous fraction and therefore the oil is termed "live". These live oils tend to have stronger, bluer, and longer-lived fluorescence lifetimes because of the dilution effect from the gaseous alkanes. This creates a difficulty in creating calibration correlations with bulk crude oils that have been topped prior to compositional analysis.[3] Therefore, assuming that fluorescence responses of "dead" oils are exactly the same as "live" HCFI can lead to incorrect conclusions. Another consideration for HCFI studies is that the reference oils should be sourced from the same locality [31, 41], to avoid the wide fluorescence variances found between oils from different sources which have similar API gravity [24].

Recently, the use of diamond anvil cells [42] coupled with fluorescence detection has allowed crude petroleum oils to be subjected to typical geological burial temperatures and pressures, and the changes in emission properties monitored. The method has also been used to explore the formation of petroleum oils from different sources [43, 44]. In the 2008 study [44], the authors clearly demonstrate the progressive blue shift (λ_{max}) and increase in emission intensity as the kerogen-based materials undergo thermal maturation, and petroleum oil is formed. The data presented suggest that the λ_{max} of spectra for inclusion oils will shift in a similar direction despite differences in composition or source kerogen. It is hoped that these studies will be extended to the collection of fluorescence lifetime data which could give a more complete insight to both oil formation and maturation processes. The method looks very promising for the development of reproducible oil maturity calibration models that than can be applied to HCFI analysis.

Another approach to the development of standards for HCFI fluorescence analysis has been the use of synthetic fluid inclusions. Synthetic hydrocarbon-bearing fluid inclusions can be generated using a variety of materials including KCl [23], NaCl [31], phosphates [45], and quartz [46]. For soluble salts, the general procedure is to add some petroleum oil to a supersaturated solution of the salt followed by a slow evaporation to form crystals [31]. Unfortunately, the process of evaporation, crystal growth, and inclusion formation is not conducive to the homogeneous

[3] Topped oils are those in which the light fraction is intentionally removed by heating to 60°C (typically) prior to analysis.

trapping of the petroleum fluids. The fluid can be partitioned into different fractions according to changes in affinity for crystal surfaces and the aqueous layer. An additional factor to consider is the fact that the petroleum fluid entrapped in synthetic fluid inclusions will be a "dead" oil, lacking the lighter gaseous hydrocarbons.

Sample Preparation

Accessing HCFI for optical analysis generally involves the preparation of thin "wafer like" sections of the sample [19]. For rock samples, whole wafers are prepared, whereas with drill core cuttings the material is often comprised of small flakes/chips of rock. For these drill cuttings, it is more practical to mount the chips in resin prior to wafer preparation. Unlike conventional rock slabbing, where a fast rotating diamond-tipped saw is used to cut through the sample, fluid inclusion saws rotate at much lower revolutions and the sample requires cooling. This is critical because the temperatures generated by these blades can compromise fluid inclusions that have low homogenisation temperatures. A cutting oil is commonly mixed with water to lubricate as well as cool the blade, and therefore one should be aware of potential damage to and contamination of wafers and their inclusions. Spurious apparent HCFI can be created by the lubrication fluid but these false fluorescent targets can usually be recognised by the fact that the fluorescence is localised at the wafer surface. After the first cut is made, the wafer is ground and then polished to generate a good-quality flat surface. Resin can then be applied to the polished surface, so that it may be mounted on a glass slide. A second parallel cut is then carefully made to produce a wafer that is several hundred microns in thickness, depending on the thickness required. For high-quality optical measurements, thicknesses of $\sim$150 μm are ideal. Grinding and polishing of the new surface results in a doubly polished wafer that is suitable for microscopic examination. It should be noted that curing of the resin needs to be at low temperature so as to avoid potential HCFI decrepitation [47]. For friable samples, McNeil and Morris [48] detail a method that employs resin impregnation into the sample that produces a polyester cast suitable for thin wafer generation. One should, however, take care with the type of resins being used, as many of these can be fluorescent under UV illumination, which can cause problems for fluorescence-based analysis methods.

Epifluorescence Microscopy

The initial step in recognising HCFI in geologic sections is by observing fluorescence using an epifluorescence microscope with UV illumination. HCFI have apparent (visually judged) fluorescent colours covering the full visible spectrum from blue to red in addition to white. Because the fluorescence of hydrocarbon based fluids is composition dependent, the apparent fluorescent colour is used to distinguish different HCFI generations within the same and different wafers. Unfortunately, the human eye lacks the ability to accurately discriminate discrete wavelengths, and therefore reporting visual fluorescence colours is at best a qualitative technique and does not provide quantitative results [20, 49, 50].

The implementation of epifluorescence microscopy for HCFI analysis has to be considered very carefully since we know that emission properties (λ_{max}, intensity, and lifetime) are all very dependant on the excitation wavelength [24]. The instrumentation is nonetheless, relatively simple and low cost, requiring a microscope, UV light source (usually a high-pressure Hg lamp, typically fitted with a bandpass filter to transmit a specific excitation light wavelength, often ~366 nm), and a filter-cube arrangement. The filter cube houses a dichroic mirror that is used to direct the UV light onto the sample and the resultant fluorescence to the operator. A barrier filter (commonly 420 nm long pass) prevents harmful UV light being observed by the observer. The quality of this filter is important, because many light oils fluoresce in the violet region, having small Stokes shifts. If the filter is not sufficiently steep edged at the cut off position, then it will either pass some of the excitation light and/or filter out a substantial amount of the petroleum fluorescence, thus distorting the emission spectrum and colour. The detection system is typically visual via binocular tubes, or video/still image recording by digital or analogue sensors. A selected list of studies that have used epifluorescence for the examination of HCFI are provided in Table 1 along with comments (if available) relating to the wavelengths of the excitation source and the range of wavelengths (by filter or spectrometer) sampled from the fluorescence emission. It is important to note the wide variety of excitation wavelengths and different bandpass filters results in the excitation of different populations of fluorophores, generating intrinsically different emission profiles. Furthermore, the fluorescence emission is then sampled in a wide variety of ways using disparate optical configurations in the emission path. This highlights the potential for problems in assessing HCFI fluorescence colour accurately, particularly if only truncated regions of the fluorescence emission are sampled or if the spectra are not corrected for instrument response. Despite this, epifluorescence remains the most broadly used first-pass technique for identifying and establishing the HCFI paragenesis.

The fluorescent colours of HCFI have been used as a guide to estimating the maturity and API gravity (a density measurement)[4] of migrating oil [21, 31, 52, 53, 103, 104]. Generally blue fluorescence is associated with oils of high-API gravity whereas inclusions with red fluorescence are linked to less mature oils and lower-API gravity [20]. However, George et al. [3] show that blue fluorescing HCFI have thermal maturities anywhere within the oil window and that fluorescent colours do not have a reliable correlation with maturity. The charge history of a reservoir, based on fluorescent colours, may be misinterpreted owing to the potential colour populations to represent one single event but that slightly different conditions, mineralogical boundaries [3], or inclusion thickness [105] may change the apparent fluorescence colour. Furthermore, the human perception of colour varies from individual to individual, particularly when the fluorescence intensity of the

[4]API gravity = ((141.5/specific gravity at 15.6°C)–131.5).

Table 1 Literature survey (not comprehensive) of the use of epifluorescence microscopy for HCFI analysis. What is very clear is lack of consistency in regard to the specific excitation wavelengths used (centre wavelength and bandpass). There is also a lack of information as to the optical configuration (filters, etc.) in the emission path

Method	Excitation and filters	Observed wavelength	References
Epifluorescence	UV	Not specified	Burruss et al. [37]
Epifluorescence	UV, 366 nm	Not specified	Burruss [8]
Unknown	Unknown	Unknown	Mukhopadhyay and Rullkotter [51]
Epifluorescence, Steady state	366 nm (epi) 270–366 nm	Not specified Not specified	McLimans [52]
Epifluorescence	Hg lamp	366 nm long pass	Bodnar [53]
Epifluorescence	UV	Not specified	Levine et al. [54]
Epifluorescence	UV	Not specified	Moser et al. [55]
Epifluorescence	UV	Not specified	Newell et al. [56]
Epifluorescence	365 nm	430–750 nm	Pironon et al. [57]
Epifluorescence	365 nm	430–750 nm	Pironon et al. [58]
Epifluorescence	UV	Not specified	Parnell et al. [59]
Epifluorescence	Not specified	Not specified	Xiao et al. [60]
Epifluorescence	UV	Not specified	George et al. [61]
Epifluorescence	UV, not specified	Not specified	Parnell et al. [62]
Epifluorescence	UV	Not specified	George et al. [63]
Epifluorescence	330–380 nm	Not specified	Dutkiewicz et al. [12]
Epifluorescence	365 nm	450 barrier	O'Reilly et al. [64]
Epifluorescence	UV	Not specified	Tseng et al. [65]
Epifluorescence	UV, not specified	Not specified	Smale et al. [66]
Epifluorescence	395 nm	440 nm barrier	Rantitsch et al. [67]
Epifluorescence	UV	Not specified	Parnell et al. [68]
Epifluorescence	UV, probably 365 nm	Not specified	Cesaretti et al. [69]
Epifluorescence	UV	Not specified	Lonnee & Al-Aasm [70]
Epifluorescence	UV	Not specified	Marchand et al. [71]
Epifluorescence	Ref. Pironon 1998	Not specified	Thiéry et al. [72]
Epifluorescence	UV	Not specified	Parnell et al. [73]
Epifluorescence	UV, not specified	Not specified	Middleton et al. [74]
Epifluorescence	~365 nm, 330–385 nm	420 barrier filter	George at al. [20]
Epifluorescence	UV, not specified	Not specified	Lavoie et al. [75]
Epifluorescence	Hg lamp, 330–380 nm	Nikon UV-2A block, ~420 nm longpass	Parnell et al. [21]
Epifluorescence	UV, not specified	Not specified	Ceriani et al. [76]
Epifluorescence	UV	Not specified	England et al. [13]
Epifluorescence	UV	Not specified	Rossi et al. [77]
Epifluorescence	365 and 410 nm	Not specified	Lisk et al. [78]
Epifluorescence	366 nm	Not specified	Mauk and Burruss [79]
Epifluorescence	Hg lamp, 362 nm	Blue and green filters	Volk et al. [80]
Epifluorescence	UV, 366 nm	Not specified	Tseng and Pottorf. [81]
Epifluorescence	UV, 365 nm	Not specified/digital camera	Dutkiewicz, Ridley, et al. [82]
Epifluorescence	UV	Not specified	Dutkiewicz and Ridley [83]
Epifluorescence	UV	Not specified	Dutkiewicz, Volk, et al. [84]

Table 1 (continued)

Method	Excitation and filters	Observed wavelength	References
Epifluorescence	UV	Not specified	Feely and Parnell [85]
Epifluorescence	Hg lamp	Not specified	Boles et al. [86]
Epifluorescence,	365 nm (12 nm bandpass)	397 long pass	Gonzalez-Partida et al. [10]
Epifluorescence	UV	Not specified	Martinez-Ibarra et al. [87]
Epifluorescence	Hg lamp	Not specified, photography	Parnell et al. [16]
Epifluorescence	UV	Not specified	Dutkiewicz et al. [88]
Epifluorescence	UV	Not specified, photography	Jonk et al. [89]
QGF	254 nm	300 – 600 nm	Liu and Eadington [34]
Epifluorescence	UV	Not specified	Volk et al. [90]
Epifluorescence	UV	Not specified	Dutkiewicz et al. [91]
Epifluorescence	UV	Not specified	Hanks et al. [92]
Epifluorescence	Not stated	Not specified	Brincat et al. [93]
Epifluorescence	330–385 nm	Not specified	Rott and Qing [94]
Epifluorescence	Blue-light	Not specified	Wierzbicki et al. [95]
Epifluorescence	UV (Leitz Ortholux II)	Not specified	Wilkinson et al. [96]
Epifluorescence	Hg lamp, 365 nm	420 nm longpass/ digital camera	Baron et al. [97]
Epifluorescence	UV, 330–385 nm	420 nm longpass	Dutkiewicz et al. [98]
Epifluorescence	365 nm	Not specified	Higgs et al. [99]
Epifluorescence	340–380 nm	425 nm barrier	Schubert et al. [100]
Epifluorescence	365 nm	420 longpass	Baron et al. [101]
Epifluorescence	UV, 360–370 nm	LP400 longpass	Bourdet et al. [102]

emission from the inclusion is very low. Coupling this human variability with the non-standard equipment used for epifluorescence imaging demonstrates the need for care in the interpretation of fluorescence colour results between laboratories. Because of these shortcomings, in many cases, the generational data produced by epifluorescence microscopy has to be validated using additional methods such as petrographic studies to elucidate cross-cutting relationships [1, 52, 106], microthermometry [2, 100, 107], or fluorescence lifetime measurements [40, 105] to accurately discriminate the different generations of inclusions.

Conventional epifluorescence microscopy has, however, two significant limitations for the accurate analysis of HCFI. First, standard epifluorescence imaging has very poor spatial resolution along the axial direction (through the sample), and as such cannot easily be used for uncovering the precise spatial arrangement of HCFI within a wafer. The out of focus light often blurs the inclusion images making it also very difficult to measure volumes of inclusions accurately. Second, the characterisation of HCFI based on apparent fluorescent colour is very subjective and at best just a qualitative guide. For more accurate and robust analyses, one must consider alternate microscopy techniques with higher spatial resolution and spectroscopic methods that yield more comprehensive data about the composition of the petroleum fluids.

High Spatial Resolution Microscopy for HCFI Analysis

While most researchers use conventional epifluorescence microscopy for routine HCFI analysis to provide information about their general location within the sample, the precise location of HCFI in space is often uncertain. Furthermore, while volume determinations may be performed by visual estimation, irregular inclusion shapes can generate large errors. There is a need therefore to make use of techniques capable of providing higher spatial resolution particularly in the axial direction from which quantitative volume and intensity information can be extracted. The ideal solution is to use confocal microscopy, and Pawley [108] gives a detailed overview of the area, the specific methods available, and the hardware required. In many cases, this drive for higher spatial resolution has been driven by applications in the life sciences. Most universities will now have a minimum of one or more confocal microscopes, which should easily be available to most inclusion researchers.

Conventional Confocal Fluorescence Microscopy

The high axial resolution in standard confocal microscopy is achieved by the use of high Numerical Aperture (NA) lenses combined with pinholes in the optical path between the sample and the detector. Stray and out-of-focus light is physically rejected whereas light originating from the desired focal plane is allowed to pass to the detector. Typically, the detector is a single channel (narrow wavelength range) unit, and therefore, to achieve a 2-D image of a target, the excitation laser spot must be scanned (or rastered) across the sample. This can be achieved in a number of ways either by laser scanning using mirrors (Confocal Scanning Laser Microscopy, CSLM) or by moving the sample stage (using piezo-driven units or stepper motors). Either way, a 3-D image stack is generated by acquiring 2-D optical slices through a sample at spacing comparable to the axial resolution, commonly 1–2 μm. These 3-D images can be used to easily discriminate between the fluorescent oil and non-fluorescent host mineral or vapour bubble. This allows both the spatial locations of HCFI to be determined with relatively high accuracy and the volumes of the fluorescent components of the inclusions to be estimated (Fig. 2) [109–112]. The accuracy of the method is constrained by a number of factors that include the inherent resolution of confocal microscopes (typically $\sim$0.3 μm – lateral and $>$1.4 μm – axial with 488 nm excitation), variations in refractive index of the sample, focal depth effects, and mineral birefringence [111]. These factors, coupled with the need to correctly estimate the thresholds for determining fluorescent and non-fluorescent voxels, can lead to significant errors in volume estimation ($\sim$5% according to [102] and [110]).

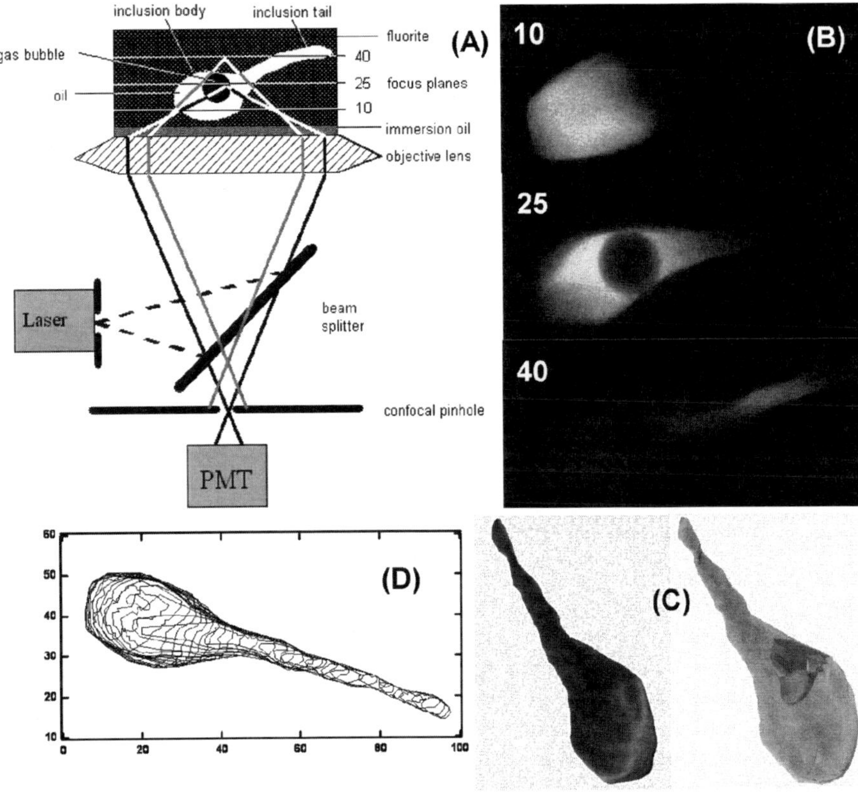

Fig. 2 Clockwise from top left: (**A**) Schematic view of the optical path during confocal recording of the image planes of a two-phase oil inclusion; (**B**) Three two-dimensional CSLM images of the oil inclusion. Two successive planes are separated by 1 μm; (**C**) 3-D reconstructions of an oil inclusion by surface modelling (GOCAD computer program); (a) external surface and (b) external surface and vapour-bubble surface in transparency. (**D**) Superimposed contours of the binarised images of an oil inclusion after thresholding. Scales in micrometer. Adapted from Pironon et al. [110] and reproduced with permission. © *European Journal of Mineralogy*

Another factor in determining the accuracy of this method (and HCFI petrographic/spectroscopic analysis in general) is the type of objective lens used. For volume measurements, accuracy is largely determined by axial and lateral resolution, and therefore one must use high NA lenses. However, the use of high NA objectives also reduces the effective working distance (typically 150–170 μm) making it difficult to analyse accurately inclusions deeply located within wafers. Aplin [111] used an oil immersion ×60 lens for this initial work, but erred in ascribing the vertical (z-axis) resolution as being ~0.1 μm and that it was limited by the stepper motor stage. In fact the vertical resolution is considerably worse than this (most certainly >0.5 μm) and limited by the numerical aperture [108]. Pironon et al. [110] also used a ×60 oil objective with an NA of 1.4, and estimated an axial resolution of 0.5 μm.

It is interesting to note that most studies use 488 nm excitation, whereas many newer confocal systems are equipped with solid state 405-nm laser diodes [102] which can be focussed to a smaller spot size thus potentially improving the accuracy of volume measurements. Kihle [113] used a $\times 100$ oil immersion objective (Olympus D Apo UV, NA $= 1.3$) for spectroscopic measurements but the system was not utilised in a confocal mode. Musgrave et al. [114] used a $\times 40$ glycerine immersion objective (Zeiss Ultrafluar D NA $= 0.6$, working distance of 360 μm) for spectroscopic measurements. For standard epifluorescence imaging and spectral analysis, the use of standard air-spaced objectives is widespread, from $\times 40$ [41] to $\times 50$ power [22]. Water immersion (WI) lenses have also been employed, Li et al. 2006 used an epiplanneofluor $\times 40$ WI objective (NA $= 0.95$).

Constraining the pressure and temperature conditions during fluid inclusion trapping is a highly desirable information for the petroleum exploration industry and can only be achieved by a combination of homogenisation temperature data and volumetric data (liquid versus vapour bubble ratios, i.e. degree of fill) [110]. Fluid inclusion micro-thermometry, a well-established methodology which is described in Roedder [1] and Shepherd et al. [19], is used to generate the homogenisation data. CLSM can be used to calculate the volume of the vapour and liquid in HCFI by surface modelling and voxel counting [110, 111]. By combining the volume estimates with the micro-thermometry data in PVT modelling software, one is able to determine the bulk composition, phase envelope, isochor, and some physical properties of the oil as explained by Aplin [111]. From examination of several coeval inclusions (inclusions that formed during the same event), it may be possible to obtain information regarding the oil's molar volume, saturation pressure, density, viscosity, and surface tension [111]. The method is now quite widely used (see Table 2) for examples of routine HCFI analysis.

However, despite its demonstrated utility for spatial and volume measurements, there are some drawbacks associated with standard confocal microscopy. First,

Table 2 Selected examples from the literature on the use of confocal microscopy for the three-dimensional analysis of hydrocarbon-bearing fluid inclusions

Method	Measurement	Excitation wavelength	Emission wavelength	References
Confocal	Volume	488 nm laser	520 ± 32 nm	Pironon et al. [110]
Confocal	Volume	488 nm laser	Not specified	Aplin et al. [111]
Confocal	Volume	488 nm laser	Not specified	Aplin et al. [115]
Confocal	Volume	Not specified	Not specified	Swarbrick et al. [116]
Confocal	Volume	Not specified	Not specified	Thiéry et al. [117]
Confocal;	Volume	488 nm laser	Not specified	Tseng and Pottorf [81].
Multi-photon (Confocal)	Volume	800 nm fs pulsed laser	Schott glass BG39	Stoller et al. [118]
Structured-light illumination	Morphology	470–490 nm; 540–550 nm	515–550 nm; 575–625 nm	Blamey et al. [119]
Confocal	Volume	405/488 nm laser	Not specified	Bourdet et al. [102]

the instrumentation in most cases can be very costly and complex, making it an expensive proposition for most geological laboratories. Second, it cannot simultaneously image non-fluorescent Aqueous Fluid Inclusions. Third, most of the potential information in the fluorescence emission is discarded because most CLSM-based methods simply utilise fluorescence intensity as a contrast agent for imaging rather than as an analytical measure of petroleum composition.

Structured-Light Illumination Microscopy

The Structured-Light Illumination (SLI) technique is a combination of an optical method and computational image analysis that gives comparable results to confocal microscopy by enhancing the axial resolution to provide high-quality optical slices [120]. The optical slices may then be merged to create 3-D visualisations of samples such as HCFI. Unlike traditional confocal microscopy, SLI is relatively inexpensive, can have faster acquisition times, and does not require laser excitation sources. Applications to HCFI studies is a relatively new development [119], and it has been shown that SLI microscopy can generate 3-D images of HCFI equivalent to that generated by conventional laser scanning methods. Unlike traditional confocal microscopy methods, an intrinsic characteristic of the methodology also allows for bubbles within aqueous fluid inclusions to be detected and imaged. This allows the pinpointing of the position of aqueous fluid inclusions together with HCFI in the same 3-D visualisation, enabling the rapid and facile determination of paragenetic relationships between HCFI and AFI.

Multi-Photon Excitation

A recent development in fluid inclusion microscopy is the application of second harmonic non-linear effects to the study of aqueous fluid inclusions through the use of pulsed femtosecond lasers [118]. The high intensity of the laser pulses generates a very weak second harmonic at the interface between a fluid inclusion and the host mineral that is half the wavelength of the laser's excitation wavelength. In essence, this allows a high-resolution confocal-like mapping of the aqueous fluid inclusion's boundaries in 3-D. However, the application of this method for volume measurements to HCFI is not feasible because the generated second harmonic light will induce fluorescence in the petroleum fluid, thus removing the potential for image contrast based on second harmonic light.

These microscopy techniques while providing increased spatial resolution to better view HCFI and the 3-D arrangement within the sample do not generate significant information about the chemical composition of HCFI. In fact, these methods tend to discard the valuable information inherent in the fluorescence emission because they only measure fluorescence at a single wavelength (or at most 2–5 individual wavelengths). To obtain useful information about the chemical composition of the fluids entrapped within the HCFI, one must adapt confocal microscopes for spectroscopic measurements.

Micro-spectroscopy of HCFI

Extraction of information from the fluorescence emission of HCFI requires the use of various spectroscopic techniques to acquire either steady-state and time-resolved emission parameters. There are no major obstacles to undertaking these types of measurements apart from interfacing the spectroscopic components to the microscope and implementing a rigorous calibration protocol.

Fluorescence Emission Micro-spectroscopy

The first obvious step in increasing the discriminating power of fluorescence is to measure the fluorescence emission spectrum. As outlined earlier, heavy (low-API gravity) oils have broad, weak, and red-shifted emission bands [25, 26, 121] while lighter oils (higher-API gravity) tend to have narrower, more intense, and bluer spectra. For spectroscopic measurements, HCFI are illuminated with a monochromatic light source (lamp, laser, or Light Emitting Diode), and the fluorescence intensity as a function of wavelength is measured. There are a number of different hardware approaches that can be taken to make these measurements. In 1994, Musgrave et al. [114] described a home-assembled system using separate lamp and scanning monochromator and photomultiplier tube detector modules. This system was then corrected for non-uniform response across the measured spectrum using a variety of standard samples including uranium glass. It is also possible using fibre optics to couple a standard fluorometer to a microscope [113], thus reducing the complexity of the overall system. Other options include integrated spectrometer/Photomultiplier Tube detector units dedicated for microscopy [31, 100], compact minispectrometers with photodiode array detector [41], or more recently Acousto-Optical Tuneable Filter (AOTF) based systems [122]. Using a Raman spectrometer (spectrograph with a multi-channel CCD detector) is another option [123, 124] that has been used for HCFI analysis. However, the use of Raman systems for fluorescence studies is typically limited by the fact that the lasers used tend to be visible/NIR (488, 532, 785 nm) rather than UV, thus lowering fluorescence intensity making it more difficult to observe some types of HCFI. Another unhelpful factor is that if these visible laser sources are used then the apparent fluorescence colour will be very different than those recorded using UV excitation.

Lasers are preferred as excitation sources for microscope-based emission spectroscopy because of the very narrow bandpass and long coherence length, which allows for more precise focussing into HCFI. This in turn ensures that the spectroscopic data is collected only from the inclusion and not from the surrounding mineral. However, to achieve the maximum benefit from laser sources one should make use of confocal optics to obtain the best spatial resolution and minimise spurious signals from the host matrix. Unfortunately, in the fluorescence HCFI studies reviewed here we have been unable to find any such combination in the literature, although most Raman spectroscopy studies of aqueous inclusions routinely make use of confocal optics to improve spatial resolution. It is vital that HCFI studies be precise in regard to the reporting of optical conditions and filters used for HCFI fluorescence analysis so that data can be compared between different laboratories. However, reviewing the literature (Table 3) show that there is generally no consistency between the different inclusion research laboratories.

Another consideration, when attempting to measure emission spectra through microscopes, is the need to undertake some form of spectral correction to account for material imperfections, light absorption, and other optical effects in the excitation and emission light paths. A variety of methods have been used, for example Blanchet et al. [41] used the spectral radiance of a reference quartz-iodine lamp illuminating an opal glass which diffuses by transmission according to Lambert's Law [134]. Musgrave et al. [114] used a variety of materials including uranium glass, whole crude oil, and pure organic materials as standards. The spectra of these materials were collected using a factory calibrated spectrometer, and then used to correct the data collected on the microscope. Unfortunately no further specific details were provided. Pironon et al. [58] used a LEITZ uranyl glass (no details available) standard to normalise emission flux measurements to calculate Qf_{535} ratios. A black body curve (using a standard tungsten lamp (3100 K) was used as a reference radiator) was used to correct spectra by Li et al. [133]. Huang's group used FluoRef™ slides for spectral correction when studying oil generation from kerogens in a diamond anvil cell [44]. This last option would be the best for HCFI analysis; however, we cannot find any validation studies on the accuracy of these slides, unlike other fluorescence standard materials [135]. It is important to note that spectral calibration is neither routine nor is there an agreed standard, easily implemented method for HCFI analysis.

Despite the assertion of Barres et al. [136], the fluorescence emission spectrum is a useful source of information on petroleum composition. The use of fluorescence spectroscopy for the analysis of HCFI was first briefly mentioned by McLimans in 1987, and detailed reports started to appear in the early 1990s [22, 23]. In these early studies, UV lamps (~365 nm centre wavelength) were used to excite the HCFI and the fluorescence spectra were measured using scanning monochromators. Technological advances in spectroscopic hardware and more affordable microscopes have opened this method to more general use since then and several publications are listed in Table 3. As regards a standard method for evaluating the fluorescence emission spectra from HCFI, there is no set procedure. For example, Munz et al. [131] collected emission spectra using the same instrumental configuration described by Kihle [113]. They correlated peak wavelength with homogenisation temperature and were able to discriminate two types of inclusion on the basis of micro-thermometry and the maximum of the fluorescence emission. The $Q_{R/G}$ (red–green quotient) proposed by Hagemann and Hollerbach in 1986 was first applied to inclusion analysis by Jochum et al. in 1995 [125]; however, there are no details provided as to instrument response correction or to the specific excitation wave-

Table 3 Literature survey of spectral methods used for HCFI analysis. What is clearly evident is the wide variation in instrumentation, excitation wavelengths, and detection systems

Method/application	Instrument	Excitation	Emission	References
Steady state spectra	Zeiss Zonax	270–366 nm	Unknown	McLimans [52]
Lifetime	Unknown	355 nm	400–600 nm	McLimans [52]
Steady state spectra	Leitz MPV III	365 nm	430–750 nm Monochromator	Pradier et al. 1990 [33]
Steady state spectra	Leitz MPV III	365 nm	430–750 nm Monochromator	Guilhaumou et al. [22]
Steady state spectra	Leitz MPV III	365 nm	430–750 nm Monochromator	Pironon and Pradier [23]
Synchronous scan	Various (FTI)	Xe lamp, Variable, Monochromator	GG385 long pass & variable, Monochromator	Musgrave et al. [114]
Steady state excitation and emission spectra	ZEISS, Photomicroscope III	Blue (not specified)	Not specified	Jochum et al. [125]
FLEEMS (synchronous)	Perkin Elmer LS50 spectrometer	Variable, Monochromator	Variable, Monochromator	Kihle [113]
Steady state spectra	Leitz MPV III	365 nm	430–750 nm Monochromator	Pironon et al. [57]
Steady state spectra	365 nm, filter	Zeiss Ultraviolet G 365 nm excitation filter 420 nm barrier,	400–700 nm (Continuous Filter Monochromator)	Stasiuk and Snowden [31]
Steady state spectra	Zeiss II photomicroscope	Hg Lamp; 368 nm filter	400–700 nm	Chi et al. [126]
Steady state spectra	As per Ref. [31]	Not specified	400–700 nm	Lavoie et al. [75]
Steady state spectra API estimation	Zeiss	Hg lamp; 368 nm filter	400–700 nm; 10 nm increments	Kirkwood et al. [127]

Table 3 (continued)

Method/application	Instrument	Excitation	Emission	References
Steady state	Zeiss MPM II	Not specified	400–700 nm	Morrow et al. [128]
Steady state spectra	Modified Zeiss type Axioplan II microscope	Laser; 360 nm	400–671 nm; OMA 2000 EG&G model 1421	Blanchet et al. [41]
Steady state spectra	Leitz MPV-III	UV; 365 nm	400–700 nm	Li and Parnell. [129]
Steady state spectra	Zeiss UMSP 50 spectrometer	Hg lamp (not specified, <395 nm)	395 nm barrier filter; 400–700 nm spectrometer	Alderton et al. [130]
Steady state spectra	As per Ref. [113]	UV, 365 nm	No barrier filter 400–700 nm	Munz et al. [131]
QGF	QFT-II	254 and 260 nm	300 – 600 nm	Liu and Eadington [34]
Steady state spectra	As per Ref. [31]	Not specified	Not specified	Lavoie et al. [132]
Lifetime	PicoQuant & Olympus	405 nm	450–800 nm, monochromator	Ryder et al. [40]
Steady state spectra	Zeiss Axioplan II microscope	Hg lamp; 365 nm filter	420–720 nm, Zeiss MPM 200/650 photometer and continuous filter monochromator	Li et al. [133]

length (they say blue light!!). They did attempt to correlate their data with the data previously collected on bulk oils by Hagemann and Hollerbach [30]; however the correlation is tentative at best.

A more detailed study, which sought to correlate crude oil and fluid inclusion $Q_{R/G}$ ratios with chemical composition was published by Stasiuk and Snowdon [31] in 1997. The values of L_{max} and $Q_{R/G}$ were shown to vary with aromatic and saturate content and a red shift correlated with increasing aromatic and NSO (Nitrogen-Sulphur-Oxygen containing species, i.e. polar fraction) concentrations. The authors also correlated the spectra obtained from HCFI with data collected from crude oils sourced from carrier beds and reservoir rocks of the Upper Devonian Birdbear Formation in Canada. There are several reports of studies which use fluorescence spectral data to estimate the API gravity of the entrapped fluids according to the Stasiuk and Snowdon correlations [126, 127]. The $Q_{F\text{-}535}$ ratio proposed by Guilhaumou, Szydlowski, and Pradier in 1990 has not been widely adopted; most studies in the literature are from the original authors or collaborators [23, 33, 57, 137, 138]. As with the other method, the $Q_{F\text{-}535}$ ratio decreases as oils become lighter. In the original work [22], the authors were also able to observe changes in fluorescence emission spectra (a red shift and an increase in the emission band width) when inclusions were overheated to 300°C in a pressure cell at ~40 MPa. The experiment seems to indicate that the changes in emission were induced by thermal quenching and photochemical reactions, not surprising since the studies involved prolonged UV exposure. Pironon and Pradier [23] also investigated heating effects on HCFI (both natural and artificial) using fluorescence and observed a similar irreversible red shift in the emission spectra and an increase in the $Q_{F\text{-}535}$ ratio. This indicates that the heating to high temperatures induces a chemical change in the entrapped fluid. The more recent work with diamond anvil cells [42–44] gives a clearer picture with regard to the evolution of petroleum oils from source kerogen and the influence of temperature.

Attempts to better visualise the variations in HCFI fluorescence saw the application of chromaticity diagrams to quantitatively assess fluorescence colour using data obtained from fluorescence spectra. The analysis of fluorescence colour variations can be improved by plotting the CIE (International Commission on Illumination 1971 [139], 1986 [140]) chromaticity coordinates obtained from fluorescence spectra on a bivariate graph. Segments of the graph are subdivided on the basis of colour and thus the method offers an improvement over purely visual characterisation of colours by individuals. Hagemann and Hollerbach in 1986 plotted spectral data from crude oils onto CIE diagrams while McLimans [52] demonstrated the analysis of single inclusions where the spectral data collected using 366-nm excitation and a Zeiss Zonax system. Specific details of the hardware are scanty but it seems to have been a reflectance measurement system. The information garnered from this type of analysis indicated that as the oils increase in maturity the x and y coordinates decrease, i.e. the fluorescence emission moves towards the blue corner of the graph. Further examples of crude oils and HCFI that have been analysed using these diagrams and comments on the method are detailed by Oxtoby [49], Alderton [130], and Schubert [100]. Blanchet et al. [41] give a detailed analysis of fluid inclusions from the North Sea and calibration oils using CIE diagrams. They observed that the fluorescence colour of inclusion oils is variable even within a single fluid inclusion assemblage and they advanced a number of explanations for this based on geological processes. There could have been a progressive entrapment of oil with API gravity increasing through time, or fractionation of the original petroleum fluid could have occurred. Alternatively, since the HCFI occur mostly at grain-overgrowth boundaries and rarely within overgrowths, this could be attributed to the weakness and permeability of the interface, which would infer that the trapping of secondary fluids is possible. Their results also demonstrated that there were significant differences between the fluorescence colour of the inclusion oils and present-day reference oils. This "red shift" is probably due to the loss of very light hydrocarbon fractions in the reference oils compared to the inclusion trapped oils. However, Blanchet [41] states that this loss of light fraction has no effect, but crucially the reference cited to support this claim has not been published or subjected to peer-review [141]. This contention is in direct opposition to the observations made when crude oils are diluted [25, 26, 29, 142].

Figure 3 shows an example of the CIE method applied to HCFI analysis which depicts a well-defined evolution in the fluorescence emission from successive hydrocarbon generations in a sample (SzD-11). For example, the CIE diagram, Fig. 3d, plots the early primary inclusions ($HC1_L$) in a narrow field ($x = 0.3$ to 0.34, $y = 0.35$ to 0.4), while the slightly younger intermediate primary inclusions ($HC2_L$) show a slight red shift ($x = 0.35$ to 0.4, $y = 0.38$ to 0.43) along the McLimans maturity curve [103]. The late-primary inclusions ($HC3_L$) are widely scattered along the same trend ($x = 0.29$ to 0.5, $y = 0.36$ to 0.46), indicating a more heterogeneous oil composition. The secondary $HC4_L$ (from SzD-11) inclusions have a narrow range between $x = 0.25$ to 0.3 and $y = 0.35$ to 0.4, similar to other $HC4_L$ inclusions sourced elsewhere in the Szeghalom Dome region (Fig. 3e). This example shows how the CIE method can be used for discriminant analysis of multiple inclusion generations across a relatively large geographical location.

In summary, one can see that emission spectroscopy is reasonably well-established methodology for HCFI analysis, but that it suffers from the use of a wide diversity of instrumentation and variances in specific methods. This makes it difficult to compare data between different laboratories. This seems to have arisen from the historical use of home-built apparatus and the relative costs of the instrumentation required. However, with the recent advent of low cost, reliable UV and violet (375 and 405 nm) solid-state lasers, and integrated CCD based spectrometers, one could suggest the possibility for developing and implementing a standard HCFI measurement and analysis methodology based on fluorescence spectroscopy.

Fluorescence Excitation-Emission Matrix (EEM) Spectroscopy

Unlike epifluorescence and steady-state emission spectroscopy where narrow band or monochromatic UV light is employed, for multidimensional fluorescence mea-

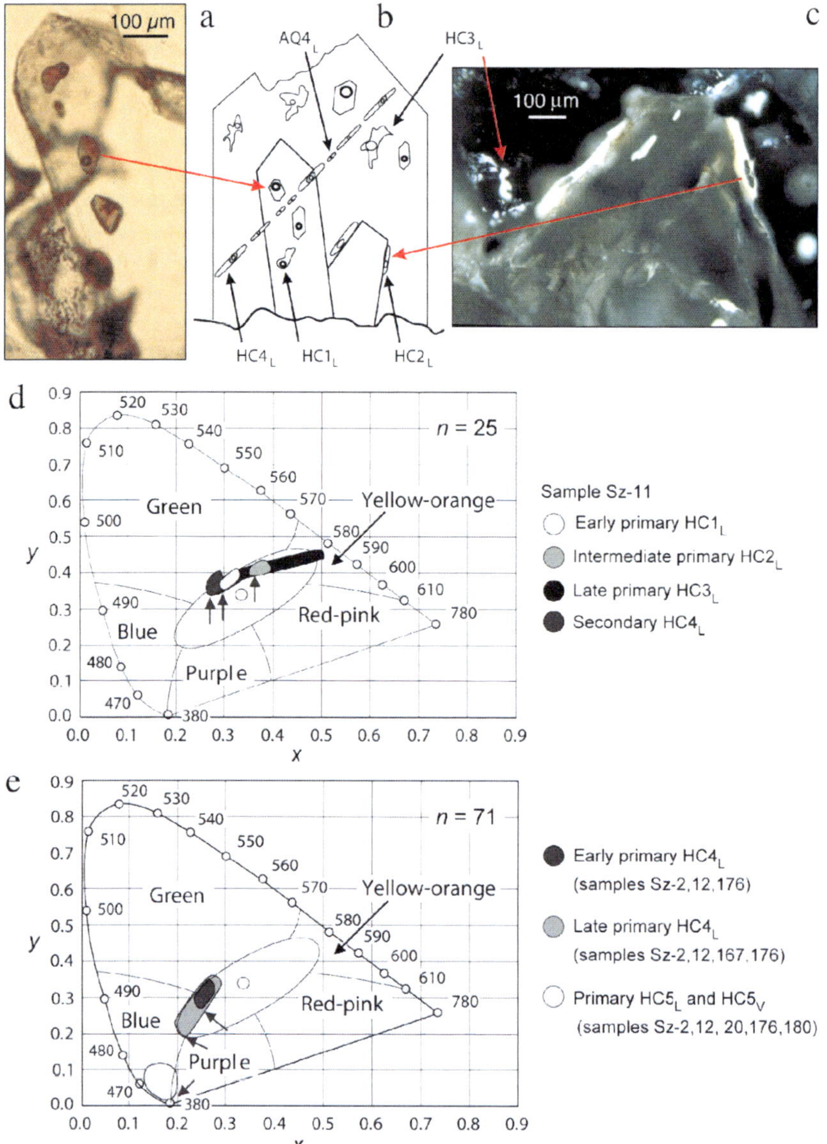

Fig. 3 (**a**) Early idiomorphic crystal of quartz from the Sz-11 well, containing primary $HC1_L$ inclusions, viewed in plane-polarised transmitted light. (**b**) Schematic petrographic relationships: fluid generations 1–3 are represented by a succession of primary inclusions ($HC1_L$, $HC2_L$ to $HC3_L$). Fluid generation 4 is represented by coexisting petroleum ($HC4_L$) and aqueous ($AQ4_L$) inclusions in a healed fracture. (**c**) Example of $HC2_L$ and $HC3_L$ inclusions viewed in UV epi-illumination. (**d**) UV-fluorescence colours of 25 inclusions from sample Sz-11, plotted in the CIE-1931 chromaticity diagram. Each petroleum generation displays slightly different colours. (**e**) UV-fluorescence colours of 71 primary $HC4_L$, $HC5_L$ and $HC5_V$ petroleum inclusions from various samples, displayed in the CIE-1931 chromaticity diagram. Reprinted with permission from *Chemical Geology*, Reference 100, © *2007, Elsevier*

surements like total synchronous fluorescence scanning (TSFS) or excitation-emission matrix (EEM) spectroscopy, a series of excitation wavelengths are used from the UV to the visible region of the spectrum. The rationale behind the use of EEM and TSFS is that these measurements explore the complete emission space, producing a map of the effect of excitation wavelength on emission properties from which energy transfer processes can be observed [143, 144]. Light oils have relatively narrow, intense fluorescence emission bands with small Stokes' shifts in comparison to heavy oils where the emission band tends to be weaker, broader, and red shifted [143]. For the instrumentation required, one can either develop a home built system with emission and excitation monochromators [114] or combine a standard fluorometer to a microscope using fibre optics [113]. For EEM measurements, the emission spectrometer is scanned over a defined emission wavelength range for a series of excitation wavelengths. In contrast, TSFS measurements involve the synchronous scanning of the excitation and emission monochromators at a variety of fixed wavelength separations [113]. In both cases, the excitation source is typically a Xenon lamp which provides a stable light output from the UV into the visible. Studies of HCFI using this technique have shown the ability to discriminate different HCFI populations based on the quantitative parameters of optimum excitation wavelength, optimum emission wavelength, and Stokes' shift [113]. However, despite the high-information content in EEM or TSFS spectra, the method has not been widely adopted for inclusion studies. This is probably due to the relatively high cost of the equipment, complexity of integration, difficulties in signal calibration and correction, and the difficulty in easily analysing the complex data generated. EEM and TSFS data are very sensitive to the collection conditions and there are no defined standard methods as yet for signal correction in microscopy systems.

The main disadvantages of steady-state spectroscopic methods for HCFI analysis are: the lack of standardisation, the difficulty in obtaining absolute intensity values and corrected emission spectra, the negative impacts of sample turbidity and opaqueness, and the possibility of photobleaching/photoalteration [23] if long exposure times are used with UV excitation [20, 24, 145]. The measurement of absolute emission intensities can be compounded by fluctuations in excitation source, detector electronics, photobleaching, sample turbidity, and absorption characteristics of the host mineral [24]. Most fluorescence measurements are made in a single channel mode, and therefore do not include a reference channel. Therefore implementation of reference measurements in order to acquire accurate intensity-based measurements is often a time-consuming process particularly when dealing with complex samples such as HCFI. An additional and significant problem lies in the wide variety of optical configurations that have been used for spectroscopic analysis (Table 2). This multiplicity of the optical configurations can lead to alterations in the true emission spectra owing to irregularities in the absorption and reflection of optical emission, interference, and dichroic filters. These factors are much less constricting when working with time-resolved measurements, particularly fluorescence lifetimes, where the lifetime is largely independent of the emission intensity.

Fluorescence Lifetime Microscopy (FLIM)

The fluorescence lifetime of a molecule can be regarded as the average time it spends in the excited state after absorbing a photon of light. For individual molecules in dilute solutions, the lifetimes can be easily ascribed to specific excited states, and one can correlate lifetime changes with specific environmental factors [146]. The factors that influence the experimentally measured value of the lifetime include molecular structure, environment, fluorophore concentration, and interacting species concentrations. In the context of complex fluids like petroleum, the interacting species can be divided into two broad classes, those that cause fluorescence quenching (non-radiative) and those that promote energy transfer which results in radiative processes like emission [147]. The radiative processes are largely determined by molecular structure and energy transfer, whereas the non-radiative processes are largely governed by collisional, static, and sphere of action quenching. On average, heavy oils (low-API gravity) have shorter average lifetimes than light oils [28, 29, 147, 148]. The average fluorescence lifetime also depends on the specific emission wavelength because different regions of the emission spectrum represent different populations of emitting fluorophores. The excitation wavelength also affects the average lifetime with a reduction in lifetime occurring as the excitation wavelength increases [24, 28, 29]. Qualitative correlations of fluorescence lifetime data with compositional parameters such as API gravity, polar concentrations, and corrected alkane concentrations have been reported [28, 40, 121, 147]. Although accurate quantitative correlations have not been possible with fluorescence lifetimes, the method can be exploited for the accurate characterisation and discrimination of different petroleum oil types, particularly when trapped in HCFI.

There are two principal methods of measuring average fluorescence lifetimes, they are the time domain (TD) and frequency domain (FD) methods [146]. In TD a pulsed laser is used as the excitation source, and the fluorescence decay is measured by either a Time Correlated Single Photon Counting (TCSPC) detection system [28, 146], a time-gated camera [149], or a streak camera [150]. A real decay profile (a combination of the Instrument Response Function (IRF) and the sample fluorescence decay) is generated that enables the fluorescence lifetime to be recovered by various mathematical deconvolution procedures. However, accurate deconvolution and recovery of fluorescence lifetimes requires the collection of an IRF which is typically a non-fluorescent scattering sample like Ludox. It is important that the IRF be collected using the same optical path length and conditions as for the sample measurement. The deconvolution of the decay curve can be accomplished using a variety of different models, depending on the sample under investigation. For simple mixtures (1–4 fluorophores), discrete lifetimes can be extracted, while for more complex situations (e.g. where a single fluorophore exists in multiple environments) lifetime distribution models (Gaussian, Lorentzian, or stretched exponentials) can be used. However, for complex fluids like crude petroleum oil, it has been found experimentally that it is best to report the intensity averaged lifetime [121]. Coupling TCSPC to microscopes for measurement of lifetimes from HCFI is straightforward

[151] and now there are many commercial systems available either for retrofitting to existing microscopes or as complete integrated systems.

FD measurement methods now normally use laser or Light Emitting Diode (LED) excitation where the output is modulated with a sinusoidal waveform and phase sensitive detection of the fluorescence emission [146]. The modulated excitation generates a modulated fluorescence emission from the sample at the same frequency, but it is demodulated and phase shifted with respect to the input according to the fluorescence lifetime. For nanosecond lifetimes in the 100 to ~1 ns range, modulation frequencies of 1 to 500 MHz range can be used. As the lifetimes decrease below ~1 ns, it is advantageous to use frequencies in the GHz range [146]. For accurate lifetime measurements of complex systems which contain more than one fluorophore, one must measure the phase and demodulation data at a range of frequencies and then fit the data to extract the lifetime information. In the case of crude oils, 40 frequencies are typically required [152] to obtain reliable and reproducible lifetime data. On the other hand, if one simply needs to rapidly discriminate oils of different lifetime, one can use a single modulation frequency. The FD principle has also been applied to standard epifluorescence wide-field imaging [153] to produce a methodology capable of rapid FLIM imaging. More recent developments have coupled this with spinning disc technologies to deliver rapid confocal-FLIM [154]. However, to date (mid 2008) there have been no reports of its use for HCFI studies. FD has an advantage over TD in that no deconvolution is required to calculate the lifetime. However, unlike the TD method, fluorescent standards of precisely known lifetime are needed to calibrate the phase-shift and demodulation scales [155]. In practical terms when studying crude petroleum oils or HCFI where there are large variations in lifetime with emission wavelength there is a need to employ several calibration standards [152, 156, 157]. Another factor which is particularly relevant to HCFI studies is the need to acquire the calibration data with similar fluorescence intensities to that being measured from the sample [155]. It should also be noted that the standards currently used are usually solutions of organic fluorophores with well-defined single-exponential lifetimes, so care needs to be exercised to ensure that fresh solutions are used, and that the lifetimes are validated using TCSPC or equivalent measurements.

When comparing FD versus TD methods for HCFI analysis, there are pros and cons for each method. One significant disadvantage of the FD methods is that the technique is much less sensitive than TCSPC, and therefore requires that the HCFI be illuminated at relatively higher light intensities and for longer periods of time. On the other hand, the method (in single frequency mode) is very rapid for imaging purposes, suitable for initial screening of HCFI. Another advantage of the FD procedure is that no deconvolution is required, making data analysis comparatively rapid. Measuring fluorescence lifetimes using TD methods requires the use of fast-pulsed lasers and sophisticated detection electronics or detectors. This contributes to overall system cost and complexity, and while TCSPC systems have become more economical, robust, and very easy to use, TD lifetime systems are still relatively expensive. The correlation between frequency domain and time domain measurements for crude oil lifetimes has been carefully evaluated [152].

The study indicated that the choice of model (discrete, Lorentzian or Gaussian) used to calculate the average lifetime is very important in both TD and FD methods.

There are several advantages of fluorescence lifetime measurements over conventional epifluorescence imaging or steady-state emission measurements. Fluorescence lifetime measurements are much less sensitive to source and detector fluctuations, light scattering, mineral absorption, and photobleaching than other techniques. In addition, the measurement is quantitative, repeatable, and easily validated using lifetime standards available for nearly all emission and excitation wavelengths to be used for HCFI analysis [146]. When lifetime measurements are implemented in confocal microscopes, measurements can easily be made on HCFI as small as 1 μm. McLimans in 1987 suggested the potential in using time-resolved spectroscopy for the analysis of HCFI. Data was presented showing a tentative negative correlation between maturity and fluorescence lifetime for several crude oils. However, this correlation does not match the observations of any subsequent studies where it is clear that the fluorescence lifetime increases with increasing API gravity and maturity [28, 29, 40, 121, 147, 152]. However, McLimans did not present any data collected from HCFI in this 1987 publication. To the best of our knowledge, the first detailed reports of fluorescence lifetime measurements on HCFI were published by our group in Galway [40, 158, 159]. These studies presented lifetime data for both crude petroleum oils and HCFI, measured using the TCSPC method. The key observation was that many of the HCFI demonstrated significantly longer average fluorescence lifetimes than the calibration oils studies, indicating that the dilution effect arising from the light hydrocarbons still present in the trapped inclusion fluids have a significant impact on emission properties.

For discriminating between multiple HCFI assemblages, one can use FLIM in a number of different modes. First, for the rapid screening of HCFI (and generation of FLIM images), the fluorescence lifetime can be measured over a wide emission range (typically using just a longpass filter) to yield a single lifetime value [105, 107, 157]. These single lifetime value measurements, using 405 nm excitation, have been used to discriminate multiple HCFI generations within the same HCFI sample [105]. This measurement mode also has the advantage of using all the emission light, so can be used to measure weakly emitting HCFI. Second, lifetimes can be measured at a range of discrete emission wavelengths. Lifetime–wavelength (τ-λ) plots can then be generated for the accurate discrimination of closely related petroleum fluids [40, 107]. This however relies on the collection of data at a range of wavelengths (using monochromator or filters) and is therefore relatively slow.

Concluding Remarks

The study of Hydrocarbon-bearing Fluid Inclusions (HCFI) is important from a geologic viewpoint because it offers a unique insight into the formation and development of petroleum basins. A combination of the relative chronology of inclusions (both aqueous and HCFI) as well as the composition and temperature-pressure-volume characteristics of individual inclusions provides valuable information about the thermal history and migration of hydrocarbons in petroleum basins [1]. Understanding these mechanisms that ultimately result in the creation of economic recoverable crude oil is vital to petroleum exploration because it provides predictive models and key indicators that may lead to further discoveries or to better reservoir management.

The use of fluorescence microscopy and spectroscopy in a wide variety of methodologies for the characterisation and analysis of HCFI is well established. The petroleum fluids entrapped in HCFI fluoresce because of the presence of polycyclic aromatic hydrocarbons. However, it is the complex chemical composition of the entrapped oils that ultimately determines the precise absorption and fluorescence characteristics of HCFI. HCFI analysis by fluorescence can be subdivided into two main categories, first estimating the hydrocarbon composition and second establishing spatial relationships between and volumes within HCFI. HCFI composition can be estimated (with varying degrees of accuracy) from a variety of emission parameters including colour, chromaticity diagrams, Stokes shifts, emission fluxes, emission intensity ratios, and fluorescence lifetimes. Confocal Microscopy allows for the facile determination of the spatial arrangement, relationships, and volumes of HCFI. This spatial information yields valuable data regarding the timing between HCFI assemblage formation, while the volume estimates of the internal phases, combined with microthermometry, provides information regarding density, pressure, and temperature at the time of trapping.

However, there are problems with the use of fluorescence methods, particularly in the areas of instrumentation, calibration, and validation. The variety of instrumentation and protocols used make it difficult to accurately compare results from various laboratories. This is compounded by the lack of a set of petroleum oils and HCFI standards, suitable for both instrument calibration and validation. A set of petroleum oils with which one can produce an accurate and reproducible calibration plot for estimating HCFI composition is of particular importance. A wide variety of very disparate petroleum oils have been used by different research groups to generate different calibration plots, based on different emission parameters, recorded on different instrumentation, and this compromises the utility of the data. The development of standard methods (defined excitation wavelengths, emission optics, and instrument calibration protocol) suitable for implementation in any inclusion laboratory should also be explored. In conclusion, while fluorescence provides a very powerful and effective tool for the analysis of fluid inclusions, there is still scope for much improvement.

Acknowledgments This research was funded by the Science Foundation of Ireland's Research Frontiers Programme Grant (05/RFP/GEO0002).

References

1. E. Roedder, Fluid inclusions. Mineralogical society of America. *Rev. Mineral.*, **12**, 1–644, (1984).

2. I.A. Munz, Petroleum inclusions in sedimentary basins: systematics, analytical methods and applications. *Lithos*, **55**(1–4), 195–212, (2001).
3. S.C. George, H. Volk, and M. Ahmed, Geochemical analysis techniques and geological applications of oil-bearing fluid inclusions, with some Australian case studies. *J. Petrol. Sci. Eng.*, **57**(1–2), 119–138, (2007).
4. I.A. Munz, K. Iden, H. Johansen, and K. Vagle, The fluid regime during fracturing of the Embla field, Central Trough, North Sea. *Mar. Pet. Geol.*, **15**(8), 751–768, (1998).
5. O. Walderhaug, Temperatures of quartz cementation in Jurassic sandstones from the Norwegian continental shelf—evidence from fluid inclusions. *J. Sediment. Petrol.*, **64**(2), 311–323, (1994).
6. O. Walderhaug, and P.A. Bjorkum, The effect of stylolite spacing on quartz cementation in the lower Jurassic Stø Formation, southern Barents Sea. *J. Sediment. Res.*, **73**(2), 146–156, (2003).
7. N.H. Oxtoby, A.W. Mitchell, and J.G. Gluyas, The filling and emptying of the Ula Oilfield: fluid inclusion constraints. In: *The Geochemistry of Reservoirs: Special Publication*, (Eds. Cubitt, J. M., and England, W. A.), Geological Society, London, **86**, 141–157, (1995).
8. R.C. Burruss, K.R. Cercone, and P.M. Harris, Fluid inclusion petrography and tectonic-burial history of the Al Ali No. 2 well; evidence for the timing of diagenesis and oil migration, northern Oman Foredeep. *Geology*, **11**(10), 567–570, (1983).
9. J. Jensenius and R.C. Burruss, Hydrocarbon-water interactions during brine migration: Evidence from hydrocarbon inclusions in calcite cements from Danish North Sea oil fields. *Geochim. Cosmochim. Acta*, **54**(3), 705–713, (1990).
10. E. Gonzalez-Partida, A. Carrillo-Chavez, J.O.W. Grimmer, J. Pironon, J. Mutterer, and G. Levresse, Fluorite deposits at Encantada-Buenavista, Mexico: products of Mississippi Valley type processes. *Ore Geol. Rev.*, **23**(3–4), 107–124, (2003).
11. M.A. Kendrick, R. Burgess, R.A.D. Pattrick, and G. Turner, Hydrothermal fluid origins in a fluorite-rich Mississippi Valley-Type district: Combined noble gas (He, Ar, Kr) and halogen (Cl, Br, I) analysis of fluid inclusions from the South Pennine Ore Field, United Kingdom. *Econ. Geol.*, **97**(3), 435–451, (2002).
12. A. Dutkiewicz, B. Rasmussen, and R. Buick, Oil preserved in fluid inclusions in Archaean sandstones. *Nature*, **395**, 885–888, (1998).
13. G.L. England, B. Rasmussen, B. Krapez, and D.I. Groves, Archaean oil migration in the Witwatersrand Basin of South Africa. *J. Geol. Soc.*, **159**(2), 189–201, (2002).
14. B. Rasmussen, Evidence for pervasive petroleum generation and migration in 3.2 and 2.63 Ga shales. *Geology*, **33**(6), 497–500, (2005).
15. V. Lüders and K. Rickers, Fluid inclusion evidence for impact-related hydrothermal fluid and hydrocarbon migration in Cretaceous sediments of the ICDP-Chicxulub drill core Yax-1. *Meteorit. Planet. Sci.*, **39**(7), 1187–1197, (2004).
16. J. Parnell, G.R. Watt, D. Middleton, J. Kelly, and M. Baron, Deformation band control on hydrocarbon migration. *J. Sediment. Res.*, **74**(4), 552–560, (2004).
17. J.M. Peter, B.R.T. Simoneit, O.E. Kawka, and S.D. Scott, Liquid hydrocarbon-bearing inclusions in modern hydrothermal chimneys and mounds from the southern trough of Guaymas Basin, Gulf of California. *Appl. Geochem.*, **5**(1–2), 51–63, (1990).
18. J.M. Hunt, *Petroleum Geochemistry and Geology*. W.H. Freeman and Company, San Francisco, (1979).
19. T.J. Shepherd, A.H. Rankin, and D.H.M. Alderton, *A Practical Guide to Fluid Inclusion Studies*, Blackie and Son, Glasgow, (1985).
20. S.C. George, T.E. Ruble, A. Dutkiewicz, and P.J. Eadington, Assessing the maturity of oil trapped in fluid inclusions using molecular geochemistry data and visually-determined fluorescence colours. *Appl. Geochem.*, **16**(4), 451–473, (2001).
21. J. Parnell, D. Middleton, C. Honghan, and D. Hall, The use of integrated fluid inclusion studies in constraining oil charge history and reservoir compartmentation: examples from the Jeanne d'Arc Basin, offshore Newfoundland. *Mar. Pet. Geol.*, **18**(5), 535–549, (2001).

22. N. Guilhaumou, N. Szydlowskii, and B. Pradier, Characterization of hydrocarbon fluid inclusions by infra-red and fluorescence microspectroscopy. *Mineral. Mag.*, **54**(375), 311–324, (1990).
23. J. Pironon and B. Pradier, Ultraviolet-fluorescence alteration of hydrocarbon fluid inclusions. *Org. Geochem.*, **18**(4), 501–509, (1992).
24. A.G. Ryder, Analysis of crude petroleum oils using fluorescence spectroscopy. In: C.D. Geddes and J.R. Lakowicz, Editors, *Reviews in Fluorescence 2005*, Springer New York, 169–198, (2005).
25. T.D. Downare and O.C. Mullins, Visible and near-infrared fluorescence of crude oils. *Appl. Spectrosc.*, **49**(6), 754–764, (1995).
26. C.Y. Ralston, X. Wu, and O.C. Mullins, Quantum yields of crude oils. *Appl. Spectrosc.*, **50**(12), 1563–1568, (1996).
27. O.C. Mullins and E.Y. Sheu, *Structure and Dynamics of Asphaltenes*, Plenum Press, New York, 21–77, (1998).
28. A.G. Ryder, T.J. Glynn, M. Feely, and A.J.G. Barwise, Characterization of crude oils using fluorescence lifetime data. *Spectrochim. Acta (A)*, **58**(5), 1025–1038, (2002).
29. X. Wang and O.C. Mullins, Fluorescence lifetime studies of crude oils. *Appl. Spectrosc.*, **48**(8), 977–984, (1994).
30. H.W. Hagemann and A. Hollerbach, The fluorescence behaviour of crude oils with respect to their thermal maturation and degradation. *Org. Geochem.*, **10**(1–3), 473–480, (1986).
31. L.D. Stasiuk and L.R. Snowdon, Fluorescence micro-spectrometry of synthetic and natural hydrocarbon fluid inclusions: crude oil chemistry, density and application to petroleum migration. *Appl. Geochem.*, **12**(3), 229–241, (1997).
32. L.D. Stasiuk, T. Gentzis, and P. Rahimi, Application of spectral fluorescence microscopy for the characterization of Athabasca bitumen vacuum bottoms. *Fuel*, **79**(7), 769–775, (2000).
33. B. Pradier, C. Largeau, S. Derenne, L. Martinez, P. Bertrand, and Y. Pouet, Chemical basis of fluorescence alteration of crude oils and kerogens–I. Microfluorimetry of an oil and its isolated fractions; relationships with chemical structure. *Org. Geochem.*, **16**(1–3), 451–460, (1990).
34. K.Y. Liu and P. Eadington, Quantitative fluorescence techniques for detecting residual oils and reconstructing hydrocarbon charge history. *Org. Geochem.*, **36**(7), 1023–1036, (2005).
35. S. Gong, S.C. George, H. Volk, K. Liu, and P. Peng. Petroleum charge history in the Lunnan low uplift, Tarim basin, China - Evidence from oil-bearing fluid inclusions. *Org. Geochem.*, **38**(8), 1341–1355, (2007).
36. R.C. Murray, Hydrocarbon fluid inclusions in quartz. *Amer. Assoc. Pet. Geol. Bull.* **41**(5), 950–956, (1957).
37. R.C. Burruss, D.J. Toth, and R.H. Goldstein, Fluorescence microscopy of hydrocarbon fluid inclusions: relative timing of hydrocarbon migration events in the Arkoma Basin, NW Arkansas. *EOS* **61**, 400, (1980).
38. R.C. Burruss, Hydrocarbon fluid inclusions in studies of sedimentary diagenesis. *Mineral. Assoc. Canada*, Short Course Handbook 6, 138–156, (1981).
39. R.C. Burruss, Practical aspects of fluorescence microscopy of petroleum fluid inclusions. Luminescence microscopy and spectroscopy: Qualitative and quantitative applications. In: Barker, C.E., and Kopp, O.C. (Eds.), SEPM Short Course 25, 1–7, (1991).
40. A.G. Ryder, M.A. Przyjalgowski, M. Feely, B. Szczupak, and T.J. Glynn, Time-resolved fluorescence microspectroscopy for characterizing crude oils in bulk and hydrocarbon bearing fluid inclusions. *Appl. Spectrosc.*, **58**(9), 1106–1115, (2004).
41. A. Blanchet, M. Pagel, F. Walgenwitz, and A. Lopez, Microspectrofluormetric and microthermometric evidence for variability in hydrocarbon fluid inclusions in quartz overgrowths: implications for inclusion trapping in the Alwyn North field, North Sea. *Org. Geochem.*, **34**(11), 1477–1490, (2003).

42. W.-L. Huang and G.A. Otten, Cracking kinetics of crude oil and alkanes determined by diamond anvil cell-fluorescence spectroscopy pyrolysis: technique development and preliminary results. *Org. Geochem.*, **32**(6), 817–830, (2001).
43. R.-F. Weng, W.-L. Huang, C.-L. Kuo, and S. Inan, Characterization of oil generation and expulsion from coals and source rocks using diamond anvil cell pyrolysis. *Org. Geochem.* **34**(6), 771–787, (2003).
44. Y.-J. Chang and W.-L. Huang, Simulation of the fluorescence evolution of "live" oils from kerogens in a diamond anvil cell: application to inclusion oils in terms of maturity and source. *Geochim. Cosmochim. Acta*, **72**(15), 3771–3787, (2008).
45. J. Kihle and H. Johansen, Low-temperature isothermal trapping of hydrocarbon fluid inclusions in synthetic-crystals of KH_2PO_4. *Geochim. Cosmochim. Acta*, **58**(3), 1193–1202, (1994).
46. S. Teinturier, M. Elie, and J. Pironon, Oil-cracking processes evidence from synthetic petroleum inclusions. *J. Geochem. Explor.*, **78-79**, 421–425, (2003).
47. A.M. Van den Kerkhof and U.F. Hein, Fluid inclusion petrography. *Lithos*, **55**(1–4), 27–47, (2001).
48. B. McNeil and E. Morris, The preparation of double-polished fluid inclusion wafers from friable, water-sensitive material. *Mineral. Mag.*, **56**(382), 120–122, (1992).
49. N.H. Oxtoby, Comments on: assessing the maturity of oil trapped in fluid inclusions using molecular geochemistry data and visually-determined fluorescence colours. *Appl. Geochem.*, **17**(10), 1371–1374, (2002).
50. S.C. George, T.E. Ruble, A. Dutkiewicz, and P.J. Eadington, Reply to comment by Oxtoby on "Assessing the maturity of oil trapped in fluid inclusions using molecular geochemistry data and visually-determined fluorescence colours". *Appl. Geochem.*, **17**(10), 1375–1378, (2002).
51. P.K. Mukhopadhyay and J. Rullkotter, Quantitative microscopic spectral fluorescence measurement of crude oil, bitumen, kerogen and coal. *AAPG Bull.*, **70**(5), 624, (1986).
52. R.K. McLimans, The application of fluid inclusions to migration of oil and diagenesis of in petroleum reservoirs. *Appl. Geochem.*, **2**(5–6), 585–603, (1987).
53. R.J. Bodnar, Petroleum migration in the Miocene Monterey Formation, California, USA: constraints from fluid inclusion studies. *Mineral. Mag.*, **54**(375), 295–304, (1990).
54. J.R. Levine, I.M. Samson, and R. Hesse, Occurrence of fracture-hosted impsonite and petroleum fluid inclusions, Quebec City region, Canada. *AAPG Bull.*, **75**(1), 139–155, (1991).
55. M.R. Moser, A.H. Rankin, and H.J. Milledge, Hydrocarbon-bearing fluid inclusions in fluorite associated with the Windy Knoll Bitumen Deposit, UK. *Geochim. Cosmochim. Acta*, **56**(1), 155–168, (1992).
56. K.D. Newell, R.C. Burruss, and J.G. Palacas, Thermal maturation and organic richness of potential petroleum source rocks in Proterozoic Rice Formation, North American Mid-Continent Rift System, northeastern Kansas. *AAPG Bull.*, **77**(11), 1922–1941, (1993).
57. J. Pironon, M. Pagel, M.H. Leveque, and M. Moge, Organic inclusions in salt .1. Solid and liquid organic-matter, carbon-dioxide and nitrogen species in fluid inclusions from the Bresse Basin (France). *Org. Geochem.*, **23**(5), 391–402, (1995).
58. J. Pironon, M. Pagel, F. Walgenwitz, and O. Barres, Organic inclusions in salt .2. Oil, gas and ammonium in inclusions from the Gabon Margin. *Org. Geochem.*, **23**(8), 739–750, (1995).
59. J. Parnell, P.F. Carey, and B. Monson, Fluid inclusion constraints on temperatures of petroleum migration from authigenic quartz in bitumen veins. *Chem. Geol.*, **129**(3–4), 217–226, (1996).
60. X.M. Xiao, D.H. Liu, and J.M. Fu, Multiple phases of hydrocarbon generation and migration in the Tazhong petroleum system of the Tarim Basin, People's Republic of China. *Org. Geochem.*, **25**(3–4), 191–197, (1996).
61. S.C. George, F.W. Krieger, P.J. Eadington, R.A. Quezada, P.F. Greenwood, L.I. Eisenberg, P.J. Hamilton, and M.A. Wilson, Geochemical comparison of oil-bearing fluid inclusions

and produced oil from the Toro Sandstone, Papua New Guinea. *Org. Geochem.*, **26**(3–4), 155–173, (1997).

62. J. Parnell, P. Carey, and W. Duncan, History of hydrocarbon charge on the Atlantic margin: evidence from fluid-inclusion studies, West of Shetland. *Geology*, **26**(9), 807–810, (1998).
63. S.C. George, M. Lisk, R.E. Summons, and R.A. Quezada, Constraining the oil charge history of the South Pepper oilfield from the analysis of oil-bearing fluid inclusions. *Org. Geochem.*, **29**(1–3), 631–648, (1998).
64. C. O'Reilly, P.M. Shannon, and M. Feely, A fluid inclusion study of cement and vein minerals from the Celtic Sea Basins, offshore Ireland. *Mar. Pet. Geol.*, **15**(6), 519–533, (1998).
65. H.-Y. Tseng, R.C. Burruss, T.C. Onstott, and G. Omar, Paleofluid-flow circulation within a Triassic rift basin: Evidence from oil inclusions and thermal histories. *Geo. Soc. Am. Bull.*, **111**(2), 275–290, (1999).
66. D. Smale, J.L. Mauk, J. Palmer, R. Soong, and P. Blattner, Variations in sandstone diagenesis with depth, time, and space, onshore Taranaki wells, New Zealand. *New Zeal. J. Geol. Geop.*, **42**, 137–154, (1999).
67. G. Rantitsch, J. Jochum, R.F. Sachsenhofer, B. Russegger, E. Schroll, and B. Horsfield, Hydrocarbon-bearing fluid inclusions in the Drau Range (Eastern Alps, Austria): implications for the genesis of Bleiberg-type Pb-Zn deposits. *Mineral. Petrol.*, **65**(3–4), 141–159, (1999).
68. J. Parnell, P.F. Carey, P. Green, and W. Duncan, Hydrocarbon migration history, west of Shetland; integrated fluid inclusion and fission track studies. Petroleum Geology of Northwest Europe: *Proc. Geol. Soc. London Conf.*, **5**, 613–625, (1999).
69. N.N. Cesaretti, J. Parnell, and E.A. Dominguez, Pore fluid evolution within a hydrocarbon reservoir: Yacoraite formation (upper Cretaceous), northwest basin, Argentina. *J. Pet. Geol.*, **23**(4), 375–398, (2000).
70. J. Lonnee and I.S. Al-Aasm, Dolomitization and fluid evolution in the Middle Devonian Sulphur Point Formation, Rainbow South Field, Alberta: petrographic and geochemical evidence. *Bull. Can. Pet. Geol.*, **48**(3), 262–283, (2000).
71. A.M.E. Marchand, R.S. Haszeldine, C.I. Macaulay, R. Swennen, and A.E. Fallick, Quartz cementation inhibited by crestal oil charge: Miller deep water sandstone, UK North Sea. *Clay Miner.*, **35**(1), 201–210, (2000).
72. R. Thiéry, J. Pironon, F. Walgenwitz, and F. Montel, PIT (Petroleum Inclusion Thermodynamic): a new modeling tool for the characterization of hydrocarbon fluid inclusions from volumetric and microthermometric measurements. *J. Geochem. Explor.*, **69**, 701–704, (2000).
73. J. Parnell, C. Honghan, D. Middleton, T. Haggan, and P. Carey, Significance of fibrous mineral veins in hydrocarbon migration: fluid inclusion studies. *J. Geochem. Explor.*, **69**, 623–627, (2000).
74. D. Middleton, J. Parnell, P. Carey, and G. Xu, Reconstruction of fluid migration history in Northwest Ireland using fluid inclusion studies. *J. Geochem. Explor.*, **69**, 673–677, (2000).
75. D. Lavoie, G. Chi, and M.G. Fowler, The Lower Devonian Upper Gaspé Limestones in eastern Gaspé: carbonate diagenesis and reservoir potential. *Bull. Can. Pet. Geol.*, **49**(2), 346–365, (2001).
76. A. Ceriani, A. Di Giulio, R.H. Goldstein, and C. Rossi, Diagenesis associated with cooling during burial: an example from Lower Cretaceous Reservoir Sandstones (Sirt Basin, Libya). *AAPG Bull.*, **86**(9), 1573–1591, (2002).
77. C. Rossi, R.H. Goldstein, A. Ceriani, and R. Marfil, Fluid inclusions record thermal and fluid evolution in reservoir sandstones, Khatatba Formation, Western Desert, Egypt: A case for fluid injection. *AAPG Bull.*, **86**(10), 1773–1799, (2002).
78. M. Lisk, G.W. O′Brien, and P.J. Eadington, Quantitative evaluation of the Oil-Leg potential in the Oliver Gas Field, Timor Sea, Australia. *AAPG Bull.*, **86**(9), 1531–1542, (2002).
79. J.L. Mauk and R.C. Burruss, Water washing of Proterozoic oil in the Midcontinent rift system. *AAPG Bull.*, **86**(6), 1113–1127, (2002).

80. H. Volk, B. Horsfield, U. Mann, and V. Suchy, Variability of petroleum inclusions in vein, fossil and vug cements – a geochemical study in the Barrandian Basin (Lower Palaeozoic, Czech Republic). *Org. Geochem.*, **33**(12), 1319–1341, (2002).
81. H.-Y. Tseng and R.J. Pottorf, Fluid inclusion constraints on petroleum PVT and compositional history of the Greater Alwyn-South Brent petroleum system, northern North Sea. *Mar. Pet. Geol.*, **19**(7), 797–809, (2002).
82. A. Dutkiewicz, J. Ridley, and R. Buick, Oil-bearing CO_2-CH_4-H_2O fluid inclusions; oil survival since the Palaeoproterozoic after high temperature entrapment. *Chem. Geol.*, **194** (1–3), 51–79, (2003).
83. A. Dutkiewicz and J. Ridley, Hydrocarbon pseudo-inclusions in barite: how to recognize and avoid artifacts. *J. Sediment. Res.*, **73**(2), 171–176, (2003).
84. A. Dutkiewicz, H. Volk, J. Ridley, and S.C. George, Biomarkers, brines, and oil in the Mesoproterozoic, Roper Superbasin, Australia. *Geology*, **31**(11), 981–984, (2003).
85. M. Feely and J. Parnell, Fluid inclusion studies of well samples from the hydrocarbon prospective Porcupine Basin, offshore Ireland. *J. Geochem. Explor.*, **78-79**, 55–59, (2003).
86. J.R. Boles, P. Eichhubl, G. Garven, and J. Chen, Evolution of a hydrocarbon migration pathway along basin-bounding faults: Evidence from fault cement. *AAPG Bull.*, **88**(7), 947–970, (2004).
87. R. Martinez-Ibarra, J. Tritlla, E. Cedillo-Pardo, J.M. Grajales-Nishimura, and G. Murillo-Muneton, Brine and hydrocarbon evolution during the filling of the Cantarell oil field (Gulf of Mexico). *J. Geochem. Explor.*, **78-79**, 399–403, (2003).
88. A. Dutkiewicz, H. Volk, J. Ridley, and S.C. George, Geochemistry of oil in fluid inclusions in a middle Proterozoic igneous intrusion: implications for the source of hydrocarbons in crystalline rocks. *Org. Geochem.*, **35**(8), 937–957, (2004).
89. R. Jonk, J. Parnell, and A. Whitham, Fluid inclusion evidence for a Cretaceous-Palaeogene petroleum system, Kangerlussuaq Basin, East Greenland. *Mar. Pet. Geol.*, **22**(3), 319–330, (2005).
90. H. Volk, S.C. George, A. Dutkiewicz, and J. Ridley, Characterization of fluid inclusion oil in a mid-Proterozoic sandstone and dolerite (Roper Superbasin, Australia). *Chem. Geol.*, **223**(1–3), 109–135, (2005).
91. A. Dutkiewicz, H. Volk, S.C. George, J. Ridley, and R. Buick, Biomarkers from Huronian oil-bearing fluid inclusions: an uncontaminated record of life before the Great Oxidation Event. *Geology*, **34**(6), 437–440, (2006).
92. C.L. Hanks, T.M. Parris, and W.K. Wallace, Fracture paragenesis and microthermometry in Lisburne Group detachment folds: implications for the thermal and structural evolution of the northeastern Brooks Range, Alaska. *AAPG Bull.*, **90**(1), 1–20, (2006).
93. M. Brincat, A. Gartrell, M. Lisk, W. Bailey, L. Johnson, and D. Dewhurst, An integrated evaluation of hydrocarbon charge and retention at the Griffin, Chinook, and Scindian oil and gas fields, Barrow Subbasin, North West Shelf, Australia. *AAPG Bull.*, **90**(9), 1359–1380, (2006).
94. C.M. Rott and H. Qing, Analysis of Mississippian anhydrite by fluorescence microscopy – implications for the origin of oil-bearing anhydrite. In *Summary of Investigations 2006, Volume 1, Saskatchewan Geological Survey, Sask,* Report 2006–4.1, 1–11, (2006).
95. R. Wierzbicki, J.J. Dravis, I. Al-Aasm, and N. Harland, Burial dolomitization and dissolution of Upper Jurassic Abenaki platform carbonates, Deep Panuke reservoir, Nova Scotia, Canada. *AAPG Bull.*, **90**(11), 1843–1861, (2006).
96. M. Wilkinson, R.S. Haszeldine, and A.E. Fallick, Hydrocarbon filling and leakage history of a deep geopressured sandstone, Fulmar Formation, United Kingdom North Sea. *AAPG Bull.*, **90**(12), 1945–1961, (2006).
97. M. Baron and J. Parnell, Relationships between stylolites and cementation in sandstone reservoirs: examples from the North Sea, U.K. and East Greenland. *Sed. Geol.*, **194**(1–2), 17–35, (2007).

98. A. Dutkiewicz, S.C. George, D.J. Mossman, J. Ridley, and H. Volk, Oil and its biomarkers associated with the Palaeoproterozoic Oklo, natural fission reactors, Gabon. *Chem. Geol.*, **244**(1–2), 130–154, (2007).
99. K.E. Higgs, H. Zwingmann, A.G. Reyes, and R.H. Funnell, Diagenesis, porosity evolution, and petroleum emplacement in tight gas reservoirs, Taranaki Basin, New Zealand. *J. Sediment. Res.*, **77**(11–12), 1003–1025, (2007).
100. F. Schubert, L.W. Diamond, and T.M. Toth, Fluid inclusion evidence for petroleum migration through a buried metamorphic dome in the Pannonian Basin, Hungary. *Chem. Geol.*, **244** (3–4), 357–381, (2007).
101. M. Baron, J. Parnell, D. Mark, A. Carr, M. Przyjalgowski, and M. Feely, Evolution of hydrocarbon migration style in a fractured reservoir deduced from fluid inclusion data, Clair Field, west of Shetland, UK. *Mar. Pet. Geol.*, **25**(2), 153–172, (2008).
102. J. Bourdet, J. Pironon, G. Levresse, and J. Tritlla, Petroleum type determination through homogenization temperature and vapour volume fraction measurements in fluid inclusions. *Geofluids* **8**(1), 46–59, (2008).
103. R.K. McLimans, Studies of reservoir diagenesis, burial history, and petroleum migration using luminescence microscopy. In Barker, C.E., Kopp, O. (Eds.), *Luminescence Microscopy: Qualitative and Quantitative Applications, (SEPM) Short Course 25*, 97–106, Society for Sedimentary Geology, Tulsa, USA, (1991).
104. R.C. Burruss, K.R. Cercone, and P.M. Harris, Timing of hydrocarbon migration: evidence from fluid inclusions in calcite cements, tectonics and burial history. In: N. Schneidermann, and P.M. Harris (eds.), *Carbonate Cements*, Special Publication – Society of Economic Paleontologists and Mineralogists, **36**, 277–289, (1985).
105. N.J.F. Blamey, A.G. Ryder, M. Feely, and P. Owens, Fluorescence lifetime analysis of single hydrocarbon-bearing fluid inclusions – A paragenetic perspective. 23rd *IMOG*, Torquay, UK, 669–670, (2007).
106. S.C. George, M. Ahmed, K. Liu, and H. Volk, The analysis of oil trapped during secondary migration. Org. Geochem., **35** (11–12), 1489–1511, (2004).
107. J. Conliffe, M. Feely, J. Parnell, N.J.F. Blamey, and A.G. Ryder, Unpublished work.
108. J.B. Pawley (Ed.). Handbook of Biological Confocal Microscopy, 2nd ed. Plenum Press, New York, (1995).
109. G. Macleod, S.R. Larter, A.C. Aplin, K.S. Pedersen, and T.A. Booth. Determination of the effective composition of single petroleum inclusions using Confocal Scanning Laser Microscopy and PVT simulation. In P.E. Brown, S.G. Hagemann (Eds.), *Biennial Pan-American Conference on Research on Fluid Inclusions (PACROFI VI)* Madison Wisconsin, USA, 81–82, (1996).
110. J. Pironon, M. Canals, M. Dubessy, F. Walgenwitz, and C. Laplace-Builhe, Volumetric reconstruction of individual oil inclusions by confocal scanning laser microscopy. *Eur. J. Mineral.*, **10**(6), 1143–1150, (1998).
111. A.C. Aplin, G. Macleod, S.R. Larter, K.S. Pedersen, H. Sørensen, and T. Booth, Combined use of confocal laser scanning microscopy and PVT simulation for estimating the composition and physical properties of petroleum in fluid inclusions. *Mar. Pet. Geol.*, **16**(2), 97–110, (1999).
112. R. Thiéry, J. Pironon, F. Walgenwitz, and F. Montel, Individual characterization of petroleum fluid inclusions (composition and P-T trapping conditions) by microthermometry and confocal laser scanning microscopy: inferences from applied thermodynamics of oils. *Mar. Pet. Geol.*, **19**(7), 847 -859, (2002).
113. J. Kihle, Adaptation of fluorescence excitation-emission micro-spectroscopy for characterization of single hydrocarbon fluid inclusions. *Org. Geochem.*, **23**(11–12), 1029–1042, (1995).
114. J.A. Musgrave, R.G. Carey, D.R. Janecky, and C.D. Tait, Adaption of Synchronously Scanned Luminescence Spectroscopy to organic-rich fluid inclusion microanalysis. *Rev. Sci. Instrum.*, **65**(6), 1877–1882, (1994).

115. A.C. Aplin, S.R. Larter, M.A. Bigge, G. Macleod, R.E. Swarbrick, and D. Grunberger. Confocal microscopy of fluid inclusions reveals fluid-pressure histories of sediments and an unexpected origin of gas condensate. *Geology*, **28**(11), 1047–1050, (2000).
116. R.E. Swarbrick, M.J. Osborne, D. Grunberger, G.S. Yardley, G. Macleod, A.C. Aplin, S.R. Larter, I. Knight, and H.A. Auld, Integrated study of the Judy Field (Block 30/7a) — an overpressured Central North Sea oil/gas field. *Mar. Pet. Geol.*, **17**(9), 993–1010, (2000).
117. R. Thiéry, J. Pironon, F. Walgenwitz, and F. Montel, Individual characterization of petroleum fluid inclusions (composition and P-T trapping conditions) by microthermometry and confocal laser scanning microscopy: inferences from applied thermodynamics of oils. *Mar. Pet. Geol.*, **19**(7), 847–859, (2002).
118. P. Stoller, Y. Krüger, J. Rička, and M. Frenz, Femtosecond lasers in fluid inclusion analysis: Three-dimensional imaging and determination of inclusion volume in quartz using second harmonic generation microscopy. *Earth Planet. Sci. Lett.*, **253**(3–4), 359–368, (2007).
119. N.J.F. Blamey, A.G. Ryder, M. Feely, P. Dockery, and P. Owens, The application of structured-light illumination to hydrocarbon-bearing fluid inclusions. *Geofluids*, **8**(2), 102–112, (2008).
120. M.A.A. Neil, R. Juškaitis, and T. Wilson, Method of obtaining optical sectioning by using structured light in a conventional microscope. *Opt. Lett.* **22**(24), 1905–1907, (1997).
121. A.G. Ryder, Quantitative analysis of crude oils by fluorescence lifetime and steady state measurements using 380-nm excitation. *Appl. Spectrosc.*, **56**(1), 107–116, (2002).
122. E.S. Wachman, W.-H. Niu, and D.L. Farkas, AOTF Microscope for imaging with increased speed and spectral versatility. *Biophys. J.*, **73**(3), 1215–1222, (1997).
123. A. Feofanov, S. Sharonov, P. Valisa, E. Dasilva, I. Nabiev, and M. Manfait, A new confocal stigmatic spectrometer for micro-Raman and microfluorescence spectral imaging analysis – design and applications. *Rev. Sci. Instrum.*, **66**(5), 3146–3158, (1995).
124. E.A.J. Burke, Raman microspectrometry of fluid inclusions. *Lithos*, **55**, 139–158, (2001).
125. J. Jochum, G. Friedrich, D. Leythaeuser, R. Littke, and B. Ropertz, Hydrocarbon-bearing fluid inclusions in calcite-filled horizontal fractures from mature Posidonia Shale (Hils Syncline, NW Germany). *Ore Geol. Rev.*, **9**(5), 363–370, (1995).
126. G. Chi, D. Lavoie, and R. Bertrand, Regional-scale variation of characteristics of hydrocarbon fluid inclusions and thermal conditions along the Paleozoic Laurentian continental margin in eastern Quebec, Canada. *Bull. Can. Petrol. Geol.*, **48**(3), 193–211, (2000).
127. D. Kirkwood, M.M. Savard, and G. Chi, Microstructural analysis and geochemical vein characterization of the Salinic event and Acadian Orogeny: evaluation of the hydrocarbon reservoir potential in eastern Gaspé. *Bull. Can. Petrol. Geol.*, **49**(2), 262–281, (2001).
128. D.W. Morrow, M. Zhao, and L.D. Stasiuk, The gas-bearing Devonian Presqu'ile Dolomite of the Cordova embayment region of British Columbia, Canada: Dolomitization and the stratigraphic template. *AAPG Bull.*, **86**(9), 1609–1638, (2002).
129. R. Li and J. Parnell, In situ microanalysis of petroleum fluid inclusions by Time of Flight-Secondary Ion Mass Spectrometry as an indicator of evolving oil chemistry: a pilot study in the Bohai Basin, China. *J. Geochem. Explor.*, **78-9**, 377–384, (2003).
130. D.H.M. Alderton, N.H. Oxtoby, H. Brice, N. Grassineau, and R.E. Bevins, The link between fluids and rank variation in the South Wales Coalfield: evidence from fluid inclusions and stable isotopes. *Geofluids*, **4**(3), 221–236, (2004).
131. I.A. Munz, M. Wangen, J-P. Girard, J-C. Lacharpagne, and H. Johansen, Pressure-temperature-time-composition (P-T-t-X) constraints of multiple petroleum charges in the Hild field, Norwegian North Sea. *Mar. Pet. Geol.*, **21**(8), 1043–1060, (2004).
132. D. Lavoie, G. Chi, P. Brennan-Alpert, A. Desrochers, and R. Bertrand, Hydrothermal dolomitization in the Lower Ordovician Romaine Formation of the Anticosti Basin: significance for hydrocarbon exploration. *Bull. Can. Pet. Geol.*, **53**(4), 454–471, (2005).
133. M. Li, L. Stasiuk, R. Maxwell, F. Monnier, and O. Bazhenova, Geochemical and petrological evidence for Tertiary terrestrial and Cretaceous marine potential petroleum source rocks in the western Kamchatka coastal margin, Russia. *Org. Geochem.*, **37**(3), 304–320, (2006).

134. R. Baranger, L.Martinez, J.-L. Pittion, and J. Pouleau, A new calibration procedure for fluorescence measurements of sedimentary organic matter. *Org. Geochem.* **17**(4), 467–475, (1991).
135. U. Resch-Genger, K. Hoffmann, and A. Hoffmann, Standardization of fluorescence measurements – criteria for the choice of suitable standards and approaches to fit-for-purpose calibration tools. *Ann. NY Acad. Sci.*, **1130**, 35–43, (2008)
136. O. Barres, A. Burneau, J. Dubessy, and M. Pagel, Application of micro-FT-IR spectroscopy to individual hydrocarbon fluid inclusion analysis. *Appl. Spectrosc.*, **41**(6), 1000–1008, (1987).
137. N. Guilhaumou, J.C. Touray, V. Perthuisot, and F. Roure, Palaeocirculation in the basin of southeastern France sub-alpine range: a synthesis from fluid inclusions studies. *Mar. Pet. Geol.*, **13**(6), 695–706, (1996).
138. N. Guilhaumou, N. Ellouz, T.M. Jaswal, and P. Mougin. Genesis and evolution of hydrocarbons entrapped in the fluorite deposit of Koh-i-Maran, (North Kirthar Range, Pakistan). *Mar. Pet. Geol.*, **17**(10), 1151–1164, (2000).
139. Commission Internationale de l'E ́clairage, Colorimetry. Publication No. 15. Commission Internationale de l'E ́clairage, Paris, (1971).
140. Commission Internationale de l'E ́clairage, Standard on Colorimetric Observers. CIE S002. Commission Internationale de l'E ́clairage, Vienna, (1986).
141. S. Mazères, Mise en oeuvre d'un microspectrofluorimètre pour l'étude de mircroéchantillons en fluorescence stationnaire et résolue dans le temps. In: Biophysique, Université Paul Sabatier, Toulouse, pp. 200, (1997).
142. A. G. Ryder and P. Owens, manuscript in preparation.
143. A.G. Ryder, Assessing the maturity of crude petroleum oils using Total Synchronous Fluorescence Scan Spectra. *J. Fluoresc.*, **14**(1), 99–104, (2004).
144. O. Abbas, C. Rébufa, N. Dupuy, A. Permanyer, J. Kister, and D.A. Azevedo, Application of chemometric methods to synchronous UV fluorescence spectra of petroleum oils. *Fuel*, **85**(17–18), 2653–2661, (2006).
145. G. Ellingsen and S. Fery-Forgues, Application of fluorescence spectroscopy to the study of petroleum: challenging complexity. *Revue De L Institut Francais Du Petrole*, **53**(2), 201–216, (1998).
146. J.R. Lakowicz, *Principles of Fluorescence Spectroscopy*. 3rd edition, Springer, New York, (2006).
147. A.G. Ryder, Time-resolved fluorescence spectroscopic study of crude petroleum oils: influence of chemical composition. *Appl. Spectrosc.*, **58**(5), 613–623, (2004).
148. A.G. Ryder, T.J. Glynn, and M. Feely. Influence of chemical composition on the fluorescence lifetimes of crude petroleum oils. *Proc SPIE – Int. Soc. Opt. Eng.*, **4876**, 1188–1195, (2003).
149. K. Dowling, M.J. Dayel, S.C.W. Hyde, P.M.W. French, M.J. Lever, J.D. Hares, and A.K.L. Dymoke-Bradshaw. High resolution time-domain fluorescence lifetime imaging for biomedical applications. *J. Mod. Opt.*, **46**(2), 199–209, (1999).
150. M.F. Quinn, A.S. Al-Otaibi, A. Abdullah, P.S. Sethi, F. Al-Bahrani, and O. Alameddine, Determination of intrinsic fluorescence lifetime parameters of crude oils using a laser fluorosensor with a streak camera detection system. *Instrum. Sci. Tech.* **23**(3), 201–215, (1995).
151. A.G. Ryder, T.J. Glynn, M. Przyjalgowski, and B. Szczupak. A compact violet diode laser based fluorescence lifetime microscope. *J. Fluoresc.*, **12**(2), 177–180, (2002).
152. P. Owens, A.G. Ryder, and N.J.F. Blamey. Frequency domain fluorescence lifetime study of crude petroleum oils. *J. Fluoresc.*, **18** (5), 997–1006, (2008).
153. T.W.J. Gadella, T.M. Jovin, and R.M. Clegg, Fluorescence lifetime imaging microscopy (FLIM) – spatial-resolution of microstructures on the nanosecond time-scale. *Biophys. Chem.*, **48**(2), 221–239, (1993).
154. E.B. van Munster, J. Goedhart, G.J. Kremers, E.M.M. Manders, and T.W.J. Gadella Jr., Combination of a spinning disc confocal unit with frequency-domain fluorescence lifetime imaging microscopy. *Cytometry Part A*, **71A**(4), 207–214, (2007).

155. K. Nithipatikom and L.B. McGown, Factors affecting calibration for phase-modulation fluorescence lifetime determinations. *Appl. Spectrosc.* **40**(4), 549–553, (1986).
156. N.J.F. Blamey, J.F. Conliffe, J. Parnell, A.G. Ryder, and M. Feely, unpublished results.
157. N.J.F. Blamey, A.G. Ryder, P. Owens, and M. Feely, unpublished results.
158. M.A. Przyjalgowski, A.G. Ryder, M. Feely, and T.J. Glynn, Analysis of hydrocarbon-bearing fluid inclusions (HCFI) using time-resolved fluorescence spectroscopy. *Proc. SPIE-Int. Soc. Opt. Eng.*, **5826**, 173–184, (2005).
159. M.A. Przyjalgowski, Time-resolved fluorescence spectroscopic analysis of petroleum oils and hydrocarbon bearing fluid inclusions (HCFI), *Ph.D. Thesis*, National University of Ireland, Galway, Galway, Ireland (2006).

Photophysics and Biophysical Applications of Benzo[*a*]phenoxazine Type Fluorophores

Paulo J.G. Coutinho

Introduction

In recent years, the application of photoluminescence methods to biomedical sciences has proved to be very successful, as shown by the number of publications in the last decade. The high sensitivity of fluorescence to the local molecular environment makes it possible to probe complex mediums and/or materials from a wide range of aspects: local polarity effects, specific physical and chemical interactions, and also morphological and topological constraints [1–3].

Nile Red (9-(diethylamino)-5H-benzo[*a*]phenoxazin-5-one), belongs to the class of benzo[*a*]phenoxazines, has a hydrophobic nature and shows low solubility and fluorescence quantum yield in water (Fig. 1).

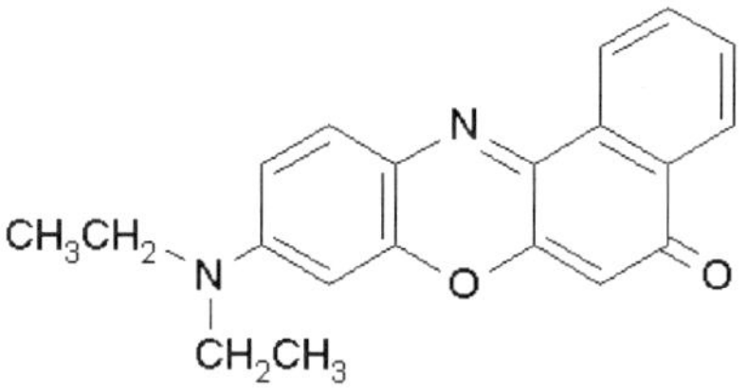

Fig. 1 Chemical structure of Nile Red

Nile Red has been extensively used to study biological membranes [4, 5], proteins [6–8], micelles [9, 10], γ-cyclodextrins [11], non-ionic surfactant microemulsions [12], Langmuir–Blodgett films [13], zeolites [14], ionic liquids [15], electrospray droplets [16] and liquid crystals [17]. Further, it has been successfully utilized in studies of membranes heterogeneity [5], in the detection of sol–gel transition [18], in protein–drug interactions [19], and formation of dendrimer–surfactant supramolecular assemblies [20].

P.J.G. Coutinho (✉)
Centro de Física, Universidade do Minho, Campus de Gualtar, 4710-057 Braga, Portugal

C.D. Geddes (ed.), *Reviews in Fluorescence 2007*, Reviews in Fluorescence 2007,
DOI 10.1007/978-0-387-88722-7_14,

The wide range of Nile Red applicability is due to its good solvatochromism (Figs. 2 and 3), and both steady-state and time-resolved emission properties are strongly medium dependent [21–28]. It usually exhibits an increase in fluorescence yield with decreasing solvent polarity accompanied by a blue shift in the peak emission.

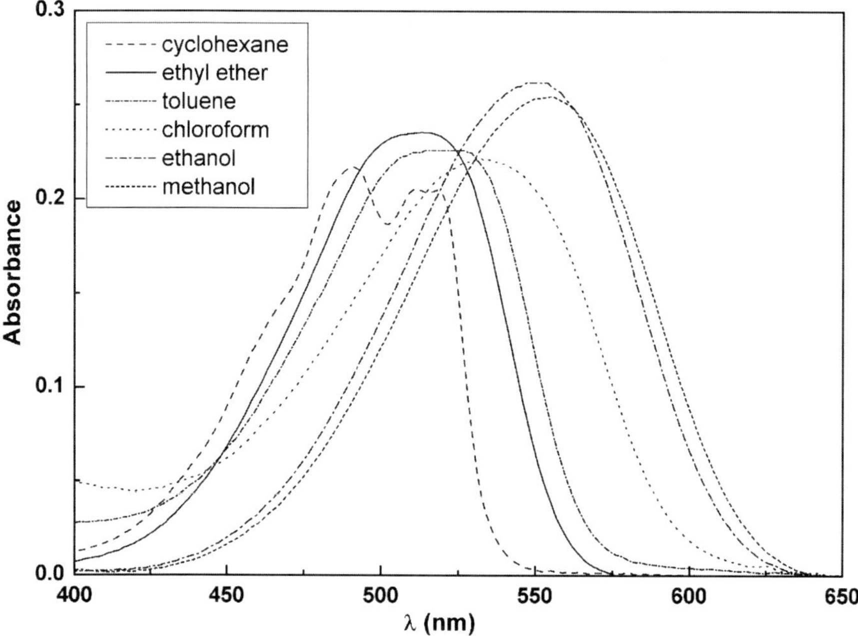

Fig. 2 Nile Red absorption spectra in several solvents

A Twisted Intramolecular Charge Transfer (TICT) state has been proposed to account for the polarity-sensitive fluorescence of Nile Red [21, 26, 27]. In this molecule, TICT would be possible due to the presence of the flexible diethylamino end group attached to the rigid structure of the molecule. However, unlike most other TICT molecules, dual fluorescence was not observed, despite the report of its existence in methanol–water mixtures [27]. Another possible explanation for TICT presence was a proposal for a non-emissive TICT state [28]. Krishna [29] reported on the observation of wavelength-dependent fluorescence intensity decay of Nile Red in egg PC vesicles, SDS micelles, and viscous alcohols, with negative amplitudes for a short-lifetime component. This indicates an excited-state dynamics between at least two emitting excited states. A much higher dependence on solvent viscosity than on polarity ruled out the TICT hypothesis and allowed the explanation based on a two-state model of excited-state solvent relaxation process of Nile Red.

This same two-state model has been used in our research group to study various microheterogeneous systems with relevance for biophysical membrane models [25, 30, 31] using fluorescence as well as fluorescence anisotropy of Nile Red.

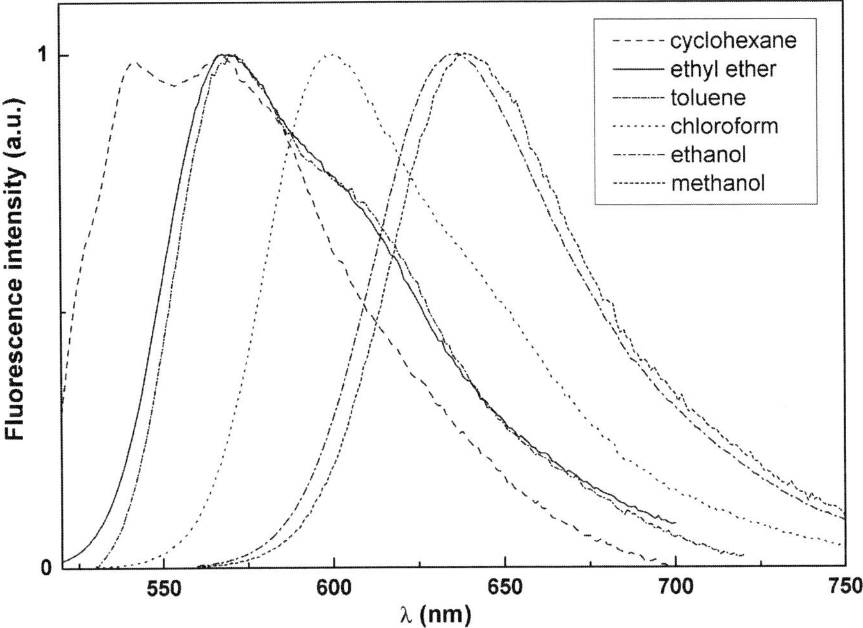

Fig. 3 Emission spectra of Nile Red in several solvents

These results will be presented in "Nile Red as a Solvatochromic and Fluorescence Anisotropy Probe" .

It was also found that the fluorescence lifetime of Nile Red is not sensitive to dielectric solvent–solute interactions but markedly decreases with the increase of the hydrogen bond donating ability in alcohols [24]. A very slight increase of fluorescence lifetime from 4.05 to 4.37 ns when the viscosity increased from 20 to 125000 cP, indicates that twisting of the diethylamino moiety of Nile Red has a very little contribution for the dissipation of the excitation energy [24]. This further indicates that the initially proposed TICT mechanism is not operative in Nile Red with the dual fluorescence in water–methanol mixtures [27] arising from aggregation effects.

Another group of benzo[a]phenoxazine type compounds are the cationic benzo[*a*]phenoxazinium derivatives bearing different substituents in the amine/imine groups and having the general structure shown in Fig. 4.

Nile Blue, Cresyl Violet, and Oxazine 720 are some of the usual dyes belonging to this group of compounds.

This type of molecules may exist as neutral molecules, monocations, dications, or even trications, depending on the environment and have been used in several applications ranging from biological stains and fluorescence standards to laser dyes in tuneable dye lasers in the range 600–900 nm [32]. In the case of laser action monocationic forms are found to be particularly efficient laser dyes.

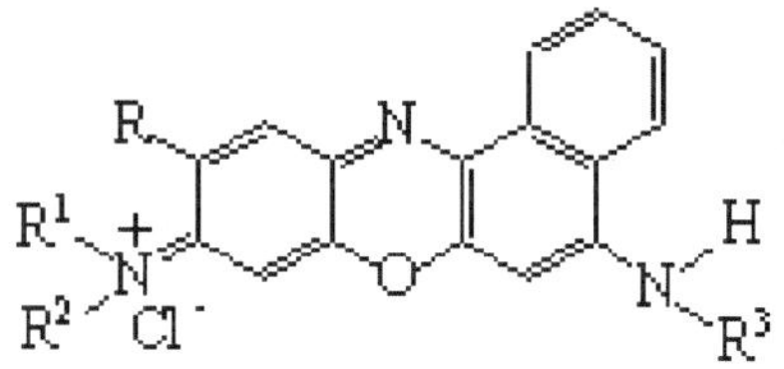

Fig. 4 General chemical structure of cationic benzo[*a*]phenoxazinium compounds

Nile Blue (NB) is a well-known DNA binding probe [33] and it was found to be localized selectively in animal tumours, as well as other similar benzo[*a*]phenoxazinium dyes [34]. It is expected that the planar hydrophobic phenoxazine moiety of NB easily intercalates between the DNA bases with the positive charge being neutralized by DNA's phosphate groups. Compared to the other conventional intercalators like Ethidium Bromide, the use of NB has a merit of having low toxicity and comparable sensitivity for DNA quantification [33]. Recently, the interaction of NB with micelles and genomic DNA has been reported taking advantage of NB photophysical behaviour [35]. Other previous studies of NR in systems of biophysical interest included TX-100 reverse micelles [36], sodium dodecylbenzenesulfonate (SDBS) micelles [37], DNA on gold electrodes with possibility of DNA sensor development [38], and DNA intercalation with the possible application in electrophoresis gel detection [39]. The main conclusions were that NR shows acid–base equilibrium with coexistence of a neutral and basic form [36], it frequently aggregates in negative surfaces [40, 37] and it interacts with DNA via both intercalative and electrostatic binding [38]. Another interesting field of application is the generation of second harmonic by aggregates of NB (and other similar dyes) at the air–water interface [41].

In pure solvents, the photophysics of Nile Blue is well documented in the literature [42–44] and also there exists already some quantum mechanical calculations [45]. Both steady-state and ultrafast time-resolved spectroscopic studies allowed the conclusion that NB either shows an intermolecular electron transfer process in electron donating solvents [43] in 100 fs or an intermolecular proton transfer process in hydrogen-bond accepting solvents [42] in a few ps.

Normal radiationless deactivation of cationic benzo[*a*]phenoxazinium derivatives was found to depend dramatically on the level of substitution of the amino groups [46]. If each amine has at least one H, the excited state deactivation proceeds via N-H vibrations and is independent of temperature. If only one or both of the amines are fully alkylated, the excited state deactivation occurs mainly through rotation of the amino groups with a corresponding dependence on viscosity and on the temperature.

The photophysical properties of new cationic benzo[*a*]phenoxazinium compounds [47–49, 40] will be presented in "Photophysical Studies of Cationic Benzo[*a*]phenoxazinium Derivatives" and interpreted in the light of the just mentioned behaviour of Nile Blue and related compounds. Preliminary results on the application of a long alkyl side chain derivative [50] to follow the micellization process will also be presented.

Nile Red as a Solvatochromic and Fluorescence Anisotropy Probe

$C_{12}E_7$ ($C_{12}H_{25}(OCH_2CH_2)_7OH$) is a non-ionic surfactant with a critical micelle concentration (CMC) of 6.9×10^{-5} M [51].

Using Nile Red fluorescence, the $C_{12}E_7$/water system was monitored at several surfactant concentrations below and above the CMC. In Fig. 5, the various fluorescence spectra of Nile Red in the $C_{12}E_7$/water system are plotted. It can be seen an increase in fluorescence intensity with increasing surfactant concentration with a concomitant blue shift of the maximum emission wavelength.

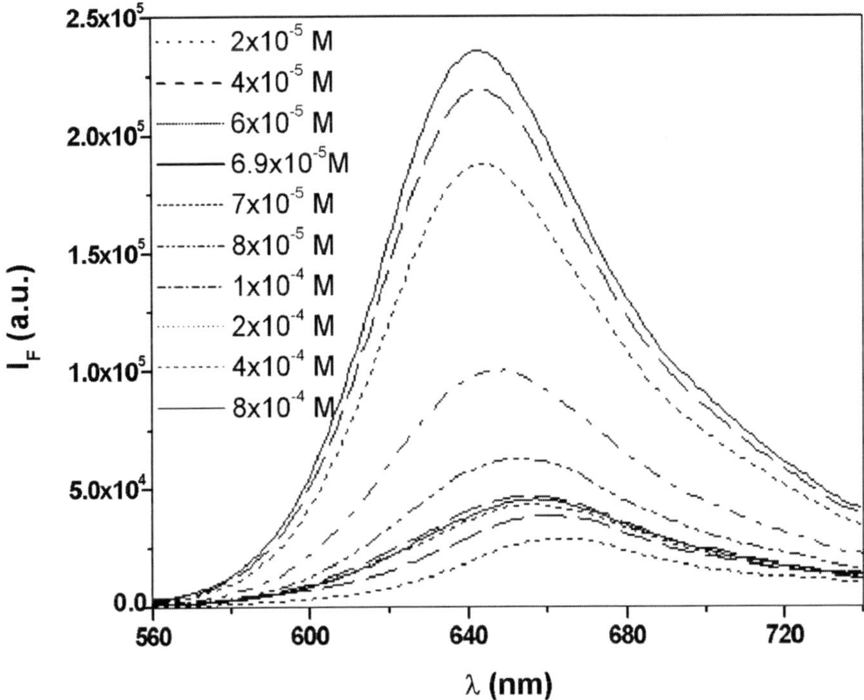

Fig. 5 Fluorescence emission spectra of Nile Red in the $C_{12}E_7$/water system as a function of $C_{12}E_7$ concentration ($\lambda_{exc} = 550$ nm) [25]

The CMC value corresponds to the start of the fluorescence intensity increase (Fig. 6). The Nile Red fluorescence lifetime markedly decreases with the increase in the hydrogen-bonding capability of the medium [24]. This fact explains the sudden rise in fluorescence intensity upon the formation of micelles, because then Nile Red has an opportunity to be more protected from water by locating itself inside the micelle.

As the maximum wavelength of the Nile Red emission spectrum increases with the polarity of the environment (see Fig. 6), the observed blue shift with increasing

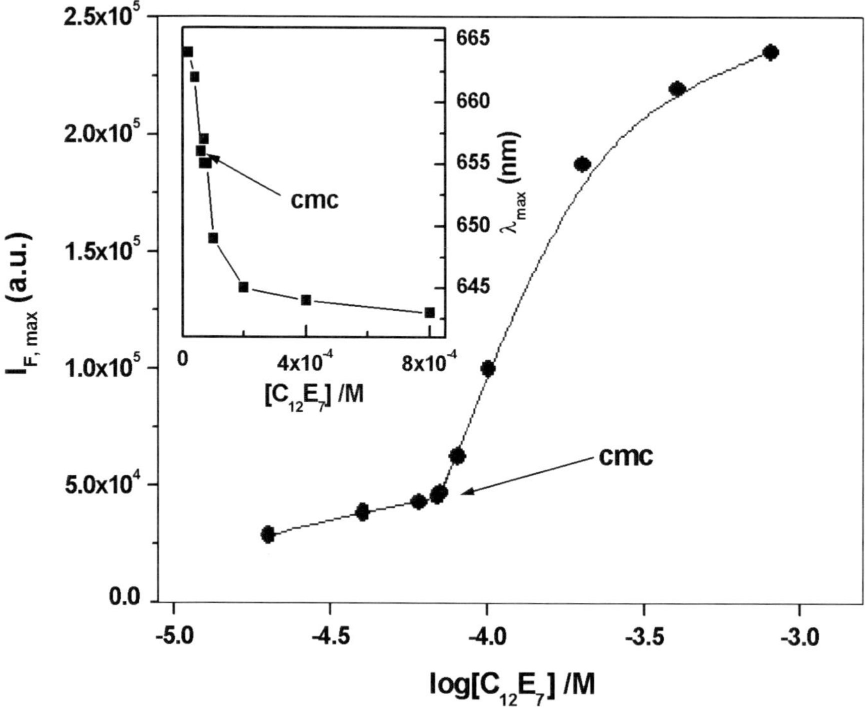

Fig. 6 Maximum emission intensity of Nile Red in the $C_{12}E_7$/water system as a function of $C_{12}E_7$ concentration. The *inset* shows the maximum emission wavelength of Nile Red in the $C_{12}E_7$/water system as a function of $C_{12}E_7$ concentration (λ_{exc}=550 nm) [25]

surfactant concentration also reflects an increased protection from water upon micellization. This in turn allows the determination of the CMC value as an inflection point on the wavelength–maximum plot (see inset of Fig. 6).

Surfactants can dissolve fat. Thus they are often used in membranology as they are able to solubilize phospholipid molecules forming mixed micelles. The amount of surfactant needed to completely disrupt a biological membrane depends on the surfactant structure and on the membrane composition, and usually various types of mixed surfactant/lipid aggregates can exist/coexist depending on the surfactant to lipid molar ratio. One way to follow variations in aggregate structures is to use fluorescence probes.

Dipalmitoylphosphatydilcholine (DPPC) is a very common and abundant phospholipid in biological membranes. Using DPPC/$C_{12}E_7$ as a model system, a concentration of 10^{-4} M, well above the CMC, was used with increasing amounts of DPPC. In the characterization of lipid/micelle interactions, the surfactant molar ratio x_s is usually considered [52]:

$$x_s = \frac{[C_{12}E_7]}{[C_{12}E_7] + [DPPC]}.$$

The solubilization of liposomes by surfactants usually occurs in three steps: mixed vesicles are formed up to a critical surfactant concentration, x_{sat}; between x_{sat} and x_{sol}, there is a coexistence of saturated mixed vesicles and saturated mixed micelles with, respectively, x_{sat} and x_{sol} surfactant contents; above x_{sol}, there are only mixed micelles [52].

Fluorescence of Nile Red in these mixed phospholipids/surfactant systems is able to detect the x_{sat} and x_{sol} critical molar ratios (Fig. 7).

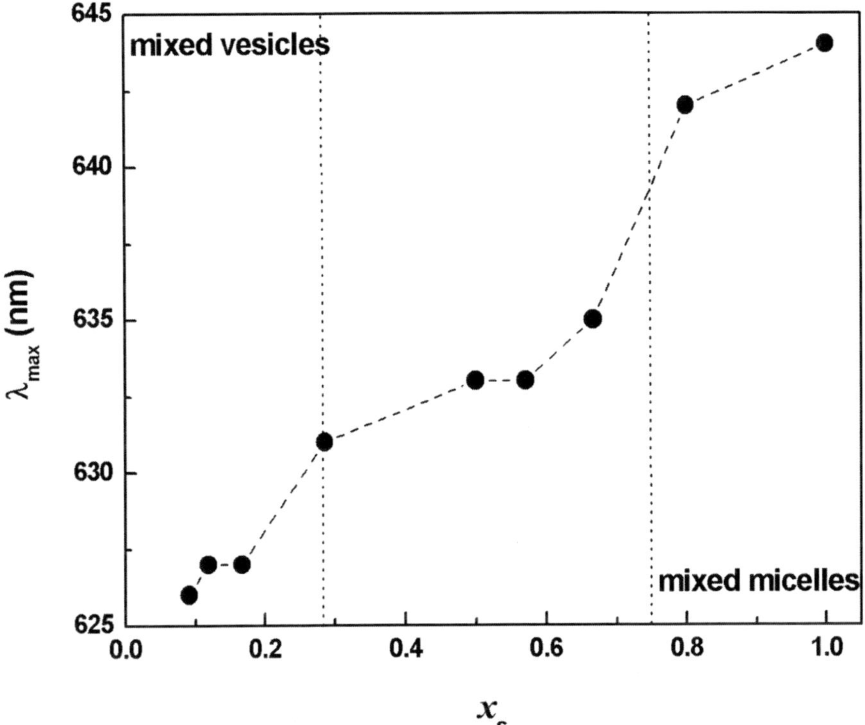

Fig. 7 Maximum emission wavelength of Nile Red in the $C_{12}E_7$/DPPC mixed system (λ_{exc} = 550 nm) as a function of the surfactant molar ratio x_s showing the three regions with mixed vesicles, both mixed vesicles and mixed micelles, and only mixed micelles [25]

Fluorescence anisotropy gives information on the local viscosity felt by the excited molecule during its lifetime. Thus this property should be useful to follow the liposomal solubilization process. Using linearly polarized filters, the usual emission spectra is recovered using the relation

$$I_{total} = I_{VV} + 2GI_{VH} \tag{2}$$

with

$$G = \frac{I_{HV}}{I_{HH}}, \tag{3}$$

where G corrects variations of the instrumental response with the polarization of light, the first subscript defines the vertical (V) or horizontal (H) polarization of the excitation light source and the second subscript defines the polarization of the detected emission.

Figure 8 represents the normalized total emission spectra of Nile Red in DPPC/$C_{12}E_7$ mixed systems.

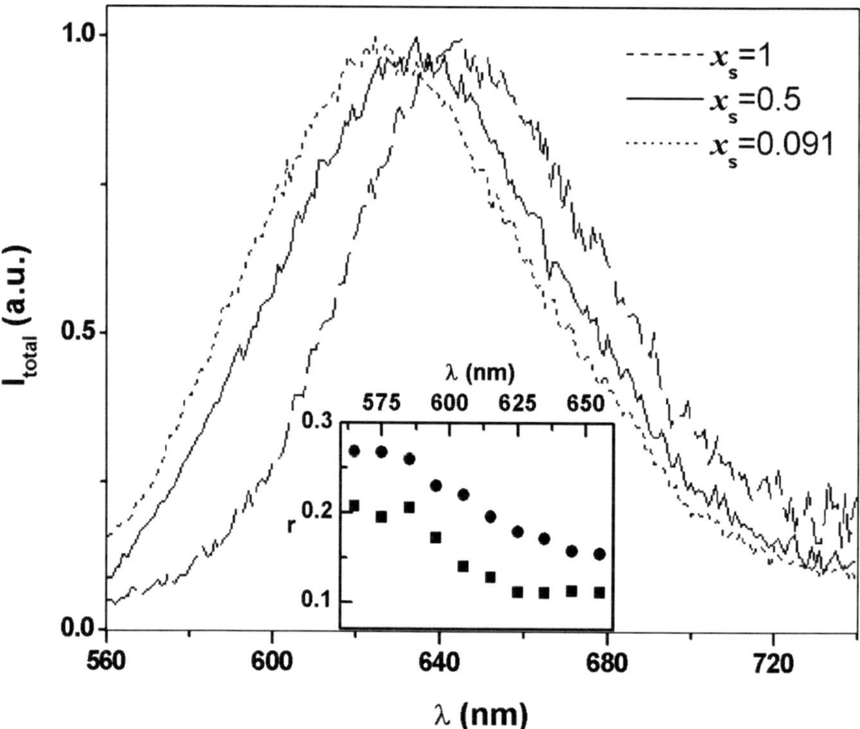

Fig. 8 Normalized total-emission spectra ($I_{VV} + 2GI_{VH}$) of Nile Red for different x_s values (0.091, 0.5, and 1). The *inset* shows the anisotropy variation with emission wavelength for x_s=0.67 (•) and 1 (■)

A significant enlargement in the blue side of the spectrum is readily observed that increases with DPPC content (lower x_s). The inset in Fig. 8 shows that the steady-state anisotropy, r, clearly varies with the emission wavelength, decreasing from a constant value on the blue side to a lower constant value on the red side of the emission spectra. Both this variations are compatible with the two-state model involving an initially excited state and a solvent relaxed state proposed by Krishna [29] (Scheme 1).

If the two states have different anisotropies then variations of the spectral weights of each state justifies the observed variation of fluorescence anisotropy with emission wavelength.

Scheme 1 Two-state model for excited state dynamics of Nile Red

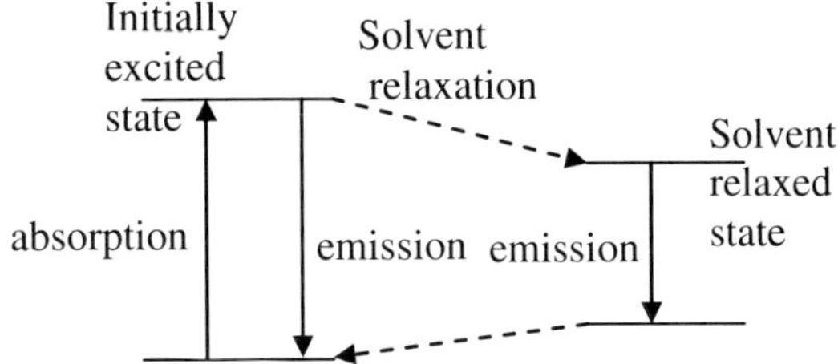

The spectral shape of the emission from any of the two states should be independent of the polarization. Only the overall emission intensity of each state should depend on the polarization of light. The I_{VV} and GI_{VH} fluorescence spectra were thus simultaneously fitted using two sets of spectral parameters where only the intensity maximum was allowed to depend on the whether the polarization was H or V. Figure 9 shows the result of this fitting procedure using one lognormal function [53] for each emitting state.

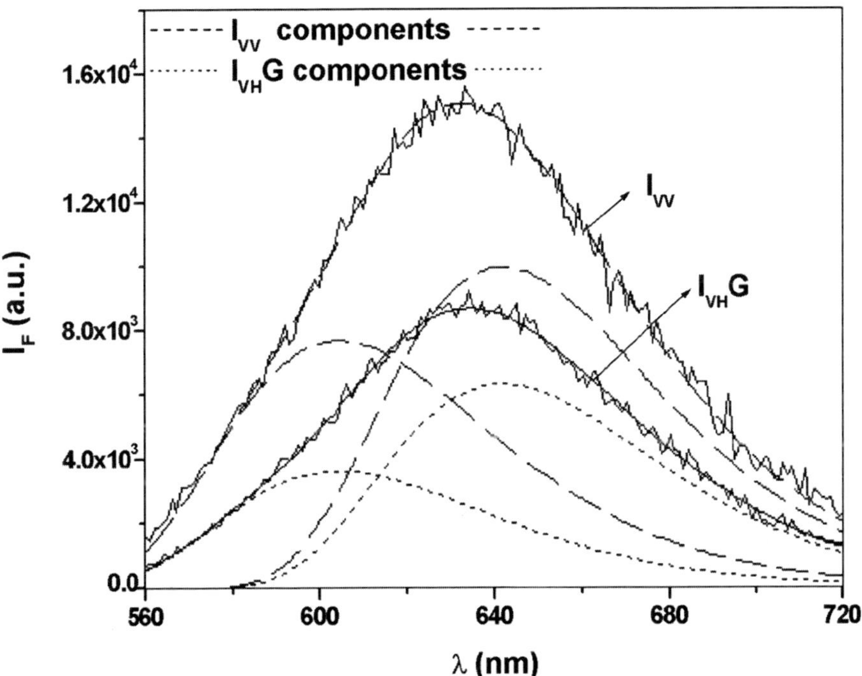

Fig. 9 Decomposition of Nile Red polarized-emission spectra (I_{VV} and GI_{VH}) with two log-normal functions. Fitting curves to experimental results (-), I_{VV} fitting components (- -), and GI_{VH} fitting components (- -)

The steady-state fluorescence anisotropies are then calculated by

$$r_1 = \frac{(A_{VV})_1 - (A_{VH})_1}{(A_{VV})_1 + 2(A_{VH})_1} \qquad r_2 = \frac{(A_{VV})_2 - (A_{VH})_2}{(A_{VV})_2 + 2(A_{VH})_2}, \tag{4}$$

where $(A_{VV}/_{VH})_1$ and $(A_{VV}/_{VH})_2$ stands for the emission maximum intensity of component 1 (initially excited state) and 2 (solvent relaxed state) obtained for the polarizer combinations VV and VH. Consequently the emission intensity fraction of the initially excited state is given by

$$f_1 = \frac{(A_{VV})_1 + 2(A_{VH})_1}{(A_{VV})_1 + 2(A_{VH})_1 + (A_{VV})_2 + 2(A_{VH})_2} \tag{5}$$

In Fig. 10, the quantities f_1, r_1, r_2 and λ_{max} for both states are plotted against the surfactant molar ratio x_s. The critical molar ratios x_{sat} and x_{sol} are represented by vertical dashed lines and correspond to variations in the trends of the plotted curves. It can be seen that the anisotropy of the initially excited state (1) is higher than that of the relaxed state (2) and is more insensitive to the structure and composition of the mixed aggregate. This is an expected result as the lifetime of the initially excited state of Nile Red is much shorter [29] than that of the relaxed state. With the rise of DPPC content (decreasing x_s), the local viscosity should increase. This results in a rise of the anisotropy of the relaxed state as can be observed in Fig. 10a. The higher local viscosity also induces a slower solvent relaxation and should thus result in a higher emitting fraction from the initially excited state, measured by the fraction f_1. This is indeed observed in Fig. 10c.

It can also be observed that the increase in DPPC content causes a blue shift of the relaxed state emission (Figs. 7, 8 and 10b). This behaviour has a general explanation because bilayer structures (vesicles) usually tend to accommodate guest molecules in their interiors, whereas micelles tend to locate them at the surface [54].

The same procedure of using Nile Red photophysical behaviour to study $C_{12}E_7$/DPPC mixed systems shown above was applied to mixed vesicle systems composed of dioctadecyldimethilammonium bromide (DODAB) and dioleoylphosphatydilcholine (DOPE) or soybean phosphatydilcholine (PC) or cholesterol (Ch) [30, 31]. The chemical structures of these lipids are represented in Fig. 11.

The fluorescence spectra of Nile Red incorporated in these vesicular systems as well as the variation of fluorescence intensity with temperature are shown in Fig. 12.

The expected effect of the presence of neutral phospholipids (PC, DOPE) in a DODAB positive bilayer is a reduction of polarity and probably a lower level of hydration in the vesicle interface, where NR is located. The Nile Red fluorescence spectra in Fig. 12 show a corresponding blue shift and an increase in fluorescence intensity. The effect of Ch is distinct and, instead of a blue shift, a huge enhancement on the blue side of the spectrum is observed accompanied by an overall decrease in the fluorescence quantum yield.

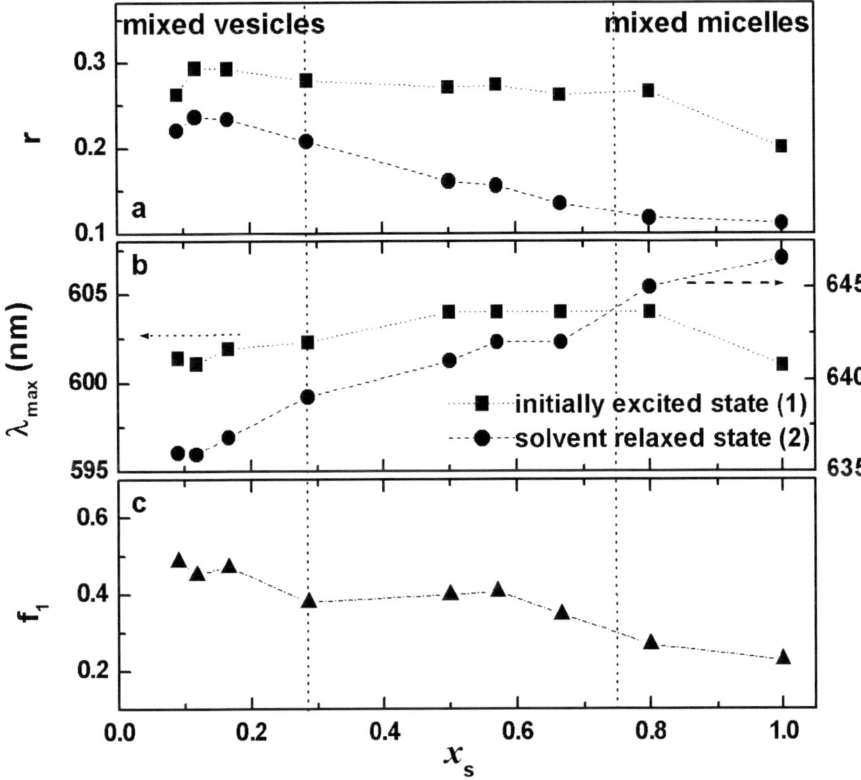

Fig. 10 (**a**) Calculated steady-state anisotropy (equation (4)) of the initially excited state (■) and solvent relaxed state (•) of Nile Red as a function of surfactant molar ratio, x_s. (**b**) Maximum emission wavelength of the initially excited state (■) and solvent relaxed state (•) as a function of x_s. (**c**) Emission intensity fraction f_1 of the initially excited state vs. x_s [25]

At room temperature (~22°C) DODAB bilayers are in the gel phase. Upon temperature increase, a phase change occurs that leads to a more fluid liquid-crystalline phase of the DODAB bilayer. This is a well-known phenomenon of pure lipid bilayers [55].

The effect of increasing temperature, given in the inset of Fig. 1, shows that, in pure DODAB, a maximum Nile Red fluorescence intensity is observed at the phase transition temperature (ca. 45°C), after which a steady decrease in intensity is seen. From these variations it is possible to infer that Nile Red molecules relocate as the temperature increases in the gel state to a less polar environment (more to the interior), giving rise to an increase in fluorescence intensity, due to less water penetration. In the liquid-crystalline phase, the normal trend of an increase in non-radiative processes with temperature (decrease in fluorescence quantum yield) is observed so that the environment in this bilayer phase seems to remain unchanged. This last behaviour is also seen in the other mixed vesicle systems although for

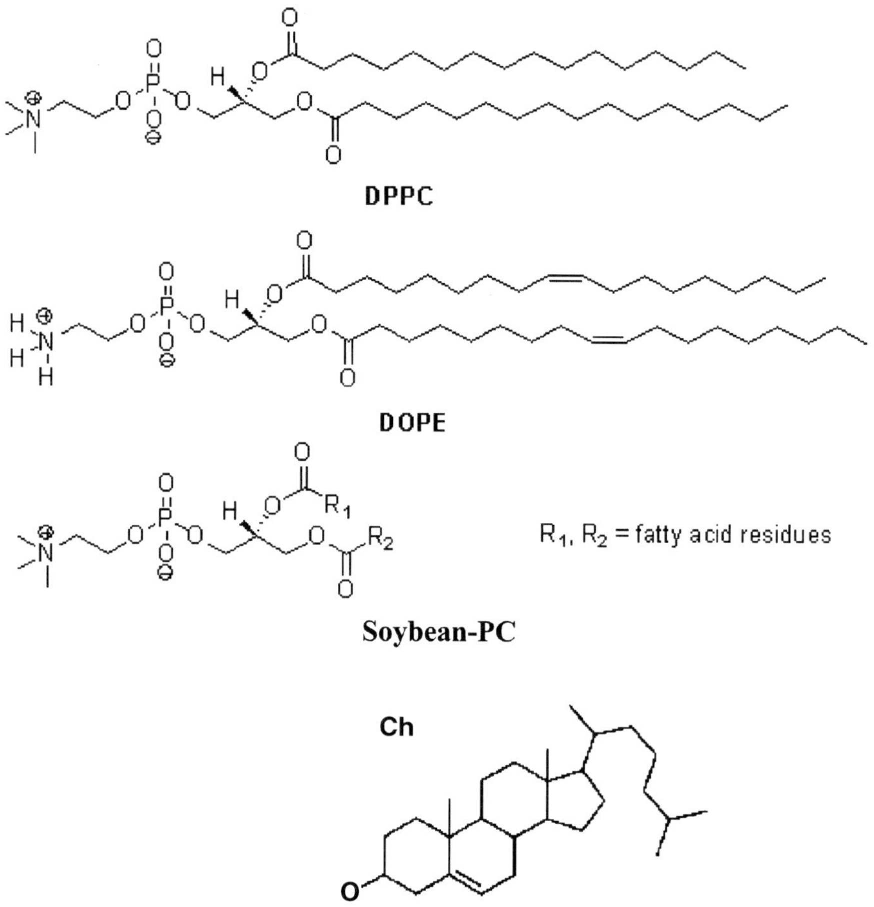

DPPC

DOPE

R1, R2 = fatty acid residues

Soybean-PC

Ch

Fig. 11 Chemical structures of DODAB, DOPE, PC, and cholesterol

cholesterol there is a peculiar behaviour. This means that all studied mixed vesicular systems are in a liquid-crystalline phase.

As in the case of C12E7/DPPC mixed systems, the steady-state anisotropy of Nile Red in the different systems, shown in Fig. 13, shows a clear wavelength dependency.

Using the same two-state model for Nile Red excited state dynamics (Scheme 1) considering a dual emission from an initially excited state and a solvent relaxed state [25], the results obtained are represented in Fig. 14 and in the inset of Fig. 13.

The inset of Fig. 13 shows the fraction of the state 1 (f_1) for the different samples. There is only a slight increase in f_1 upon addition of DOPE or PC to DODAB and the values obtained are close to those seen using NR in egg PC [29]. Only in the system with cholesterol a huge increase of f_1 is observed at a molar ratio of 1:1. If Nile Red in these systems, especially in the presence of cholesterol, feels more

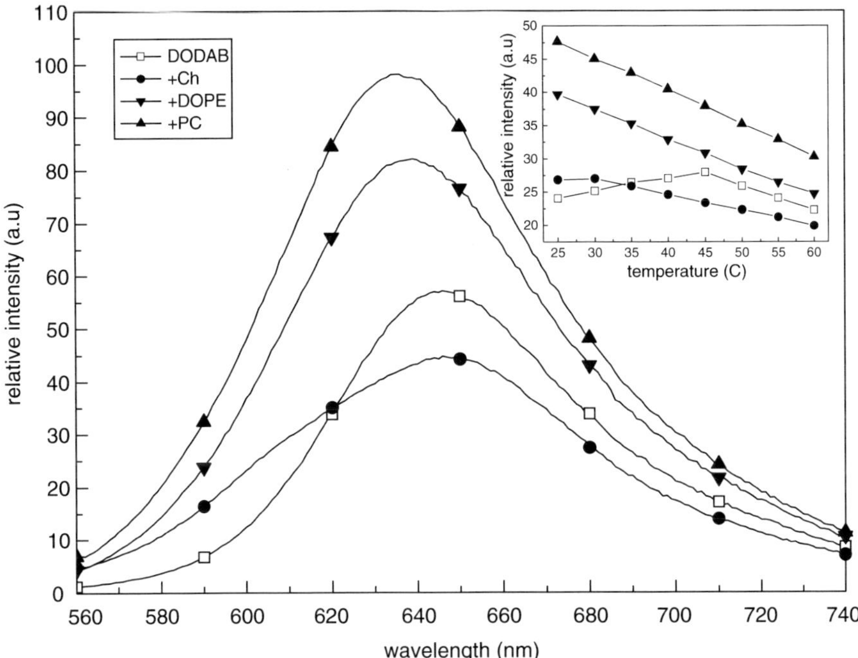

Fig. 12 Emission spectra of Nile Red in the DODAB+ (Ch, DOPE or PC) systems for a molar ratio 1:1 ($\lambda_{exc} = 550$ nm, the peak absorption). The *inset* shows the total fluorescence intensity with temperature

than one average environment, each polarized emission spectra is expected to be composed of two components, corresponding to states 1 and 2, per environment. In that case, the recovered parameters from the anisotropy analysis reflect and average over all possible environments where Nile Red is located.

In Fig. 14, both DOPE and PC follow similar trends: the increase of additive content induces a decrease to a plateau of the peak emission wavelength accompanied, only in the case of the initially excited state, by a reduction in anisotropy. This blue shift has already been explained by the decreasing polarity of the interface. A slight increase of the lifetime of state 1 could explain the decrease in r_1. For cholesterol, the situation is again very different. Not only the decrease does not reach a plateau but an inversion occurs from 33 to 50% cholesterol content with the anisotropy of state 1 following an opposite variation and being always higher than for DOPE or PC.

The higher value of r_1 for DODAB+Ch system suggest either some resistance to rotational relaxation motion (more viscous environment) or a restricted environment for Nile Red. The above observations can be explained, if initially there is an even distribution of Ch between the lipids, which on Ch enrichment leads to the formation of lipid–cholesterol condensed complexes [56]. Overall this would explain the blue shift in the Nile Red emission spectrum as the lipid packing changes and the dye is

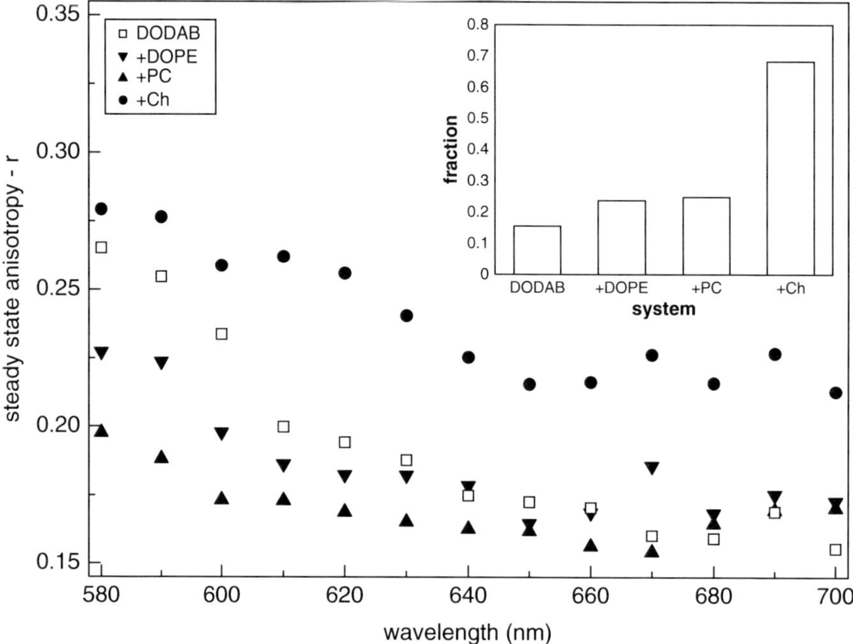

Fig. 13 Variation in Nile Red anisotropy with emission wavelength for NR in the DODAB+ (Ch, DOPE, or PC) systems for a molar ratio 1:1. *Inset* is the fraction of fluorescence attributed to excited state 1

less accessible to the aqueous region. At the higher Ch concentrations, the cholesterol can self associate to form crystalline domains [57, 58], which expel the probe to a more polar environment. This effect is more pronounced in the wavelength data, but is also perceptible in the anisotropy data.

Although simplistic this form of analysis shows its merits in elucidating information concerning these difficult systems and could report very distinct structural changes in membranes induced by different lipid compositions.

Photophysical Studies of Cationic Benzo[*a*]phenoxazinium Derivatives

New cationic phenoxazine derivatives were synthesized in collaboration with the chemical department [47–49]. In this section, an overview of some of its photophysical properties will be presented.

In Fig. 15, the chemical structure of some of the synthesized compounds is shown.

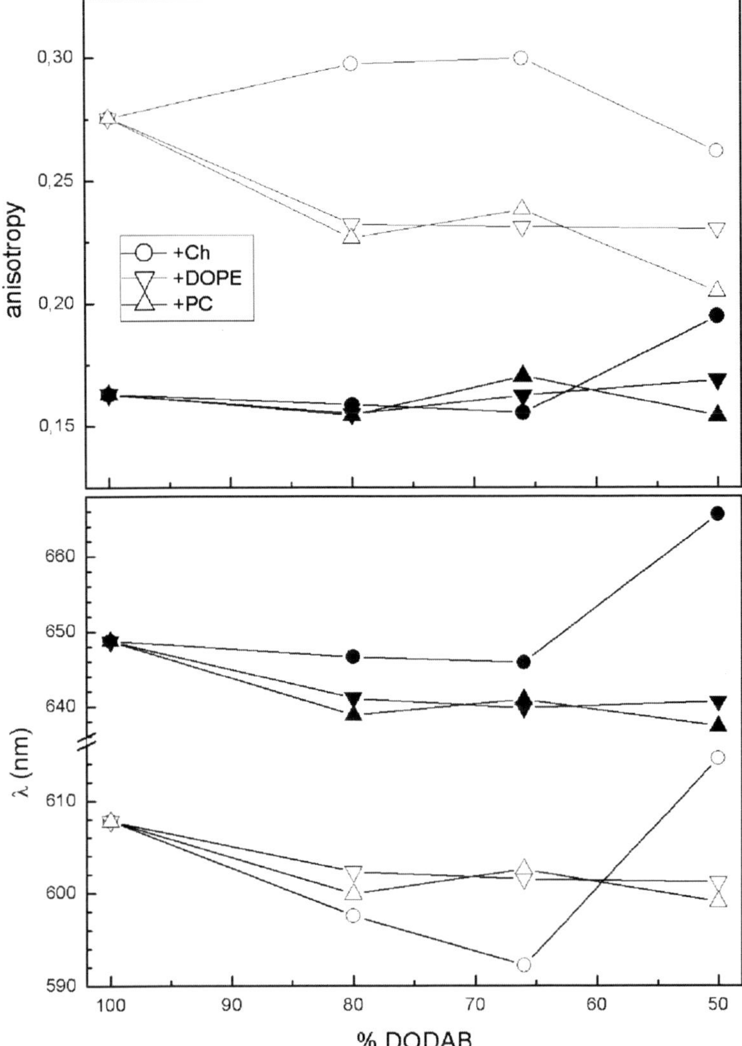

Fig. 14 Position of the peak emission and associated anisotropies for Nile Red in DODAB with different quantities of additives. *Open symbols* relate to state 1 and *closed* to state 2

In Table 1, the wavelength of absorption and emission maximum, and the fluorescence quantum yield, Φ_F, are tabulated for both ethanol and water at pH 7.4.

In aqueous solutions at physiological pH (pH 7.4, adjusted with HCl and NaOH), the position of the absorption maximum shows a huge bathochromic shift for compound **1a** from 500 nm (in ethanol) to 640 nm (in water). This fact could be explained by the common solvatochromic effect on $\pi\pi^*$ electronic transitions. But

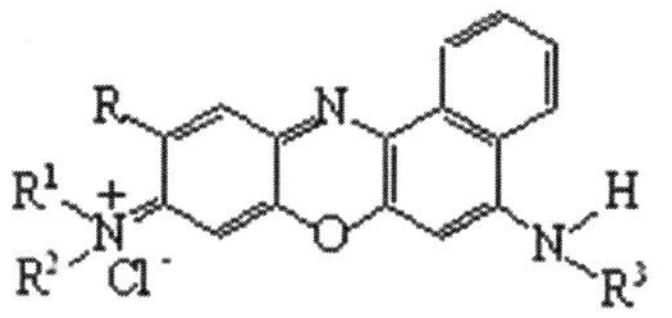

1a $R = H, R^1 = R^2 = CH_3, R^3 = (CH_2)_2CO_2CH_3$
1b $R = H, R^1 = R^2 = CH_3, R^3 = (CH_2)_2CO_2H$
1c $R = H, R^1 = R^2 = CH_3, R^3 = (CH_2)_2CO_2CH_2CH_3$
1d $R = H, R^1 = R^2 = CH_3, R^3 = Ph$
1e $R = H, R^1 = R^2 = CH_3, R^3 = H$
1f $R = H, R^1 = R^2 = CH_2CH_3, R^3 = (CH_2)_2CO_2CH_3$
1g $R = H, R^1 = R^2 = CH_2CH_3, R^3 = (CH_2)_2CO_2H$
1h $R = H, R^1 = H, R^2 = (CH_2)_5CH_3, R^3 = Ph$
1i $R = CH_3, R^1 = H, R^2 = CH_2CH_3, R^3 = (CH_2)_2CO_2H$
1j $R = CH_3, R^1 = H, R^2 = CH_2CH_3, R^3 = (CH_2)_2CO_2CH_2CH_3$
1l $R = CH_3, R^1 = H, R^2 = CH_2CH_3, R^3 = (CH_2)_2CO_2CH_3$
1m $R = CH_3, R^1 = H, R^2 = CH_2CH_3, R^3 = Ph$

Fig. 15 Chemical structure of the studied compounds

the observed value of the shift (140 nm) is too large and also not all compounds show maximum absorbance near 500 nm in ethanol (only **1a**, **1d**, **1g** and **1m**) which now appears in the 600 nm region.

In water, charged dyes are known to aggregate due to the high dielectric constant of water reducing the electrostatic repulsion. The oxazine dyes, which belong to the

Table 1 Photophysical properties of cationic phenoxazine derivatives in ethanol and in water at pH 7.4

	Maximum absorption wavelength (nm)		Maximum emission wavelength (nm)		Fluorescence quantum yield, Φ_F	
Compound	Ethanol	Water	Ethanol	Water	Ethanol	Water
1a	500	640	612	682	0.051	0.10
1b	625	635	669	682	0.110	0.094
1c	633	640	620	678	0.053	0.080
1d	515		643		0.0017	
1e	625		661		0.19	
1f	638	650	618	685	0.049	0.065
1g	635	645	666	684	0.225	0.080
1g	520		622		0.0022	
1i	625	625	644	652	0.44	0.28
1j	630	625	643	654	0.50	0.32
1l	615	620	643	650	0.49	0.19
1m	523					

same class of compounds, are known to form H-aggregates [40]. This type of dimers has an absorption band to the blue of the monomer and a very low fluorescence quantum yield.

The observed differences in absorption maximum can then be due to variations of dimerization equilibrium constants with the solvent. In order to test this hypothesis, absorption spectra of compound **1a** at different concentrations (Fig. 16) in ethanol and in water were measured. The formation of H-aggregates in water is confirmed with a separation of ~40 nm between the absorption maxima of the monomer and the dimer. The extent of this dimerization depends on the substituents of the benzo[*a*]phenoxazine compounds studied in this work as can be seen in Fig. 17.

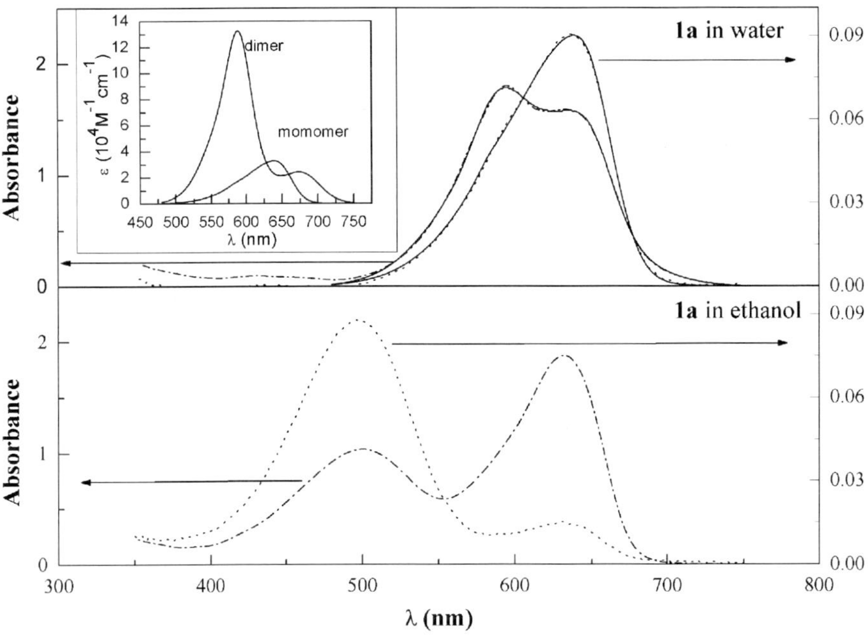

Fig. 16 Absorption spectra of **1a** at 2.75×10^{-6} M (*dotted line*) and 5×10^{-5} M (*dash–dotted line*) concentration, in water (pH 7.4) (**A**) and in ethanol (**B**). *Solid lines* in **A** are the fitted spectra. The *inset* in **A** plots the monomer molar absorptivity and half the dimer molar absorptivity

Pure monomer spectra can be obtained from the excitation spectra because H-aggregates are non-fluorescent. In order to estimate the dimerization equilibrium constant, both absorption spectra, at high and low concentrations, were fitted using the following equations:

$$A(\lambda) = \alpha \, \mathrm{Ex}(\lambda) \, f_{\mathrm{M}} C + \frac{\varepsilon_{\mathrm{D}}(\lambda)}{2} \, (1 - f_{\mathrm{M}}) C, \tag{6}$$

$$k_{\mathrm{D}} = \frac{1 - f_{\mathrm{M}}}{2C \, f_{\mathrm{M}}^{2}}, \tag{7}$$

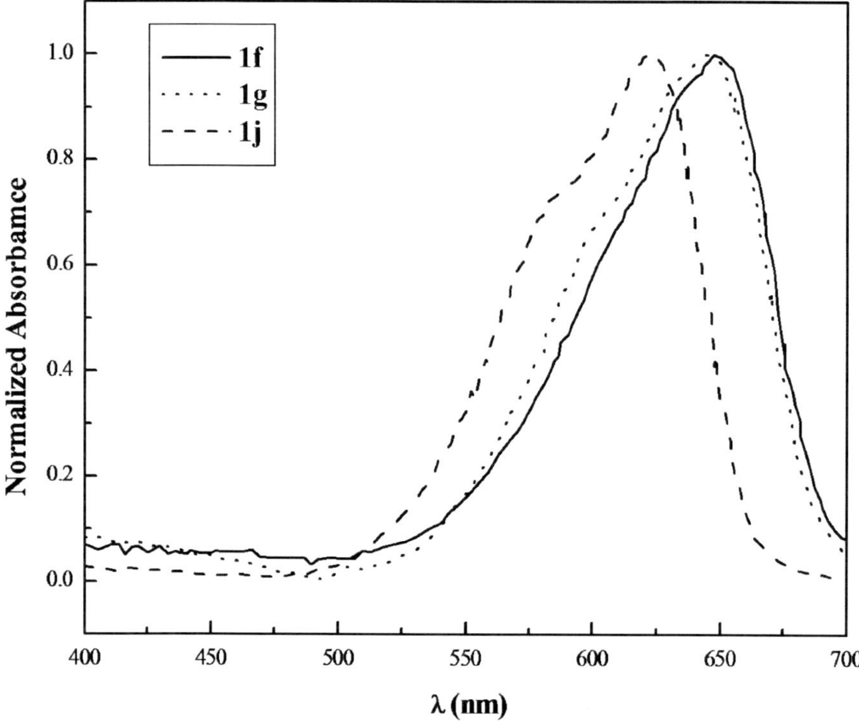

Fig. 17 Absorption spectra of **1f** at 1.3×10^{-6} M,**1g** at 1.5×10^{-6} M, and **1j** at 2.3×10^{-6} M in water (pH 7.4)

where k_D is the dimerization equilibrium constant, C is dye concentration, f_M is the mole fraction of monomer, Ex(λ) represents the excitation spectrum, α is a proportionality constant, and $\varepsilon_D(\lambda)$ is the dimer absorption spectrum taken as a sum of three Gaussian functions.

In Fig. 16A, the solid lines represent the result of this fitting procedure. A value of $k_D=2.09\times10^3 M^{-1}$ was then obtained for the dimerization equilibrium constant. This estimate is in accordance with a recent k_D determination of an oxazine derivative where double lysine labelling was used in order to obtain a pure dimer absorption spectrum [59]. From the described fitting procedure, it was also possible to obtain the dimer absorption spectrum which is plotted in the inset of Fig. 16A.

In the case of ethanol, not only the wavelength separation of the two bands is much greater (>100 nm), but also the weight of the blue band increased upon dilution (Fig. 16B). This is a clear indication that no dimerization is occurring in ethanol. Another possibility is an interaction of the compounds with the solvent (possibly with the labile hydrogen atoms in the amine groups), which gains importance with dilution. If the complete dissociation of the labile proton is considered, the effect can be explained by the influence of concentration in the ionization equilibrium of weak acids:

$$HA^+ \rightarrow H^+ + A, \tag{8}$$

where HA+ represents the cationic phenoxazine derivative. If the acid/base ionization constant is 10^{-5} M, the ratio $[A]/[HA^+]$ goes from 2.7 to 0.56 when the HA^+ concentration increases from 5 to 50 μM.

In order to further elucidate this solvent interaction, absorption spectra of compounds **1a**, **1c**, **1f**, **1g**, **1j**, and **1m** were run in eight more solvents of different polarity and proton donor ability such as 1,4-dioxane, chloroform, ethyl acetate, dichloromethane, acetone, methanol, DMF and acetonitrile. Figure 18 shows absorption spectra of compound **1a** in various solvents.

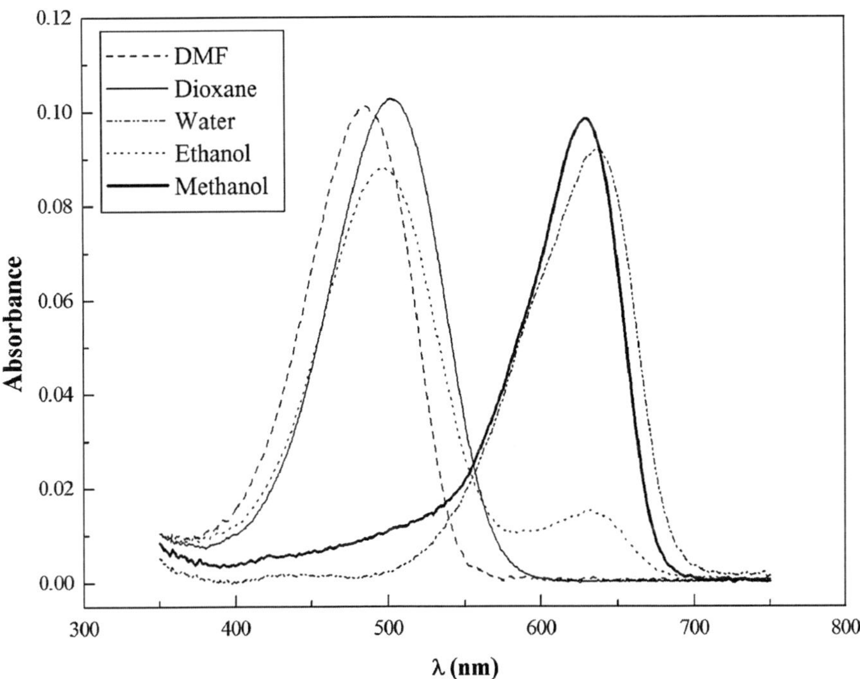

Fig. 18 Absorption spectra of **1a** at 5.0×10^{-6} M in DMF, 1,4-dioxane, water (pH 7.4), ethanol and methanol

One main absorption in the 500 nm region and other near 600 nm are observed. In the case of proton accepting solvents, this bands can then correspond to the deprotonated and protonated form of the phenoxazine derivatives.

Some representative wavelength maxima (λ_1 and λ_2) and molar absorptivities (ε_1 and ε_2) of the two bands observed for compounds **1a**, **1c**, **1f**, **1g**, **1j**, and **1m** are listed in Table 2.

The relative weight of the two bands depends not only on the solvent but also on the substituents of the benzo[*a*]phenoxazine moiety. Also included in Table 2 are the Kamlet–Taft solvent parameters [60] π^*, β, and α. Parameter π^* takes into

Table 2 The absorption spectra data of compounds **1a**,**1c**,**1f**, **1g**,**1j**, and**1m** in selected solvents from Ref. [47]

Solvent		Ethyl acetate		Dichloromethane		Acetone		Ethanol		Methanol	
$\varepsilon^a \mid \pi^*$		6.08	0.55	8.93	0.82	21.0	0.71	25.3	0.54	33.0	0.6
$\beta \mid \alpha$		0.45	0	0	0.3	0.48	0.08	0.77	0.83	0.62	0.93
	$\lambda_1 \mid \lambda_2$	489		505	629	492		500	634	500	629
1a	$\varepsilon_1 \mid \varepsilon_2$	1.3		1.2	3.4	3.5		1.7	0.3	0.2	2.0
	Conc.[c]	7.2		3.9		4.5		5.0		5.0	
	$\lambda_1 \mid \lambda_2$	489	611	503	638	494			633		629
1c	$\varepsilon_1 \mid \varepsilon_2$	1.5	0.3	0.8	2.2	1.5			2.3		2.2
	Conc.[c]	7.0		1.9		4.7		2.8		4.0	
	$\lambda_1 \mid \lambda_2$	499	621	514	635	505		512	638		637
1f	$\varepsilon_1 \mid \varepsilon_2$	2.1	0.6	1.1	3.7	3.1		1.9	4.3		4.6
	Conc.[c]	4.8		4.5		4.2		9.9		4.6	
	$\lambda_1 \mid \lambda_2$	494		539	634	505			635		632
1g	$\varepsilon_1 \mid \varepsilon_2$	2.1		0.4	0.6	1.4			2.2		2.2
	Conc.[c]	1.9		3.8		3.1		5.6		4.1	
	$\lambda_1 \mid \lambda_2$	495			620	496			630		625
1j	$\varepsilon_1 \mid \varepsilon_2$	0.9			2.8	3.4			3		6.8
	Conc.[c]	4.9		2.0		2.9		4.8		4.8	
	$\lambda_1 \mid \lambda_2$	509		510		514		523		527	634
1m	$\varepsilon_1 \mid \varepsilon_2$	2.2		4.8		3.8		0.3		3.1	1.4
	Conc.[c]	2.5		2.0		3.2		20.4		13.7	

[a]Dielectric constant,[b]value for distilled water, [c]concentrations in μM, $\lambda_{max,}$in nm and ε in $10^4 M^{-1} cm^{-1}$.

account the solvent capability of stabilizing ionic solutes (dipolarity/polarizability). Parameter β evaluates the ability of a solvent to accept a proton (donate an electron pair) in a solvent to solute hydrogen bond. Parameter α evaluates the ability of a solvent to donate a proton (accept an electron pair) in a solvent to solute hydrogen bond.

In solvents with high hydrogen bond accepting capability (acetone, DMF, dioxane, and ethyl acetate), the predomination of the blue (~500 nm) absorption band can be seen. It is then possible to conclude that the solvent interacts with benzo[*a*]phenoxazinium dyes by either accepting a hydrogen bond or by completely removing the amine proton thus originating the corresponding base. Douhal [42] showed that for Nile Blue (NBA: benzo[*a*]phenoxazinium salt with $R^1 = R^2 = CH_2CH_3$ and $R^3 = H$) the hydrogen bond interaction results in a ~5 nm blue shift, and the complete removal of a proton from the amine (forming Nile Blue base – NBB) results in the formation of a new band shifted ~100 nm to the blue. Thus, the band near 500 nm can be assigned to the basic form of the studied compounds.

Among the used solvents, only chloroform and dichloromethane cannot accept a proton. But in some compounds the band corresponding to the basic form is observed. This can be explained by the presence of trace amount of water in these solvents that can then accept a proton from the red absorbing acid form of these dyes. Douhal [42] also observed a small fluorescence from NBB in a solution of NBA in neat 1-chloronaphthalene, which decreased upon drying of the solvent.

In both DMF and acetone, only the basic form is observed for all compounds. This is due to the strong proton accepting nature of these solvents.

Coming from ethanol to water, the proton accepting capability is decreasing while the proton donor strength is increasing. In accordance with these variations, the basic form only is important in ethanol and completely vanishes in water. It is worth mentioning that in methanol the absorbance spectrum shows a tail to the blue which is completely absent in water solutions (see Fig. 18).

The R^1, R^2, and R^3 substituents also affect the relative amount of basic and acid forms as can be seen from Table 2. This probably occurs through inductive/chemical effects on the acidity of the amino group of the benzo[*a*]phenoxazinium derivatives. For example, in compounds **1a** and **1c**, only an extra methylic group in the R^3 ester substituent decreases the acidity of the amino group leading to the appearance of acid form in ethyl acetate. Using geometry optimization with semi-empirical quantum methods within the PM3 framework by ArgusLab software, it is possible to conclude that the resonance structure with a double bond in the N atom with the R^3 substituent establishes an hydrogen bond with the ester group (N···H distance of 1.02 Å and H···O distance of 1.83 Å). The slightly stronger electron donating capability of the C_2H_5 group in the ester leads to a stronger hydrogen bond and a corresponding reduced acidity of the amine. Thus, the absorbance spectrum of these oxazine dyes can be tuned by the type of substituents and are very sensitive to the proton accepting capability of the solvent.

In the case of compounds **1a** and **1f** which differ only in an extra methyl group in R^1 and R^2positions, the inductive effect in the proton in the other side of the molecule is expected to be very small. Yet the dominant form in the absorption spectrum changes from basic to acid in ethanol solutions. This variation can be mainly due to a duplication of concentration from **1a** to **1f**. If the acid/base dissociation constant is 4×10^{-6}M, the ratio [A]/[HA^+] goes from 1.47 to 0.9 when the HA^+ concentration increases from 5 μM to 10 μM.

Compounds **1j** and **1m** have two labile protons but the absorbance characteristics are not much different from those observed for the other compounds which only have one removable proton.

The fluorescent properties of benzo[*a*]phenoxazinium salts **1a–m** measured in absolute ethanol, using Oxazine 1 as standard (fluorescence quantum yield, Φ_F = 0.11 [61] in ethanol), are indicated in Table 1. When the basic form is dominant (compounds **1a**, **1c**, **1d**, **1f**, **1g**, and **1m**) the fluorescence occurs with a large Stokes shift (102–128 nm) with emission maximum in the 612–643 nm region. When the acid form is more important (compound **1b**, **1e**, **1g**, **1i**, and **1j**), the emission maximum occurs in 643–669 nm with a smaller Stokes' shift (53–81 nm). Douhal [42] observed a similar behaviour for Nile Blue. The emissions in the 612–643 nm and 643–669 nm regions can be assigned to the basic and acid forms of benzo[*a*]phenoxazinium dyes, respectively.

As the Stokes' shift of the basic form is larger than that of the acid one, the interaction with the solvent seems more important for the unprotonated form than for the protonated one. Also from Table 1, the fluorescence quantum yield, ϕ_F, for the basic form is one order of magnitude lower than for the acid one. Thus, the stronger interaction of the basic form with the solvent seems to introduce an enhanced non-radiative deactivation of the excited state. A parallelism can be made here with Nile Red, the neutral phenoxazine derivative presented in "Nile Red as a Solvatochromic and Fluorescence Anisotropy Probe" as these interactions originate probably from H-bond type interactions.

When R^3 is a phenyl group (compounds **1d**, **1g**, and **1m**), the absorption maximum of the basic form shifts 10 nm to the red (Table 1) and Φ_F decreases another order of magnitude. When the R group is changed from H to CH_3, a ~4 times increase in the fluorescence quantum yield (**1b/1i**) is observed. It seems that electron donating groups in the R position enhances the fluorescence of this compounds while electron donating groups in the R^3 position originates a quenching effect.

In aqueous media, only the acid form of all benzo[*a*]phenoxazinium compounds is observed. For those compounds that were excited in ethanol in the basic form (**1a**, **1c**, **1d**, and **1f**), an increase of quantum yield is observed in water as the acid form is more fluorescent than the basic one (see Table 1). For the other compounds, it seems that the acid form is less fluorescent in water than in ethanol. This can be the result of the formation of non-fluorescent H-aggregates in aqueous media. This phenomena still occurs in the diluted solutions used in fluorescent studies as is apparent from Fig. 18.

In order to use benzo[*a*]phenoxazinium compounds as probes in biophysical models of membranes, a derivative was synthesized with a long alkyl side-chain (Fig. 19) [50].

This compound was incorporated in aqueous solutions of either a neutral (Triton® X-100) or a cationic (CTAB) surfactant (Fig. 20).

As shown in Figs. 21 and 22, this compound is able to detect the critical micelles concentration (CMC) of both neutral and cationic surfactants.

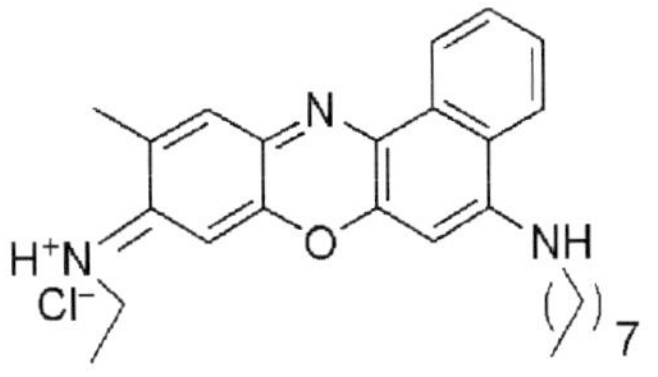

Fig. 19 Chemical structure of C8 – benzo[*a*]phenoxazinium derivative

$H_3C(H_2C)_{15}-N^+(CH_3)_3\ Br^-$

CTAB

n ~ 10

Triton® X-100

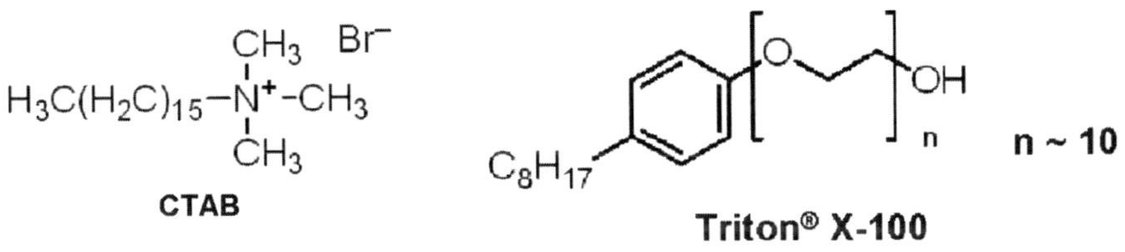

Fig. 20 Chemical structure of the surfactants used

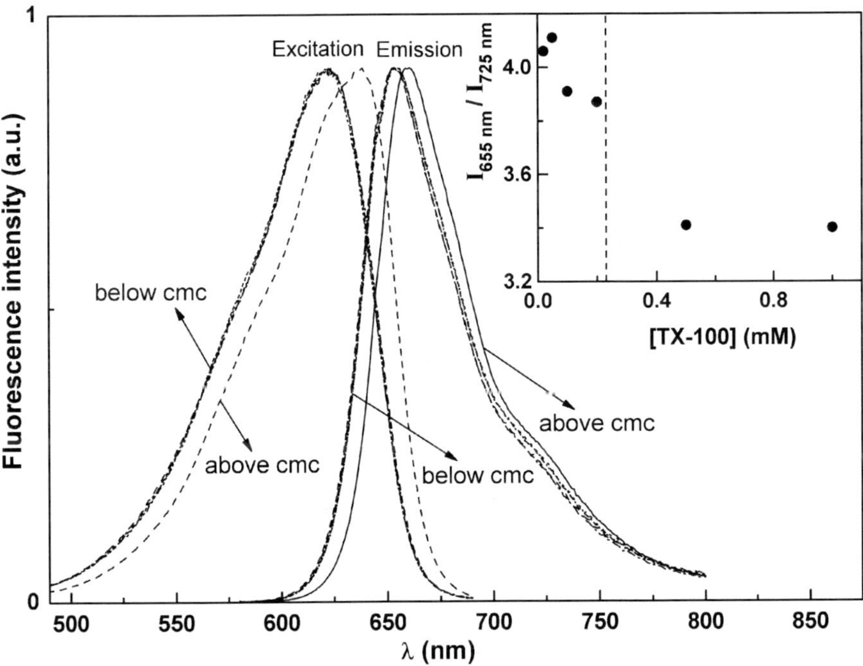

Fig. 21 Excitation and emission spectra of C8-benzo[*a*]phenoxazinium derivative in aqueous solutions of TX-100. The *inset* quantifies the observed variation of spectral shape and the CMC is indicated by a *vertical dotted line*

In Triton® X-100 (TX-100), the compound fluorescence can detect the formation of micelles through two spectral features. One is the spectral shape where a shoulder at 725 nm increases with the concentration of TX-100. The other is the maximum fluorescence wavelength that shows a sudden shift to the red when CMC is reached (Fig. 22).

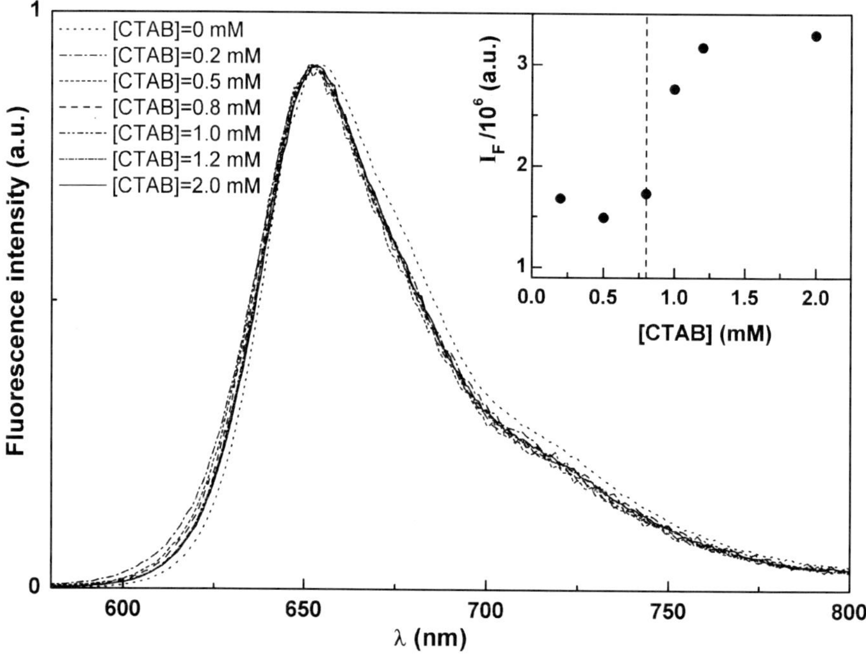

Fig. 22 Emission spectra of C8-benzo[*a*]phenoxazinium derivative in aqueous solutions of CTAB. The *inset* shows the maximum intensity as function of CTAB concentration and the CMC is indicated by a *vertical dotted line*

In CTAB, the micellization process is still detectable, although both the compound and CTAB are positively charged. This indicates that the C8 side chain acts as an anchor to the micellar or pre-micellar aggregates. The shape of the emission spectra does not change significantly, but the fluorescence intensity increases with the concentration of CTAB allowing the determination of its CMC.

A further confirmation that the C8-side chain derivative associates with micelles can be obtained from anisotropy studies. In Fig. 23, fluorescence anisotropy spectra are shown the studied compound in water and in TX-100 aqueous solutions above the CMC.

The high anisotropy observed in TX-100 reveals a higher local viscosity and indicates that the C8-side chain derivative is associated with micelles. In water, a fluorescence anisotropy value of 0.05 indicates that the compound is slightly associated, but not as H-aggregates which, as mentioned previously, have very low fluorescence quantum yield. As for Nile Red probe, an initial decrease of anisotropy

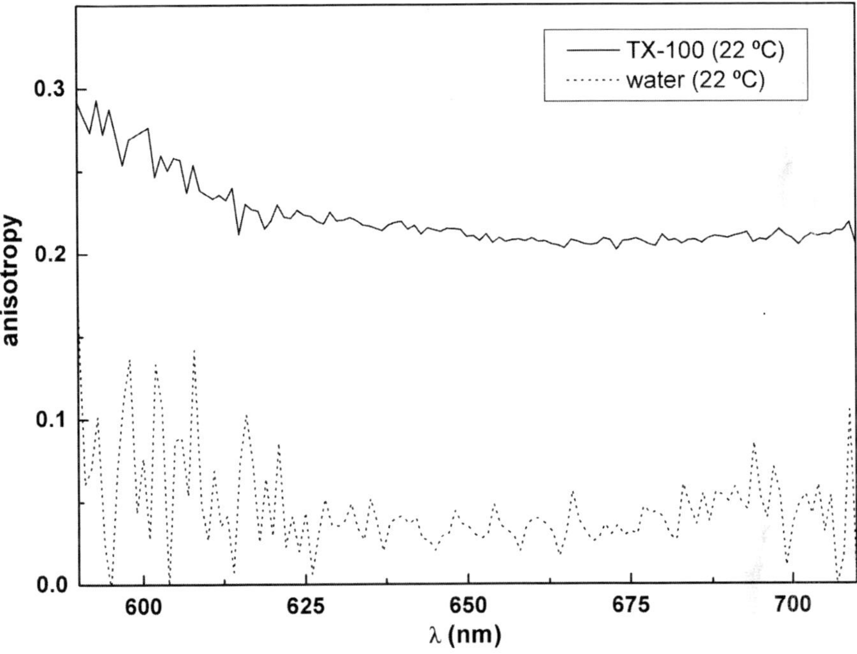

Fig. 23 Fluorescence anisotropy spectra of C8-benzo[*a*]phenoxazinium derivative in water and in TX-100 micelles

with wavelength is observed when the synthesized probe is associated with TX-100 micelles. This can be explained either by the previously mentioned two-state model of excited state solvent relaxation [29, 47] or by a possible contribution of the neutral form of the C8-side chain benzo[*a*]phenoxazinium derivative. The presence of the neutral basic form has been observed by K. Das et al. [36] in TX-100 reverse micelles. The neutral form could feel a more viscous environment if it locates more to the interior of the TX-100 micelles.

References

1. J. N. Miller, Fluorescence energy transfer methods in bioanalysis, *Analyst*, 130, 265–270, 2005.
2. K. Suhling, J. Siegel, PMP Lanigan, S. Lévêque-Fort, SED Webb, D. Phillips, DM Davis and PMW French, Time-resolved fluorescence anisotropy imaging applied to live cells, *Optics Letters*, 29, 584–586, 2004.
3. M. Antiaa, L. D. Islasa, D. A. Bonessc, G. Baneyxa and V. Vogel, Single molecule fluorescence studies of surface-adsorbed fibronectin, *Biomaterials*, 27, 679–690, 2006.
4. P. Greenspan and S. D. Fowler, Spectrofluorometric studies of the lipid probe, Nile Red, *Journal of Lipid Research*, 26, 781–789, 1985.
5. Ira and G. Krishnamoorthy, Probing the Link between Proton Transport and Water Content in Lipid Membranes, *Journal of Physical Chemistry B*, 105, 1484–1488, 2001.

6. M. Mazumdar, P. K. Parrack and B. Bhattacharyya, *European Journal of Biochemistry*, 204, 127–132, 1992,
7. D. L. Sackett and J. Wolff, Nile Red as a polarity-sensitive fluorescent probe of hydrophobic protein surfaces, *Analytical Biochemistry*, 167, 228–234, 1987.
8. D. M. Davis and D. J. S. Birch, Extrinsic fluorescence probe study of human serum albumin using Nile Red , *Journal of Fluorescence*, 23–32, 1996.
9. N. C. Maiti, M. M. G. Krishna, P. J. Britto, and N. Periasamy, Fluorescence dynamics of dye probes in micelles, *Journal of Physical Chemistry B*, 101, 11051–11060, 1997.
10. A. Datta, D. Mandal, S. Kumar Pal and K. Bhattacharyya, Intramolecular charge transfer processes in confined systems. Nile Red in reverse micelles, *Journal of Physical Chemistry B*, 101, 10221–10225, 1997.
11. V. J. P. Srivatsavoy, Enhancement of excited state nonradiative deactivation of Nile Red in γ-cyclodextrin: evidence for multiple inclusion complexes, *Journal of Luminescence*, 82, 17–23, 1999.
12. G. Hungerford, E. M. S. Castanheira, M. E. C. D. Real Oliveira, M. da Graca Miguel and H. D. Burrows, Monitoring ternary systems of C12E5/water/tetradecane via the fluorescence of solvatochromic probes, *Journal of Physical Chemistry B*, 106, 4061–4069, 2002.
13. A. Kumar Dutta, K. Karnada and K. Ohta, Langmuir-Blodgett films of Nile Red: a steady-state and time-resolved fluorescence study , *Chemical Physics Letters*, 258, 369–375, 1996.
14. S. Uppili, K. J. Thomas, E. M. Crompton, and V. Ramamurthy, Probing zeolites with organic molecules: supercages of X and Y zeolites are superpolar, *Langmuir*, 16, 265–274, 2000.
15. A. J. Carmichael and K. R. Seddon, Polarity study of some 1-alkyl-3-methylimidazolium ambient-temperature ionic liquids with the solvatochromic dye, Nile Red, *Journal of Physical Organic Chemistry*, 13, 591–595, 2000.
16. S. Zhou and K. D. Cook, Probing solvent fractionation in electrospray droplets with laser-induced fluorescence of a solvatochromic dye, *Analytical Chemistry*, 72, 963–969, 2000.
17. M. Choi, D. Jin, H. Kim, T. J. Kang, S. C. Jeoung, and D. Kim, Fluorescence Anisotropy of Nile Red and Oxazine 725 in an isotropic liquid crystal, *Journal of Physical Chemistry B*, 101 , 8092–8097, 1997.
18. K. Matsui and K. Nozawa, Molecular probing for the microenvironment of photonics materials prepared by the Sol–Gel process, *Bulletin of the Chemical Society of Japan*, 70, 2331–2335, 1997.
19. M. B. Brown, J. N. Miller and N. J. Seare, An investigation of the use of Nile Red as a long-wavelength fluorescent probe for the study of α1-acid glycoprotein-drug interactions, *Journal of Pharmaceutical and Biomedical Analysis*, 13, 1011–1017, 1995.
20. D. M. Watkins, Y. Sayed-Sweet, J. W. Klimash, N. J. Turro, and D. A. Tomalia, Dendrimers with hydrophobic cores and the formation of supramolecular dendrimer-surfactant assemblies, *Langmuir*, 13, 3136–3141, 1997.
21. G. B. Dutt, S. Doraiswamy, and N. Periasamy, Molecular reorientation dynamics of polar dye probes in tertiary-butyl alcohol–water mixtures, *Journal of Chemical Physics*, 94, 5360–5368,1991.
22. G. B. Dutt and S. Doraiswamy, Picosecond reorientational dynamics of polar dye probes in binary aqueous mixtures, *Journal of Chemical Physics*, 96, 2475–2491, 1992.
23. J. F. Deye, T. A. Berger, and A. G. Anderson, Nile Red as a solvatochromic dye for measuring solvent strength in normal liquids and mixtures of normal liquids with supercritical and near critical fluids, *Analytical Chemistry*, 62, 615–, 1990.
24. A. Cser, K. Nagy and L. Biczók, Fluorescence lifetime of Nile Red as a probe for the hydrogen bonding strength with its microenvironment, *Chemical Physics Letters*, 360, 473–478, 2002.
25. P. J. G. Coutinho, E. M. S. Castanheira, M. C. Rei and M. E. C. D. Real Oliveira, Nile Red and DCM fluorescence anisotropy studies in C12E7/DPPC mixed systems, *Journal of Physical Chemistry B*, 106, 12841–12846, 2002.
26. N. Ghoneim, Photophysics of Nile Red in solution. Steady state spectroscopy, *Spectrochimica Acta Part A*, 56, 1003–1010, 2000.

27. A. K. Dutta, K. Kamada and K. Ohta, Spectroscopic studies of Nile Red in organic solvents and polymers, *Journal of Photochemistry and Photobiology A: Chemistry*, 93, 57–64, 1996.
28. N. Sarkar, K. Das, D. N. Nath, and K.Bhattacharyya, Twisted charge transfer processes of Nile Red in homogeneous solutions and in faujasite zeolite, *Langmuir*, 10, 326–329, 1994.
29. M. M. G. Krishna, Excited-state kinetics of the hydrophobic probe Nile Red in membranes and micelles, *Journal of Physical Chemistry A*, 103, 3589–3595, 1999.
30. G. Hungerford, E. M. S. Castanheira, A. L. F. Baptista, P. J. G. Coutinho and M. E. C. D. Real Oliveira, Domain formation in DODAB–cholesterol mixed systems monitored via Nile Red anisotropy, *Journal of Fluorescence*, 15(6), 835–840, 2005.
31. G. Hungerford, A. L.F. Baptista, P. J.G. Coutinho, E. M.S. Castanheira and M. E. C. D. Real Oliveira, Interaction of DODAB with neutral phospholipids and cholesterol studied using fluorescence anisotropy, *Journal of Photochemistry and Photobiology A: Chemistry*, 181, 99–105, 2006.
32. R. Gvishi, R. Reisfeld and M. Eisen, Structures, spectra and ground and excited state equilibria of polycations of oxazine-170 , *Chemical Physics Letters*, 161, 455–460, 1989.
33. Q-Y. Chen, D-H. Li, Y. Zhao, H-H. Yang, Q-Z. Zhua and J-G. Xu, Interaction of a novel red-region fluorescent probe, Nile Blue, with DNA and its application to nucleic acids assay, *The Analyst*, 124, 901–907, 1999.
34. H. J. van Staveren, O. C. Speelman, M. J. H. Witjes, L. Cincotta, W. M. Star, Fluorescence imaging and spectroscopy of ethyl Nile Blue A in animal models of (pre)malignancies, *Photochemistry and Photobiology*, 73, 32–38, 2001.
35. R. K. Mitra, S. S. Sinha and S. K. Pal, Interactions of Nile Blue with micelles, reverse micelles and a genomic DNA, *Journal of Fluorescence*, 18, 423–432, 2008.
36. K. Das, B. Jain and H. S. Patel, Nile Blue in Triton-X 100/benzene–hexane reverse micelles: a fluorescence spectroscopic study, *Spectrochimica Acta Part A*, 60, 2059–2064, 2004.
37. H-W. Gao, Q-S. YE and W-G. Liu, Langmuir aggregation of Nile Blue and Safranine T on sodium dodecylbenzenesulfonate surface and its application to quantitative determination of anionic detergent, *Analytical Sciences*, 18, 455–459, 2002.
38. H. Ju, Y. Ye and Y. Zhu, Interaction between Nile Blue and immobilized single- or double-stranded DNA and its application in electrochemical recognition, *Electrochimica Acta*, 50, 1361–1367, 2005.
39. Y-I. Yang, H-Y. Hong, I-S. Lee, D-G. Bai, G-S. Yoo, and J-K. Choi, Detection of DNA using a visible dye, Nile Blue, in electrophoresed gels, *Analytical Biochemistry*, 280, 322–324, 2000.
40. C. Nasr and S. Hotchandani, Excited-state behavior of Nile Blue H-aggregates bound to SiO_2 and SnO_2 colloids, *Chemical Materials*, 12, 1529–1535, 2000.
41. D. A. Steinhurst and J. C. Owrutsky, Second harmonic generation from oxazine dyes at the air/water interface, *Journal of Physical Chemistry B*, 105, 3062–3072, 2001.
42. A. Douhal, Photophysics of Nile Blue A in pronton-accepting and electron-donating solvents, *Journal of Physical Chemistry*, 98, 13131–13137, 1994.
43. T. Kobayashi, Y. Takagi, H. Kandori, K. Kemnitz and K. Yoshihara, Femtosecond intermolecular electron transfer in diffusionless, weakly polar systems: Nile Blue in aniline and N, N dimethylaniline, *Chemical Physics Letters*, 180, 416–422, 1991.
44. A. Grofcsik, M. Kubinyi and W. J. Jones, Intermolecular photoinduced proton transfer in Nile Blue and Oxazine 720, *Chemical Physics Letters*, 250, 261–265, 1996.
45. A. Grofcsika, M. Kubinyia, A. Ruzsinszkya, T. Veszprémi and W. J. Jones, Quantum chemical studies on excited state intermolecular proton transfer of oxazine dyes, *Journal of Molecular Structure*, 555, 15–19, 2000.
46. R. Sens and K. H. Drexhage, Fluorescence quantum yield of oxazine and carbazine dyes, *Journal of Luminescence*, 25/25, 709–712, 1981.
47. V. H. J. Frade, M. Sameiro T. Gonçalves, P. J.G. Coutinho and J. C.V.P. Moura, *Journal of Photochemistry and Photobiology A: Chemistry* 185, 220–230, 2007.

48. V. H. J. Frade, P. J. G. Coutinho, J. C. V. P. Moura and M. S. T. Gonçalves, Functionalised benzo[*a*]phenoxazine dyes as long-wavelength fluorescent probes for amino acids, *Tetrahedron*, 63, 1654–1663, 2007.
49. V. H. J. Frade, S. A. Barros, J. C. V. P. Moura, P. J. G. Coutinho and M. S. T. Gonçalves, Synthesis of short and long-wavelength functionalised probes: amino acids' labelling and photophysical studies, *Tetrahedron*, 63, 12405–12418, 2007.
50. P. J. G. Coutinho, C. M. A. Alves and M. S. T. Gonçalves, submitted to publication.
51. K. Meguro, M. Ueno and K. Esumi, Micelle formation in aqueous media, in Nonionic Surfactants: Physical Chemistry, M. J. Schick Ed., Surfactant Science Series, Marcel Dekker: New York, Vol. 23, pp 109–183, 1987.
52. H. Heerklotz, H. Binder, G. Lantzsch, G. Klose, and A. Blume, Lipid/detergent interaction thermodynamics as a function of molecular shape, *Journal of Physical Chemistry B*, 101, 639–645, 1997.
53. D. B. Siano and D. E. Metzler, Band shapes of the electronic spectra of complex molecules, *Journal of Chemical Physics*, 51, 1856–1861, 1969.
54. J. Shobha, V. Srinivas, and D. Balasubramanian, Differential modes of incorporation of probe molecules in micelles and in bilayer vesicles, *Journal of Physical Chemistry*, 93, 17–20, 1989.
55. T. Inoue, Interaction of Surfactants with Phospholipid Vesicles, in Vesicles, M. Rosoff Ed., Surfactant Science Series, Marcel Dekker: New York, Vol. 62, pp. 151–195, 1996.
56. H. M. McConnell and A. Radhakrishnan, Condensed complexes of cholesterol and phospholipids, *Biochimica Biophysica Acta - Biomembranes*, 1610, 159–173, 2003.
57. R. P. Masona, T. N. Tulenkob and R. F. Jacob, Direct evidence for cholesterol crystalline domains in biological membranes: role in human pathobiology, *Biochimica Biophysica Acta - Biomembranes*, 1610, 198–207, 2003.
58. S. Koronkiewicz and S. Kalinowski, Influence of cholesterol on electroporation of bilayer lipid membranes: chronopotentiometric studies, *Biochimica Biophysica Acta – Biomembranes*, 1661, 196–203, 2004.
59. Ni. Marmé, G. Habl and J-P. Knemeyer, Aggregation behavior of the red-absorbing oxazine derivative MR 121: A new method for determination of pure dimer spectra, *Chemical Physics Letters*, 408, 221–225, 2005.
60. M. J. Kamlet, J. L. M. Abboud, M. H. Abraham, and R. W. Taft, Linear solvation energy relationships. 23. A comprehensive collection of the solvatochromic parameters, .pi.*, .alpha., and .beta., and some methods for simplifying the generalized solvatochromic equation, *Journal of Organic Chemistry*, 48, 2877–2887, 1983.
61. R. Sens and K. H. Drexhage, Fluorescence quantum yield of oxazine and carbazine laser dyes, *Journal of Luminescence*, 24, 709–712, 1981.

A Fluorescence Quenching Method to Study Interactions of Hemoglobin Derivatives with Erythroid Spectrin

Abhijit Chakrabarti

Abstract We have used a simple fluorescence technique to study the interactions of erythroid spectrin with hemoglobin and the globin chains. This could be used to study interactions of other proteins with hemoglobin derivatives and other heme proteins. The concentration-dependent change in the fluorescence intensity of fluorescein-conjugated spectrin (F-spectrin) in the presence of oxy-hemoglobin and other hemoglobin derivatives, e.g. globin chains, hemoglobin variants, and hemoglobin mixtures, indicated binding with dissociation constants (K_d) ranging from of 3.3 ± 0.2 to 21.6 ± 1.6 μM at 25°C. The K_d values have been evaluated from the increase in the extent of quenching of the fluorescein fluorescence of F-spectrin by reverse titration with the increasing concentrations of different hemoglobin derivatives, e.g. normal hemoglobin HbA, the alpha- and beta-globin chains, the hemoglobin variant HbE in mixtures predominantly in oxy-form isolated from the blood samples of patients of hemoglobinopathy. Results of such studies indicated the preferential spectrin binding of HbE, and alpha-globin chains compared to the HbA and the beta-globin chains. Taken together, this simple fluorescence method could be used to study interactions of hemoglobin derivatives with spectrin and provides a quantitative estimate of such protein–protein interactions. This could explain the pathophysiological implications of spectrin–hemoglobin interactions in disorders, e.g. thalassemia.

Keywords HbE · Thalassemia · Globin subunits · Erythroid spectrin · Hemoglobin–protein interactions

Abbreviations

HbA,	Adult hemoglobin
HbF,	Fetal Hemoglobin
FITC,	Fluorescein isothiocyanate

A. Chakrabarti (✉)
Biophysics Division and Structural Genomics Section, Saha Institute of Nuclear Physics, 1/AF Bidhannagar, Kolkata 700064, India
e-mail: abhijit.chakrabarti@saha.ac.in

C.D. Geddes (ed.), *Reviews in Fluorescence 2007*, Reviews in Fluorescence 2007,
DOI 10.1007/978-0-387-88722-7_15,

F-spectrin,	Fluorescein-conjugated spectrin
HMWA,	High molecular weight aggregate of spectrin and hemoglobin
RD,	Relative pixel densities, after subtracting from the respective background
K_d,	Binding dissociation constant of hemoglobin derivative to spectrin
Mg/ATP,	ATP in 10-fold molar excess of $MgCl_2$
2,3-DPG,	2-,3-diphosphoglycerate
Mg/ATP,	ATP in presence of 10-fold molar excess of $MgCl_2$
PMB,	p-hydroxymercuribenzoic acid sodium salt
SEM,	Standard error of the mean

Introduction

Fluorescence resonance energy transfer (FRET) is the most common spectroscopic technique to study association between two proteins [1, 2]. One of the best example of FRET is the elaboration of association of the chaperonin proteins GroEL and GroES [3]. Fluorescence anisotropy has also been used in studying association of proteins, e.g. peptide binding to calmodulin [1, 4]. We have used a simple fluorescence quenching technique to study interactions of hemoglobin and its derivatives with spectrin, the major membrane skeletal protein of erythrocytes.

Hemoglobin is a soluble globular tetrameric protein composed of two identical α-like globin chains and two identical β-like globin chains found within vertebrate erythrocytes, at a very high concentration of about 5 mM [5]. Spectrin, the major protein of red cell membrane skeleton, is composed of two large, worm-like subunits (α- and β-chain), which are associated into double stranded, fiber-like flexible heterodimers about 100 nm in contour length. In normal red cells, spectrin dimers are assembled, head to head, predominantly into tetramers imparting mechanical flexibility to the erythrocytes [6–10].

Hemoglobin has been found to be associated with erythrocyte membranes prepared by hypotonic lysis of erythrocytes [11–13]. Several reports indicated the association of spectrin with hemoglobin under different experimental and physiological conditions, e.g. in senescent red blood cells and under oxidative stress [14–16] The extent of peroxide-induced hemoglobin–spectrin complexation, the effects of bound hemoglobin on conversion of spectrin dimer to the tetramer, and the identification of two types of hemoglobin–spectrin complexes within the membrane skeleton has also been evaluated [16].

Hemoglobin E is the most common hemoglobin variant in the world [17]. It is generated by a point mutation (Glu26(B8)→Lys) in the β-globin gene [18]. The primary clinical importance of HbE trait arises when the β^E allele interacts with a β-thalassaemia mutation leading to a severe anaemia known as HbEβ-thalassaemia [19]. Several blood diseases are associated with erythrocyte deformation and defects in spectrin, e.g. various types of hereditary haemolytic anaemia involve mutations in spectrin [20–22]. Various reports indicating the presence of cross-linked aggregates of hemoglobin and spectrin in thalassemic erythrocytes have been directly implicated in explaining pathophysiological symptoms like reduced deformability, enhanced rigidity, and enhanced phagocytosis by macrophages [23, 24].

Phosphate metabolites, ATP and DPG, are found to modulate different molecular and biochemical processes inside erythrocyte. It has been shown earlier that DPG affects the mobility of intrinsic membrane proteins and weakens the association between the components of the membrane skeleton [25, 26]. Both ATP and DPG bind to hemoglobin in a comparable manner [27]. Irreducible complexation of hemoglobin with spectrin is a natural phenomenon of erythrocyte aging and such complex, generated in vivo, was associated with increased red cell membrane rigidity [28, 29]. There were conflicting reports both for and against the binding of hemoglobin [30, 31] to erythroid spectrin before we measured the binding dissociation constant between them using this fluorescence technique [32, 33].

Studies from this laboratory have also shown that spectrin exhibits chaperone-like properties and could bind denatured heme proteins in an ATP-dependent manner [34–36]. This review elaborates on the fluorescence quenching method that we have adopted to study protein–protein interactions when one protein is a heme protein and the other being spectrin [32, 33]. We have also studied the spectrin binding properties of the individual globin subunits [37] and intact hemolysates with varying compositions of HbA and HbE. However, this work could be extended to any other protein that interacts with a particular heme protein for the estimation of binding constants. Results of binding studies of HbA, the variant HbE, the two globin chains, α-globin and β-globin of HbA, and mixture of HbE and HbA, purified from hemolysates of the blood samples of HbEβ-thalassemics, with spectrin in presence and absence of phosphate metabolites ATP, 2,3 DPG and high salt have been discussed. The binding affinity increased about 5-fold from HbA to HbE. The binding dissociation constant, K_d decreased to $16.1 \pm 1.4\ \mu M$ for the α-globin and remained comparable at $21.8 \pm 1.9\ \mu M$ for the β-globin with HbA. Such binding affinities decreased in presence of the phosphate metabolites and somewhat increased in presence of high salt. The spectrin binding affinity of HbE and α-globin chain has been always higher than those of HbA and the β-globin chain.

Materials and Methods

Sephadex G-50, DEAE-cellulose, CM-cellulose, Sepharose 4B, PMB, sodium azide, DTT, PMSF, Tris, KCl, ATP, and DPG were obtained from Sigma-Aldrich (St. Louis, MO). BioGel-P2 was from BioRad and Sephadex G-100 was from Amersham. Fluorescein isothiocyanate (FITC) was obtained from Molecular Probes. All other chemicals were of analytical grade and obtained locally. De-ionized water was doubly distilled on quartz before using for the preparation of buffers and all other solutions.

Collection and Isolation of Hemoglobin from Human Blood Samples

Human blood samples, taken for diagnosis from patients suffering from Eβ-thalassemia, were collected from different Thalassemia Clinics in the city of Kolkata and were characterized by the BioRad Variant HPLC system with enriched levels of HbE [38]. Samples were collected with the proper consent of patients who did not have blood transfusions. The levels of different hemoglobin variants, mainly of the normal adult HbA, HbE, and HbF in the blood samples, were estimated from the HPLC system, Variant^{TM} in the blood samples of normal and homozygous HbE individuals and Eβ-thalassemics.

Human erythrocytes, after removal of the buffy coat and plasma, were extensively washed with phosphate-buffered saline (5 mM phosphate, 0.15 M NaCl, pH 7.4). Hemoglobin was isolated from packed clean erythrocytes by osmotic lysis using three volumes of 1 mM Tris, pH 8.0, at 4°C for 1 h. The hemoglobin mixture was purified by gel filtration on Sephadex G-100 column (30×1 cm) in the same buffer containing 100 mM KCl. The hemoglobin samples were stored in oxy-form at –70°C for not more than 7 days and characterized by the measurements of absorption at 415 and 541 nm, respectively. The protein concentration was determined spectrophotometrically using the molar extinction coefficient of 125,000 $M^{-1}cm^{-1}$ at 415 nm and 13,500 $M^{-1}cm^{-1}$ at 541 nm, respectively [39, 40]. Both HbA and HbE were also purified in cyano-met form by ion-exchange chromatography on SP-Sephadex elaborated earlier [41]. ATP concentration was determined assuming a molar absorbance of 15,400 $M^{-1}cm^{-1}$ at 259 nm [42]. ATP was used in presence of 10 fold molar excess of $MgCl_2.6H_20$ (Mg/ATP).

Isolation and Purification of Spectrin

Clean, white ghosts both from ovine and human blood were prepared by hypotonic lysis in 5 mM phosphate, 1 mM EDTA containing 20 g/ml of PMSF at pH 8.0 (lysis buffer) and dimeric spectrin was isolated at 37C following the protocol of Gratzer [43, 44]. Spectrin concentration was determined spectrophotometrically using an absorbance of 10.7 at 280 nm for 1% spectrin solution [43].

Preparation of Human α- and β-Globin Chains

The PMB derivatives of HbA were prepared following the method of Bucci and Fronticelli [45]. The α-PMB and β-PMB chains were separated by following a method consisting of two-column selective ion-exchange chromatography as described earlier [46]. To obtain α-PMB, the splitting solution was equilibrated with 0.01 M phosphate buffer at pH 8.0 and passed through a DEAE-cellulose column equilibrated and eluted with the same buffer. To obtain β-PMB, the splitting solution was equilibrated with 0.01 M phosphate buffer at pH 6.6 and applied on a CM-cellulose column, equilibrated and eluted with the same buffer. The PMB was removed from the isolated α-PMB and β-PMB chains by the addition of 50 MM 2-mercaptoethanol in 0.1 M phosphate buffer, pH 7.5. The intact globin chain was purified from the mixture of globin chains and unreacted PMB by gel filtration on a BioGel P2 column. Immediately after separation, the subunits were dialyzed extensively against 0.1 M phosphate buffer, pH 7.5 [47]. The globin chains were not stored for more than 48 h at 4°C.

Fluorescein-Conjugated Spectrin

Spectrin was covalently labeled with FITC in a buffer of pH ~9.0. About 1 mg spectrin dimer was reacted with 50–100 fold molar excess of FITC, taken in a small volume of dry acetone. The labeling was carried out in a buffer containing 20 mM $NaHCO_3$ for about an hour at 4°C. The FITC-labeled spectrin (F-spectrin) was separated and purified from the reaction mixture by gel filtration on Sephadex G-50 column using the buffer containing 5 mM Tris–HCl, 50 mM KCl, pH 8 [32, 33]. The concentration of F-spectrin was determined spectrophotometrically from the absorbance at 495 nm and using the molar extinction coefficient of 76,000 $M^{-1}cm^{-1}$ [48]. The labeling ratio of fluorescein to spectrin in F-spectrin was determined to be 2 fluorescein per spectrin dimer.

Fluorescence Measurements and Quenching of F-spectrin by Hemoglobins

The steady state fluorescence measurements were performed using a FluoroMax 3 fluoremeter (Jobin-Yvon Edison, NJ) fluorescence spectrophotometer. The excitation and the emission bandpass were set to 5 nm each. F-spectrin (20–50 nM) was excited at 495 nm and the change in fluorescence emission intensity at 520 nm was monitored in presence of the hemoglobin derivatives in a buffer containing 5 mM Tris–HCl, 50 mM KCl, pH 8.0. At both the wavelengths of excitation and emission maxima, 495 and 520 nm respectively, hemoglobin absorbed minimally and corrections of intensities due to inner filter effects were not necessary up to 50 μM hemoglobin [32, 33].

The changes in the extent of fluorescence quenching as a function of increasing concentrations of hemoglobin derivatives were analyzed by a model independent method and the apparent dissociation constant of hemoglobin binding to spectrin (K_d) was determined using non-linear curve fitting analysis following the equations below. All experimental points for binding isotherm were fitted by least-square analysis using Microcal Origin software package (Version 5.0) from Microcal Software Inc., Northampton, MA.

$$K_d = [C_0 - (\Delta F/\Delta F_{max})C_0][C_L - (\Delta F/\Delta F_{max})C_0]/(\Delta F/\Delta F_{max})C_0, \quad (1)$$

$$C_0(\Delta F/\Delta F_{max})^2 - (C_0 + C_L + K_d)(\Delta F/\Delta F\text{max}) + C_L = 0, \quad (2)$$

In equations (1) and (2), ΔF is the change in fluorescence emission intensity at 520 nm ($\lambda_{ex} = 495$ nm) for each point on the titration curve. ΔF_{max} denotes the same when hemoglobin is completely bound to spectrin, C_L is the concentration of the ligand, hemoglobin, and $C_0$is the initial concentration of spectrin. From the double reciprocal plots of $1/\Delta F$ against $1/C_L$, the magnitude of ΔF_{max} was determined using equation (3):

$$1/\Delta F = 1/\Delta F_{max} + 1/[K_{app}\Delta F_{max}(C_L - C_0)]. \quad (3)$$

Since in this case $C_L >> C_0$, hence ($C_L - C_0$) C_L, the linear double reciprocal plot was obtained by plotting $1/\Delta F$ against $1/C_L$ and the plot is extrapolated to the ordinate to obtain the value of ΔF_{max} from the intercept [32–35]. The above approach is based on assumptions that emission intensity is proportional to the concentration of the ligand and the ligand concentration was in large excess than spectrin.

We have studied the effects of high salt, 0.5 M and 1.0 M KCl, to elucidate the mechanism of hemoglobin binding to spectrin and the effects of two important phosphate metabolites – Mg/ATP and DPG, at two different concentrations 0.5 and 1.0 mM. All estimated K_d values are represented as mean $\pm$ SEM employing a minimum of 4–5 sets of independent experiments from two sets of samples with nearly identical compositions in terms of HbE, HbF, and HbA. Also, to check out the statistical significance level, all the reported values were subjected to the two-tail Student's *t*-test with equal variance.

Results

Heme is a potent inhibitor of the intrinsic protein fluorescence due to its strong absorption in the Soret region with a molar extinction coefficient of 125,000 $M^{-1}cm^{-1}$ at 415 nm causing inner filter effect [1]. We have bypassed this problem by covalently modifying spectrin with FITC, and the fluorescein-conjugated spectrin (F-spectrin) was used for the study of the interaction of hemoglobin with spectrin. Hemoglobin showed minimal absorption at 495 and 520 nm, the excitation and emission maxima of F-spectrin, respectively. Figure 1 shows the fluorescence emission spectra of F-spectrin indicating quenching of the fluorescence intensity with increasing concentrations of oxy-HbA. Figure 2 shows a representative binding isotherm as a plot of $\Delta F/\Delta F_{max}$ against the concentrations of hemoglobin from the fluorescence intensity data. The inset of Fig. 2 shows the linear double reciprocal plot of $1/\Delta F$ against 1/[HbA], extrapolated to the ordinate for obtaining the value of ΔF_{max} from the intercept. The binding data were analyzed by model independent

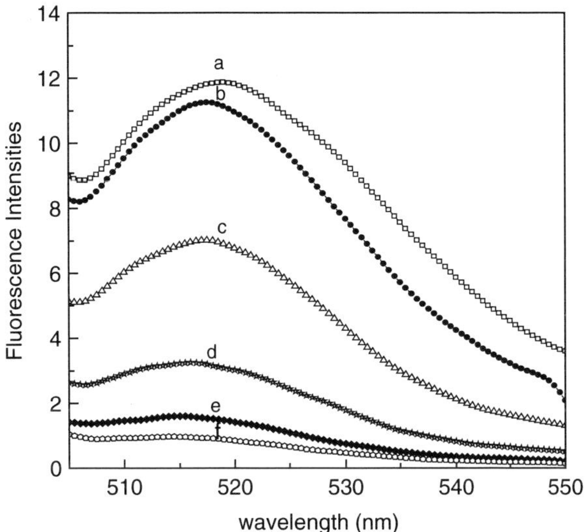

Fig. 1 Fluorescence emission spectra of F-spectrin (a) in absence and presence of (b) 5 μM; (c) 10 μM; (d) 20 μM; (e) 30 μM; and (f) 40 μM Hemoglobin. The excitation wavelength was at 495 nm

curve fitting elaborated in the previous section. The apparent dissociation constant of HbA binding to spectrin was evaluated to be 22.5±1.6 μM. Due to no inner filter effect of hemoglobin even at a concentration as high as 50 μM, 80% saturation of hemoglobin binding to spectrin could be attained. Presence of a large number of moderately high affinity hemoglobin binding sites in spectrin is probable in view of the high concentrations of hemoglobin. Identical sets of experiments were performed with all hemoglobin samples, differing in the composition of HbA, HbF, and HbE, with enriched levels of HbE. Table 1 summarizes the apparent dissociation constants of binding of hemoglobin and spectrin.

The extent of fluorescence quenching was found to increase with increasing levels of HbE in the hemoglobin mixtures. Figure 3 shows the histogram representations of the extent of fluorescence quenching ($\Delta F/F_0$) where F_0 is the fluorescence intensity of F-spectrin in the absence of hemoglobin, against the elevated levels of HbE% in the hemoglobin mixtures at three fixed concentrations of total hemoglobin at 5, 10 and 20 μM. The K_d values on the other hand decreased with increasing levels of HbE% in the hemoglobin mixtures indicating high affinity binding of HbE over HbA (Table 1).

Spectrin binding experiments were also carried out with purified globin chains from HbA. K_d decreased to 16.1±1.4 μM for α-globin and for β-globin chains remained comparable to that of HbA at 21.8±1.9 μM [37]. The binding dissociation constants for HbA and β-globin chains also remained comparable in the presence of both ATP and 2,3 DPG. However, the binding affinity decreased substantially for the α-globin chains with K_d values increasing from 16.1±1.4 μM to 38.0±3.3 μM in presence of ATP and 28.9±2.1 μM in presence of 2,3 DPG ($P<0.05$).

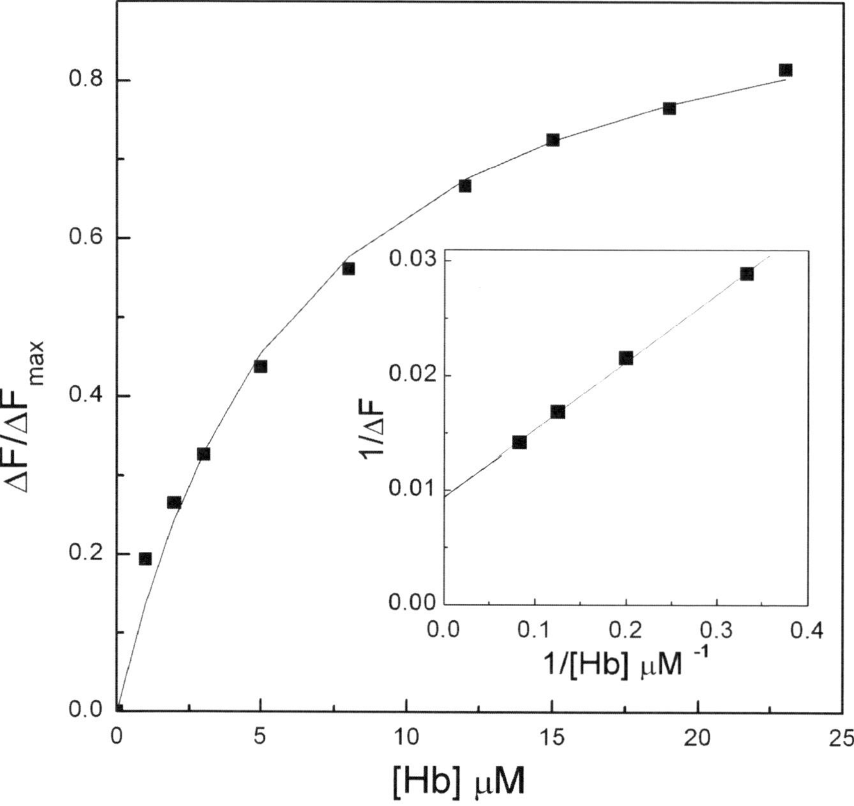

Fig. 2 The binding of hemoglobin with spectrin as reflected in the plot of $\Delta F/\Delta F_{max}$ against the hemoglobin concentrations [Hb], from the fluorescence intensity data. *Inset* shows the linear double reciprocal plot of $1/\Delta F$ against 1/[Hemoglobin], extrapolated to the ordinate for obtaining the value of ΔF_{max} from the intercept

The fluorescence quenching measurements of F-spectrin were then done with purified hemolysates having varying levels of HbE along with HbA and HbF from normal individuals with HbA ranging from 90–95% and HbE up to 90–92% from homozygous HbE patients. Only those hemolysates were used for study where the levels of HbF stayed within 5–8%. HbA (>95%) and HbE (>90%) stays predominantly in oxy-form, determined from the absorbance at 415 and 541 nm as the yardstick [40]. Three different sets of experiments were performed with hemoglobin mixtures having ~30%, ~50%, and ~70% HbE with the rest of HbA characterized by the Variant™ system. Table 1 summarizes the apparent dissociation constants of spectrin binding of the hemoglobin mixtures in the presence and absence of Mg/ATP, DPG, and KCl. The apparent dissociation constant of HbA binding to spectrin was evaluated to be 22.5±1.6 μM which decreased to 17.1±1.5 μM in presence of 30% HbE and 5.4±0.5 μM in presence of 70% HbE. In presence of both Mg/ATP and DPG, the binding affinity decreased. For example, fin case

Table 1 Effect of Mg/ATP, DPG, and KCl on the binding of spectrin and hemoglobin mixtures (K_d in μM) containing increasing levels of HbE

Hemoglobin mixtures	Control (μM)	ATP (mM) 0.5	ATP (mM) 1.0	DPG (mM) 0.5	DPG (mM) 1.0	KCl (M) 0.5	KCl (M) 1.0
HbA~95%	22.5±1.6	19.2±1.4	21.6±1.6	19.9±1.5	20.4±1.5	19.5±1.6	17.5±1.4
HbE~30%	17.1±1.5	18.5±1.4	17.6±1.3	18±1.3	18.4±1.4	16.3±1.2	14.4±1.1
HbE~50%	5.8±0.5	8.5±0.6	8.9±0.7	6.8±0.5	7.2±0.5	5±0.4	4.1±0.3
HbE~70%	5.4±0.5	7.1±0.5	8.4±0.6	6.5±0.5	7.7±0.6	5.5±0.4	4.2±0.3
HbE~90%	5.1±0.5	7.9±0.7	8.1±0.6	7.5±0.6	7.7±0.6	4.4±0.3	3.3±0.2

The error bars are mean ± SEM of 4–5 independent experiments ($P<0.05$)

of hemoglobin mixture containing ~70% HbE the K_d value increased to 8.4±0.6 μM and 7.7±0.6 μM from 5.4±0.5 μM in presence of 1 mM Mg/ATP and DPG, respectively ($P<0.05$). On the other hand, the K_d values decreased to 4.2±0.3 μM in presence of 1 M KCl (Table 1).

Figure 4 summarizes the binding affinities in form of histogram representation of spectrin complexation with HbA, HbE, α-globin, and β-globin chains of the affinity constants ($= 1/K_d$). The spectrin binding affinities are found to increase from HbA

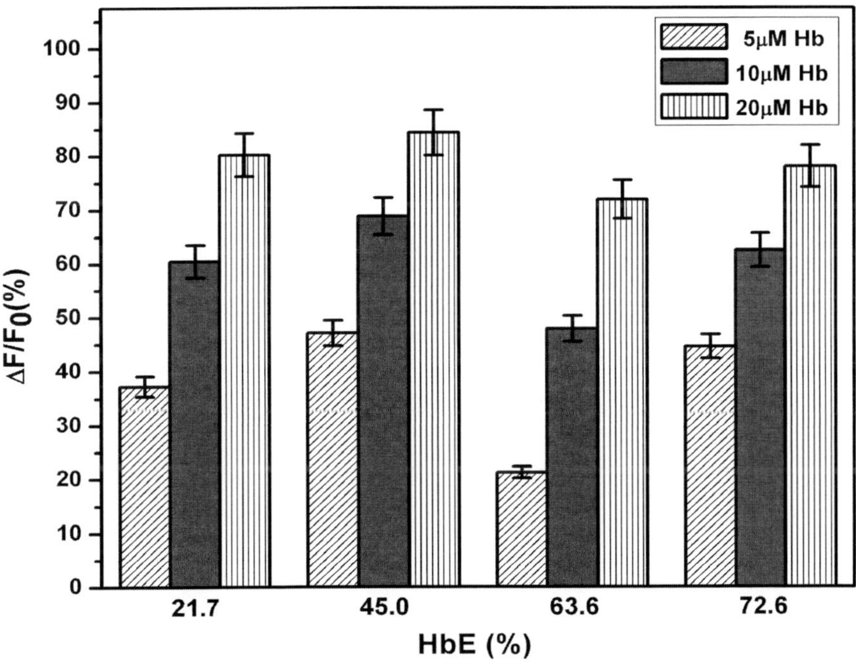

Fig. 3 Histogram representation of the extent of fluorescence quenching $\Delta F/F_0$ % at 520 nm against the level of HbE % indicated in the X-axis, three different concentrations of 5, 10, and 20 μM with respect to the total hemoglobin

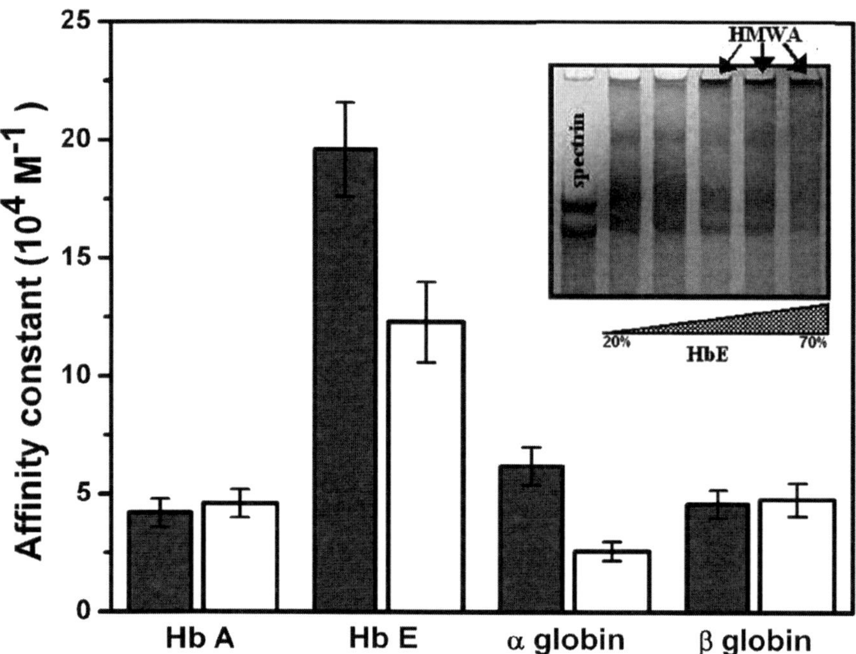

Fig. 4 Histogram representation of the affinity constants of spectrin binding to HbA, HbE, α-globin chains, and β-globin chains in the presence (*filled bar*) and absence (*empty bar*) of 1 mM Mg/ATP. The error bars are SEM of 5–10 independent experiments. *Inset* shows a representative 4% SDS–PAGE showing high molecular weight aggregates generated from 1 μM spectrin, 22.5 μM Hemoglobin, and 1 mM H_2O_2 in absence of DTT with hemolysates containing HbE starting from 20 to 90%

to HbE and α-globin to β-globin chains. It is also evident that ATP decreases the binding affinity substantially in HbE and in α-globin chains. Inset of Fig. 4 shows a representative 4% SDS–PAGE experiment showing peroxide-induced cross-linked complexes (HMWA), generated from 1 μM spectrin, 22.5 μM hemoglobin and 1 mM H_2O_2 in absence of DTT with hemolysates containing HbE starting from 20 to 90%, further indicating physical association of spectrin with globin chains of HbA [33].

Discussion

Using this simple fluorescence technique, we have determined the binding constant between spectrin and hemoglobin, for the first time, to be around 10^5 M^{-1}, two orders of magnitude higher than that between the cytoplasmic domain of band 3 and hemoglobin [32, 49]. This moderately high affinity binding of hemoglobin with spectrin could play an important role in the maintenance of structural integrity of the erythrocyte membrane. Moreover, in hemoglobinopathy, e.g. HbEβ-thalassemia, where the amount of HbE rises to more than 50% in acute cases, the stronger spectrin–hemoglobin interaction takes place as indicated by significant lowering of the dissociation constants from 22.51.6 μM to 5.40.5 μM. Implications of these results in hemoglobin disorders might go a long way in understanding the stability, deformability and cytoskeletal integrity of erythrocytes containing varying amounts of HbE. This decreasing trend in K_d with increasing HbE content was also associated with increase in the extent of fluorescence quenching of F-spectrin indicating its possible usefulness in the detection of differential spectrin interaction of HbA and HbE (Fig. 3). The maximum extent of hemoglobin-induced quenching of F-spectrin, in our opinion, could even be used in the detection of HbEβ-thalassemia and other hemoglobin disorders involving a particular hemoglobin variant out about 1000 known variants.

The spectrin–hemoglobin binding is favored in presence of high salt indicating possible role of hydrophobic interactions in such binding (Table 1). The presence of hydrophobic patches on human globin subunits could be responsible for the interaction with spectrin as predicted in an earlier report [50]. K_d values decreased to 4.2±0.3 μM in presence of 1 M KCl and the effect was more pronounced in the hemolysates with higher HbE content. The effects of Mg/ATP on the spectrin binding affinity of HbA, HbE and the globin chains along with the yield of oxidative cross-linked complexes were even more interesting [33, 37]. In presence of 1.0 mM Mg/ATP and DPG, the binding affinities were found to decrease significantly. The difference in the K_d values listed in Table 1, both in the absence and presence of ATP, DPG and KCl were found to be statistically significant ($P<0.05$). K_d values for HbA and β-globin chains remained comparable both in the presence and absence ATP and 2,3 DPG. However, the K_d increased substantially for the α-globin chains in presence of 1 mM Mg/ATP and 2,3 DPG ($P<0.05$). The histogram representations of the spectrin binding affinities between the two variants HbA and HbE and the two globin chains of HbA clearly indicates preference of HbE over HbA and α-globin chains over β-globin chains in an ATP-dependent manner.

Our earlier work revealed a large number of moderately high affinity hemoglobin binding sites in spectrin [32, 33, 37]. A rough estimate from fluorescence studies revealed ~100 hemoglobins to bind to one spectrin dimer. Stoichiometric analysis of the soluble cross-linked complexes generated between HbA and spectrin in presence of H_2O_2, after separation on Sepharose 4B, indicated 8-10 globin chains to associate with one spectrin subunit [33, 37]. A representative SDS-PAGE is shown in the inset of Fig. 4 indicating increased formation of HMWA with increase in HbE% in the hemolysates. The band seen in the middle of HMWA and the α-spectrin subunit corresponding to a molecular weight of ~400 kDa were analyzed for finding the stoichiometry of spectrin subunit with the globin chains.

Thermodynamic studies on the subunit assembly of hemoglobin showed the equilibrium constant associated with the conversion of hemoglobin from dimer to tetramer has been in the range of $1.0–1.2 \times 10^6$ M^{-1} between 20 and 27°C [51, 52]. We have chosen the hemoglobin concentration above 10^{-5} M in order to assure the presence of significant amount of hemoglobin tetramers preventing the formation of the hemoglobin dimers. The hemoglobin-induced quenching of F-spectrin is a static effect and enables one to obtain a quantitative estimate of the binding constants. Comparative experiments with different fluorescein or rhodamine labeled protein that interacts with hemoglobin can be used to measure the relative binding constant. Our measurements of the excited state lifetime of the fluorescein moiety in F-spectrin also showed no significant changes in the mean lifetime of spectrin in presence of different hemoglobin derivatives [44].

A possible mechanism involved in the interaction of HbE and β-thalassemia lies in the observation that HbE is oxidatively unstable in vitro [53]. When the level of HbE rises above ~30% (HbE carrier) basal level to ~70% in acute cases of HbEβ-thalassaemia, stronger spectrin–hemoglobin interactions take place as evidenced by significant lowering of the binding dissociation constants from 17.1±1.5 μM (~30% HbE) to 5.4±0.5 μM (~70% HbE). The preferential HbE interactions, within a mixture of HbA and HbE, with erythroid spectrin could contribute to the premature hemolysis of erythrocytes. The results of the present work are more significant in the light of erythroid spectrin exhibiting chaperone-like activity [34]. Moreover, both the binding of erythroid spectrin to a denatured heme protein and proteolysis of excess α-globin chains in β-thalassemic cells have been shown to be ATP-dependent, supporting our observations of the effects of Mg/ATP and DPG [35, 54, 55]. The susceptibility of HbE to oxidation could also play a major role governing the release of heme which strongly destabilizes lipid bilayer and eventually triggers the oxidative haemolytic effect in such hemolytic anemia [56, 57].

References

1. J R Lakowicz, (2006), Principles of Fluorescence Spectroscopy 3rd ed, Springer, Singapore.
2. R H Fairclough, C R Cantor, (1978), The use of singlet energy transfer to study macromolecular assemblies, Methods Enzymol 48, 347–379.
3. H S Rye, (2001), Application of fluorescence resonance energy transfer to the GroEL-GroES chaperonin reaction, Methods 24, 278–288.
4. T J Lukas, W H Burgess, F G Prendergast, W Lau, D M Watterson, (1986), Calmodulin binding domains: characterization of a phosphorylation and calmodulin binding site from myosin light chain kinase, Biochemistry 25, 1458–1464.
5. S H Martin, E J Jr. Benz, (2000) Pathobiology of the human erythrocyte and its hemoglobins, in Hematology, Basic Principles and Practice, R Hoffman, E J Jr. Benz, S J Shattil, B Furie, H J Cohen, L E Silberstein, P McGlave, Eds, pp 358–364, Churchill Livingstone, Pensylvania, Philadelphia.
6. S R Goodman, K Shiffer, (1983), The spectrin membrane skeleton of normal and abnormal human erythrocytes: a review, Am J Physiol. 244, C121–C141.
7. D M Shotton, B E Burke, D Branton, (1979), The molecular structure of human erythrocyte spectrin. Biophysical and electron microscopic studies, J Mol Biol. 131, 303–329.
8. K E Sahr, P Laurila, L Kotula, A L Scarpa, E Coupal, T L Leto, A J Linnenbach, J C Winkelmann, D W Speicher, V T Marchesi, P J Curtis, B G Forget, (1990), The complete cDNA and polypeptide sequences of human erythroid alpha-spectrin, J Biol Chem. 265, 4434–4443.
9. J C Winkelmann, J-G Chang, W T Tse, A L Scarpa, V T Marchesi, B G Forget, (1990), Full-length sequence of the cDNA for human beta-spectrin, J Biol Chem. 265, 11827–11832.

10. J Hartwig, (1995), Actin binding protein 1: spectrin superfamily, Protein Profile 2, 703–800.
11. S Fischer, R L Nagel, R M Bookchin, E F Jr. Roth, I Tellez-Nagel, (1975), The binding of hemoglobin to membranes of normal and sickle erythrocytes. Biochim Biophys Acta 375, 422–433.
12. N Shaklai, J Yguerabide, H M Ranney, (1977), Interaction of hemoglobin with red blood cell membranes as shown by a fluorescent chromophore. Biochemistry 16, 5585–5592.
13. P Jarolim, M Lahav, S C Liu, J Plaek, (1990), Effect of hemoglobin oxidation products on the stability of red cell membrane skeletons and the associations of skeletal proteins: correlation with a release of hemin. Blood 76, 2125–2131.
14. L M Snyder, F Garver, S C Liu, L Leb, J Trainor, N L Fortier, (1985), Demonstration of hemoglobin associated with isolated, purified spectrin from senescent human red cells, Br J Haematol. 61, 415–419.
15. N Shaklai, B Frayman, N Fortier, M Snyder, (1987), Crosslinking of isolated cytoskeletal proteins with hemoglobin: a possible damage inflicted to the red cell membrane, Biochim Biophys Acta 915, 406–414.
16. C R Kiefer, J F Trainor, J B McCanny, C R Valeri, L M Snyder, (1995), Hemoglobin-spectrin complexes: interference with spectrin tetramer assembly as a mechanism for compartmentalization of band 1 and band 2 complexes, Blood 86, 366–371.
17. D J Weatherall, J B Clegg, (1996), Thalassaemia – a global public health problem, Nature Med. 2, 847–849.
18. J Traeger, W G Wood, J B Clegg, D J Weatherall, (1980), Defective synthesis of HbE is due to reduced levels of beta E mRNA, Nature 288, 497–499.
19. S Fucharoen, P Winichagoon, P Pootrakul, A Piankijagum, P Wasi, (1987), Variable severity of Southeast Asian beta 0-thalassaemia/HbE disease, Birth Defects Orig Artic Ser. 23, 241–248.
20. J Delaunay, D Dhermy, (1993), Mutations involving the spectrin heterodimer contact site: clinical expression and alterations in specific function, Semin Hematol. 30, 21–33.
21. H Wichterle, M Hanspal, J Palek, P Jarolim, (1996), Combination of two mutant alpha spectrin alleles underlies a severe spherocytic haemolytic anaemia, J Clin Invest. 98, 2300–2307.
22 P G Gallagher, M J Petruzzi, S A Weed, Z Zhang, S L Marchesi, N Mohandas, J S Morrow, B G Forget, (1997), Mutation of a highly conserved residue of betaI spectrin associated with fatal and near-fatal neonatal haemolytic anaemia, J Clin Invest. 99, 267–277.
23. S L Schrier, E Rachmilewitz, N Mohandas, (1989), Cellular and membrane properties of alpha and beta thalassemic erythrocytes are different: implication for differences in clinical manifestations, Blood 74, 2194–2202.
24. S L Schrier, (1994), Thalassaemia: pathophysiology of red cell changes, Annu Rev Med. 45, 211–218.
25. M Schindler, D E Koppel, M P Sheetz, (1980), Modulation of membrane protein lateral mobility by polyphosphates and polyamines, Proc Natl Acad Sci USA. 77, 1457–1461.
26. M P Sheetz, J Casaly, (1980), 2,3-Diphosphoglycerate and ATP dissociate erythrocyte membrane skeletons, J Biol Chem. 255, 9955–9960.
27. H H Lo, P R Schimmel, (1969), Interaction of human hemoglobin with adenine nucleotides, J Biol Chem. 244, 5084–5086.
28. L M Snyder, L Leb, J Piotrowski, N Sauberman, S C Liu, N L Fortier, (1983), Irreversible spectrin-hemoglobin crosslinking in vivo: a marker for red cell senescence, Br J Haematol. 53, 379–384.
29. N Fortier, L M Snyder, F Garver, C Kiefer, J McKenney, N Mohandas, (1988), The relationship between in vivo generated hemoglobin skeletal protein complex and increased red cell membrane rigidity, Blood 71, 1427–1431.
30. P Chaimanee, G H Yuthavong, (1977), Binding of hemoglobin to spectrin of human erythrocytes, FEBS Lett. 78, 119–123.
31. R Cassoly, (1978), Evidence against the binding of native hemoglobin to spectrin of human erythrocyte, FEBS Lett. 85, 357–360.

32. P Datta, S B Chakrabarty, A Chakrabarty, A Chakrabarti, (2003), Interaction of erythroid spectrin with hemoglobin variants: implications in beta-thalassemia, Blood Cells Mol Dis. 30, 248–253.
33. P Datta, S Basu, S B Chakrabarty, A Chakrabarty, D Banerjee, S Chandra, A Chakrabarti, (2006), Enhanced oxidative cross-linking of hemoglobin E with spectrin and loss of erythrocyte membrane asymmetry in hemoglobin Eβ-Thalassaemia, Blood Cells Mol Dis. 37, 77–81.
34. M Bhattacharyya, S Ray, S Bhattacharya, A Chakrabarti, (2004), Chaperone activity and prodan binding at the self-associating domain of erythroid spectrin, J Biol Chem. 279, 55080–55088.
35. A Chakrabarti, S Bhattacharya, S Ray, M Bhattacharyya, (2001), Binding of a denatured haeme protein and ATP to erythroid spectrin, Biochem Biophys Res Commun. 282, 1189–1193.
36. A Chakrabarti, D A Kelkar, A Chattopadhyay, (2006), Spectrin organization and dynamics: new insights, BioSci Rep. 26, 369–386.
37. P Datta, S Chakrabarty, A Chakrabarty, A Chakrabarti, (2007), Spectrin interactions of globin chains in presence of phosphate metabolites and hydrogen peroxide: Implications in thalassemia. J Biosci. 32, 1147–1151.
38. H M Waters, J E Howarth, K Hyde, S Goldstone, K I Cinkotai, M Kadkhodaei-Elyaderani, J T Richards, (1998), An evaluation of the Bio-Rad Variant Hemoglobin Testing System for the detection of haemoglobinopathies, Clin Lab Haematol. 20, 31–40.
39. A Riggs, (1981), Preparation of blood hemoglobins of vertebrates, Methods Enzymol. 76, 5–29.
40. E E Di Iorio, (1981), Preparation of derivatives of ferrous and ferric hemoglobin, Methods Enzymol. 76, 57–72.
41. U Sen, J Dasgupta, D Choudhury, P Datta, A Chakrabarti, S B Chakrabarty, A Chakrabarty, J K Dattagupta, (2004), Crystal structures of HbA2 and HbE and modeling of hemoglobin delta 4: interpretation of the thermal stability and the antisickling effect of HbA2 and identification of the ferrocyanide binding site in Hemoglobin, Biochemistry 43, 12477–12488.
42. J Sambrook, E F Fritsch, T Maniatis, (1989), In: Molecular cloning: A laboratory manual, C Nolan, ed, C5, Cold Spring Harbor Laboratory Press, New York.
43. W B Gratzer, (1982), Preparation of Spectrin, Methods Enzymol. 85, 475–480.
44. S Ray, A Chakrabarti, (2003), Erythroid spectrin in miceller detergents. Cell Motil Cytoskeleton 54, 16–28.
45. E Bucci, C A Fronticelli, (1965), New method for the preparation of α and β subunits of human hemoglobin, J Biol Chem. 240, PC551–PC552.
46. G Geraci, L J Parkhurst, Q H Gibson, (1969) Preparation and properties of alpha- and beta-chains from human hemoglobin, J Biol Chem. 244, 4664–4667.
47. E Bucci, (1981), Preparation of Derivatives of Ferrous and Ferric Hemoglobin, Methods Enzymol. 76, 99–100.
48. R P Haughland, (1996), Handbook of Fluorescent Probes and Research Chemicals, 6th Edition, M T Z Spence, ed, Molecular Probes, Eugene, OR.
49. N Shaklai, H Abrahami, (1980), The interaction of deoxyhemoglobin with the red cell membrane, Biochem Biophys Res Commun. 95, 1105–1112.
50. A S Bhown, F Hunter, J C Bennett, (1989), Analysis of the spectrin binding domains of globin, Biochem Biophys Res Commun. 164, 894–902.
51. R Jr. Valdes, G K Ackers, (1977), Thermodynamic studies on subunit assembly in human hemoglobin. Self-association of oxygenated chains (alphaSH and betaSH): determination of stoichiometries and equilibrium constants as a function of temperature, J Biol Chem. 252, 74–81.
52. S H Ip, G K Ackers, (1977), Thermodynamic studies on subunit assembly in human hemoglobin. Temperature dependence of the dimer-tetramer association constants for oxygenated and unliganded hemoglobins, J Biol Chem. 252, 82–87.

53. H Frischer, J Bowman, (1975), Hemoglobin E, an oxidatively unstable mutation, J Lab Clin Med. 85, 531–539.
54. J R Shaeffer, (1983), Turnover of excess hemoglobin alpha chains in beta-thalassemic cells is ATP-dependent, J Biol Chem. 258, 13172–13177.
55. J R Shaeffer, (1988), ATP-dependent proteolysis of hemoglobin alpha chains in beta-thalassemic haemolysates is ubiquitin-dependent, J Biol Chem.263, 13663–13669.
56. I Solar, J Dulitzky, N Shaklai, (1990), Hemin-promoted peroxidation of red cell cytoskeletal proteins, Arch Biochem Biophys. 283, 81–89.
57. D T Chiu, J van den Berg, F A Kuypers, I J Hung, J S Wei, T Z Liu, (1996), Correlation of membrane lipid peroxidation with oxidation of hemoglobin variants: possibly related to the rates of hemin release, Free Radic Biol Med. 21, 89–95.

Photoluminescence of Pharmaceutical Materials in the Solid State. 4. Fluorescence Studies of Various Solvated and Desolvated Solvatomorphs of Erythromycin A

Harry G. Brittain

Abstract The hydrate, methanolate, ethanolate, and isopropanolate solvatomorphs of erythromycin A have been prepared and characterized as to their crystallographic and solvent content characteristics. Even though erythromycin A does not exhibit fluorescence when in a dissolved state, in the solid state the solvated materials were found to be mildly fluorescent. Differences in fluorescence intensity were noted among the solvatomorphs, which could be roughly correlated with the degree of crystallinity of the materials. Desolvation of the dihydrate phase (known to yield an isomorphic desolvate) led to only minor changes in the excitation and emission spectra, but desolvation of the alchoholate solvatomorphs (known to yield largely amorphous products) caused large decreases in the intensities of the excitation and emission spectra. Since the fluorescence properties of organic solids are critically dependent on the details of the crystal structure and the delocalization of excitation energy, the loss of crystal structure appears to suppress the degree of energy transfer and this in turn affects the fluorescence intensities.

Introduction

In the absence of excited-state reactions, the solution-phase fluorescence spectroscopy of organic compounds can usually be interpreted in terms of the energy levels of the isolated molecule, where the excitation spectrum is effectively that of the absorption spectrum. In the solid state, however, it is well established that effects associated with excited-state intermolecular energy transfer cause the fluorescence spectra of the same compound to be quite different relative to those obtained in the solution phase [10, 9]. Since the transfer of electronic energy among excited molecules would be rapid when compared to the time frame of fluorescence, excitation energy becomes delocalized and this causes the excited state molecular orbital

H.G. Brittain (✉)
Center for Pharmaceutical Physics, 10 Charles Road, Milford NJ 08848, USA
e-mail: hbrittain@centerpharmphysics.com

C.D. Geddes (ed.), *Reviews in Fluorescence 2007*, Reviews in Fluorescence 2007,
DOI 10.1007/978-0-387-88722-7_16, © Springer Science+Business Media, LLC 2009

to extend over the ensemble of molecules involved in the energy transfer. The interaction leads to splitting of the single-molecule energy levels into a bundled set of levels, with the magnitude of the splitting being determined by the strength of the coupling. This type of splitting is commonly referred to as Davydov splitting and becomes manifest in the appearance of new bands in the excitation spectrum. In addition, the theory predicts that the mean frequency of the Davydov components would be displaced to lower energies relative to that of the free molecule value as a result of cooperative interactions in the crystalline state.

It is known that as long as the energies of the molecular orbitals of a molecule are affected by differences in structures of the various crystal forms, solid-state spectroscopy can yield important information regarding the properties of polymorphic or solvatomorphic systems [21, 4, 5, 3]. Less well recognized is the fact that when the differing structural aspects of polymorphs (i.e., a substance exists in structures characterized by different unit cells, but where each of the forms consists of exactly the same elemental composition) or solvatomorphs (i.e., a substance exists in structures characterized by different unit cells, but where these unit cells differ in their elemental composition through the inclusion of one or more molecules of solvent) cause an alteration in molecular orbitals, analytical techniques that measure transitions among the electronic states derived from these orbitals can be used to obtain additional information on the systems. Since fluorescence spectroscopy is derived from transitions among molecular electronic states, it is to be anticipated that the excitation and emission spectra of luminescent molecules could be used to study the patterns of energy flow within the respective solids [8]. For example, such methodology was used to understand the solid-state fluorescence of four polymorphs of diflunisal [6] and isostructural trihydrate solvatomorphs of ampicillin and amoxicillin [7].

As part of the present work, four solvatomorphs (hydrate, methanolate, ethanolate, and isopropanolate) of erythromycin A have been prepared, and characterized by x-ray powder diffraction (XRPD), thermal analysis, and fluorescence spectroscopy. The substances were subsequently desolvated, and re-characterized so as to learn the effect of solvent removal from the solids on the physical properties of the erythromycin A solvatomorphs.

Experimental Details

Materials

Erythromycin A dihydrate was obtained from the Aldrich Chemical Company and recrystallized from water (isolation temperature of 5–8°C) prior to its solid-state characterization. To prepare the alcoholate solvatomorphs, the dihydrate was dissolved in the appropriate organic solvent, whereupon the products were obtained after sufficient evaporation of the solvent had taken place. The stoichiometry of the products was established using thermogravimetry, using an Ohaus model MB45 system where the samples were heated isothermally at 130°C until constant weight was reached. The crystallographic and spectroscopic properties of the materials that were obtained as a result of the desolvation were obtained immediately after completion of the thermogravimetric determination.

Methods

XRPD patterns were obtained using a Rigaku MiniFlex powder diffraction system, set up to operate with a horizontal goniometer in the $\theta/2$-θ mode, and using the nickel-filtered Kα emission of copper as the x-ray source. Samples were packed into an aluminum holder using a back-fill procedure, and were scanned over the range of 50 to 6 degrees 2-θ, at a scan rate of 0.5 degrees 2-θ/min. Using a data acquisition rate of 1 point per second, the scanning parameters equate to a step size of 0.0084 degrees 2-θ. Calibration of each XRPD pattern was effected using the characteristic scattering peaks of aluminum at 44.738 and 38.472 degrees 2-θ.

All solid-state fluorescence excitation and emission spectra were obtained on samples packed into 5-mm glass NMR tubes. For the erythromycin solvatomorphs, the optimal emission spectra were obtained using an excitation wavelength of 390 nm, and the optimal excitation spectra were obtained using an emission monitoring wavelength of 465 nm. The spectra were obtained on a Perkin–Elmer LS 5B luminescence spectrometer, whose sample compartment was modified to enable measurements to be made on samples contained in the NMR tubes. The liquid cell holder was removed, and replaced by an aluminum block that had a hole drilled through its length to permit kinematic placement of the sample tube. The block had an additional lateral removal of metal that permitted irradiation of the sample and fluorescence detection at right-angles.

Results

Erythromycin A is an antibiotic substance produced by a strain of *Streptomyces erythreus*, originally found in a soil sample from the Philippine Archipelago [14]. The substance has been the subject of an extensive profile [12], and has been shown to be capable of forming a number of solvatomorphic crystal forms.

The fluorescence of organic molecules in the solid state often bears little resemblance to the fluorescence of the same molecule in the solution phase, and is usually dominated by cooperative energy transfer effects. In crystalline solids, the ground electronic state remains localized on individualized molecules, while the excited electronic states of these can be so strongly interactive that excitation energy becomes delocalized among the coupled molecules. Since the transfer of electronic energy between the excited molecules is rapid relative to the time frame of fluorescence, the delocalization of excitation energy has the effect of extending the excited state molecular orbital over the molecules whose excited states are involved in the energy transfer [9]. For a pair of molecules, the interaction leads to a splitting of the single-molecule energy level into a pair of levels, with the magnitude of this splitting being determined by the strength of the coupling. This type of splitting is denoted as Davydov splitting, and its effects include the appearance of new bands in the excitation spectrum. In addition, the theory predicts that the mean frequency of the Davydov components would be displaced to lower energies from that of the free molecule value as a result of cooperative interactions in the crystalline state. The result of exciton coupling is to produce an excitation multiplet corresponding to a band of levels, each differing in energy by small amounts, with the overall spread of the excitation band being determined by the strength of the coupling.

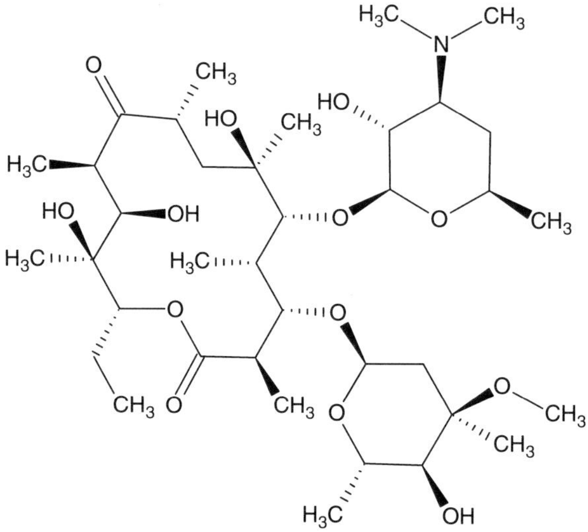

In crystals of organic molecules, the magnitude of the interaction energy and degree of exciton coupling is necessarily dependent on the relative orientation of the molecules in the crystal, as well as on their spatial arrangement. It follows that since polymorphic or solvatomorphic crystal forms are characterized by the existence of differing structural properties, the nature of the exciton coupling in the various forms would be influenced by the structural characteristics of each form. As a result, one would anticipate that the magnitude of the Davydov splitting, and the degree of shifting of the levels, would be dependent on the exact structural details existing in the different polymorphic forms.

Erythromycin A Dihydrate

The hydrate phases that can be formed by erythromycin have been studied at great length [1, 11, 2, 17, 13, 19], and it has also been established that when the dihydrate phase is desolvated, one obtains an isomorphic anhydrous crystal form [20, 15].

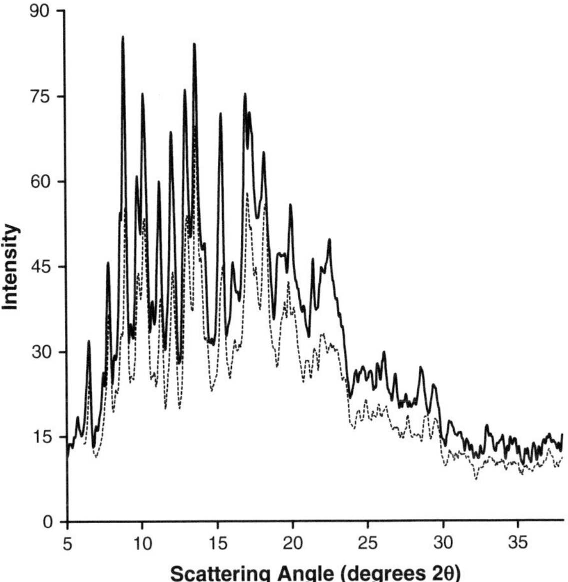

Fig. 1 X-ray powder diffraction patterns of erythromycin A dihydrate (*solid trace*) and its thermally dehydrated product (*dashed trace*)

The XRPD pattern of erythromycin A dihydrate is shown in Fig. 1, along with the XRPD pattern of the product obtained after its thermal dehydration. The empirically measured degree of thermally induced weight loss was found to be 4.75%, which agrees well with the theoretical water content of the dihydrate phase (calculated as 4.68%). The isomorphic nature of the dehydrated product with the initial dihydrate phase is evident in that both powder patterns consist of essentially the same sequence of scattering peaks. However, comparison of the relative intensities of the diffraction patterns indicates that the dehydration process causes some decrease in the overall crystallinity of the product.

The excitation and emission spectra obtained for the initial erythromycin A dihydrate are shown in Fig. 2. The excitation spectrum was found to consist of two barely resolved bands, with the main wavelength maximum being observed at 398 nm, and a shoulder located at 358 nm. Both of these features are significantly red-shifted relative to the known absorption band maxima of solubilized erythromycin A [12], indicating the existence of considerable delocalization of excitation energy in the solids. Excitation into either feature yielded a single emission band having a maximum at 484 nm.

The wavelengths of the excitation and emission spectra of the dehydrated erythromycin A dihydrate were equivalent to those of the initial substance, except for an approximate 25% decrease in overall intensity. This trend follows the comparable decrease in intensity observed in the XRPD patterns of the two materials, indicating that the degree of delocalization of energy transfer in the solids parallels the degree of crystallinity in the materials. The desolvation process was not found

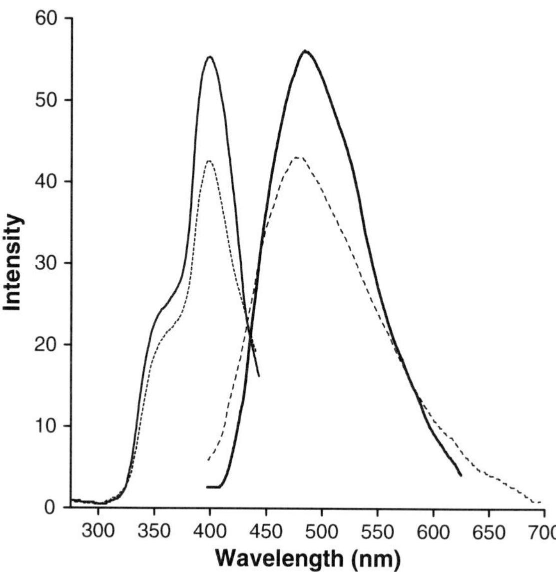

Fig. 2 Excitation (fine traces) and emission (heavy traces) spectra obtained for erythromycin A dihydrate (*solid traces*) and its thermally dehydrated product (*dashed traces*)

to affect the maximum of the excitation bands, but did cause a small blue shift in the maximum of the emission band down to 478 nm.

Erythromycin A Methanolate

The methanolate solvatomorph of erythromycin A does not appear to have been discussed to any degree in the literature, although Sarisuta et al. [18] did mention that they studied the effect of methanol on the crystalline properties of erythromycin without stating what they learned.

The XRPD pattern of the erythromycin A product obtained through crystallization from methanol is shown in Fig. 3, as is the XRPD pattern of the desolvated product obtained after its thermal desolvation. The XRPD of the methanolate product differed substantially from that of the dihydrate product, indicating the existence of a different type of solvatomorphic structure. The empirically measured degree of thermally induced weight loss was found to be 9.40%, which would be taken to indicate the existence of a di-methanolate solvatomorph (theoretical total volatile calculated to be 8.03%). Unlike the dihydrate phase, the methanolate solvatomorph of erythromycin A was found to thermally desolvate into an essentially amorphous product. A similar result has been reported in the thermal desolvation of the ethanolate solvatomorph [16].

The excitation and emission spectra obtained for the initial erythromycin A methanolate are shown in Fig. 4, where the excitation spectrum consisted of a band having a wavelength maximum of 399 nm. The excitation spectrum appeared to

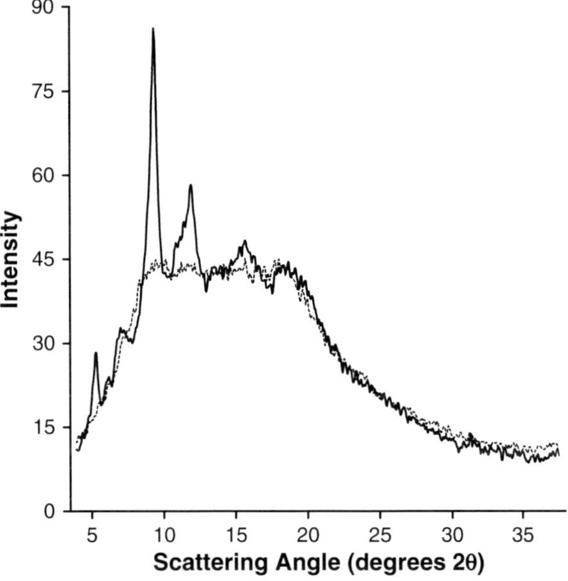

Fig. 3 X-ray powder diffraction patterns of erythromycin A methanolate solvatomorph (*solid trace*) and its thermally desolvated product (*dashed trace*)

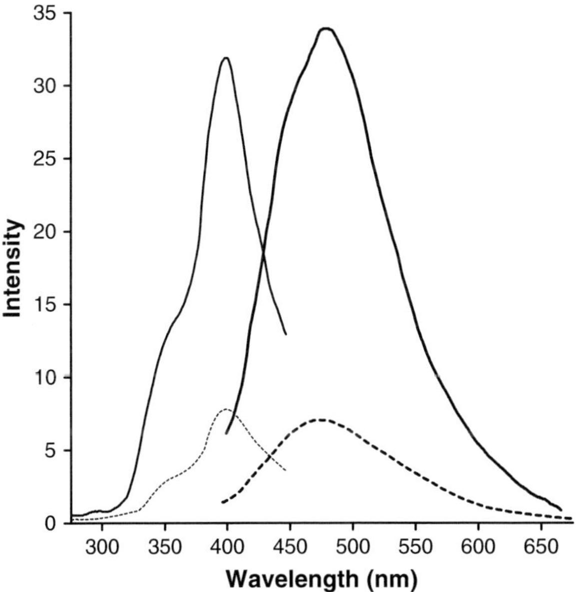

Fig. 4 Excitation (fine traces) and emission (heavy traces) spectra obtained for erythromycin A methanolate (*solid traces*) and its thermally dehydrated product (*dashed traces*)

also contain an unresolved shoulder around 350 nm, but this feature was far less visible in the excitation spectrum of the methanolate solvatomorph than it was for the dihydrate solvatomorph. The emission band was characterized by a wavelength maximum at 479 nm whose intensity was somewhat reduced relative to the intensity of the dihydrate emission band.

While the wavelengths of the excitation and emission spectra of the desolvated erythromycin A methanolate were roughly equivalent to those of the initial substance, the relative intensities of these features were 75% smaller than the corresponding features observed in the initial material. Since the desolvated product is amorphous in character, it would follow that the complete loss in crystallinity that accompanied the desolvation so strongly perturbed the degree of energy transfer that the resulting product was almost non-fluorescent.

Erythromycin A Ethanolate

The solid-state properties of the ethanolate solvatomorph of erythromycin A has been investigated by Mirza et al. [16], who reported that the isolated solid exhibited a substantially different XRPD pattern compared to that of the dihydrate phase. However, the degree of crystallinity of the ethanolate solvatomorph was shown to be much less than that of the dihydrate form.

The XRPD pattern of the erythromycin A ethanolate product obtained in the present study is shown in Fig. 5, together with the XRPD pattern of its thermally desolvated product.

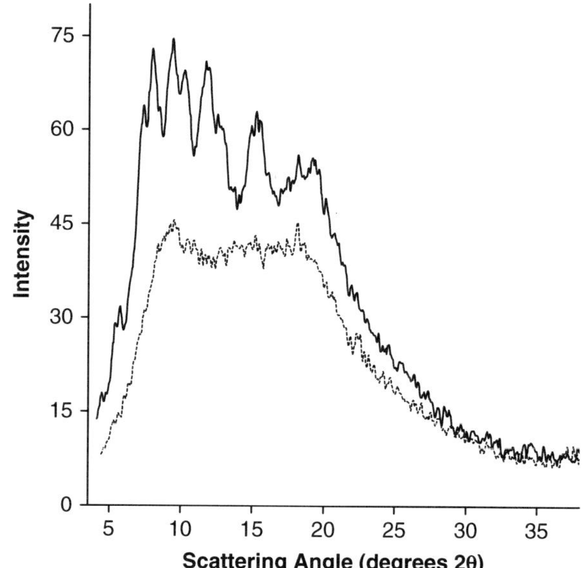

Fig. 5 X-ray powder diffraction patterns of erythromycin A ethanolate solvatomorph (*solid trace*) and its thermally desolvated product (*dashed trace*)

The XRPD of the ethanolate product was consistent with that reported by Mirza et al. [16], different from that of the methanolate solvatomorph, and also completely different from the XRPD of the dihydrate substance. This finding indicates that each solvatomorphic structure must be fundamentally different in nature. The empirically measured degree of thermally induced weight loss was found to be 12.10%, which would be taken to indicate the existence of a di-ethanolate solvatomorph (theoretical total volatile calculated to be 11.15%). As was the case for the methanolate solvatomorph, the ethanolate solvatomorph of erythromycin A was also found to thermally desolvate into an essentially amorphous product.

The excitation and emission spectra obtained for the initial erythromycin A ethanolate solvatomorph are shown in Fig. 6, where the excitation spectrum consisted of a band having a wavelength maximum at 403 nm and an unresolved shoulder around 350 nm. The emission band was characterized by a wavelength maximum at 484 nm, the intensity of which was comparable to the relative intensity of the methanolate solvatomorph emission band.

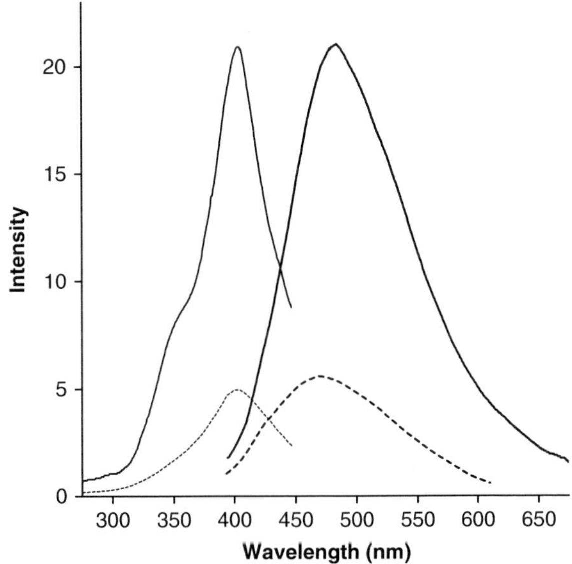

Fig. 6 Excitation (fine traces) and emission (heavy traces) spectra obtained for erythromycin A ethanolate (*solid traces*) and its thermally dehydrated product (*dashed traces*)

As had been noted for the fluorescence properties of the methanolate solvatomorph, the maxima of the excitation and emission spectra of the desolvated erythromycin A ethanolate product were roughly equivalent to those of the initial substance and approximately 75% weaker relative to the corresponding features of the initial material. Since the desolvated ethanolate product is also amorphous in character, the general trend that loss in crystallinity accompanying the desolvation process breaks up the energy transfer and delocalization of excitation energy so that the resulting product becomes almost non-fluorescent.

Erythromycin A Isopropanolate

The isopropanolate solvatomorph of erythromycin A has been studied by Sarisuta et al. [18] and Mirza et al. [16], with both groups reporting the isolation of a substantially crystalline product whose XRPD pattern was more intense and completely different relative to that of the dihydrate phase.

The XRPD pattern of the erythromycin A isopropanolate product obtained in the present study is shown in Fig. 7, together with the XRPD pattern of its thermally desolvated product.

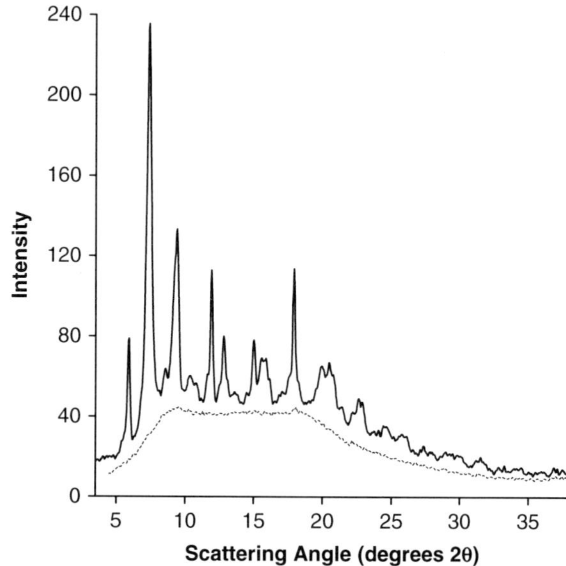

Fig. 7 X-ray powder diffraction patterns of erythromycin A isopropanolate solvatomorph (*solid trace*) and its thermally desolvated product (*dashed trace*)

The XRPD of the isopropanolate product was consistent with that reported by Mirza et al. [16], and different from the other solvatomorphs studied in this work. The empirically measured degree of thermally induced weight loss was found to be 14.75%, which would be taken to indicate the existence of a di-isopropanolate solvatomorph (theoretical total volatile calculated to be 14.07%). As was the case for the methanolate and ethanolate solvatomorphs, the isopropanolate solvatomorph of erythromycin A was observed to thermally desolvate into an amorphous product.

The excitation and emission spectra obtained for the initial erythromycin A isopropanolate solvatomorph are shown in Fig. 8, where the excitation spectrum consisted of a band having a wavelength maximum at 393 nm and an unresolved shoulder around 350 nm. The emission band was characterized by a wavelength maximum at 468 nm, and the intensity of this band was considerably higher than observed for any of the other solvatomorphs.

The maxima observed for the excitation and emission spectra of the desolvated erythromycin A isopropanolate product were roughly equivalent to those of the

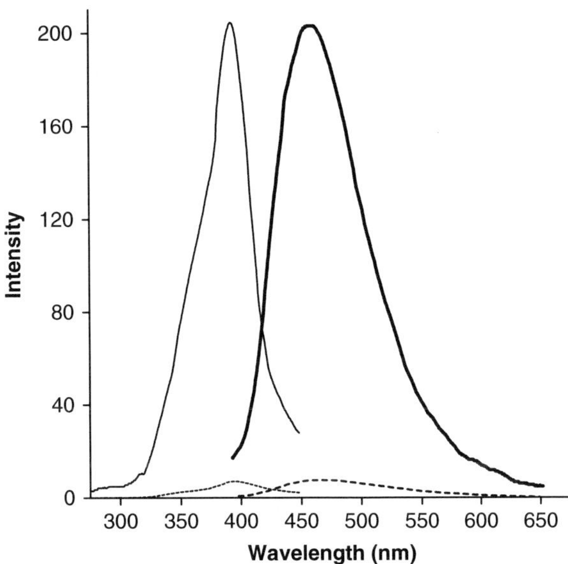

Fig. 8 Excitation (fine traces) and emission (heavy traces) spectra obtained for erythromycin A isopropanolate (*solid traces*) and its thermally dehydrated product (*dashed traces*)

initial substance, but the decrease in fluorescence intensity (approximately 95% less) that accompanied the desolvation was much higher than noted for the other solvatomorphs. This can be explained by considering that the initial substance was obtained in a highly crystalline form, and its reduction to an amorphous state strongly affects the energy transfer so that delocalization of excitation energy (and hence solid state fluorescence) becomes almost non-existent.

Discussion

If one correlates the degrees of crystallinity associated the solvated and desolvated erythromycin A solvatomorphs with the relative intensity of the observed emission band, one notes an approximate linear relationship between the intensity of the XRPD pattern with the fluorescence intensity. This observation is consistent with the development of fluorescence in the solids as resulting from the delocalization of absorbed excitation energy, and its subsequent delocalization through the crystal structure by means of energy transfer. This process would require favorable interaction of the chromophores in the solids, which could only be achieved by the existence of translational repetitiveness in the structures.

Dehydration of crystalline erythromycin A dihydrate results in the formation of a desolvate that still possesses a considerable degree of crystal structure. This isomorphic desolvate can either undergo rehydration to re-form the dihydrate phase, or undergo lattice relaxation with a concomitant reduction in unit cell volume if maintained in a dry atmosphere [20]. This latter process appears to have only a small effect on the fluorescence intensity of the dehydrated solid.

The solvatomorphs formed by the inclusion of alcohol molecules in the solids exhibit completely different behavior. When heated, these tend to lose solvent near the boiling points of the respective bulk liquids [16], which would indicate that the solvent molecules are not an integral part of the crystal lattice. When the solvent molecules are thermally expelled from the solids, the resulting amorphous solids that are obtained from the methanolate, ethanolate, and isopropanolate solvatomorphs all exhibit the same amorphous XRPD pattern, indicating that desolvation of all three alcoholic solvatomorphs yields the same substance. This has been illustrated in Fig. 9.

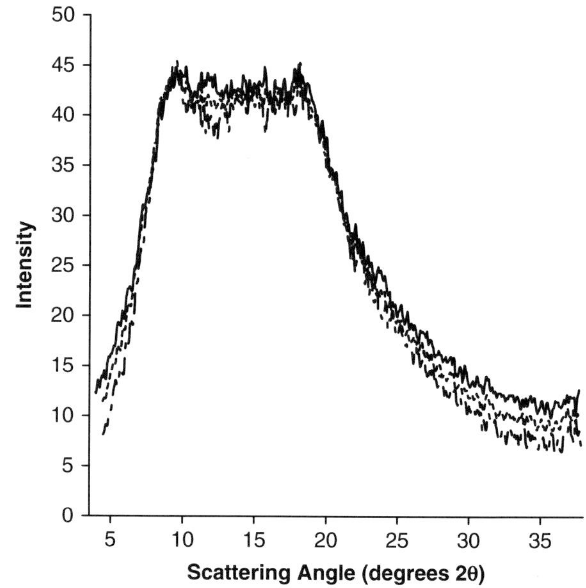

Fig. 9 X-ray powder diffraction patterns of the desolvated erythromycin A methanolate (*solid trace*), ethanolate (*long and short dashed trace*), and isopropanolate (*short dashed trace*) solvatomorphs

The fluorescence spectra resulting from the desolvated solvatomorphs was extremely weak, and each solid exhibited essentially the same fluorescence spectrum that was characterized by approximately the same relative intensity (see Fig. 10). This finding supports the deduction that the three desolvated alcoholate solvatomorphs of erythromycin A each consist of a structurally equivalent amorphous phase.

These observations are consistent with the general conclusion that fluorescence will only be of appreciable magnitude in erythromycin A solids if those materials are characterized by the presence of crystallinity.

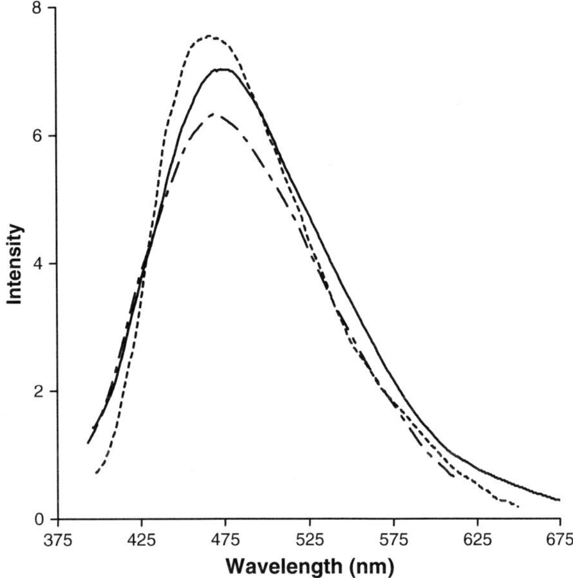

Fig. 10 Emission spectra of the desolvated erythromycin A methanolate (*solid trace*), ethanolate (*long and short dashed trace*), and isopropanolate (*short dashed trace*) solvatomorphs

References

1. Allen PV, Rahn PD, Sarapu AC, Vanderwielen AJ (1978). "Physical Characterization of Erythromycin: Anhydrate, Monohydrate, and Dihydrate Crystalline Solids." *J. Pharm. Sci.*, 67, 1087–1093.
2. Bauer J, Quick J, Oheim R. (1985). "Alternate Interpretation of the Role of Water in the Erythromycin Structure." *J. Pharm. Sci.*, 74, 899–900.
3. Bernstein J (2002). "Analytical Techniques for Studying and Characterizing Polymorphs." in *Polymorphism in Molecular Crystals*, Clarendon Press, London, pp. 94–150.
4. Brittain HG (1997). "Spectral Methods for the Characterization of Polymorphs and Solvates." *J. Pharm. Sci.*, 86, 405–412.
5. Brittain HG (1999). "Methods for the Characterization of Polymorphs and Solvates." in *Polymorphism in Pharmaceutical Solids*, Brittain HG ed.;: Marcel Dekker, New York, pp. 227–278.
6. Brittain HG, Elder BJ, Isbester PK, Salerno AH (2005a). "Solid-State Fluorescence Studies of Some Polymorphs of Diflunisal." *Pharm. Res.*, 22, 999–1006.
7. Brittain HG (2005b). "Solid-State Fluorescence of the Trihydrate Phases of Ampicillin and Amoxicillin", *AAPS PharmSciTech*, 6(3), article 55.
8. Brittain HG (2006). "Luminescence Spectroscopy." in *Spectroscopy of Pharmaceutical Solids*, Brittain HG ed., Taylor & Francis, New York, pp. 151–204.
9. Craig DP, Walmsley SH (1968). *Excitons in Molecular Crystals*. W.A. Benjamin, New York.
10. Davydov AS (1962). *Theory of Molecular Excitons*. McGraw-Hill, New York.
11. Fukumori Y, Fukuda T, Yamamoto Y, Shigitani Y, Hanyu Y, Takeuchi Y, Sato N (1983). "Physical Characterization of Erythromycin Dihydrate, Anhydrate and Amorphous Solid and their Dissolution Properties." *Chem. Pharm. Bull.*, 31, 4029–4039.
12. Koch WL (1979). "Erythromycin", in *Analytical Profiles of Drug Substances*, Florey K, ed., Academic Press, New York, pp. 159–177.

13. Laine E, Kahela P, Rajala R, Heikkila T, Saarnivaara K, Piippo I (1987). "Crystal Forms and Bioavailability of Erythromycin." *Int. J. Pharm.*, 38, 33–38.
14. *Merck Index* (2001). Merck & Co., Inc., Whitehouse Station, NJ, pp. 654–655.
15. Miroshnyk I, Khriachtchev L, Mirza S, Rantanen J, Heinamaki J, Yliruusi J (2006). "Insight into Thermally Induced Phase Transformations of Erythromycin A Dihydrate." *Cryst. Growth Design*, 6, 369–374.
16. Mirza S, Miroshnyk I, Heinamaki J, Christiansen L, Karjalainen M, Yliruusi J (2003). "Influence of Solvents on the Variety of Crystalline Forms of Erythromycin". *AAPS PharmSci*, 5(2), article 12.
17. Murthy KS, Turner NA, Nesbitt RU, Fawzi MB (1986). "Characterization of Commercial Lots of Erythromycin Base." *Drug Dev. Indust. Pharm.*, 12, 665–690.
18. Sarisuta N, Kumpugdee M, Muller BW, Puttipipatkhachorn S (1999). "Physico-Chemical Characterization of Interactions Between Erythromycin and Various Film Polymers." *Int. J. Pharm.*, 186, 109–118.
19. Stephenson GA, Stowell JG, Pascal HT, Pfeiffer RR, Byrn SR (1997). "Solid-State Investigations of Erythromycin A Dihydrate: Structure, NMR Spectroscopy, and Hygroscopicity." *J. Pharm. Sci.*, 86, 1239–1244.
20. Stephenson GA, Groleau EG, Kleemann RL, Xu W, Rigsbee DR (1998). "Formation of Isomorphic Desolvates: Creating a Molecular Vacuum." *J. Pharm. Sci.*, 87, 536–542.
21. Threlfall TL (1995). "Analysis of Organic Polymorphs – A Review." *Analyst*, 120, 2435–2460.

Index